D1556795

LADY MARGARET HALL LIBRARY
OXFORD

3063043410

**ORGANIC PHOTOCHEMISTRY:**
**A Comprehensive Treatment**

**Ellis Horwood and Prentice Hall**
are pleased to announce their collaboration in a new imprint whose list will encompass outstanding works by world-class chemists, aimed at professionals in research, industry and academia. It is intended that the list will become a by-word for quality, and the range of disciplines in chemical science to be covered are:

ANALYTICAL CHEMISTRY
ORGANIC CHEMISTRY
INORGANIC CHEMISTRY
PHYSICAL CHEMISTRY
POLYMER SCIENCE & TECHNOLOGY
ENVIRONMENTAL CHEMISTRY
CHEMICAL COMPUTING & INFORMATION SYSTEMS
BIOCHEMISTRY
BIOTECHNOLOGY

## Ellis Horwood PTR Prentice Hall
## PHYSICAL CHEMISTRY SERIES

*Series Editors:*
Ellis Horwood, M.B.E.
Professor T J Kemp, University of Warwick

**Current Ellis Horwood PTR Prentice Hall**
**Physical Chemistry Series titles**

| | |
|---|---|
| Blandamer | **CHEMICAL EQUILIBRIA IN SOLUTION: Dependence of Rate and Equilibrium Constants on Temperature and Pressure** |
| Bugayenko | **HIGH ENERGY CHEMISTRY** |
| Horspool & Armesto | **ORGANIC PHOTOCHEMISTRY: A Comprehensive Treatment** |
| Navratil | **NUCLEAR CHEMISTRY** |

# ORGANIC PHOTOCHEMISTRY: A Comprehensive Treatment

Dr WILLIAM HORSPOOL
Department of Chemistry, University of Dundee
Professor DIEGO ARMESTO
Department of Organic Chemistry,
Universidad Complutense, Madrid

**ELLIS HORWOOD PTR PRENTICE HALL**
NEW YORK LONDON TORONTO SYDNEY TOKYO SINGAPORE

First published in 1992 by
**ELLIS HORWOOD LIMITED**
Market Cross House, Cooper Street,
Chichester, West Sussex, PO19 1EB, England

A division of
Simon & Schuster International Group
A Paramount Communications Company



Printed and bound in Great Britain
by Bookcraft Ltd, Midsomer Norton, Avon

---

British Library Cataloguing in Publication Data

---

A catalogue record for this book is available from the British Library

ISBN 0–13–639477–9

---

Library of Congress Cataloging-in-Publication Data

---

Available from the Publishers

*To Una and Mercedes*

# Table of contents

# Preface

Organic photochemistry has in the last 30 or so years reached the full maturity of a scientific discipline. The steady development of the subject has led to a great wealth of novel reactions which can be of use to the synthetic chemist. In recent times most of the textbooks devoted to the subject have been of a specialist nature or have sought to give the reader an overview of the physical, inorganic, and organic areas. There are few recent textbooks on the subject directed especially at senior undergraduates and postgraduates and this present text is aimed at that particular level.

This book on organic photochemistry divides the material covered according to the chemical structure of the substrate undergoing irradiation. Thus there are major sections dealing with hydrocarbons, oxygen-containing compounds, sulphur-containing compounds, nitrogen-containing compounds and halogenated compounds. Within these major groupings there are subsections corresponding to the more familiar functional groups. For the purpose of classifying the reactions within the major groups the general rule adopted is that a substituent bearing a heteroatom is ignored if its presence does not substantially alter the photochemical reactivity of the parent molecule. Thus hexafluorobenzene and benzene are dealt with as hydrocarbons since they both undergo photoisomerization and the presence of fluorine does not alter this to any great extent. However, chlorobenzene derivatives are contained in the halogen chapter since the reactivity of these molecules is dominated by C—halogen fission. Apart from the main chapters a brief introductory chapter provides the link with the more physical aspects of the subject. The last chapter provides some simple details on experimental techniques. The authors would like to acknowledge the assistance of Dr John

Coyle who supplied the introductory chapter and some of the material used in Chapter 2.

Throughout the text references are supplied to review texts such as the annual review *Photochemistry*, published by the Royal Society of Chemistry, which provides excellent cover of all the available publications. Reference is also made to chapters in specialist texts and important primary papers. These references should provide sufficient data to permit a more systematic study of a specific area. No attempt has been made at comprehensive referencing.

All in all the purpose of the text is to provide an entry point to the blossoming area of organic photochemistry. We hope readers will be attracted to the new and often superior synthetic reactions described.

*Dundee*
*Madrid*
*October* 1992

William Horspool
Diego Armesto

# 1

# Introduction

This book deals specifically with organic photochemistry, an area of research that has been with us since before the start of the 1900s. Notable contributions were made by Ciamician and Silber [1] who published many papers in the period 1900–1915 and by Schonberg and Mustafa [2] in the 1940s. However, the major advance in understanding of the subject and a resurgence of interest did not occur until the 1950s and 1960s [3]. Since then the literature available in research publications, reviews, and monographs has increased dramatically and the list included in the reference section can only be a sample of what is available for consultation [4]. This present text highlights the synthetic value of organic photochemical reactions and, since there are sufficient texts in which the general principles underlying the reactions are dealt with thoroughly [5], only a summary treatment of the physical processes involved is included.

## 1.1 GENERAL PRINCIPLES

A photochemical reaction occurs when a molecule is raised from its electronic ground state to a higher state. In this higher state the molecule can follow a variety of reaction paths. Thus photochemistry is concerned with electronically excited molecules and the changes which they undergo. Such excited species can be produced in a variety of ways [6], but by far the most commonly used method for their generation is through absorption by a molecule of a photon of ultraviolet or visible light. As a consequence photochemists have an interest in molecular spectroscopy, and the ultraviolet/visible absorption spectrum of compounds provides essential information about the wavelengths that are absorbed by a given molecule. Absorption is a prerequisite to any photochemistry since only light which is absorbed can bring about a photochemical change. The absorption spectrum, a typical example of which is that of benzophenone shown in Fig. 1.1, also gives an indication of the wavelengths required to obtain different electronically excited states since each major absorption band corresponds to a transition to a different state. However, by no means all the transitions result in a distinct and observable absorption band.

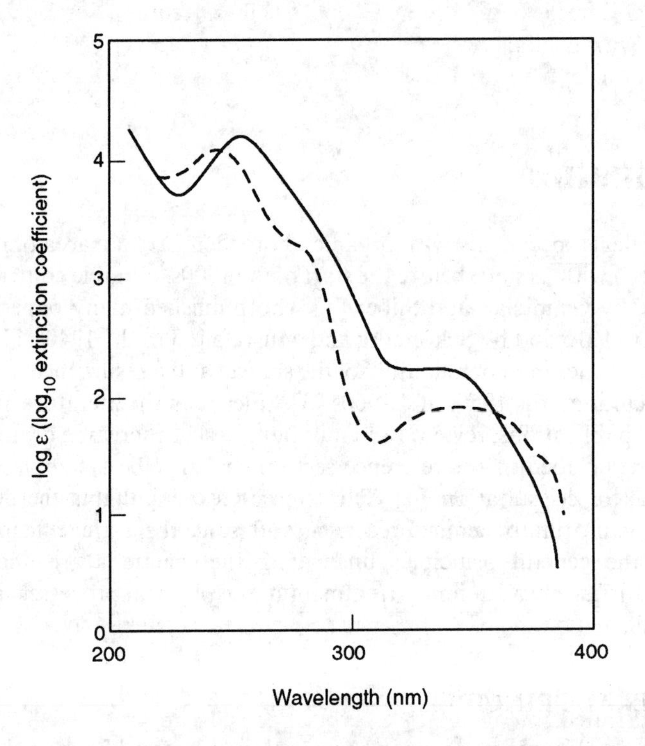


Fig. 1.1. Absorption spectrum of benzophenone in cyclohexane (---) and methanol (–) as solvents

The process by which excited states are formed, their physical properties and the paths for physical decay that are open to them are all of considerable interest. In this brief survey of the general principles governing photochemistry it is inappropriate to go into these topics in any detail [7]. However, material is included to help in developing an understanding of the mechanisms of photochemical reactions and the ways in which these reactions differ from their thermal, non-photochemical, counterparts.

## 1.2 ELECTRONICALLY EXCITED STATES AND TRANSITIONS

### 1.2.1 Orbital types

In organic photochemistry the most commonly used description of electronically excited states is based on the valence-shell molecular orbitals, and some familiarity with orbitals and their nomenclature is necessary. The non-scale representations of the $n$, $\pi$, and $\sigma$ orbital types given in Figs 1.2–1.4 represent

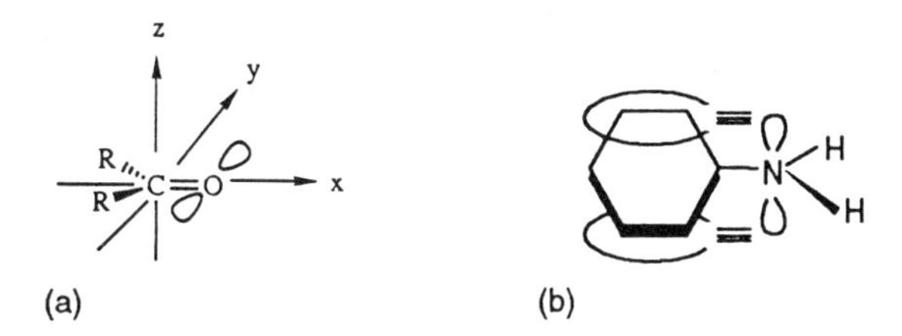


Fig. 1.2. (a) $n$-Orbital on carbonyl oxygen. (b) Interaction of nitrogen lone pair with aryl $\pi$-system.

the most likely distribution of the electrons in the orbitals concerned. The non-bonding $n$-orbitals are found in molecules containing a hetero atom and often are involved in the lowest electronic transitions. The example illustrated, a carbonyl group (Fig 1.2(a)), is for most purposes considered to be a purely $p$-orbital located on the oxygen. In some instances, e.g. in aniline (Fig. 1.2(b)), the lone pair is not completely non-bonding since it can overlap with the aromatic $\pi$-orbitals. $\pi$-Orbitals are delocalized over at least two atoms, and ethene, illustrated in Fig. 1.3(a), is typical of such an arrangement. In this the electronic distribution is evenly spread between the two atoms. This is not the case with the $\pi$-bond of a carbonyl group (Fig. 3(b)) where there is polarization of the $\pi$-bond resulting in greater electron density at the hetero atom. Corresponding to each of the $\pi$-orbitals is a $\pi^*$-anti-bonding orbital and the relationship between the two is shown pictorially in Fig. 1.3(c). $\sigma$-Orbitals (Fig. 1.4) are the skeletal orbitals of organic compounds and are

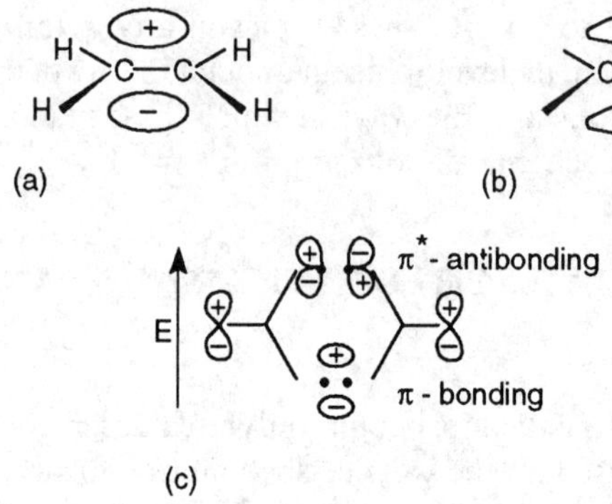

Fig. 1.3. (a) Olefinic $\pi$-bond. (b) Carbonly $\pi$-bond. (c) Linear combination of two $p$-orbitals.

much stronger than $\pi$-bonds. The orbital of the electron pair defined as a $\sigma$-bond is symmetrical around the bond axis. Again there are associated $\sigma^*$-antibonding orbitals as illustrated in Fig. 1.4.

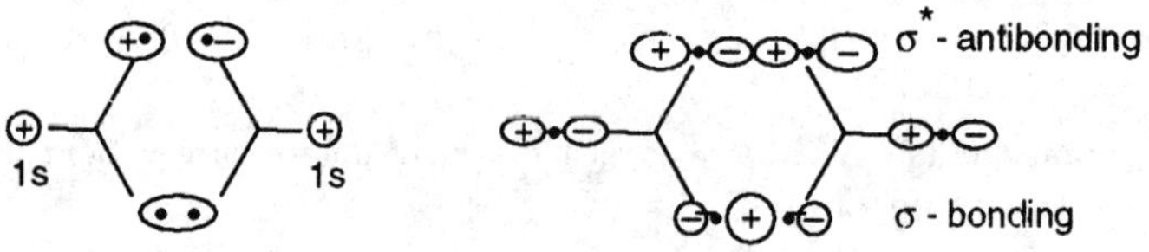

Fig. 1.4. $\sigma$ and $\sigma^*$-orbitals.

### 1.2.2 Molecular orbitals

Normally the molecular orbitals of the ground state, the unexcited state of the molecule, are used to describe and categorize the electronic transitions that can occur and the excited states they produce. In Fig. 1.5 a partial set of molecular orbitals for formaldehyde (methanal) ignoring C—H is shown. In Fig. 1.5(a) the orbitals are occupied by pairs of electrons in the lowest energy configuration $[(\pi_{CO})^2(n_O)^2(\pi^*_{CO})^0(\sigma^*_{CO})^0]$, and this is the normal, ground-state arrangement of electrons in the molecule. Excitation with ultraviolet light of wavelength $\lambda = 280$ nm can lead to a different state, of higher energy, in which an electron has been promoted from the higher of the two non-bonding ($n$) orbitals to the lowest antibonding ($\pi^*$) orbital; this is shown

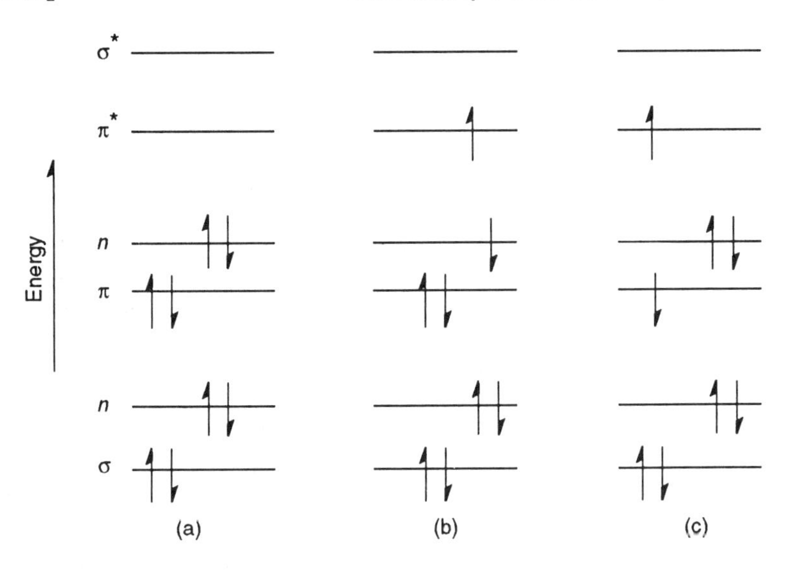


Fig. 1.5. Some of the molecular orbitals for methanal showing the electronic configuration of (a) The ground state. (b) The $n \rightarrow \pi^*$ singlet excited state. (c) The $\pi \rightarrow \pi^*$ singlet excited state.

in Fig. 1.5(b) and described by $(\pi_{CO})^2(n_O)^1(\pi^*_{CO})^1(\sigma^*_{CO})^0$. This excited state is designated as an $n\pi^*$ state, and the transition leading to its formation is an $n \rightarrow \pi^*$ transition. Radiation of shorter wavelength, around 180 nm, can cause a different electronic change to produce the $\pi\pi^*$ excited state depicted in Fig. 1.5(c) and described by $(\pi_{CO})^1(n_O)^2(\pi^*_{CO})^1(\sigma^*_{CO})^0$.

For any molecule there is a series of excited states that differ in their electronic nature and properties. These are conveniently labelled $S_1$, $S_2$, $S_3$, and so on where $S$ indicates a singlet state with an overall spin of zero and where the unpaired electrons have opposed spin. The numerical subscripts refer to increasing energy of the excited state. Molecules that contain carbon-carbon multiple bonds but have no hetero atom such as oxygen or nitrogen normally have a lowest-energy $S_1$ excited state that is $\pi\pi^*$ in character. In molecules that contain hetero atoms but are without extensive conjugation the highest energy of the filled orbitals in the ground state is often non-bonding and such molecules generally have $n\pi^*$ or $n\sigma^*$ excited states that are lower in energy than $\pi\pi^*$, $\pi\sigma^*$, or $\sigma\sigma^*$ states. This generalization is subject to some conditions. First, molecules with conjugated multiple bonds may have $\pi\pi^*$ states that are lowest in energy even if a non-bonding orbital is occupied in the ground state. This may be apparent from the electronic absorption spectrum since a $\pi\pi^*$ band is much more intense than an $n\pi^*$

band. Secondly, there are some excited states that are not well described by this simple molecular orbital model. In particular there are Rydberg states that are visualized better as arising from the promotion of an electron to an orbital that extends substantially beyond the core of nuclei and inner shell electrons and has many of the characteristics of an atomic orbit. Rydberg states can be regarded as having a large, positively charged core and an outer region of electron density, and their importance is recognized in the photochemistry of many saturated compounds such as alkanes, haloalkanes, alcohols, ethers and amines as well as that of some simple alkenes [8].

### 1.2.3 Multiplicity and lifetime

The third point to be made about the likely electronic nature of the lowest energy excited state is that much of what has been said so far refers primarily to singlet excited states in which the electron spins remain opposed. However, for most singlet excited states there is a corresponding triplet state in which the spins are parallel. Fig. 1.6 shows the occupation of the orbitals in the

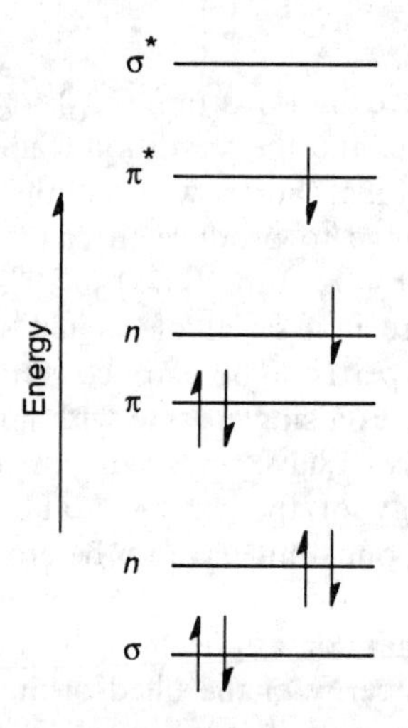


Fig. 1.6. An orbital energy diagram for the $n\pi^*$ triplet state of methanal.

$n\pi^*$ triplet state of formaldehyde. Triplet states are not readily obtained directly by absorption of a photon since such an absorption is strongly forbidden [9] and most observable spectra are singlet–singlet absorption spectra. Some triplet states can be generated indirectly by absorption of ultraviolet or visible light as a result of a photophysical process referred to as intersystem crossing from the initially generated singlet state. A triplet state always has an energy lower than a singlet state with the same electronic configuration but the singlet–triplet energy difference varies considerably. This energy gap is much higher for $\pi\pi^*$ states than for $n\pi^*$ or Rydberg states and so it is not unusual for a molecule with a lowest-energy singlet state, $S_1$, that is $n\pi^*$ in nature to have a lowest triplet state $T_1$ that is $\pi\pi^*$. The emphasis on the excited states of lowest energy arises because of the observation formulated by Kasha [10] that luminescence or photochemical reaction normally occurs from the lowest-energy singlet or triplet excited state rather than from a higher energy state. There are exceptions to Kasha's rule but it is a good starting point when considering which excited state is responsible for an observed photochemical process.

Absorption of light is an extremely rapid event and the excited state is formed initially with the same geometry as the ground state. As well as being electronically excited, this initial state is a vibrationally excited species and by rapid vibrational deactivation it relaxes to the excited state with equilibrium geometry. The relaxed excited state is a distinct species with characteristic properties such as bond lengths, angles, vibrational frequencies, and a dipole moment that can in principle be measured and is generally different from that of the ground-state molecule. The excited state also has its own characteristic range of chemical reactions and, as we shall see later, this range is much less restricted by thermodynamic constraints than is the chemistry of the ground state. This, together with the number of excited states, both singlet and triplet, that are available for each molecule, makes for a potentially very extensive array of photochemistry for each class of compound. In practice there is a considerable variety of photochemical reactions for most groups, but one severe limit on observable photochemical change is the very short lifetime of the excited state. Most lifetimes are in the range $10^{-3}$ to $10^{-12}$ s in fluid solution at room temperature [11].

### 1.2.4 Radiative and non-radiative processes

The lifetime of an excited state can be expressed as the reciprocal of the sum of first-order rate constants corresponding to all the routes by which the excited state reacts or decays. As well as chemical reactions that may have very high rate constants there are rapid physical processes by which the excited state decays to regenerate the ground state from which it was derived. For many excited states, such fast physical processes dominate the lifetime, that is they are very much faster than the chemical reaction. This provides the chief justification for Kasha's rule, since most excited states that are higher in energy than the lowest excited singlet or triplet decay very rapidly. The decay leads to the lowest excited states by a physical process known as internal conversion (if there is no change in spin) or intersystem crossing (if there is a change in spin). Decay of the lowest excited states to the ground state is much slower than the preceding process because the energy gap between the lowest states and the ground state is quite large but chemical reaction still has to compete with non-radiative decay to the ground state and also with luminescent decay [12]. The latter processes are called fluorescence if there is no change in spin and phosphorescence if there is a spin change. This means that efficient photochemical reactions are less widespread than might be anticipated. The efficiency with which absorbed radiation causes a compound to undergo a specified chemical change can be expressed in terms of the quantum yield ($\Phi$) which is defined as

$$\Phi = \frac{\text{Number of molecules of product produced}}{\text{Number of photons of light absorbed}}$$

High quantum yields are desirable for the application of photochemical reactions in organic synthesis although reactions with low or very low quantum yields can be employed if the chemical yield is high by extending the time of irradiation. This approach is of use only when the photons are wasted by physical decay of the excited state to reform the ground state.

The various physical processes that interconvert the lower-lying electronic states of a molecule can be depicted in a Jablonskii diagram as shown in Fig. 1.7. This is a state energy diagram rather than an orbital energy diagram as in Figs 1.5 and 1.6. Absorption, radiationless decay and radiative decay (luminescence) are shown using appropriate arrows. Such a diagram can be constructed for a particular compound on the basis of results from absorption spectra, from luminescence studies and from photophysical experiments [13]. Often it is not easy to acquire all the detailed quantitative information, but to an organic photochemist there is value in the qualitative information about the nature and relative energy of low-lying excited states and the photophysical processes with which any observable photochemical reaction has to compete.

## 1.3 QUENCHING AND SENSITIZATION

The descriptions so far have made no distinction between intermolecular and intramolecular processes, although many of the photophysical changes are intramolecular and do not involve specific interaction with a second molecular

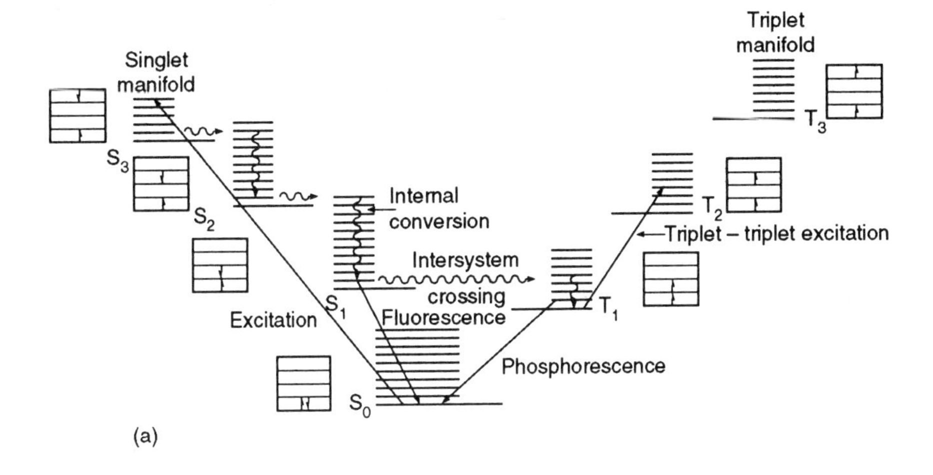


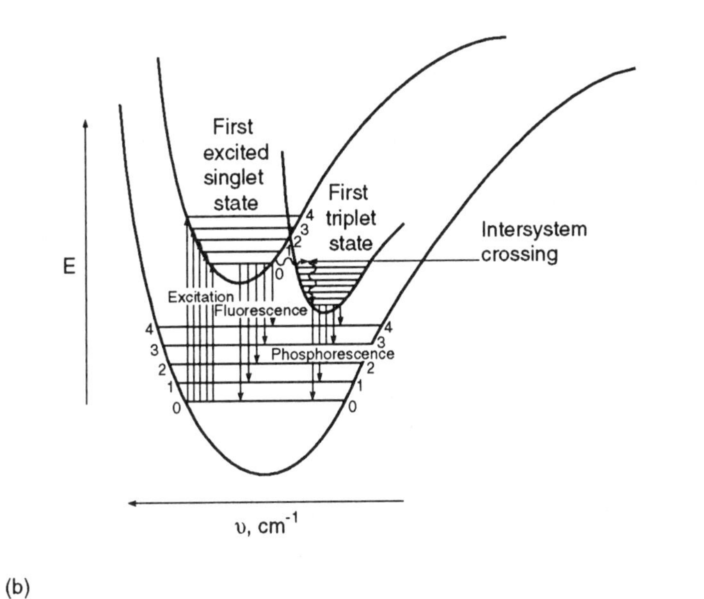


Fig. 1.7. Modified Jablonskii diagram. (b) Schematic representation of relative energies of fluorescence and phosphorescence.

species. However, intermolecular interaction between a molecule in its excited state and a molecule in its ground state can lead to deactivation of the electronically excited state and the generation of the excited state of the other molecule. This phenomenon is known generally as quenching (**1**). The reverse

$$M^* + Q \longrightarrow M + Q^*$$

(1)

process, in which a molecule in the ground state is raised to its excited state by energy transfer from another excited state molecule, is known as sensitization (**2**). Thus there is a direct relationship between the phenomena

$$M + Sens^* \longrightarrow M^* + Sens$$

(2)

referred to as quenching and sensitization. There are many different mechanisms for quenching [14]. Some do not involve close approach of the two species whereas in others an intermediate complex, an exciplex or an excimer, is formed. The details of the mechanisms for quenching do not concern us here but the outcome is a very useful technique in the hands of either the mechanistic or the synthetic photochemist. Quantitative quenching studies allow the properties of the excited states taking part in the photochemical reactions to be investigated. Thus, for example, it may be possible to determine the lifetime of an excited state which in turn provides a starting point for estimating the rate constant for the primary photochemical step. Selective quenching, used qualitatively, can remove an unwanted component from a photochemical reaction which occurs by way of two different excited states and this is particularly useful in inhibiting a triplet-state reaction in the presence of an accompanying singlet-state process.

### 1.3.1 Triplet-state sensitization

A very important use of the energy-transfer processes mentioned above is in the generation of excited states, usually denoted by M* or some such symbolism, that are not readily accessible by the normal process of light absorption. This is termed sensitization [15] and it is frequently employed to provide a route to triplet states of molecules for which the indirect route of absorption followed by intersystem crossing from singlet to triplet is very inefficient, i.e. very slow by comparison with other decay processes open to the singlet state. The aromatic ketones such as benzophenone ($Ph_2CO$) undergo intersystem crossing very efficiently: for every photon absorbed by the ketone one molecule is produced in its triplet state; i.e. $\Phi_{ISC}$, the quantum yield for intersystem crossing, is unity. In such cases there is no problem in studying the chemistry of their triplet states. However, a conjugated diene such as penta-1,3-diene does not undergo intersystem crossing readily and the triplet state chemistry of such compounds can only be carried out with the help of a triplet sensitizer. Sensitization can also introduce an element of photochemical selectivity in that a triplet state can be generated in the absence of a related singlet excited state whose chemical reactions may be undesirable. An example of the processes taking place is shown in **3** and **4**

$$\text{Donor}\,(T_1) + \text{Acceptor}\,(S_0) \longrightarrow \text{Donor}\,(S_0) + \text{Acceptor}\,(T_1)$$

(3)

$$Ph_2CO\ (T_1,\ E_T = 69\ \text{kcal mol}^{-1},\ 289.1\ \text{kJ mol}^{-1}) + \text{naphthalene}\ (S_0) \longrightarrow$$
$$Ph_2CO\ (S_0) + \text{naphthalene}\ (T_1,\ E_T = 61\,\text{kcal mol}^{-1},\ 255.6\ \text{kJ mol}^{-1}).$$

(4)

where the triplet-state donor transfers its energy to the ground-state acceptor with the resultant production of ground-state donor and triplet-state acceptor. One should note that the energy transfer has occurred with conservation of the overall spin angular momentum of the molecules involved.

The efficiency of energy transfer in a particular system depends on the concentration of the energy-accepting molecule (acting as a quencher in this instance), the rate constant for the quenching interaction, and the lifetime of the excited states in the absence of quencher. In general, triplet states are much longer-lived than singlet states and so they are quenched more readily. This is the main reason why phosphorescence is seldom observed in fluid solution at room temperature because very small amounts of quenching impurities inhibit the process. Fluorescence is not affected so markedly and it is usually studied under normal conditions, whereas phosphorescence studies are carried out in a frozen glass matrix at low temperature under conditions where the rate constant for quenching is considerably reduced. Molecular oxygen is an important, though often unwanted, triplet-state quencher in organic photochemistry. Its ground state is a triplet state and it can interact with many triplet excited states of other molecules, either by a transfer of energy which results in the production of a low energy excited singlet state of molecular oxygen termed singlet oxygen (**5**), by transfer of an

$$M\,(T_1) + O_2\,(T_0) \longrightarrow M\,(S_0) + O_2\,(S_1)$$

(5)

electron which generates the dioxygen radical anion and the radical cation of the organic compound (**6**) or by a quenching mechanism that does not

$$M(T_1) + O_2(T_0) \longrightarrow M^{+\bullet} + O_2^{-\bullet}$$

(6)

produce a different oxygen specie. The outcome is the quenching of the triplet excited state [16] and in the first two cases production of a very reactive oxygen species that may react chemically with any organic compound present. The chemical reactivity of singlet oxygen is employed usefully in reactions with a variety of unsaturated compounds, but unless this is intended the reaction with oxygen has to be regarded as an undesirable component of an organic photochemical reaction. It is common practice to remove as much oxygen as possible by carrying out reactions with a stream of oxygen-free nitrogen or some other inert gas bubbling through the irradiated solution.

The quenching of electronically excited states that involves transfer of energy or transfer of an electron is very important in many photobiological phenomena and especially in photosynthesis [17]. One of the remarkable features of photosynthesis is that all the photochemical reactions occur in the presence of a relatively high concentration of molecular oxygen. This is possible partly because the excited states involved have extremely short lifetimes and hence quenching by oxygen is less efficient than in other systems. Also present in such natural systems are the carotenes that are able to quench any singlet oxygen that is formed as a result of the primary quenching reaction and so prevent singlet oxygen from causing damage by chemical reaction.

## 1.4 PHOTOCHEMICAL REACTIONS

### 1.4.1 Kinetic versus thermodynamic control

Having looked briefly at the various photophysical pathways that are open to an electronically excited state we turn now to the question of photochemical reactions and initially to the differences that might be expected between such a reaction and a thermal ground-state process. The general kinetic model that underlies much of an organic chemist's thinking in relation to ground-state chemistry can also be used to a large extent for excited-state chemistry. One of the major differences encountered is in connection with reversibility. Thermal reactions may operate under conditions where there is effective reversibility, i.e. substrates and products are in thermodynamic equilibrium, and the extent of reaction and the ratios of products depend on the differences in standard free energy between the components. The process is said to be operating under conditions of thermodynamic control. This applies to some thermal reactions, though many are carried out under conditions where thermodynamic equilibrium is not achieved and in many instances practical reversibility is not feasible. Such reactions are said to operate under conditions of kinetic control, and product ratios are determined by relative rates of reaction. In a few cases the balance between kinetic and thermodynamic control is quite fine and the reaction time can influence the final outcome. The general situation with photochemical reactions is different in that they operate, with very few exceptions, under conditions of kinetic control. The pathway from an excited state to products does not go through an excited state of the products, i.e. the general mechanism is as shown in **7**. There is

$$\mathrm{M} \xrightarrow{h\nu} \mathrm{M}^* \longrightarrow \mathrm{P}$$

(7)

usually a considerable energy difference between the excited state of the starting material and the ground state of the product and therefore the product does not revert to the excited state of the starting material. If reversion does take place, especially if the product is a reactive intermediate rather than the final product, it returns to the ground state of the substrate rather than the excited state, and this does not represent reversibility in a thermodynamic sense.

One explanation for this departure of photochemical reactions can be found in detailed calculations of potential energy surfaces for excited state processes. It is generally found that along favoured pathways for the excited state there is an area where the excited state surface comes very close in energy to the ground state surface. At this point rapid crossing is favoured from the upper to the lower surface and so the product appears in its ground state rather than in an excited state [18]. The lack of reversibility in photochemical reactions means that product ratios are determined by kinetic factors such as relative rates of competing steps at a branching point in the mechanism whether this involves alternative routes for an excited state to follow or different fates of an intermediate species.

There are very few exceptions to the generalization that the product of a photochemical reaction is formed directly in its ground state rather than through one of its excited states. One group of reactions to which it does not apply are rapid proton transfers in which the protonated or deprotonated species formed from the excited base or acid respectively is also in an excited state. In these situations an equilibrium is achieved between two excited states and it is possible to measure, usually by indirect methods, $pK_a$ values for the excited states. For organic compounds in which the acidic or basic group is part of the chromophore, such as phenols, saturated carboxylic acids or amines, and aromatic carboxylic acids or amines, it is found that the excited-state $pK_a$ values differ substantially from those of the ground state, commonly by several units (**8** and **9**) [19].

$$M^* + H^+ \rightleftharpoons (MH^+)^*$$

(8)

$$MH^* \rightleftharpoons (M^-)^* + H^+$$

(9)

### 1.4.2 Photostationary states

Practical reversibility of a different kind can arise in some photochemical reactions when the product is capable of absorbing the irradiating light. If the product undergoes a photochemical reaction to give the substrates then a two-way photochemical reaction (**10**) can be set up. This is not an

$$M \underset{h\nu}{\overset{h\nu}{\rightleftharpoons}} P$$

(10)

equilibrium in the thermodynamic sense discussed earlier because the detailed pathways are different: the forward reaction proceeds by way of an excited state of substrate (**11**) but the reverse reaction proceeds through an excited

$$M \xrightarrow{h\nu} M^* \longrightarrow P$$

(11)

state of the product (**12**).

$$P \xrightarrow{h\nu} P^* \longrightarrow M$$

(12)

In a situation like this a photostationary state can be reached in which the rate of the forward reaction equals the rate of the reverse reaction and no change in composition occurs on further irradiation. The composition at the photostationary state can be influenced by the choice of irradiating wavelength since the relative rates of reaction depend on the relative intensities of light absorbed by the two species. Relative absorption is measured by absorption coefficients and these vary with wavelength so that a change in composition with wavelength is inevitable unless the spectra of substrate and product have exactly the same form in the spectral region considered.

### 1.4.3 Photochemical reaction versus thermal reaction

Thermodynamic considerations do have a place in an understanding of photochemical reactions in the sense that any chemical reaction proceeding to near completion must be thermodynamically favourable, that is, it must involve an overall decrease in free energy as it proceeds. The lowest excited states of organic compounds are quite high energy species, commonly in the range 200–500 kJ $mol^{-1}$ higher in energy than the corresponding ground states, and so the range of reactions that is open in principle to an excited state is far greater than that open to the ground state. The limitations that arise on account of the short lifetime of the excited state have already been discussed but, nevertheless, the principle underlies the successful application of photochemical reactions in the synthesis of small ring-compounds or compounds that are otherwise strained or of high energy.

Thermal reactions operating under conditions of kinetic control are normally discussed in mechanistic terms that are based on electron distribution within the molecule. The preferred sites of attack as a result of this electron distribution or the relative stabilities of alternative intermediate species (which are taken to reflect the relative stabilities of activated complexes and hence to be a guide to relative activation energies and rate constants) determine the resultant reaction path. With certain limitations it is possible to rationalize photochemical reactions in a similar way, bearing in mind that the electron distribution in an excited state may be different from that in the corresponding ground state. As an example, many of the thermal reactions of ketones can be understood on the basis of initial attack by a nucleophile at the partially positively charged carbon atom of the carbonyl group or initial attack by an electrophile at the partially negatively charged oxygen atom. The starting point for this rationale is the polarization of electrons in the carbon-oxygen bonds with a higher electron density near the more electronegative oxygen. The $n\pi^*$ excited state of a ketone is less strongly polarized because the electron density has been transferred from an orbital located largely on oxygen to an orbital covering the carbon and oxygen nuclei more equally. The main electronic feature of the $n\pi^*$ excited states is that the oxygen atom has odd-electron, radical-like, character as a result of the loss of an electron from the non-bonding orbital. Most of the chemistry of ketones in these excited states can be successfully rationalized on the basis of the odd-electron nature and the major reactions involve subsequent radical species rather than the electron-paired charged species associated with thermal reactions of ketones.

An example of how approaches to thermal and photochemical reactions of the same substrate differ more substantially is found in a consideration of aromatic substitution reactions. Ground-state reactions are generally rationalized on the basis of the relative stabilities of $\sigma$-bonded intermediates formed by initial attack on the aromatic ring. Such a rationalization is not tenable for excited-state reactions because it takes no account of the different electron distribution in the excited state compared with that of the ground state. Instead arguments based on preferred sites of attack are often successful, with the charge distributions in the excited state determined by calculation or by analogy with compounds for which calculated charge densities are available. As a final example of disparities resulting from electronic differences between the ground state and the excited state the effect of electronic excitation on the course of chemical reactions is dramatically demonstrated in the complete reversal of the Woodward–Hoffmann rules [20] for the preferred course of concerted photochemical as opposed to thermal pericyclic reactions.

The factors already considered make for considerable differences between the thermal chemistry and the photochemistry of many organic compounds, and the differences are enhanced as a result of another feature of excited-state properties. This relates to the ability of a molecule to donate or to accept an electron. Ground-state molecules do take part in electron-transfer reactions but this is most usually in the context of the supply or removal of electrons through agencies such as electrode processes or the use of reactive metals. It is unusual to come across a thermal electron-transfer reaction between two electron-paired molecules. However, any molecule on electronic excitation becomes both a better electron donor (because less energy is required to remove an electron from the molecule if there is already one in a higher-energy orbital) and a better electron acceptor (because there is a greater release of energy on addition of an electron to an excited state in which a half-filled orbital of lower energy is available). The result is that electron transfer (**13**) is a common first step in photochemical mechanisms

$$M^* + N \xrightarrow{h\nu} M^{+\bullet} + N^{-\bullet} \text{ or } M^{-\bullet} + N^{+\bullet}$$

(13)

and the subsequent chemistry is that of radical cations and radical anions [21]. The formation of dioxygen radical anion (**6**) is a specific example of this process and, by way of another illustration, such interaction is not uncommon in the photochemistry of ketones, with electron donor compounds like tertiary amines giving rise to ketone radical anions and amine radical cations.

Electron transfer of this general type is the first step in some photobiological processes, including, as has already been mentioned, the process of photosynthesis following the initial light-harvesting step. In synthetic or mechanistic chemistry it is often possible actively to promote reaction through a radical cation or radical anion by the use of a sensitizer known to be very strongly electron-accepting or electron-donating in its excited state. Cyano-substituted aromatic compounds such as 1,4-dicyanobenzene or 9,10-dicyanoanthracene are widely used as electron-accepting sensitizers in photochemical reactions, and methoxy-substituted aromatics such as 1,4-dimethoxybenzene as electron-donating sensitizers. In a few instances it has been demonstrated that when methoxy-substituted aromatic compounds are irradiated in the absence of a suitable acceptor an electron is ejected completely from the molecule (**14**) so that any further reaction may involve a solvated electron.

$$M^* \longrightarrow M^{+\bullet} + e^-$$

(14)

This brings us to the point where a detailed account of the photochemical reactions of organic compounds containing particular fuctional groups can be described and the general considerations covered in this introduction be appreciated and applied in the context of specific examples.

## REFERENCES

[1] G. Ciamician and P. Silber, *Ber. dtsch. Chem. Ges.*, 1909, **33**, 2911.

[2] e.g. A. Schonberg and A. Mustafa, *J. Chem. Soc.*, 1947, 997; A. Mustafa, *Chem. Rev.*, 1953, **51**, 1.

[3] An historical review of photochemistry has been published by H. D. Roth, *Angew. Chem. Int. Edn. Engl.*, 1989, **28**, 1193.

[4] The series *Photochemistry*, Eds D. Bryce-Smith and A. Gilbert, Royal Society of Chemistry, London, vols 1–23 is a particularly useful source of photochemical reactions.

[5] e.g. R. P. Wayne, *Principles and Applications of Photochemistry*, Oxford Science Publications, 1988.
[6] N. J. Turro and V. Ramamurthy, in *Rearrangements in Ground and Excited States*, ed. P. De Mayo, Academic Press, 1980, Chapter 13.
[7] J. A. Barltrop and J. D. Coyle, *Principles of Photochemistry*, Wiley, 1978; R. Devonshire, *Physical Photochemistry*, Ellis Horwood, 1986.
[8] M. B. Robin, *Higher Excited States of Polyatomic Molecules*, Academic Press, 1985, Volume 3, Chapter 1.
[9] H. H. Jaffe and M. Orchin, *Theory and Application of Ultraviolet Spectroscopy*, Wiley, 1962, Chapter 6.
[10] M. Kasha, *Radiation Research, Supplement 2*, 1960, 243.
[11] S. L. Murov, *Handbook of Photochemistry*, Marcel Dekker, 1973.
[12] R. S. Becker, *Theory and Interpretation of Fluorescence and Phosphorescence*, Wiley, 1969.
[13] W. M. Horspool, in *Synthetic Organic Photochemistry*, ed. W. M. Horspool, Plenum Press, 1984, Chapter 9.
[14] D. O. Cowan and R. L. Drisko, *Elements of Organic Photochemistry*, Plenum Press, 1976.
[15] P. S. Engel and B. M. Monroe, *Adv. Photochem.*, 1971, **8**, 245–313.
[16] H. H. Wasserman and R. W. Murray, *Singlet Oxygen*, Academic Press, 1979.
[17] G. Porter, in *Light, Chemical Change and Life*, eds. J. D. Coyle, R. R. Hill, and D. R. Roberts, Open University 1982, Chapter 6.3.
[18] N. J. Turro, *Modern Molecular Photochemistry*. Benjamin/Cummings, 1978, Chapter 4.
[19] J. F. Ireland and P. A. H. Wyatt, in *Advances in Physical Organic Chemistry*, eds V. Gold and D. Bethell, 1976, Volume 12, p. 131.
[20] T. L. Gilchrist and R. C. Storr, *Organic Reactions and Orbital Symmetry*, Second Edition, C.U.P., 1979.
[21] P. S. Mariano and J. L. Stavinoha, in *Synthetic Organic Photochemistry*, ed. W. M. Horspool, Plenum Press, 1984, Chapter 3; R. S. Davidson, *Advances in Physical Organic Chemistry*, eds V. Gold and D. Bethell, Academic Press, 1983, Volume 19, p. 1.

# 2

# Hydrocarbon systems

The photochemistry discussed in this chapter describes the reactions of what are loosely referred to as hydrocarbons. Thus the material covered ranges from alkanes, through alkenes and alkynes to arenes. In the main, reactions which occur in the solution phase are discussed since it is these which are, more often than not, of value to a chemist in synthesis.

## 2.1 ALKENES

### 2.1.1 Spectra of alkenes and dienes

The bonding electrons in alkenes are found in the $\pi$ and $\sigma$ bonds of the compounds and, as a result, photochemical excitations will be one of two types, either a $\sigma \rightarrow \pi^*$ excitation, where an electron is promoted from a $\sigma$ orbital to a vacant low-lying $\pi^*$ anti-bonding orbital, or a $\pi \rightarrow \pi^*$ excitation, involving the promotion of an electron from the ground-state $\pi$ orbital to a vacant $\pi^*$ anti-bonding orbital. The simple monoalkenes usually absorb light of wavelength shorter than 210 nm. The parent of the series, ethene, absorbs well outside the normally accessible ultraviolet region and has a maximum at 162 nm in the gas phase [1, 2]. This band, arising from an allowed $\pi\pi^*$ transition, is quite broad, extending from 145 to 190 nm with a tail up to approximately 207 nm. This absorption gives rise to the lowest excited singlet states which are of relatively high energy (about 150 kcal $mol^{-1}$; 650 kJ $mol^{-1}$) and are of little synthetic value because of their inaccessibility. The spectra also show sharp lines which are the commencement of a Rydberg series in which a $\pi$-electron is promoted into a $\sigma$-type orbital. The introduction of substituents shifts the $\pi\pi^*$ absorption towards longer wavelength and *cis*-but-2-ene, for example, shows a maximum at 174 nm while the *trans*-isomer has a maximum at 178 nm. This difference in the position of the maximum where the *trans*-isomer is at a longer wavelength is often but not always observed. In addition to this the *trans*-isomer usually has a larger extinction coefficient for the $\pi\pi^*$ band. Increasing alkyl substitution also brings about a drop in energy of the first Rydberg transition involving $\pi \rightarrow 3\sigma$ excitation. The triplet $\pi\pi^*$ for ethene and other simple alkenes is very weak [1, 2]. Typically ethene shows an absorption extending up to 350 nm for this triplet excitation with an extinction coefficient $\varepsilon$ of approximately $10^{-4}$ $dm^3$ $mol^{-1}$ $cm^{-1}$. The energy of this triplet state is around 80 kcal $mol^{-1}$; 350 kJ $mol^{-1}$. The difference in energy between the singlet and the triplet state of ethene leads to a very low intersystem crossing rate [3]. Thus the triplet-state reactivities of the simple alkenes are usually studied using triplet sensitizers.

The introduction of alkyl substituents to the simple alkenes does bring about a small shift of the absorption maximum towards longer wavelength. A small shift is also observed for non-conjugated dienes, such as hexa-1,5-diene ($\lambda_{max}$ = 178 nm, $\varepsilon$ = 26 000 $dm^3$ $mol^{-1}$ $cm^{-1}$), and the absorption maxima are similar to those for the isolated alkenes [4]. Norbornadiene (**1**)

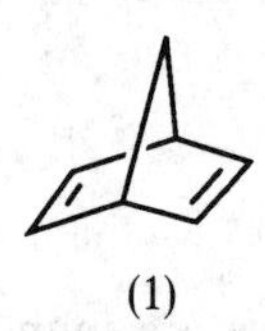

(1)

is an exception to this since the double bonds are close enough to interact and as a consequence the absorption is at longer wavelength and of greater intensity [5]. If, however, the substituent is a double bond or an aryl ring and is capable of conjugation with the alkene moiety then the ultraviolet absorption moves to longer wavelength and into the region of the spectrum which is more readily accessible to solution-phase photochemistry. Buta-1,3-diene is typical with an absorption at 210 nm [1]. The singlet-state energy of these conjugated dienes thus becomes progressively lower with the introduction of conjugation. Buta-1,3-diene has s-*cis*- and s-*trans*-conformations in the ground state and excitation will lead, therefore, to two first excited singlet states. A large singlet–triplet splitting is still present in such dienes and intersystem crossing is inefficient. Thus any study of the triplet reactivity of dienes usually requires the use of triplet sensitizers to populate the $\pi\pi^*$ state. As with the singlet state there are two possible triplet states corresponding to the s-*cis*- and the s-*trans*-geometry with triplet energies of around 53 kcal $mol^{-1}$ (225 kJ $mol^{-1}$) and 60 kcal $mol^{-1}$ (250 kJ $mol^{-1}$) respectively [6].

### 2.1.2 Excited-state geometry of alkenes

Excitation of a planar alkene like ethylene results initially in the formation of a planar excited state molecule in observance of the Franck–Condon Principle. This initially formed planar excited-state $\pi\pi^*$ species, whether as a singlet or a triplet, rapidly relaxes by rotation of the terminal groups — methylenes in the case of ethylene — through 90° around the central bond to give the lowest energy conformation possible. Spectroscopic evidence has been obtained which shows that there is additional distortion of the methylene groups from planarity as the C-atoms tend towards tetrahedral ($sp^3$) hybridization. In this energy minimum there is essentially no $\pi$-bond, the electronic interaction is at a minimum and the terminal carbon atoms are orthogonal. The orthogonal triplet state is important in the sensitized photochemical processes of alkenes since this can account readily for some of the reactions which the excited state can undergo.

### 2.1.3 Direct irradiation. Singlet state reactivity of alkenes

#### *2.1.3.1 cis–trans*-isomerization

The most commonly observed photoprocess for alkenes in solution is *cis–trans* or more generally *Z–E* isomerization which can occur by both direct and sensitized irradiation and is usually associated with $\pi\pi^*$ excited states. Direct irradiation for the simplest alkenes is difficult to achieve as a result of the high energy absorptions mentioned earlier. A disadvantage of the need to go below 200 nm to irradiate such compounds is that wavelengths in this region are not readily accessible to solution-phase photochemists. However, as the substitution is changed, so the absorption of the alkene moves towards longer wavelengths. Such is the case with the *cis*- and *trans*-stilbenes represented by **2** and **3** respectively. The direct irradiation of either of these

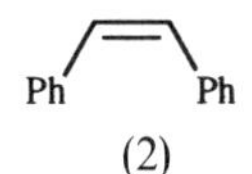


(2)

Ph

Ph

(3)

isomers at 313 nm results in the formation of a mixture of 93% *cis* and 7% *trans* irrespective of the length (within reason) of irradiation [7]. The ratio of products does not alter and such a state is referred to as a photostationary state. The photostationary-state composition is dependent upon the relative extinction coefficients of the *cis*- ($\varepsilon = 2280\ dm^3\ mol^{-1}\ cm^{-1}$) and the *trans*- ($\varepsilon = 16\ 300\ dm^3\ mol^{-1}\ cm^{-1}$) alkene at the wavelength used for the irradiation and on the partitioning ratios for the decay of each of the excited states to the ground state. In the case of stilbene the partitioning ratio in the excited states is almost unity. Thus the photostationary state is rich in the compound with the lower extinction coefficient. The direct irradiation of stilbene is still an area of considerable interest [8] along with *cis-trans* isomerization of alkenes in general and Hammond [9] estimates that there are several thousand references in the literature to such processes. This work has established the influence of substituent effects [10], temperature [11], and a variety of other parameters. The synthetic value of *trans–cis*-isomerization lies in the fact that the more stable can be readily converted into the less stable isomer. Some examples of the result of direct irradiation of alkenes are shown in **4** [12–15].

Ph $\underset{285\ nm}{\overset{h\nu}{\rightleftarrows}}$ Ph

$\phi_{c\text{-}t} = 0.33$
$\phi_{t\text{-}c} = 0.22$

Ref [12]

$CO_2H$ $\xrightarrow{h\nu}$ $HO_2C$

74%

Ref [13]

Ph Bu$^t$ Bu$^t$ Ph $\xrightarrow{h\nu}$ Ph Ph Bu$^t$ Bu$^t$

26%

Ref [14]

(4)

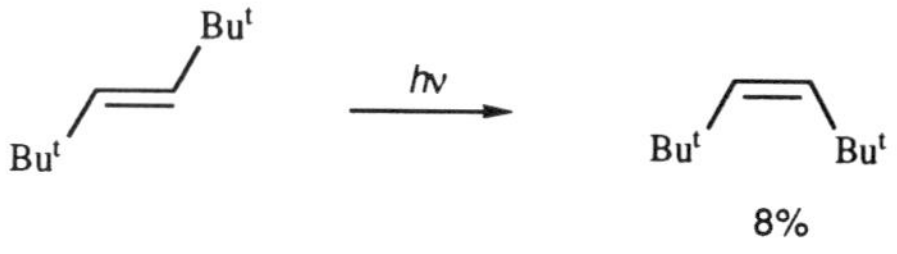

8% Ref [15]

(4 continued)

### *2.1.3.2 Cumulenes*

By analogy with alkene photochemistry the most efficient process in allene and cumulene photochemistry is isomerization by rotation around the $\pi$-bond [16]. This is exemplified by the racemization of the allene (**5**) [17]. Another

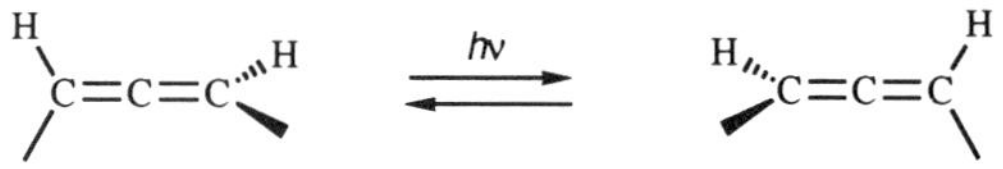

(5)

example has shown that the irradiation of partially resolved 1,3-di-*t*-butylallene resulted in its photoracemization [16]. Studies have been carried out on a cyclic allene (**6**) where it was discovered that benzene-sensitized

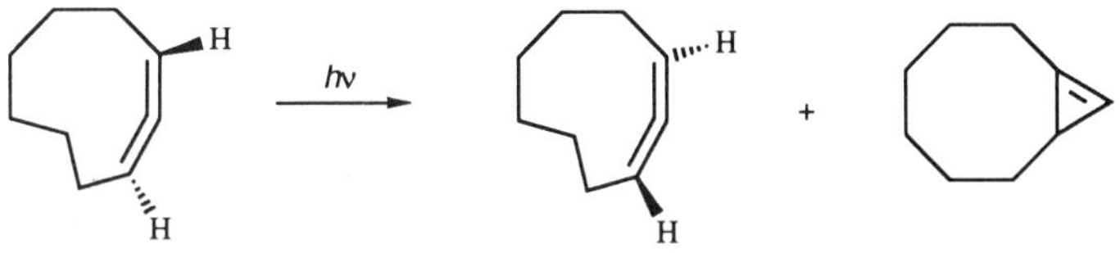

(6)

irradiation affords a racemized product although this is accompanied by rearrangement [18]. Other research has sought to provide a route to photoresolution of allenes but so far little success has been achieved with only a 3.4% enantiomeric excess being obtained from the irradiation of penta-2,3-diene (**5**) with an optically active sensitizer [19].

### *2.1.3.3 Rydberg reactivity*

As pointed out earlier, in addition to the $\pi \rightarrow \pi^*$ excitation path for simple alkenes Rydberg transitions also occur [1, 20]. Generally the excited state can be pictured as that shown in **7** where 2,3-dimethylbut-2-ene has undergone

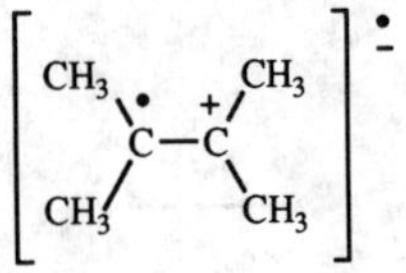

(7)

excitation which has resulted in a $\pi \rightarrow 3s$ transition producing the species shown. Such an intermediate readily accounts for the nucleophilic trapping which is encountered in the direct irradiation of alkenes in alcoholic media. A typical example of this behaviour is shown in **8**, where the intermediate

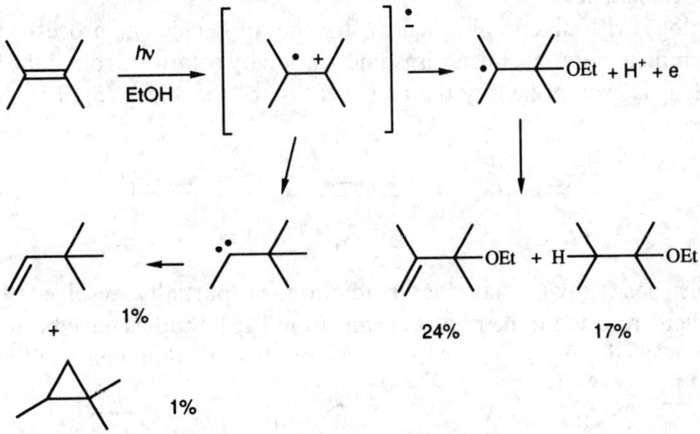

hν EtOH

OEt + H⁺ + e

OEt + H OEt

1%

+

24% 17%

1%

(8)

is trapped nucleophilically by ethanol [21]. A proton and the loosely bound Rydberg electron are lost to the medium. The radical formed disproportionates to afford the two ethers. Alternatively, the intermediate undergoes methyl migration to afford a carbene which gives low yields of 3,3-dimethylbut-1-ene and 1,1,2-trimethylcyclopropane by processes typical of a carbene intermediate. Reactions of this type have been reviewed [22] and some examples of this nucleophilic trapping are illustrated in **9**, where it can be seen that the substitution pattern on the alkene plays an important role in determining the outcome of the reaction [21, 23].

In the absence of a nucleophilic solvent, irradiation of the alkenes can still produce carbenes via the Rydberg state by a 1,3-hydrogen migration in competition with the other principal photoprocesses such as *cis–trans* isomerization or double-bond migration. Thus in pentane, for example,

hν / MeOH

$OCH_3$ + $OCH_3$ + $OCH_3$

11%

22%

Ref [23]

hν / MeOH

$CH_3O$ + $OCH_3$

19% 16%

+ $CH_3O$ + $OCH_3$

23% 7%

Ref [21,23]

hν / MeOH

+ + + $OCH_3$

5% 26% 21% 1%

$CH_3O$ $CH_3O$

+ + + $OCH_3$

13% 4% 1% Ref [21]

(9)

irradiation of the alkene (**10**) affords the carbene, which either undergoes a

hν / $C_5H_{12}$

18%

hν $C_5H_{12}$

17% 24%

(10)

H-migration yielding 3,3-dimethylbut-1-ene or else cyclizes to yield 1,1,2-trimethylcyclopropane as well as affording 2,3-dimethylbut-1-ene [24]. This is common behaviour and some other examples are illustrated in **11**

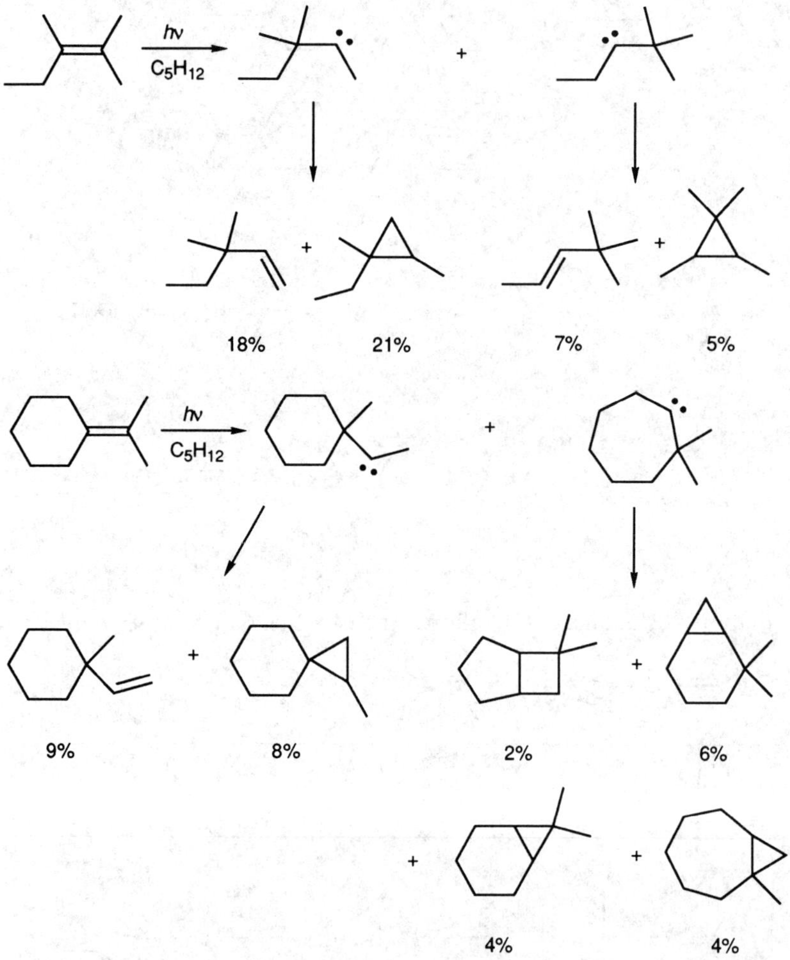

$h\nu$
$C_5H_{12}$

18% 21% 7% 5%

$h\nu$
$C_5H_{12}$

9% 8% 2% 6%

4% 4%

(11)

where it can be seen that in unsymmetrically substituted alkenes the possible carbenes are formed by group migrations that even result in ring expansion [25]. When the alkene is within a ring (**12**) the carbene formed on excitation

hν
$C_5H_{12}$
20%
Ref [25]

CH
hν/185 nm
$C_5H_{12}$
+
$CH_2$
+
22%
(12)
43%
Ref [26]

and group migration is the result of ring contraction rather than alkyl group migration [25] and even with cyclohexene there is evidence that the cyclopentyl-substituted carbene, formed by ring contraction, is involved in competition with the alternative carbene intermediate, obtained by a 1,2-hydrogen migration [26].

### *2.1.3.4 Dimerization*

#### *2.1.3.4.1 Singlet-state reactivity*

Direct irradiation can also bring about dimerization, as with the irradiation of neat 2,3-dimethylbut-2-ene (**13**) when the dimer octamethylcyclobutane is

hν
neat
75%

(13)

formed accompanied by the usual monomeric products illustrated in **8** [27]. The dimerization reaction is stereospecific and this is demonstrated by the irradiation of *cis*-but-2-ene (**14**) which yields the all *cis*-dimer and the

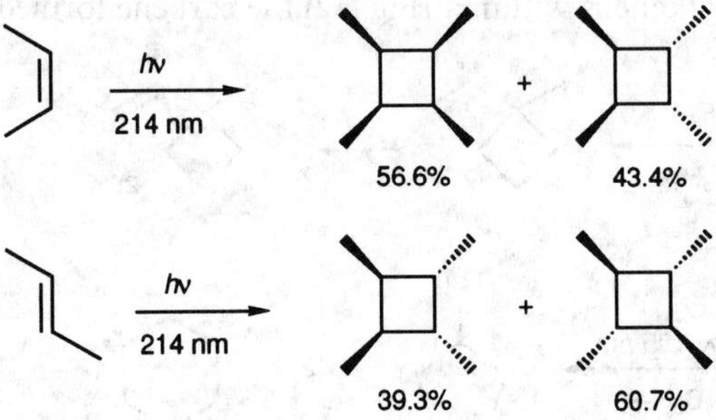

(14)

*cis-anti-cis*-dimer while *trans*-but-2-ene affords the *cis-anti-cis*-dimer and the *trans-anti-trans*-dimer [28]. Cyclohexene also dimerizes, yielding a mixture of dimeric products [29] while norbornene (**15**) affords two dimers in a ratio of 1 : > 10 [30].

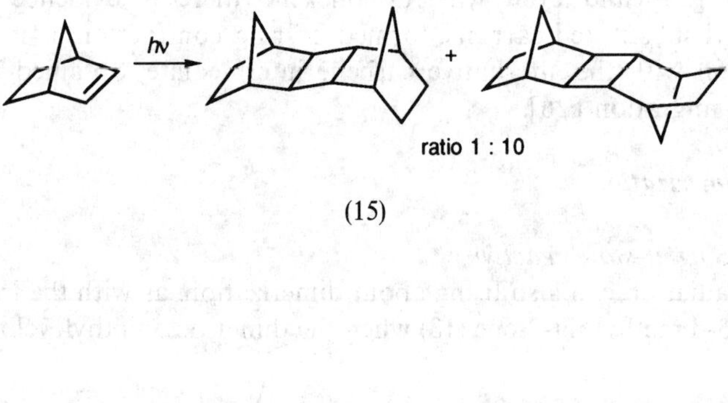

(15)

*2.1.3.4.2 Triplet-state reactivity*

Dimerization of alkenes can also be effected *via* the triplet state. However, the majority of cases reported involve cyclic alkenes and there are few, if any, examples involving open-chain systems. Dimerization appears to favour the formation of the *cis-anti-cis*-arrangement of rings in the dimer. Examples of dimerization of cyclopropenes (**16**) [31], cyclobutenes (**17**) [32], cyclopen-

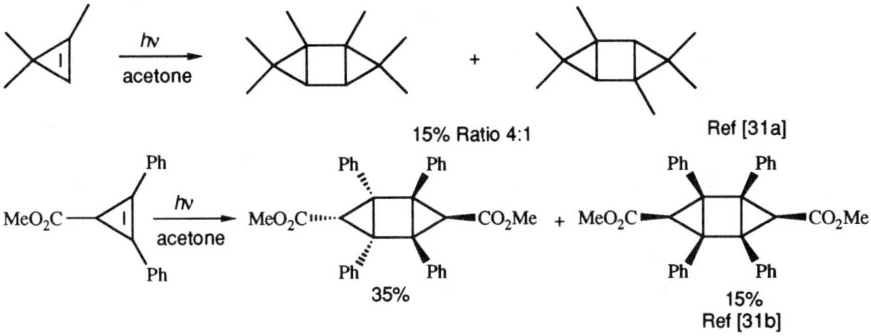
hν
acetone
+
15% Ratio 4:1
Ref [31a]
Ph
MeO2C
Ph
hν
acetone
Ph
Ph
MeO2C
CO2Me
Ph
Ph
35%
+
MeO2C
Ph
Ph
CO2Me
Ph
Ph
15%
Ref [31b]

(16)

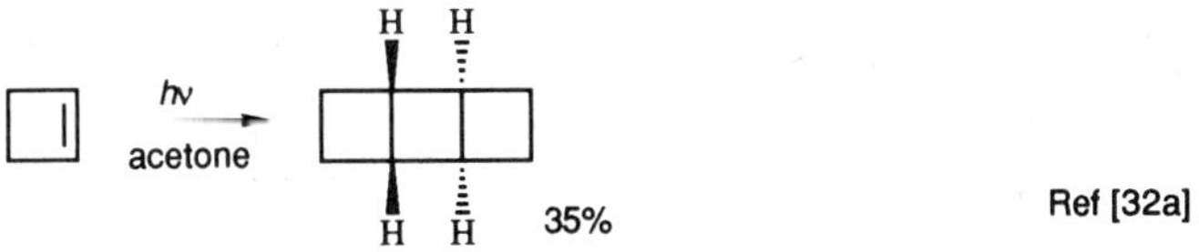
hν
acetone
H H
H H
35%
Ref [32a]

(17)

tenes (**18**) [33], and cyclohexenes (**19**) [34] have been reported. Norbornene

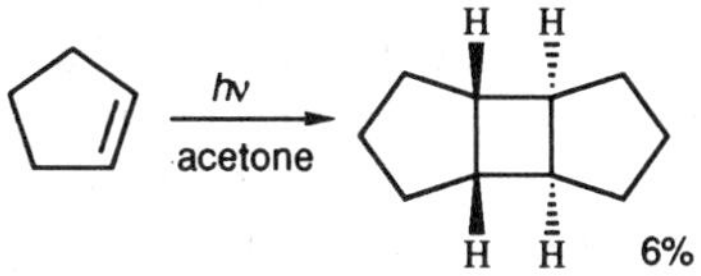
hν
acetone
H H
H H
6%

(18)

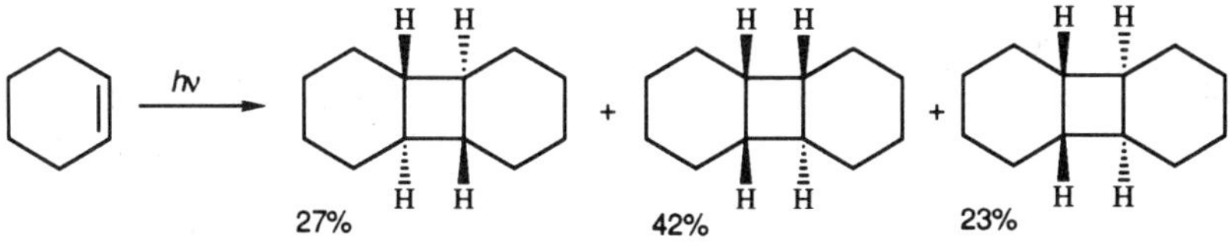
hν
H H
H H
27%
+
H H
H H
42%
+
H H
H H
23%

(19)

can also be made to dimerize on sensitization and there is a preference for the formation of the *exo-trans-endo*-dimer. A variety of sensitizers have been used for this and in addition the influence of solvent has been studied [30, 35]. Some use can be made of such additions as in the synthesis of cage compounds shown in **20** [36–38].

hν acetone

62%

Ref [36]

MeO OMe Cl Cl Cl Cl N $CO_2Et$ N $CO_2Et$

hν acetone

MeO OMe Cl Cl Cl Cl N $CO_2Et$ N $CO_2Et$

Ref [37]

O

hν acetone

O

32%

Ref [38]

(20)

*2.1.3.4.3 Copper(I)-catalysed addition reactions*
Interestingly dimerization can also be brought about in the presence of copper(I) salts such as Cu(I) triflate (trifluoromethanesulphonate) which has been shown to be the most effective. The reaction under these conditions often yields different results from those obtained under sensitized conditions. The dimerization of norbornene illustrates this by the predominant formation of the *exo-trans-exo*-dimer (**21**) [39]. The route by which the dimerization

hν Cu(I)triflate

$Cu^+$ + $Cu^+$

exo-exo-complex

10% 90% exo-trans-exo

(21)

takes place involves the formation of complexes between the alkene component and the highly electrophilic Cu(I) ion [40]. Photoexcitation of the complex yields the products. In the case of norbornene it has been suggested that two complexes are involved with a preference for the formation of the thermodynamically favoured *exo-exo*-complex. Many examples of the use of copper complexes have appeared over the years and a few are shown in **22** [40–43].

hν, Cu(I)triflate

R= H

R = (2-(CH$_2$Pr$^i$)-3,5,6... aryl: OMe, CH$_2$Pr$^i$, MeO, OMe)

Ref [41]

hν, Cu(I)triflate

| $R^1$ | $R^2$ | $R^3$ | % |
|---|---|---|---|
| H | H | OEt | 74 |
| H | Me | OEt | 60 |
| Me | H | OEt | 76 |

Ref [42]

hν, Cu(I)triflate

Total yield 80%

Ref [43]

(22)

### *2.1.3.5 Sensitized irradiation of alkenes. Triplet-state reactivity*

Sensitized *cis–trans*-isomerization of an alkene can be brought about by the use of a triplet energy sensitizer such as a ketone. The sensitized isomerization of stilbene (**2, 3**) is typical of the type of study carried out. In this system, as in others, the photostationary state composition is dependent upon the triplet energy of the sensitizer employed. Some sensitizers may have an energy which is too low to populate the vertical *cis* or *trans* triplet state but may still bring about isomerization. This occurs because the sensitizer in question has been able to populate the twisted triplet state directly, by-passing the vertical excited state. A typical energy diagram (Fig. 2.1) for stilbene shows the

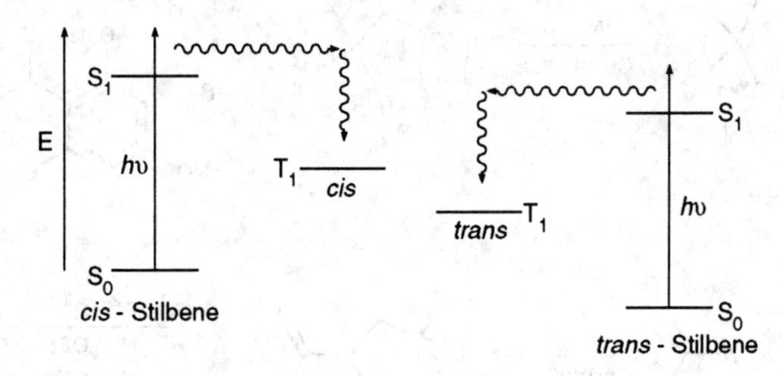


Fig. 2.1. Relative relationship of excited states of *cis*- and *trans*-stilbene.

relationships of these energy levels. The following summarize the reactivity:

(1) The use of high-energy sensitizers can readily populate both excited triplet states provided the available triplet energy is greater than the energy of the *cis*- or the *trans*-excited states.
(2) The use of low-energy sensitizers will be able to populate the non-planar triplet state provided that the energy available is less than the energies

**Table 2.1.** Results from the sensitized isomerization of pent-2-ene [45]

| Photostationary state composition (*trans*:*cis*) | Sensitizer (energy kJ mol$^{-1}$) |
|---|---|
| 1.0 | Benzene (351) |
| 1.8 | Acetone (334) |
| 3.9 | Acetophenone (309) |

of both the vertical excited states but greater than the energy of the non-planar excited states.

(3) The use of sensitizers with an energy intermediate between that of the *cis*- and the *trans*-excited states can populate the *trans*-excited state preferentially. This will lead to a photostationary state rich in the *cis*-isomer by selective isomerization of the *trans*-isomer.

Typical examples of these effects applied to pent-2-ene are shown in Table 2.1 [45] where decreasing the energy of the sensitizer increases the amount of *trans*-isomer in the photostationary state.

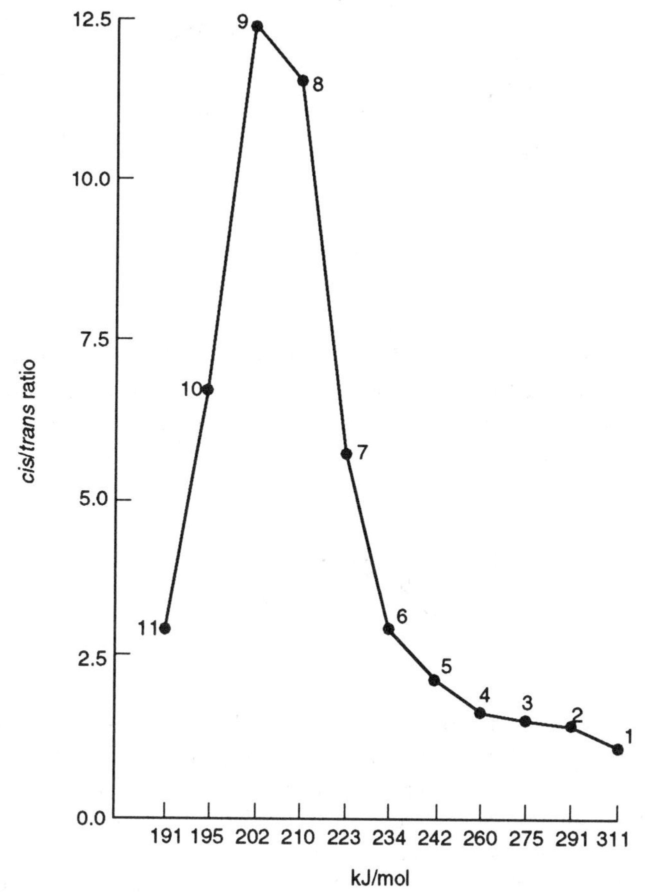


Fig. 2.2. Photosensitized isomerization of stilbene. 1, Acetophenone. 2, Benzophenone.3, Thioxanthone. Michler's ketone. 5, Acetonaphthone. 6, Chrysene. 7, Fluorenone. 8, Benzil. 9, Pyrene. 10, Benzanthrene. 11, 3-Acetylpyrene. (Graph drawn from recalculated data from [46])

More information regarding this has been assembled for the sensitized isomerization of the stilbenes and the effect of sensitizer energy on this is represented in Fig. 2.2 [46]. Some examples showing sensitized irradiation are illustrated in **23** [47–49].

hν
acetone

Ref [47]

hν
sens.

Ref [48]

(23)

*2.1.3.5.1 Photoprotonation of alkenes*
Sensitized reactions of the cycloalkenes, e.g. cyclohexenes, cycloheptenes and cyclo-octenes, have also been shown to effect *cis–trans*-isomerization using aromatic hydrocarbon sensitizers such as xylene, toluene, or benzene [22]. In the case of cyclo-octene (**24**) the isolation and identification of the

hν

(24)

*trans*-isomer had been demonstrated [22]. With the increase in strain from the $C_8$ to the $C_7$ and $C_6$ cyclic compounds the identification of the corresponding *trans*-cycloheptene and *trans*-cyclohexene relied on the reactions observed and on spectroscopy [50]. When the reactions are carried out in the presence of a proton source such as an alcohol or an acid, irradiation is followed by conversion into an isomerized product and/or an ether. The reaction mechanism for the processes encountered in such sensitized irradiations is illustrated in (**25**), whereby excitation by xylene-sensitized irradiation

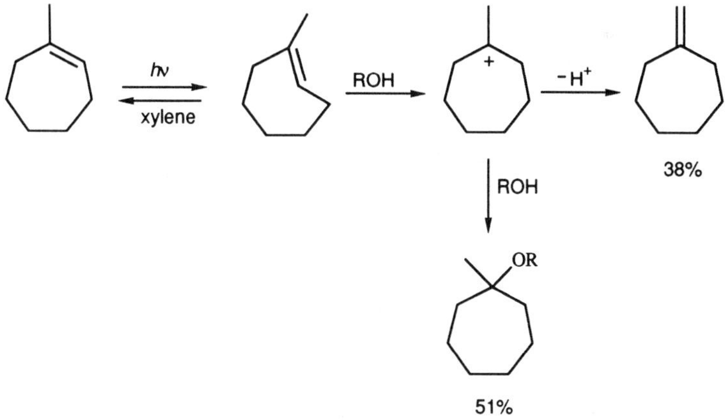

(25)

affords the *trans*-isomer, which is then protonated by solvent such as an alcohol or an acid. The resultant carbocation either deprotonates to afford an alkene or is trapped by solvent to yield an ether. In some cases these products are accompanied by a reduction product (**26**). These reduction

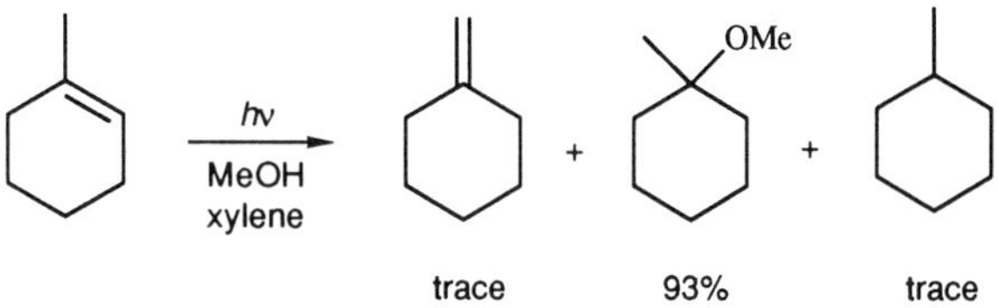

(26)

products are formed by hydride transfer to the carbocation, a mechanism which can be proven using deuterated alcohols [51]. In one case, the xylene-sensitized irradiation of the octalin (**27**), reduction is the dominant

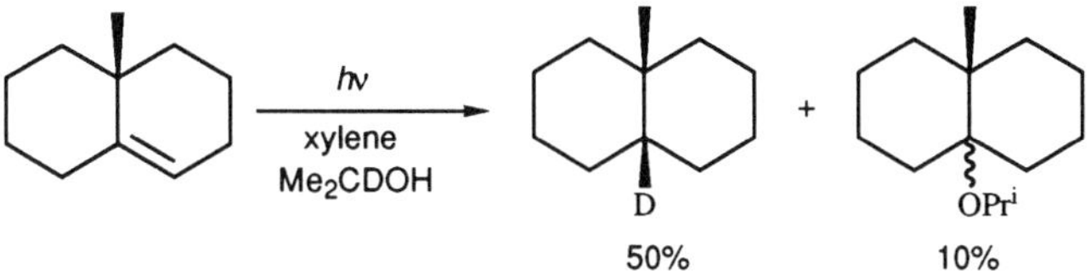

(27)

reaction [51]. This, however, is a special case, since the alcohol, propan-2-ol, is bulky, and nucleophilic trapping of a tertiary centre is difficult. Furthermore hydride transfer from propan-2-ol is a facile process. The evidence for the involvement of the cation is compelling since the addition follows the expected path involving the more stable ion. The deprotonation and addition reactions are typical for such a species, as is the rearrangement which is sometimes observed. The reactions discussed above show some dependence on substitution, and cycloheptene [52] and cyclohexene [52] undergo protonation less readily than the substituted derivatives illustrated. Cycloheptene [52] does not undergo addition of alcohol when irradiated with xylene as the sensitizer but instead affords adducts with the sensitizer (see later). Apparently the *trans*-cycloheptene is insufficiently basic to abstract a proton from the alcohol, and to achieve addition a mineral acid (0.25% sulphuric acid) is required to yield the corresponding cycloheptyl ether. Indeed most of the 1-alkylcycloalkenes yield a *trans*-isomer which is basic enough to protonate in methanol or ethanol but will require mineral acid to achieve the protonation step with less acidic alcohols such as *t*-butanol. Thus there is a fine balance between the need for mineral acid or not. Cyclohexene [52], however, adds methanol readily without the need for added mineral acid. This reaction is complicated by the fact that *trans*-cyclohexene readily undergoes a thermal addition with the *cis*-isomer to yield the 2 + 2 dimers (see 2.1.3.4.2). The scope of the addition reaction mode is illustrated in **28** [21, 52–55].

hν
xylene
MeOH
OMe MeO
36% 26% 39%
Ref [53, 54]

hν
xylene
MeOH
MeO OMe
37% 10% 21%
Ref [21]

(28)

OMe

hν xylene $MeOH/H_2SO_4$

59% Ref [52]

$CO_2Me$ hν $C_6H_6$ MeCOOH AcO $CO_2Me$ + $CO_2Me$ OAc

58% 2% Ref [55]

(28 continued)

### 2.1.4 Photoreactions of dienes and trienes

#### *2.1.4.1 Conjugated dienes*

The first electronic transition in dienes is $\pi\pi^*$ and in butadiene this broad band is centred at 210 nm [1]. Evidence has been reported which shows that irradiation at 254 nm can bring about the reaction of buta-1,3-diene [56]. Like the alkenes, 1,3-dienes undergo *trans–cis*-isomerization on sensitization. Penta-1,3-diene has been studied in some detail [57] and again the outcome of the reaction is dependent upon the energy of the sensitizer employed, with the complication, compared with the stilbenes, that the energy of the vertical *cis*-triplet state is lower than that of the *trans*-triplet state. Thus the use of high-energy sensitizers leads to a photostationary state composition (*trans*:*cis*) of 1.23. When the triplet energy of the sensitizer is between that of the *trans*-triplet and that of the *cis*-triplet the outcome of the reaction is variable.

#### *2.1.4.2 Electrocyclic reactions of conjugated dienes*

Electrocyclic reactions are common photochemical processes of 1,3-dienes. Typical of this is the cyclization of the *trans-trans*-diene (**29**) which on

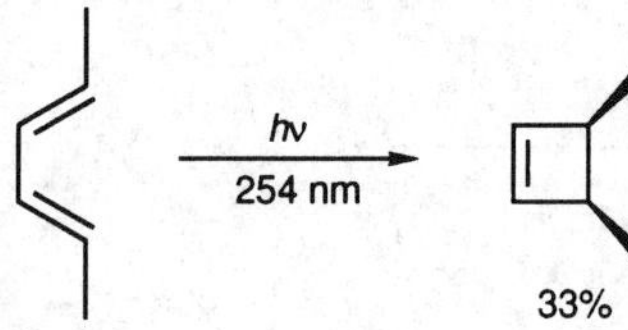

(29)

irradiation at 254 nm affords the cyclobutene specifically [58]. This reactivity is quite general and cyclic alkenes, e.g. **30**, also undergo this specific reaction

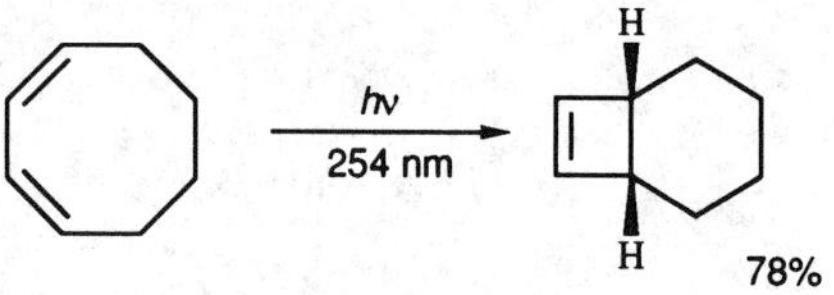

(30)

path to yield the bicyclic product [59]. These are examples of pericyclic processes where the outcome of the reaction is governed by the symmetry of the electronic wavefunctions or orbitals of the starting material. In the excited singlet-state reactions of the type illustrated above the outcome of the reaction, in simple terms, is controlled by the symmetry of the lowest unfilled molecular orbital (LUMO) (**31**) although steric factors can also

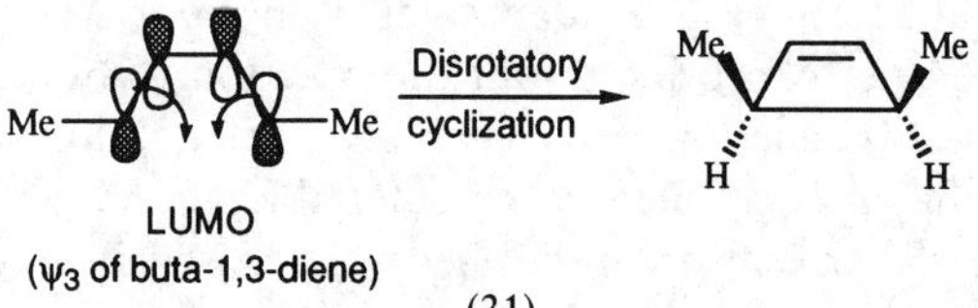

(31)

exercise some control over the outcome of the reaction [58]. The cyclization of a diene involves a four-electron $\pi$-system (fitting into the general reactivity of a $4n$ process) and will occur by a disrotatory mode of the LUMO (i.e. $\Psi_3$ of buta-1,3-diene) placing the methyl groups of **29** or the ring residue of **30** *cis* to each other. This rule is part of a set developed by Woodward and Hoffman and many texts have been devoted to this subject [60]. In theory the forward (illustrated) and backward reactions should occur with equal facility and should be controlled by the same factors. However, the cyclobutenes formed are normally transparent to ultraviolet light at normal wavelengths and the reverse reaction is not often observed. Another example of the control exercised by orbitals on the outcome of photochemical reactions is seen with the *trans-cis-trans*-triene (**32**). The photoreactions of this type of

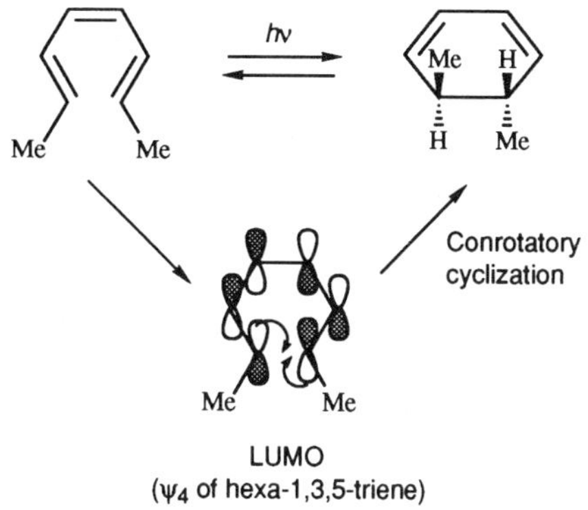

Conrotatory cyclization

LUMO
($\psi_4$ of hexa-1,3,5-triene)

(32)

compound fall into another class, that of the $4n + 2$ processes, since the reaction involves six electrons ($6\pi$). In simple terms the ring-closing reaction is controlled by the symmetry of $\Psi_4$ of the hexa-1,3,5-triene and follows a conrotatory path to yield the cyclohexadiene [61]. This process is well exemplified in the literature and such ring closures or ring openings operate readily with cyclohexadienes (**33**) and dihydronaphthalenes such as (**34**) [62].

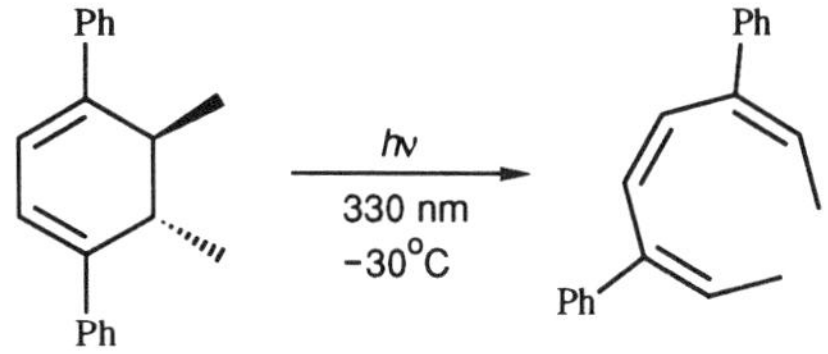

(33)

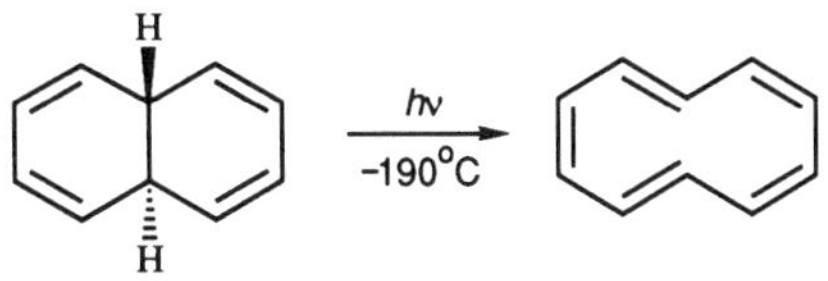

(34)

Clearly the conrotatory process will only take place when steric factors do not prevent it. Thus the *trans*-dihydronaphthalene (**34**) undergoes ring opening to yield the [10]-annulene but the anhydride (**35**) fails to react via

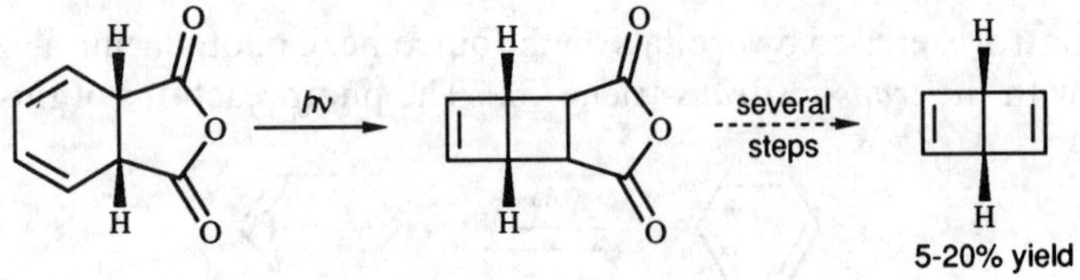

(35)

the ring opening path. In cases like this the four-electron disrotatory closure can operate and yields the bicyclic compound which was used as a starting point for the synthesis of *Dewar* benzene [63]. Indeed there is often a fine balance between the ring opening of a cyclohexadiene and the disrotatory cyclization of the diene moiety. There are some elegant studies on such reactions and the photoreactions illustrated by the vitamin D series (**36**) is

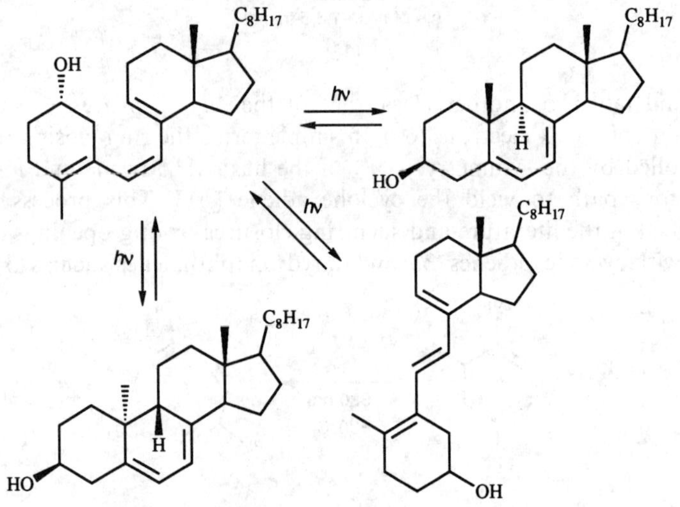

(36)

a fine example showing the operation of the two possible conrotatory cyclizations [64]. Other examples of 4e and 6e processes are included in (**37**) [65–67].

### *2.1.4.3 Bicyclobutane formation*

The electrocyclic processes described above for the 1,3-dienes are accompanied by another cyclization reaction which affords bicyclobutanes. The route to these products is thought to involve biradicals and be a non-concerted process and arises on direct irradiation [55] as well as under conditions of mercury sensitization [68]. Thus excitation affords a singlet state which ring

Ph Ph hν direct H Ph Ph H 59% Ref [65]

O O hν H O O H Ref [66]

HOOC H hν HOOC H 50% Ref [67a]

O hν O 59% Ref [67b]

O O hν H O O H 60% Ref [67c]

(37)

closes to a biradical and ultimately the bicyclobutane as shown in **38**. The

hν

$S_1$

(38)

reaction can be synthetically useful and can be used for the formation of bicyclobutanes or else, in the presence of a protic solvent, ring opening can yield alkoxycyclopropane derivatives. Some applications are shown in (**39**) [69]

hν
cyclohexane
Vycor filter

hν
cyclohexane
Vycor filter

Ref [69a]

hν
ROH

OR
40%

Ref [69b]

hν
quartz

MeOH

OMe

30%

Ref [69c]

(39)

### *2.1.4.4 1,4-Dienes*

*2.1.4.4.1 The di-π-methane rearrangement*

The reaction which falls under this categorization is remarkably widespread and involves the photochemical rearrangement of penta-1,4-dienes with di-substitution at the central carbon into vinylcyclopropanes [70]. There are other variants of this reaction; the oxa-di-$\pi$-methane process is dealt with in Chapter 3 and the aza-di-$\pi$-methane reaction in Chapter 5. In the absence of dimethyl substitution at C-3, (2 + 2)-cycloaddition results on irradiation affording the products shown in **40** [71]. The di-$\pi$-methane reaction can be

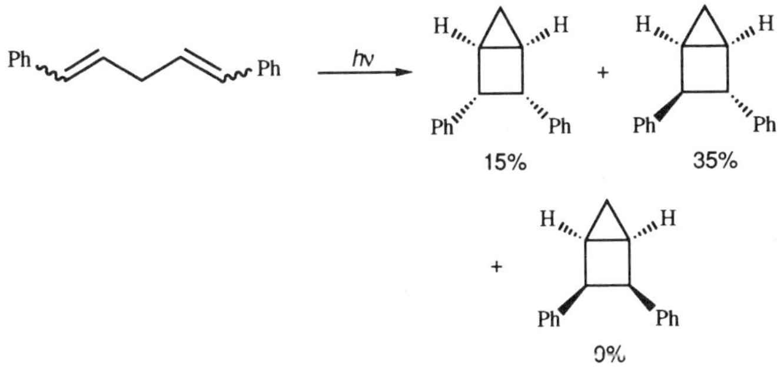

(40)

interpreted as a 1,2-migration followed by cyclization to yield the product as shown in **41** and this rationalization provides a useful predictive approach. However, the biradicals represented in such a scheme (**41**) do not represent

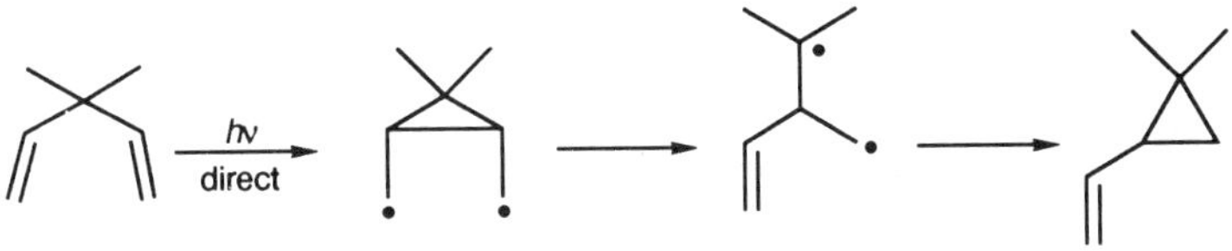

(41)

discrete intermediates and are incorporated to make the understanding of the process simpler. The di-$\pi$-methane reaction of these open-chain dienes generally involves the singlet excited state (sensitization usually brings about *trans–cis*-isomerization (**42**) [72] or occasionally group migration (**43**) [73]),

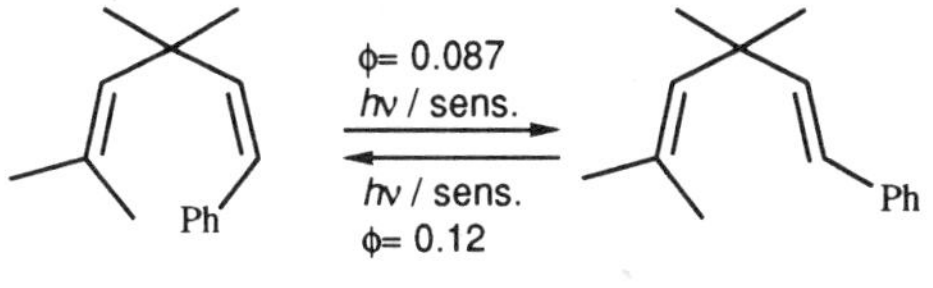

(42)

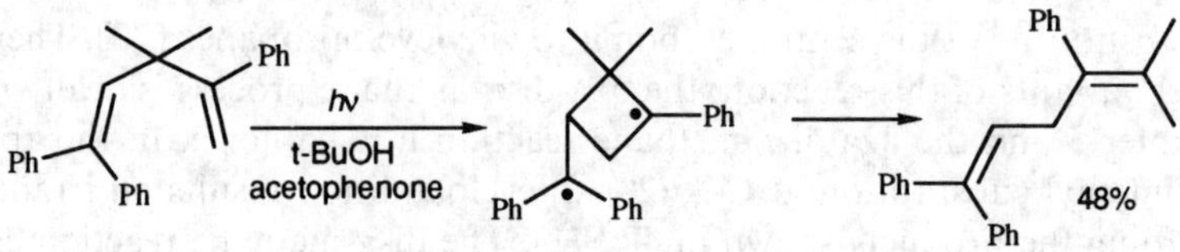

(43)

is concerted and is stereospecific at all centres, particularly retaining the original geometric arrangement at the terminal carbons of the diene. In addition the reaction shows remarkable regioselectivity and the simplistic more stable biradical approach can be used again to predict the outcome of the reaction. Thus for example the diene (**44**) on direct irradiation affords

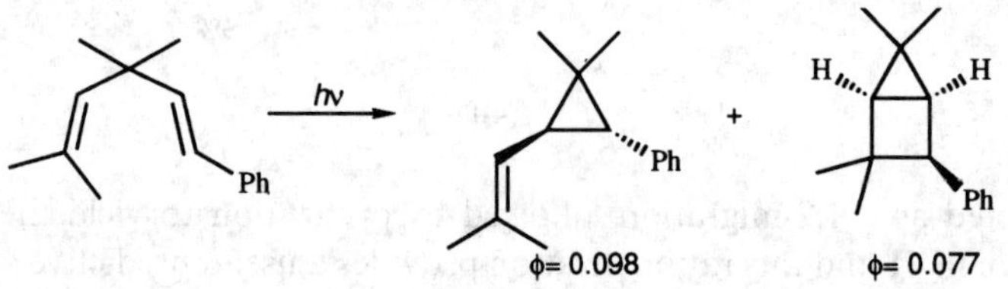

(44)

exclusively the vinyl cyclopropane shown [72]. Other examples illustrating the process are given in **45** [74]. The reaction also arises when one of the

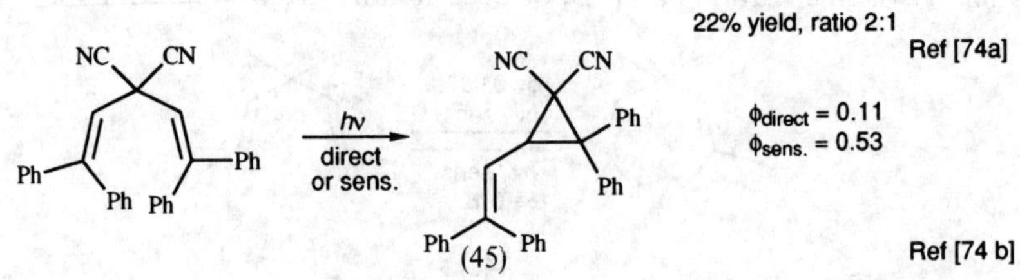

(45)

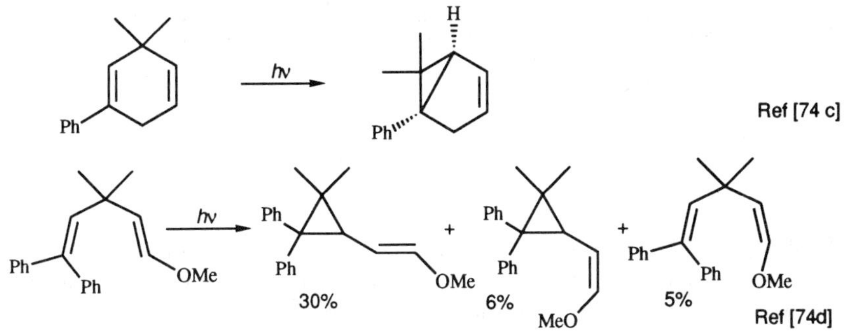

Ref [74 c]

Ref [74d]

(45 continued)

alkene components is replaced by an allene (**46**) [75], although in this instance

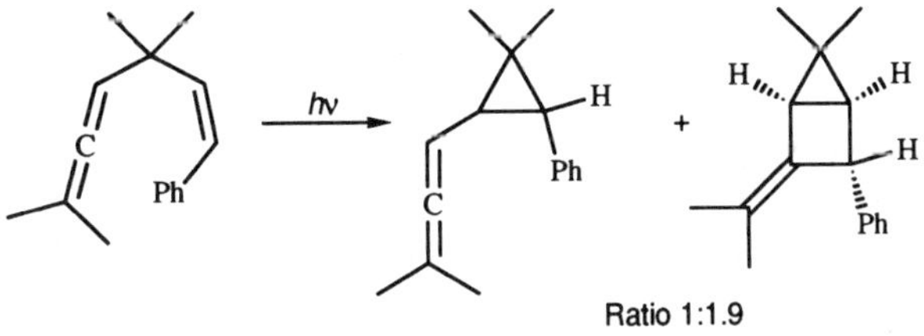

Ratio 1:1.9

(46)

the preference is for (2 + 2)-cycloaddition rather than rearrangement, or an alkyne (**47**) [76] when again cyclopropanes are obtained. It should be noted

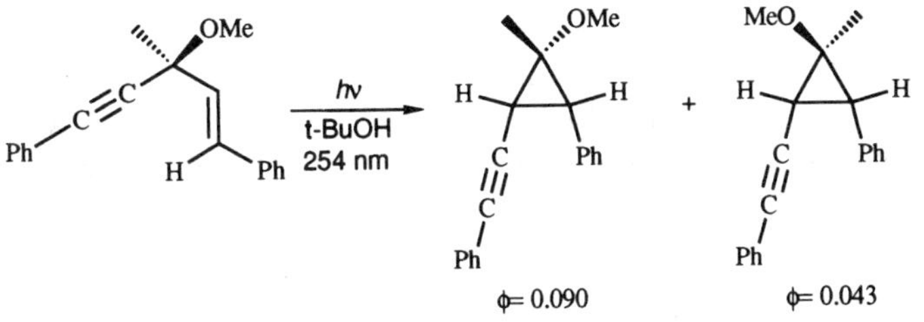

$\phi = 0.090$ $\phi = 0.043$

(47)

that in some cases the acyclic 1,4-dienes do undergo the di-$\pi$-methane process from the triplet state. This reactivity is observed when the diene is substituted with electron-withdrawing groups at positions 1 or 3 at the diene. Usually the triplet state is reactive for a diene constrained within a ring.

A variant of the di-$\pi$-methane reaction involves an aryl vinyl interaction as in the conversion of the propene derivative into cyclopropanes (**48**) [77].

Ph, OMe, Ph, Ar $\xrightarrow[C_6H_6]{h\nu,\ 254\ nm}$ Ph, Ph, OMe, Ar

Ar = $p$-$ClC_6H_4$ 60%
Ar = $p$-$MeC_6H_4$ 60%
Ar = $p$-$FC_6H_4$ 65%

(48)

In this case bridging occurs between a phenyl group and the alkene and results in phenyl group migration prior to the formation of the three-membered ring. Further examples are illustrated (**49**) [78].

Ph, Ph $\xrightarrow[t\text{-}BuOH]{h\nu}$ H, Ph, Ph, H 40% + Ph, Ph 7% Ref [78a]

$p$-$MeOC_6H_4$, Et, Ph, Ph $\xrightarrow{h\nu}$ Et, H, Ph, $p$-$MeOC_6H_4$, Ph $\phi$= 0.070 + Et, H, Ph, $p$-$MeOC_6H_4$, Ph $\phi$= 0.023 Ref [78b]

(49)

As mentioned above in constrained 1,4-dienes the triplet state is reactive. A typical example of di-$\pi$-methane reactivity in such a system is the conversion of barrelene (**50**) into the fluxional hydrocarbon semibullvalene [79]. Arene

$\xrightarrow[acetone]{h\nu}$ 40%

(50)

moieties can replace any of the vinyl components of the triene and sensitization is again effective in the conversions shown in (**51**) [80]. In some

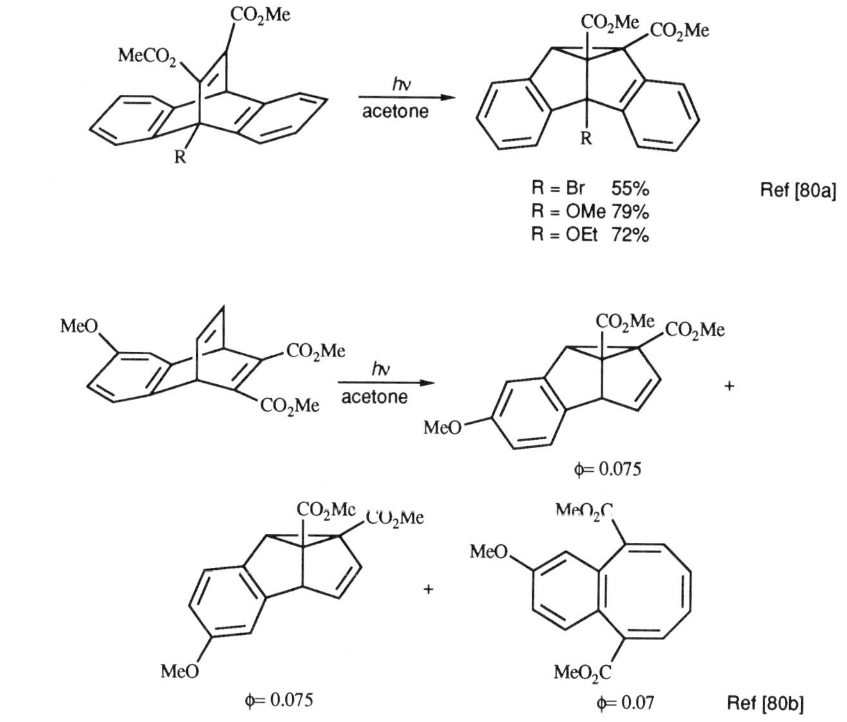

R = Br 55%
R = OMe 79%
R = OEt 72%

Ref [80a]

$\phi$= 0.075

$\phi$= 0.075

$\phi$= 0.07

Ref [80b]

(51)

cases, a singlet-state reaction competes with the triplet reaction and yields cyclo-octatetraene derivatives. Interest has also been shown in the photochemical reactivity of dibenzobarrelenes in the solid state, particularly as a method of controlling the rearrangement and obtaining products with high enantiomeric excess [81]. The aryl vinyl interactions are also seen in the photochemical rearrangements of benzonorbornadienes. In this situation, when a substituent is present on the arene or the norbornadiene, there can be a choice of rearrangement path and the outcome of the reaction depends on the nature of the substituent and the influence it exerts on the stability of the intermediates (**52**) [82]. Many systems have been shown to undergo

hν sens. CN 100% via CN Ref [82a]

$CO_2Me$ MeO hν sens. $CO_2Me$ MeO 100% via $CO_2Me$ MeO Ref [82b]

N MeO CN hν sens. acetophenone OMe N 100% via N MeO Ref [82c]

(52)

this variant of the di-$\pi$-methane process and some of the uses to which it has been put are shown in **53** [83].

O N Ph hν sens. acetone Ph N O 100% via O N Ph Ref [83a]

OMe N hν sens. OMe N 85% + N OMe 15% Ref [83b]

N N O N O hν sens. N N O N O 40% Ref [83c]

(53)

*2.1.4.4.2 Norbornadiene–quadricyclane conversion*

The benzophenone-sensitized conversion of norbornadiene to quadricyclane is another efficient reaction of a 1,4-diene and provides a facile entry to highly strained polycyclic compounds (**54**) [84]. Since the discovery of the initial

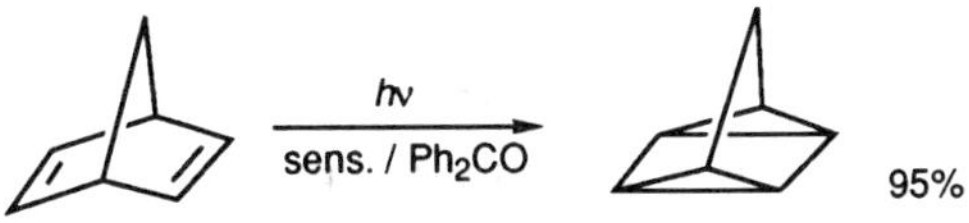

(54)

reaction many examples have been reported (**55**) [85]. Interest has been

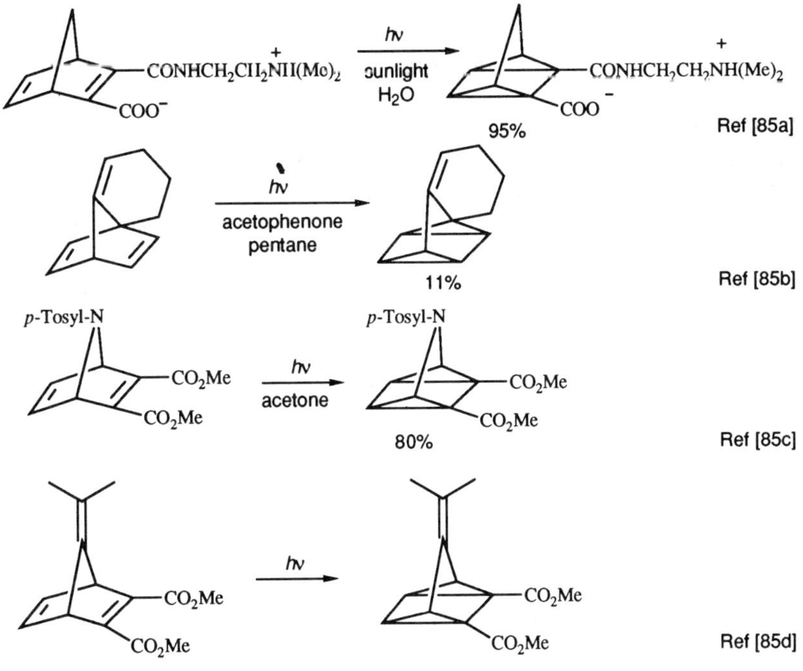

(55)

shown in such strained compounds as energy storage systems [86] and, for example, the photoconversion of the norbornadienes to quadricyclanes can be reversed thermally by silver ion (**56**) [87].

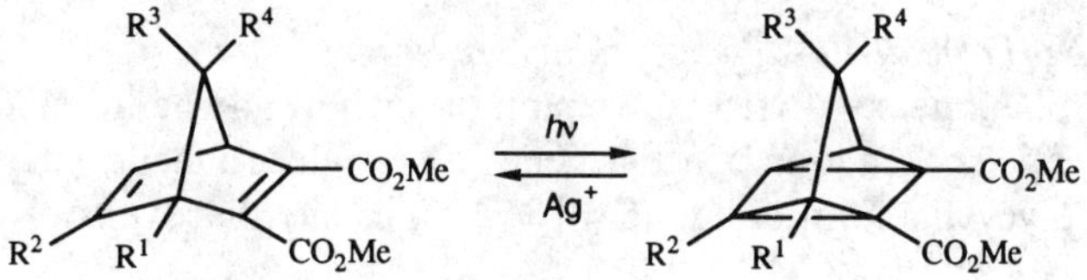

| $R^1$ | $R^2$ | $R^3$ | $R^4$ |
|---|---|---|---|
| H | H | H | H |
| Me | H | H | H |
| H | Me | H | H |
| H | H | Me | H |
| H | H | H | Me |
| H | H | Me | Me |

(56)

### *2.1.4.5 Miscellaneous reactions of dienes and alkenes*

*2.1.4.5.1 Cycloaddition reactions*

Cycloaddition reactions can be brought about on sensitized irradiation of conjugated dienes. Thcsc can be synlhetically useful and provide a route to the formation of cyclobutanes. Thus the acetophenone-sensitized irradiation of buta-1,3-diene in the presence of 2-acetoxyacrylonitrile affords the isomeric cyclobutane adducts (**57**) in a total yield of 34% [88]. A similar cycloaddition

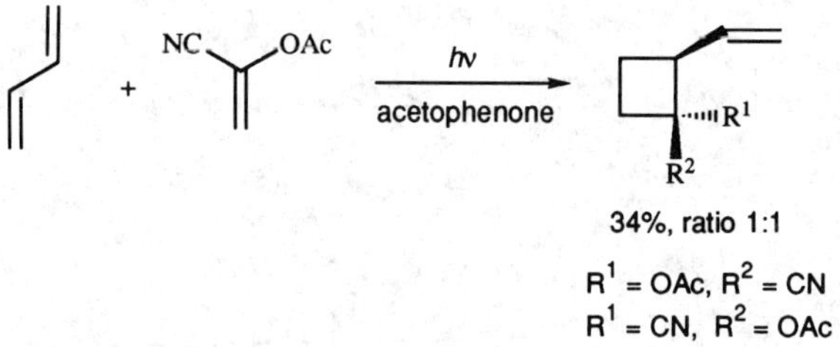

34%, ratio 1:1

$R^1$ = OAc, $R^2$ = CN
$R^1$ = CN, $R^2$ = OAc

(57)

reaction has also been used as a key step in a synthesis of *grandisol* (**58**) [89].

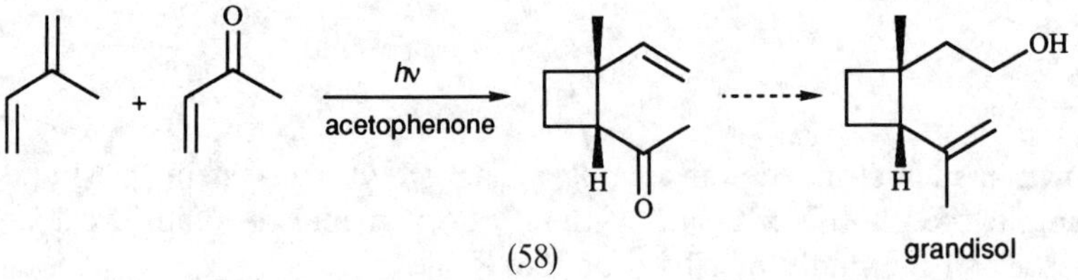

grandisol

(58)

The synthetic value of (2 + 2)-cycloadditions to alkenes is also illustrated by the acetone-sensitized addition of 6-methoxydihydropyran to an imide which yields a (2 + 2)-adduct subsequently converted into *biotin* (**59**) [90]. Another

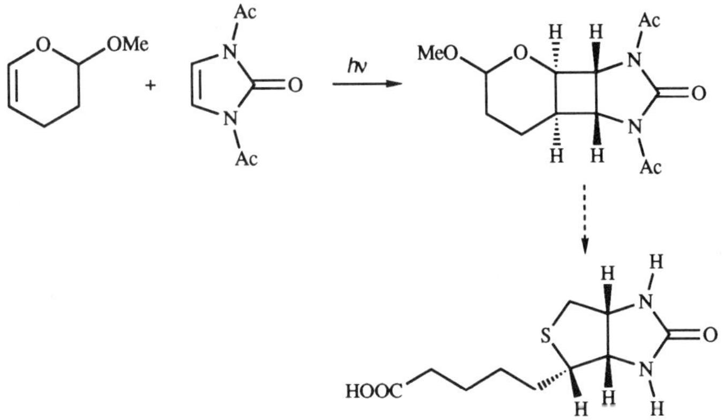

(59)

efficient example is the cycloaddition of dichlorovinylene carbonate (**60**) to

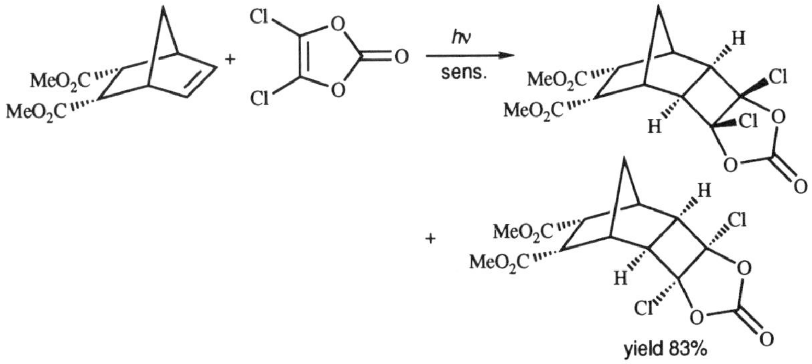

(60)

afford the two adducts in high yield [91].

*2.1.4.5.2 Hydrogen abstraction reactions*

Alkenes in their triplet states can undergo hydrogen abstraction reactions. Thus the acetone-sensitized irradiation of the cyclic alkene affords two products formed by cyclization within a biradical arising by a 1,6-hydrogen transfer (**61**) [92]. A more conventional hydrogen abstraction is also reported

hν
acetone
+
H
H
71% (total yield)

(61)

following the irradiation of derivatives of 1,1-diphenylethene whereby hydrogen transfer affords an *o*-xylylene which can be trapped by maleic anhydride (**62**) [93]. Reactions of this type are reminiscent of the Norrish

hν
Ph
Ph
Ph
Ph

(62)

Type II process of ketones (see Chapter 3).

## 2.2 AROMATIC COMPOUNDS

### 2.2.1 Spectra and excited states

The absorption and emission spectra of aromatic molecules and the photophysical properties of their excited states have been the subjects of considerable study [94] and only a short account is given here. The ultraviolet absorption spectrum of benzene shows two intense bands below 220 nm and a much weaker structured band centred around 255 nm. A simple molecular orbital model of the electronic structure of benzene based on the linear combinations of six carbon 2*p* orbitals has a degenerate pair of highest filled bonding orbitals, and a degenerate pair of lowest unfilled antibonding orbitals. This would lead to a fourfold degeneracy for the lowest energy transition, but when electron–electron interaction is taken into account a

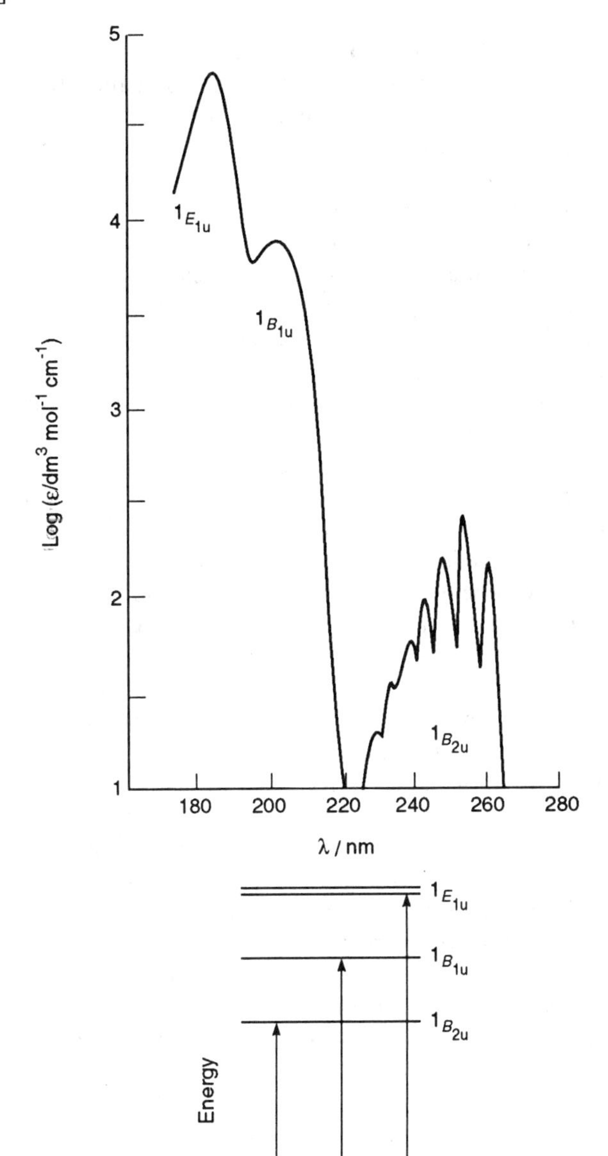
(a)
Log (ε/dm³ mol⁻¹ cm⁻¹)
5
4
3
2
1
180
200
220
240
260
280
λ / nm
1E1u
1B1u
1B2u
(b)
Energy
1E1u
1B1u
1B2u
1B1g (ground state)

Fig. 2.3.

simple state energy level diagram (Fig. 2.3) can be constructed. Only the transition to the $^1E_{1u}$ state is fully allowed and the transition to the lowest excited-state singlet, $^1B_{2u}$, is strongly forbidden. Its observed intensity arises from vibronic coupling due to lowering of the hexagonal symmetry. Calculations account nicely for the observed bands in the absorption spectrum.

The lowest triplet state of benzene is probably that corresponding to the $^1B_{1u}$ state. The transition $S_0$ to $T_1$ cannot be observed in normal absorption studies, although weak bands around 340 nm are seen in oxygen-perturbation experiments using a low-temperature oxygen matrix [95]. A $T_1$ to $T_n$ absorption spectrum has been recorded in flash photolysis experiments [96]. Benzene phosphorescence, however, is very efficient in a low-temperature matrix and the yield of triplet states is high in fluid solution or in the vapour phase at room temperature.

Substituents on the benzene ring, especially those capable of $\pi$-orbital overlap with the ring, can influence the absorption and emission properties. Such effects have a direct bearing on synthetic applications of aromatic photochemistry. A knowledge of these spectral changes permits the correct selection of exciting wavelength to use. Furthermore a substituent can affect the electronic nature and energy of the excited state that is lowest in energy and so a reaction that is observed for benzene itself may not occur for certain substituted benzenes. For example, the many photo-ring-isomerization reactions described in this section seem to be restricted to aromatic compounds carrying alkyl, fluoro, chloro, perfluoroalkyl or cyano groups.

As the extent of the conjugation is increased in fused polycyclic aromatic hydrocarbons, so the absorption and emission maxima shift to longer wavelengths and the lowest excited states decrease in energy. Emission from the second excited singlet state of azulene was once thought to be exceptional but it is now established that weak $S_2$ to $S_0$ fluorescence accompanies $S_1$ to $S_0$ fluorescence for many aromatic hydrocarbons, including naphthalene [97]. The mechanism for this emission involves thermal repopulation of $S_2$ from $S_1$ [98].

Heteroaromatic systems can have unshared pairs of electrons that influence the photophysical properties. The absorption spectrum of pyridine shows a weak $n\pi^*$ band at around 270 nm as a shoulder on the more intense $\pi\pi^*$ band. The $\pi\pi^*$ bands for pyridine are quite similar to those for benzene, although of greater intensity. However, the lowest excited singlet state is clearly $n\pi^*$ and weak fluorescence ($\Phi < 10^{-4}$) from this state has been reported [99]. The triplet $n\pi^*$ and $\pi\pi^*$ states of pyridine are close together in energy. Increasing the number of nitrogen atoms in the ring lowers the energy of the $n\pi^*$ singlet state and for pyridazine the $n\pi^*$ absorption band is at about 340 nm.

### 2.2.2 Photoisomerization

Many isomers of benzene can be obtained by photochemical isomerization [100]. In the vapour phase, high-energy radiation of $\lambda < 200$ nm converts benzene into fulvene (**63**) and further reaction occurs to give *cis-*

hν
vapour
fulvene
hν
+

(63)

and *trans*-isomers of hexadienyne. Fulvene is stable in dilute solution but it polymerizes when the solution is concentrated. In the liquid phase using the same wavelengths, benzene produces fulvene and two other isomers, benzvalene and bicyclo[2.2.0]hexadiene (*Dewar* benzene) (**64**). With 254 nm light in

hν
liquid
+
+
benzvalene

(64)

the liquid phase, benzvalene is the major product ($\Phi_{\text{initial}} = 0.18$) and this together with the observation that its formation is quenched by alkenes suggests that it arises from the $S_1$ state. However, excess vibrational energy must be required to form benzvalene from the $S_1$ benzene because in the gas phase only light of shorter wavelengths within the $S_0 \rightarrow S_1$ absorption band is effective [101]. A similar line of argument suggests that the bicyclohexadiene arises from the $S_2$ state.

Benzvalene absorbs strongly at 254 nm and is converted to benzene. As a result of this it is impossible to obtain mixtures richer than about 1% benzvalene in benzene. The synthesis of benzvalene and of the naphthalene analogue is best carried out by a thermal method [102], although the photochemical route is preparatively useful for other aromatic systems. A further isomer of benzene is prismane (**65**). This has not been prepared from

$h\nu$

prismane

(65)

benzene itself but its derivatives are formed from some substituted benzenes. Prismane arises by way of (2 + 2)-photocycloaddition reactions of *Dewar* benzene (bicyclo[2.2.0]hexadiene).

The formation of these isomers of benzene has been treated theoretically [103] in terms of orbital symmetry correlations with intermediates having biradical or zwitterionic character. Prefulvene (**66**) correlates with the $S_1$ state

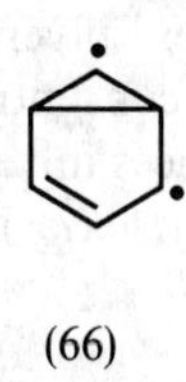

(66)

of benzene and this offers a plausible route to fulvene by subsequent carbon–carbon bond cleavage and a 1,2-hydrogen shift or to benzvalene by coupling of radical centres. A biradical (**67**) derived by way of a different

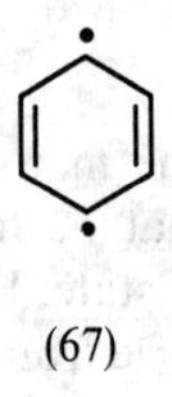

(67)

bonding pattern correlates with the $S_2$ state of benzene and could be a precursor to bicyclohexadiene. The validity of this orbital symmetry approach has been challenged [104], and a possible alternative *Mobius* benzene (*cis,cis,trans*-cyclohexatriene) has been suggested [105].

The formation of similar isomers has been reported for a number of substituted benzenes, although only a restricted range of substituents is involved and some of these are shown in **68** and **69** [106, 107]. Groups that

F F F F F F

$h\nu$

F F F F F F

$\phi = 0.003–0.08$

Ref [106]

$CF_3$ $F_3C$ $CF_3$ $F_3C$ $CF_3$ $CF_3$

$h\nu$ Pyrex

$CF_3$ $CF_3$ $CF_3$ $CF_3$ $CF_3$ $CF_3$

~100%

Ref [106a]

$h\nu$ quartz

$CF_3$ $CF_3$ $F_3C$ $CF_3$ $CF_3$ $CF_3$

~100%

Ref [106b]

(68)

$h\nu$

65%

Ref [107a]

$SiMe_3$ $SiMe_3$ $Me_3Si$ $SiMe_3$

$h\nu$

$SiMe_3$ $SiMe_3$ $Me_3Si$ $SiMe_3$

up to 25%

+

$SiMe_3$ $SiMe_3$ $Me_3Si$ $SiMe_3$

50%

Ref [107b]

(69)

$MeO_2C$ → $h\nu$ → $MeO_2C$ → $h\nu$ → $MeO_2C$

Ref [107c]

(69 continued)

are effective in promoting the formation of isolable benzene isomers are either bulky or strongly electron-withdrawing and most examples involve *t*-butyl, fluoro- or perfluoroalkyl. The effectiveness of the $CF_3$ group has been attributed [108] to both electronic and steric effects.

A number of general comments can be made about these reactions. Alkyl and related substituents normally lead to benzvalene formation if steric interaction is not too great. When severe steric crowding occurs, bicyclohexadiene or prismane formation predominates. Fluoro-substituted benzenes generally give rise to bicyclohexadienes and this may be the result of the $^1B_{2u}$ and $^1B_{1u}$ states being much closer together in energy than for benzene itself. Finally the ratio of isomers may vary with time, reflecting the photochemical reversibility of interconversion between the different products. Prismanes are relatively photostable, however, and they can be the predominant product on prolonged irradiation with short-wavelength light. Of relevance to the effect of bulky substituents is that a few heavily substituted bicyclohexadienes have been madc by a thermal route that are more stable thermodynamically than the isomeric benzenes [109]. On irradiation these bicyclic compounds can be converted to the corresponding benzene.

Photoisomerization to give bicyclic or polycyclic products is also a feature of the chemistry of heteroaromatic compounds, which are described in Chapters 4 and 5, and a few reactions of this type have been reported for five-membered ring systems involving trifluoromethylthiophenes (**70**) [110].

$CF_3$, S, $CF_3$ → $h\nu$ vapour → $CF_3$, S, $CF_3$ + $F_3C$, $CF_3$, S + $F_3C$, S, $CF_3$

(70)

A number of photoisomers related to bicyclohexadiene or benzvalene have been isolated starting from polycyclic aromatic hydrocarbons, including *t*-butylnaphthalenes e.g. **71** and **72** [111, 112] 9-*t*-butyl anthracene [113],

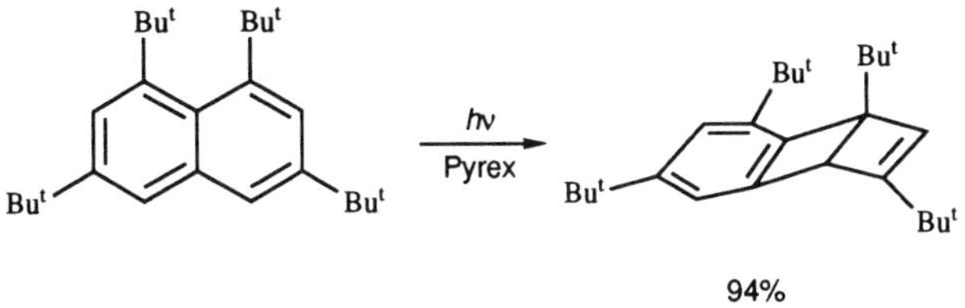

Bu$^t$ Bu$^t$ Bu$^t$ Bu$^t$ $h\nu$ Pyrex Bu$^t$ Bu$^t$ Bu$^t$ Bu$^t$

94%

(71)

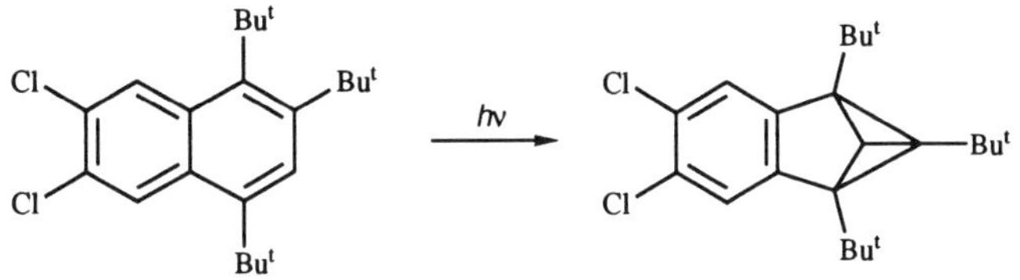

Cl Cl Bu$^t$ Bu$^t$ Bu$^t$ $h\nu$ Cl Cl Bu$^t$ Bu$^t$ Bu$^t$

(72)

and decamethylanthracene (**73**)[114]. Relief of steric crowding again seems

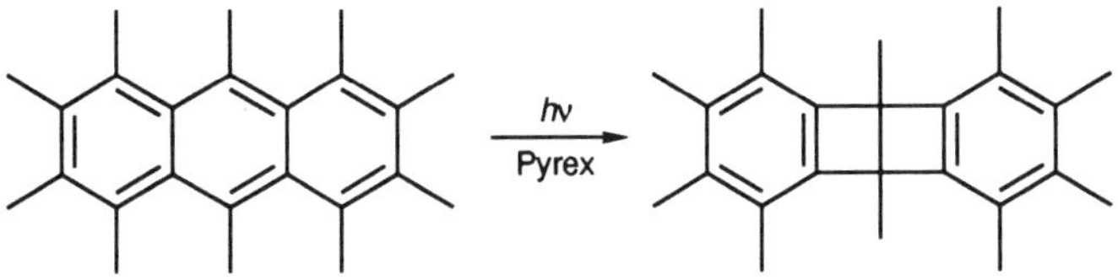

$h\nu$ Pyrex

(73)

to play a part in the formation of these as isolable products.

Benzvalenes and related compounds, like all bicyclobutanes and azabicyclohexadienes, are sensitive to attack by proton acids and this accounts for the photochemical reactions of several aromatic compounds in aqueous conditions. The initial products from benzene are bicyclo[3.1.0]hexenes oxygenated at C-2 (**74**) and these undergo sensitized isomerization to isomers

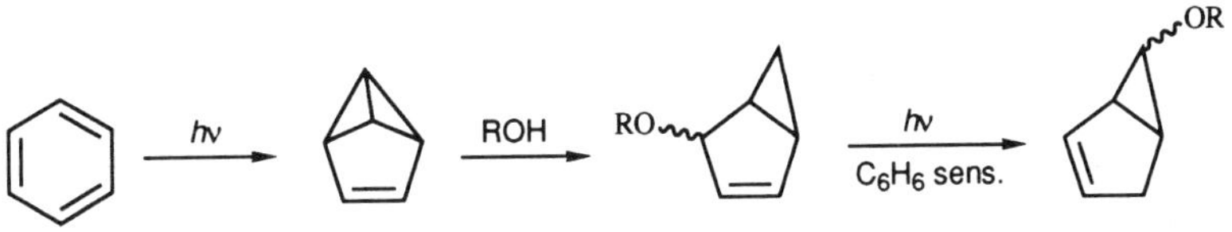

$h\nu$ ROH RO $h\nu$ $C_6H_6$ sens. OR

(74)

oxygenated at C-6 [115]. Deuterium labelling studies [116] support the intermediacy of benzvalene in this reaction. 4-Pyrones give rise to dihydroxycyclopentenones (**75**) [117]. The preparation [118] of bicyclo[3.1.0]hex-

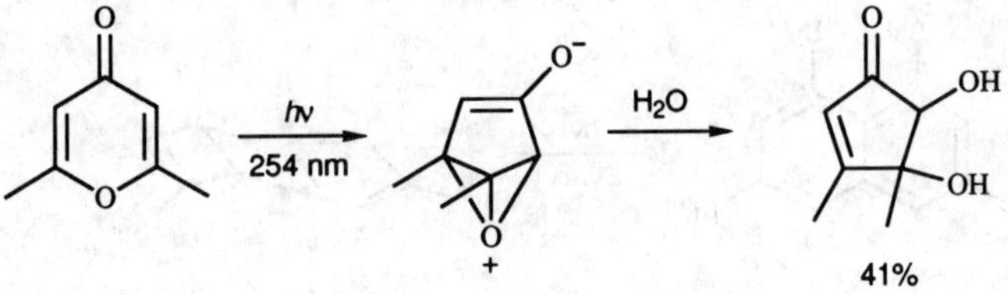

(75)

3-en-2-one from phenol by irradiation in strong acid bears some resemblance to the reactions of 4-pyrones, and both are related to the photorearrangement of cyclohexa-2,5-dienones (see Chapter 3).

The isolation of polycyclic photoisomers is a feature of the photochemistry of some aromatic systems, particularly those with *t*-butyl, F or $C_nF_{2n+1}$ substituents, but for other compounds the photoisomer cannot be isolated but leads to products with the same ring system and with certain atoms of the ring transposed. *o*-Xylene gives the *meta* and the *para*-isomers sequentially (**76**) [119], and *o*-di-*t*-butylbenzene reacts similarly [120]. A study of the

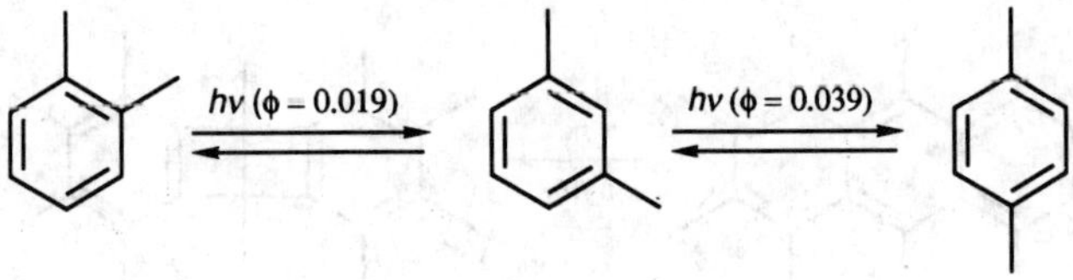

(76)

$^{14}$C-labelled mesitylene showed that the ring atoms were transposed (**77**) and

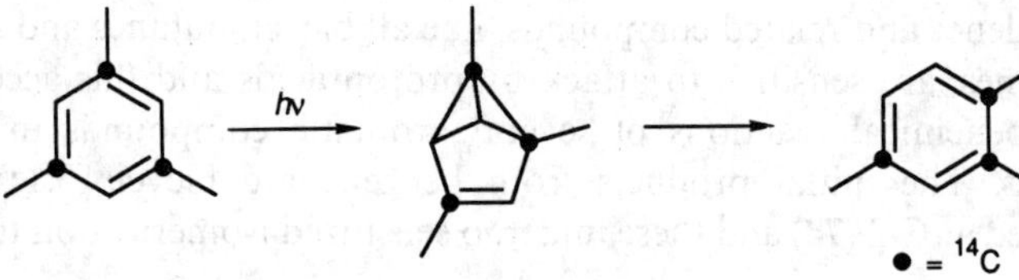

(77)

the mechanism did not involve cleavage of the bond to the alkyl group [121]. This led to the acceptance of a mechanism for the 1,2-shifts based on the formation and subsequent re-aromatization of a benzvalene intermediate. In principle other mechanisms may operate, such as the formation and further rearrangement of an intermediate bicyclohexadiene followed by re-aromatization or the photochemical interconversion of isomeric benzvalenes which is known to occur in one tri-*t*-butylbenzene system.

The fact that several pathways for group rearrangement could lead to the same overall result, especially in systems where some ring atoms are indistinguishable because they carry the same substituents, led to an analysis [122] based on 12 possible permutation patterns for the transformation of a six-membered ring. The value of this analysis is that it allows a permutation pattern to be identified from the results based on systems where fully labelled rings are not used. For example the reaction of 4-hydroxy-2,6-dimethyl-pyrylium (**78**) conforms to a specific identifiable pattern while that of

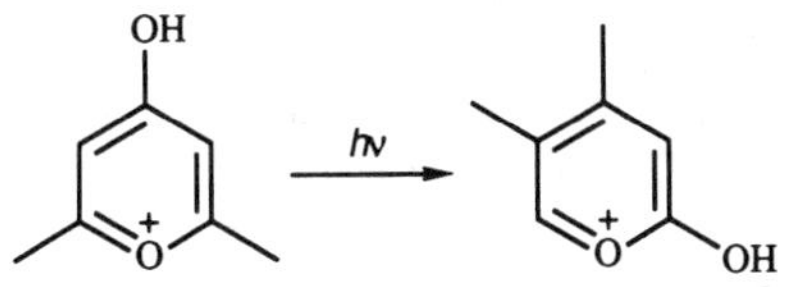

(78)

2-ethyl-4-hydroxy-3-methylpyrylium (**79**) follows two different paths to afford

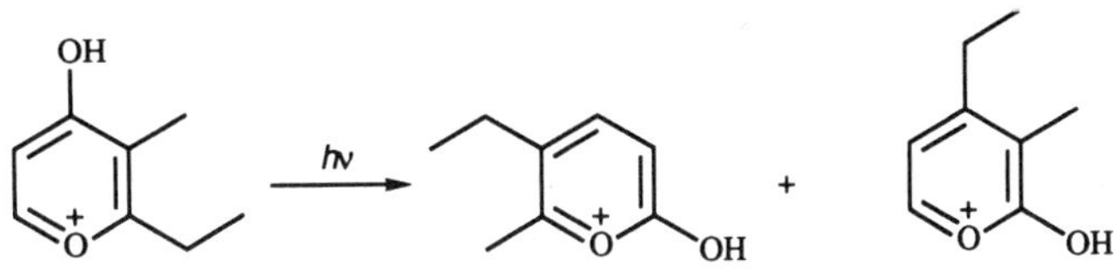

(79)

the two products obtained [123]. For the pyrylium ions a series of oxabicyclo[3.1.0]hexenium ions were proposed (**80**), although in sulphuric

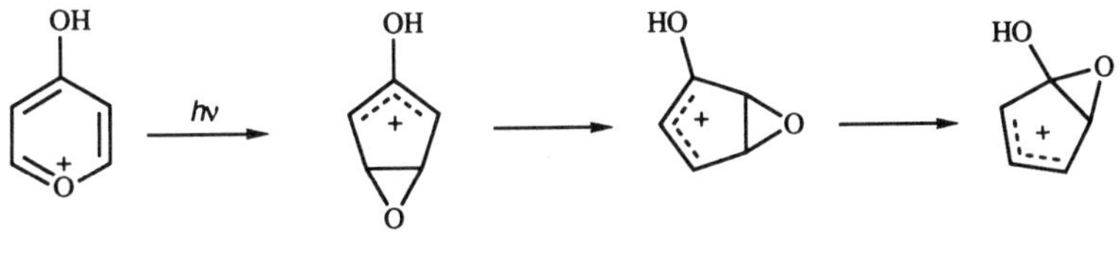

(80)

acid a cyclic sulphate intermediate can also be isolated [123]. Further studies on pyrylium salts have included the photo-isomerization of 2-hydroxy-pyrylium systems [124] and of pyrylium compounds bearing only alkyl groups [125]. In these systems the discovery that successful reaction depends on the presence of alkyl groups on C-3 and C-5 led to a modification of the original mechanistic proposal.

Five-membered heterocyclic aromatic compounds undergo a number of photochemical transformations that involve transposition of the ring atoms [126]. One difference between five- and six-membered systems is that for the smaller rings bicyclic isomers have been isolated in only a few cases. However, such intermediates have been widely postulated and there is a diversity of mechanisms reported in the literature. Some of the diversity may be accounted for by alternative ways of representing the structure of intermediates, particularly when sulphur is involved. Furans offer a relatively simple picture, and photochemical isomerization involving transposition of ring atoms 2 and 3 is general for 2-alkylfurans (**81**) [127]. The most likely mechanism

hν

(81)

involves cleavage of the weaker of the two C—O bonds of the ring followed by reclosure to give an acylcyclopropene. Such cyclopropenes have been isolated for the alkylfurans [128] and for tetrakis(trifluoromethyl)furan (**82**)

$F_3C$ $CF_3$ $F_3C$ $CF_3$ hν $F_3C$ $CF_3$ $CF_3$ $F_3C$ O

20%

(82)

[129]. The cyclopropenes undergo photochemical reaction to give the original furan or its ring-transposed isomer. This mechanism accounts for some of the other products that are formed on irradiation of furans, such as the cyclopropene by loss of carbon monoxide from the acylcylopropene or low yields of pyrroles if the reaction is carried out in the presence of a primary amine (**83**) [130].

hν
PrNH$_2$

ratio 8:1

(83)

2-Arylthiophenes are converted [131] to the 3-aryl isomers in high yield on irradiation (**84**) although yields are lower for 2-alkyl analogues. A complex

hν
quartz

(84)

mechanistic picture emerges from studies on deuterium-labelled compounds [132] because the label is found in all positions of both recovered starting material and product. Several mechanisms have been proposed [126] but in some cases the differences may be more apparent than real. Some of the differences could be in the representation of the intermediate as a tricyclic species (**85**) or as an equilibrating system of isomers (**86**) (see also Chapter

(85)

(86)

4). The intermediacy of 5-thiabicyclopentene is supported by the isolation of such a compound (**87**) albeit in low yield on the irradiation of 3-cyano-2(or

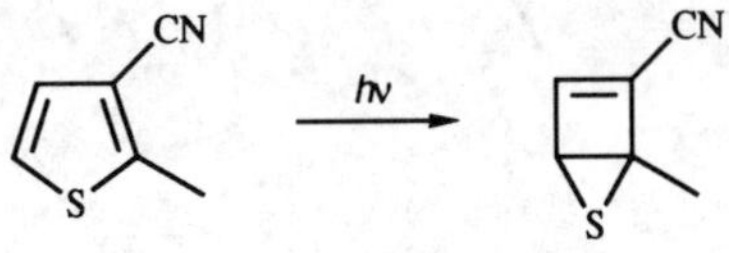

(87)

4)-methylthiophene [133]. As with furans, thiophenes also give rise to pyrroles when they are irradiated with primary amines [130], but a mechanism involving nucleophilic attack on a thiabicyclopentene [134] is preferred to one involving a thioacyclopropene. In all of the reactions of furans, pyrroles, and thiophenes there is a tendency for isomerization to occur from 2-substituted to 3-substituted compounds but not vice versa. An attempt to account for this has been made [135] based on the involvement of two different types of excited state for phenyl-substituted heteroaromatics.

### 2.2.3 Photochemical ring-opening

Photochemical reactions can occur that lead to ring-opening of six-membered aromatic systems by way of attack on a first-formed bicyclic photoisomer. The photohydrolysis of pyridine is one example of this and related to this process is the formation of 5-aminopenta-2,4-dienenitrile (**88**) when pyridine

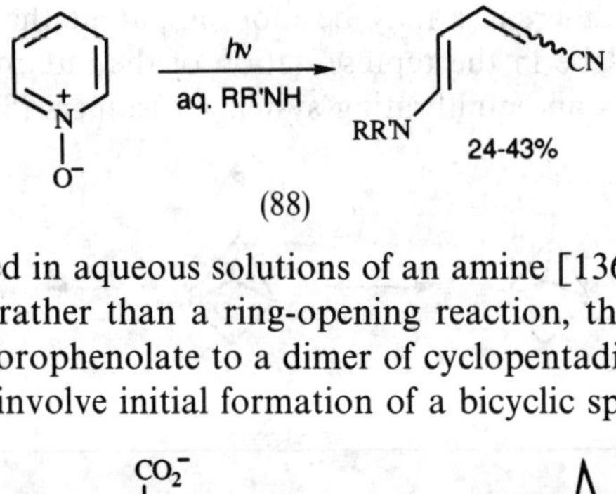

(88)

*N*-oxide is irradiated in aqueous solutions of an amine [136]. Although it is a ring-contraction rather than a ring-opening reaction, the photochemical conversion of 2-chlorophenolate to a dimer of cyclopentadiene-2-carboxylic acid (**89**) may also involve initial formation of a bicyclic species [137].

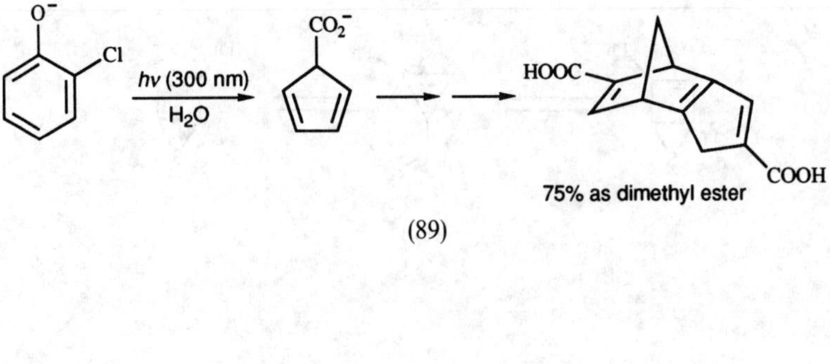

(89)

Photolysis of certain types of five-membered heterocyclic aromatic compounds leads to extrusion of small molecules such as $N_2$ or $CO_2$. This leads to the formation of reactive species that may be employed in the subsequent synthesis of other heterocyclic systems. Sydnones lose carbon dioxide on irradiation (**90**) and the nitrilimines so generated can be trapped by dimethyl

$R^1-C\equiv\overset{+}{N}-\overset{-}{N}-R + CO_2$ ; $MeO_2CC\equiv CCO_2Me$ ; 10-84%

(90)

acetylenedicarboxylate, affording pyrazoles [138], or some other trapping agent. Benzonitrile sulphide can be formed by photolysis of 1,2,3,4-thiatriazoles (**91**) and this provides a route to isothiazoles by reaction with alkynes

$PhC\equiv\overset{+}{N}-\overset{-}{S}$ ; $MeO_2CC\equiv CCO_2Me$

(91)

[139]. 1,2,3-Thiadiazoles have been used as precursors to thiirenes, and the parent compound has been observed by photolysis in a low-temperature matrix [140]. In the presence of electron-deficient alkynes at normal temperatures the products are thiophenes (**92**) [141]. In the absence of added

$F_3CC\equiv CCF_3$ ; 12%

(92)

alkyne these thiadiazoles give a variety of sulphur heterocycles on irradiation and this has been employed as a means of cross-linking vinyl polymers [142].

### 2.2.4 Photosubstitution

Electrophilic attack on the aromatic ring plays a very large part in the ground-state chemistry of benzene and other aromatic compounds. The excited-state chemistry is quite different and photochemical electrophilic substitution has not been widely reported. The absence of this type of reactivity may be due in part to the ease with which many electrophiles quench the excited states rather than reacting chemically with them. On the other hand, photochemical substitution of aromatic compounds by nucleophilic reagents has provided a wealth of valuable mechanistic and synthetic results and most of this section is devoted to such processes. There is no discussion here of aromatic substitution by radical mechanisms where the radical is generated independently of any light absorption by the aromatic compound.

Early results on photochemical isotope exchange between aromatics such as toluene, anisole and nitrobenzene and $CF_3COOD$ were not reproducible. Later work [143] using tritiated trifluoroacetic acid confirmed that quantum yields for exchange were quite low (less than 0.01 for anisole) and for mechanistic studies the results of protodeuteration of deuterated aromatics in acetic acid proved more useful. It is, however, not clear if the orientational preferences are substantially different from those for ground state reactions [144]. This route to deuterated aromatic compounds is not synthetically viable. Much better routes are available such as the photolysis of aryl bromides in the presence of deuterated trialkylstannanes [145] or $d_4$-methanol (**93**) [146]. The only other photochemical electrophilic substitution

Br
X
hν
$CD_3OD$
D
X
> 70%

(93)

to be noted here is the reaction between anthracene and aroyl chlorides (**94**)

hν
ArCOCl
COAr
+
COAr
20-30%
30-60%

(94)

that gives 2- and 9-substituted Friedel–Crafts products in reasonable yields [147].

Photochemical nucleophilic substitution reactions in aromatic compounds have been reported extensively [148] and many of the reactions are preparatively useful. Most of the examples involve systems where there is activation by an electron-withdrawing group such as nitro, and some of these are illustrated (**95**) [148–151]; a mechanism is proposed in which the

$NO_2$, OMe $\xrightarrow[HO^-]{h\nu}$ $NO_2$, OH

$\phi = 0.22$ Ref [148]

$NO_2$, OMe $\xrightarrow[CN^-]{h\nu}$ $NO_2$, CN

44% Ref [149]

$NO_2$, OMe $\xrightarrow[NH_3]{h\nu}$ $NO_2$, $NH_2$

89% Ref [150]

$NO_2$, OMe $\xrightarrow[OH^-]{h\nu}$ $NO_2$, OH + OH, OMe

$\phi = 0.020$ $\phi = 0.091$ Ref [151]

(95)

nucleophile attacks the excited state of the aromatic compound directly. Full details remain to be elucidated, although an exciplex has been detected in the reaction of 3,5-dinitroanisole with hydroxide ion [152]. The entries in **95** show that there is strong activation by a nitro group when there is a leaving group in the *meta* position (compare the *ortho/para* directing influence in the ground state). Other studies have shown that the effect is transmitted readily to the second ring in naphthalene derivatives. When there is no good leaving group or when the leaving group is *ortho* or *para* to the nitro group, the reaction is less efficient and other processes compete, such as replacement of the nitro group itself. Rationalizations have been proposed based on static reactivity indices such as electron density or on dynamic reactivity calculations (potential energy surfaces). The former are easier to work out and, despite inconsistencies in detail, the qualitative agreement with experimental results is satisfactory.

It has become increasingly apparent that many of the observed aromatic photosubstitution reactions do not conform to expectations based on this first mechanism, and two other major groups of photochemical nucleophilic substitutions have emerged. One involves the aromatic radical cation and the other the aromatic anion. For the first group of reactions it is proposed that the excited state of the aromatic compound loses an electron and the radical cation is attacked by the nucleophile. These reactions are more efficient for aromatics with electron-donating substituents such as methoxy. The electron-donating groups are *ortho/para* directing, and quantum yields often do not depend upon the nature of the added nucleophile; results match those of similar electrochemical anodic reactions. Evidence for the radical cations and solvated electrons has been obtained by flash spectroscopy and by scavenging of the electrons in irradiated solutions of anisole [153]. Many of the reported reactions in this group employ cyanide ion as the nucleophile, as seen in **96** and **97** [148, 154–157], and, in the presence of oxygen, photocyanation can be highly efficient, even when the leaving group is a

OMe, OMe → ($h\nu$, $CN^-$) → OMe, CN 73% Ref [148]

OMe, OMe → ($h\nu$, $HO^-$) → OMe, OH 52% Ref [148]

(96)

OMe, OMe (1,3-dimethoxybenzene) $\xrightarrow[CN^-]{h\nu}$ OMe, OMe, CN 40% + NC, OMe, OMe, CN 7% Ref [148]

OMe, OMe (1,4-dimethoxybenzene) $\xrightarrow[CN^-]{h\nu}$ OMe, CN 22% Ref [148]

OMe, Cl $\xrightarrow[CN^-]{h\nu}$ OMe, CN 95% Ref [148]

OMe, Cl $\xrightarrow[CN^-]{h\nu}$ OMe, Cl, CN 54% Ref [148]

OMe, F $\xrightarrow[CN^-]{h\nu}$ OMe, CN 94% $\phi = 0.53$ Ref [148]

OMe $\xrightarrow[CN^-/O_2 \text{ absent}]{h\nu}$ CN 3% Ref [154]

(96 continued)

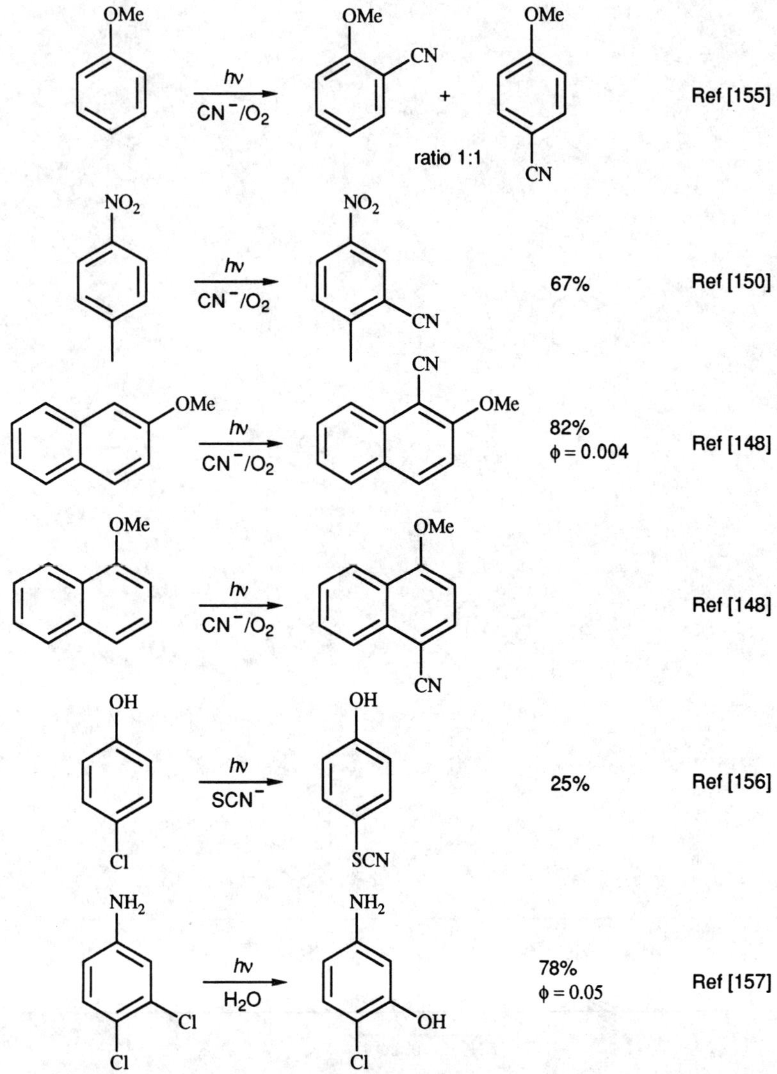

OMe → hν, $CN^-/O_2$ → OMe, CN + OMe, CN

ratio 1:1

Ref [155]

$NO_2$ → hν, $CN^-/O_2$ → $NO_2$, CN

67%

Ref [150]

OMe → hν, $CN^-/O_2$ → CN, OMe

82%
$\phi = 0.004$

Ref [148]

OMe → hν, $CN^-/O_2$ → OMe, CN

Ref [148]

OH, Cl → hν, $SCN^-$ → OH, SCN

25%

Ref [156]

$NH_2$, Cl, Cl → hν, $H_2O$ → $NH_2$, OH, Cl

78%
$\phi = 0.05$

Ref [157]

(97)

hydride ion. Water is also effective in replacing halide, as is nitrite ion (**98**)

$$\text{4-ClC}_6\text{H}_4\text{NMe}_2 \xrightarrow[\text{NaNO}_2]{h\nu} \text{4-O}_2\text{NC}_6\text{H}_4\text{NMe}_2 \quad 60\%$$

$$\text{4-ClC}_6\text{H}_4\text{NMe}_2 \xrightarrow[\text{NaNO}_2]{h\nu\text{/sensitizer}} \text{4-Cl-2-O}_2\text{NC}_6\text{H}_3\text{NMe}_2$$

(98)

[158], although in the latter case a different product is formed on direct or sensitized irradiation and it is thought that the direct reaction may not involve a radical cation.

Hydrocarbons such as naphthalene (**99**) [159], biphenyl or phenanthrene

$$\text{naphthalene} \xrightarrow[\text{CN}^-/\text{O}_2]{h\nu} \text{1-cyanonaphthalene} \quad 65\%$$

(99)

react photochemically with cyanide to give cyano-substituted derivatives and, as expected on the basis of initial formation of the aromatic radical cation, these reactions are enhanced in the presence of a good electron acceptor such as *p*-dicyanobenzene (**100**) [160]. The reaction with anisole is also

$$\text{phenanthrene} \xrightarrow[p\text{-C}_6\text{H}_4(\text{CN})_2]{h\nu\text{/CN}^-} \text{9-cyano-9,10-dihydrophenanthrene (49\%)} + \text{9-cyanophenanthrene (6\%)}$$

(100)

enhanced under these conditions [161], and the products (61%) are *o*- and *p*-cyanoanisoles in a ratio of 1:1.

Photosubstitution reactions of heteroaromatic compounds have not been widely reported [162] and those for nitrogen systems are exemplified in Chapter 5. Illustration **101** [163] contains examples for furans and

$NO_2$ hν CN$^-$ CN
54%
$\phi = 0.51$

$NO_2$ hν MeO$^-$ OMe
41%
$\phi = 0.16$

(101)

thiophenes.

### 2.2.5 Photoaddition

Two types of non-cyclic addition reactions to aromatic rings have already been exemplified. These are the ones which occur by way of attack on a first-formed benzvalene or related intermediate and lead to products that incorporate a hydroxylic addend (**74, 75**) and those that involve the formation of dihydro products accompanying the photosubstitution reactions of aromatic hydrocarbons with cyanide. Amongst other photochemical reactions that give rise to dihydro compounds, the best known are those of amines or of hydride reducing agents with aromatics.

Benzene reacts photochemically with primary or secondary amines [164] to give 1,2- and 1,4-dihydro adducts (**102**) in which the 1,4-adducts predominate.

+ RR'NH hν NRR' + NRR'

(60%, R= $C_6H_{11}$, R'= H)

(102)

nate. Gram quantities of substituted cyclohexa-1,4-dienes can be made in this way. Substituted benzenes such as toluene, chlorobenzene or trifluoromethylbenzene react in a similar manner [165]. Tertiary amines react by way of the α-position in one of the alkyl groups, as in the reaction of naphthalene with triethylamine (**103**) [166]. It is thought that in these

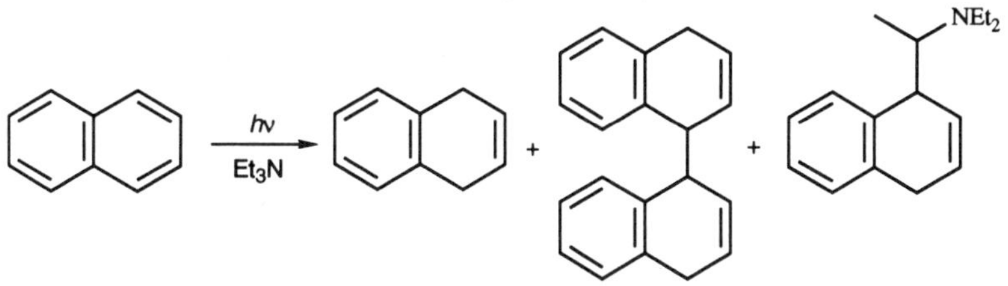

(103)

process the initial reaction of the excited state is to accept an electron from the amine, and the aromatic radical anion then abstracts a proton from the nitrogen of the amine or from the α-position.

Intramolecular examples of this reaction are known (**104**) [167], and can

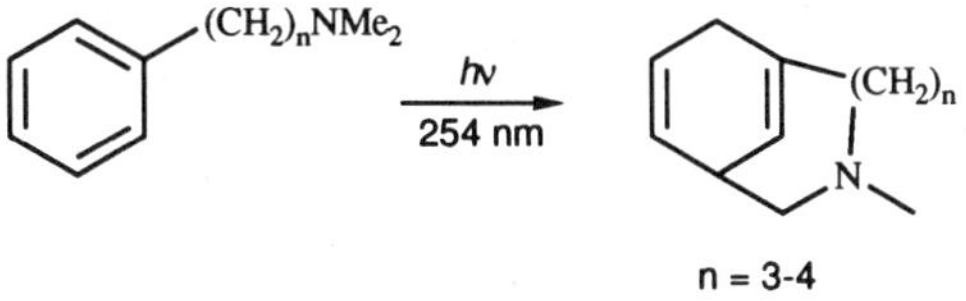

(104)

be employed in the formation of spiro products in the phenanthrene system (**105**) [168]. Anthracene reacts in a different way with *N*,*N*-dimethylaniline

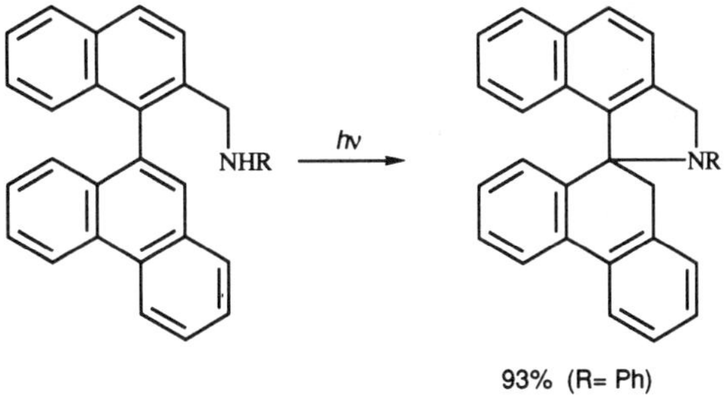

(105)

[169] producing an adduct linked to the ring of the aromatic amine rather than to one of the α-carbons of the alkyl groups (**106**). That a radical anion

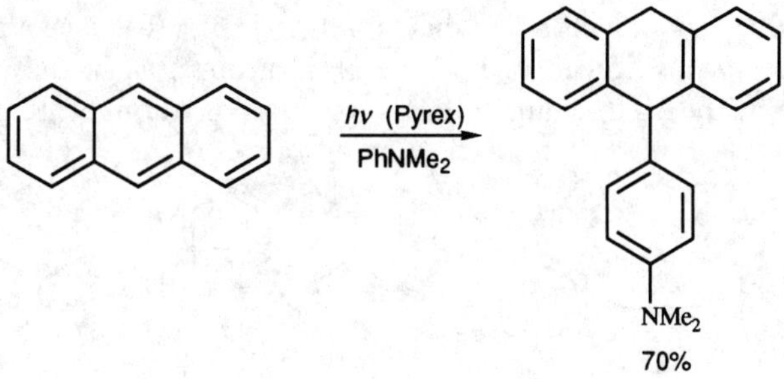

*hν* (Pyrex)
$PhNMe_2$

$NMe_2$
70%

(106)

is very likely to be involved when the excited phenanthrene interacts with an amine is suggested by the ready formation of 9,10-dihydrophenanthrene-9-carboxylic acid (**107**) when the reaction is carried out in the presence of

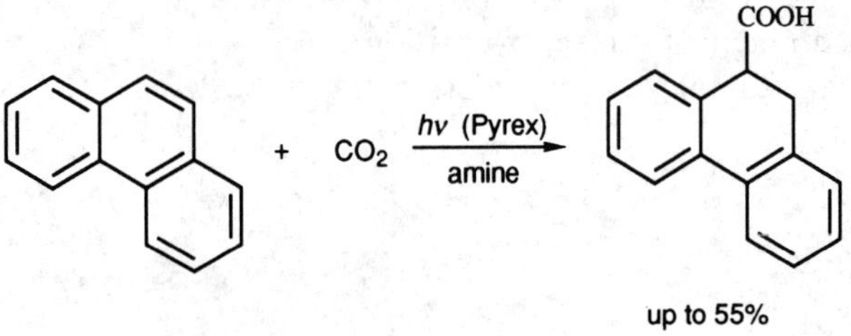

COOH

+ $CO_2$ *hν* (Pyrex) amine

up to 55%

(107)

carbon dioxide [170].

Sodium borohydride is an effective reducing agent for aromatic compounds substituted with electron-withdrawing groups (**108**) [171], and also for

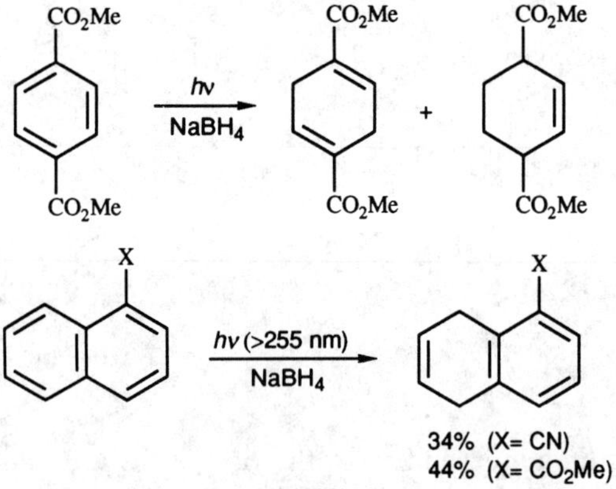

$CO_2Me$ *hν* $NaBH_4$ $CO_2Me$ + $CO_2Me$

$CO_2Me$ $CO_2Me$ $CO_2Me$

X *hν* (>255 nm) $NaBH_4$ X

34% (X= CN)
44% (X= $CO_2Me$)

(108)

aromatic hydrocarbons or electron-rich aromatics in the presence of an electron acceptor (**109**) [172]. Deuterium-labelling studies have pointed to

*hv*, $NaBH_4$
*p*-$C_6H_4(CN)_2$

59%

(109)

the existence of at least two mechanistic pathways [171], one involving an electron transfer and the other hydride transfer to the excited-state aromatic compound. The successful use of electron acceptors as co-reagents suggests that radical cations might also play a part in these systems.

There is also one report of photochemical addition of an alcohol to a benzene ring [173], and this involves reduction of a substituted phthalimide (**110**). Both of the bridgehead hydrogens are derived from the solvent hydroxyl

*hv*, $NaBH_4$
ROH

14-31%

(110)

group, and the carbonyl groups undoubtedly play some part in the process.

### 2.2.6 Photocycloaddition

The photochemical reactions of benzenoid compounds with alkenes give an array of cycloadducts that arise by way of addition across the 1,2-, 1,3-, or 1,4-positions of the aromatic ring (**111**). One of the first reported reactions

\+ $C_2H_4$ *hν*

1,2- 1,3- 1,4-

(111)

was that of benzene with maleic anhydride [174], which leads to a 2:1-adduct (**112**). This arises by way of a Diels–Alder reaction between a first-formed

88%

(112)

1,2-cycloadduct and a second molecule of maleic anhydride. It became apparent as more substrates were studied that the predominant products in many processes with simple addends were tricyclo[3.2.1.0$^{2,8}$]oct-3-enes, and a typical example of this involves benzene and cyclobutene (**113**) [175].

$\phi = 0.10$

(113)

1,4-Cycloadducts are not generally formed with mono-alkenes, although this is a major mode of addition [176] for many allenes (**114**).

$H_2C{=}C{=}CH_2$

(114)

With the wealth of data that are now available it is possible to make a few generalizations about photocycloadditions to benzene [177]. First, although most alkenes give mixtures of products, 1,3-cycloadducts usually predominate if the ionization potential of the alkene (8.65 eV < I.P. < 9.6 eV) is not very different from that of benzene (9.24 eV). If the alkene has a relatively low ionization potential (e.g. maleic anhydride) or a relatively high one (e.g. dimethoxyethylene) then the major adducts are often 1,2-cycloadducts. In some cases the 1,2-cycloadducts are labile and undergo a subsequent photochemical rearrangement to give a 1,4-adduct (**115**) [178]. Although this

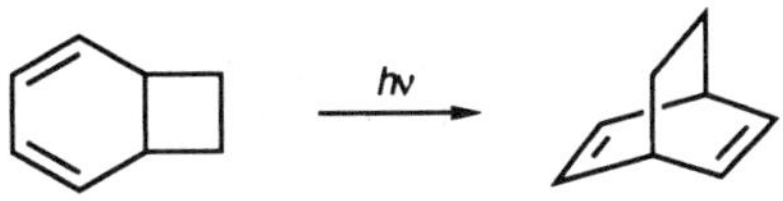

(115)

consideration of ionization potentials may imply that exciplex intermediates or even radical ions are involved in the mechanism of the cycloaddition, evidence for such intermediates is not easy to obtain. Nonetheless the idea provides a useful rule-of-thumb guideline. By an extension of this yardstick, electron-deficient aromatics such as hexafluorobenzene (**116**) [179] or ben-

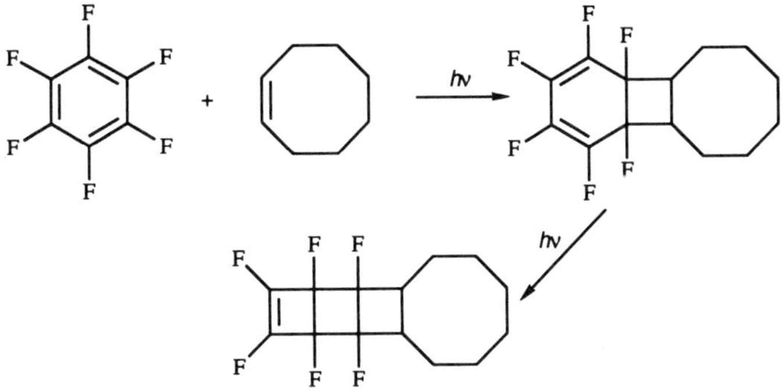

(116)

zene-1,3,5-tricarboxylates [180] give products arising mainly by way of initial 1,2-cycloaddition when irradiated with simple alkenes. Under these conditions 1,3-cycloadducts are formed as minor products.

The stereochemistry of the original alkene is preserved in all of the types of cycloadduct, but there is no single pattern of *exo*/*endo* product formation. With some electron-rich alkenes exclusive *exo* product formation is observed (**117**) [181], but in other instances mixtures of *exo* and *endo* isomers are

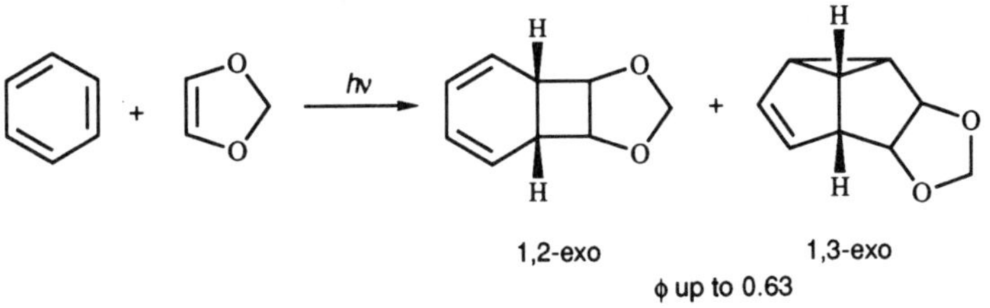

1,2-exo 1,3-exo

$\phi$ up to 0.63

(117)

formed. With substituted benzenes the question of regioselectivity arises. Many mono-substituted benzenes show a preference for 2,6-adduct formation (**118**) [182], but this is not exclusive, as shown by the reactions of benzonitrile

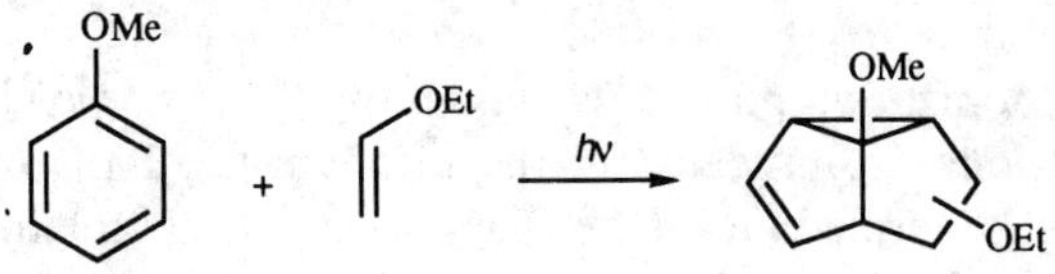

(118)

(**119**) [183, 184]. Other examples of the addition process are shown in (**120**) [178, 185–189] and (**121**) [190–195].

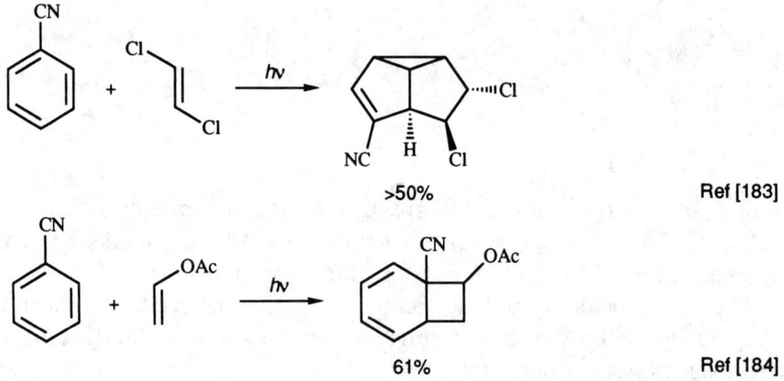

>50% Ref [183]

61% Ref [184]

(119)

$+ H_2C{=}CH_2 \xrightarrow{h\nu}$

ϕ up to 0.11

Ref [178]

$\xrightarrow{h\nu}$

ϕ ~0.25

Ref [185]

$\xrightarrow{h\nu}$

minor

major

Ref [186]

$\xrightarrow{h\nu}$

H H O

H H

ϕ up to 0.76

Ref [187]

$\xrightarrow{h\nu}$

O O O O

ϕ = 0.3

Ref [188]

$CO_2Me$

$\xrightarrow{h\nu}$

H H $CO_2Me$ H H $CO_2Me$

Ref [189]

(120)

OMe

$\xrightarrow{h\nu}$

OMe

(121)

Ref [190]

OMe + $h\nu$ → OMe

30% Ref [191]

OMe + CN $h\nu$ → OMe CN

73% Ref [192]

R (R= alkyl) + O O O $h\nu$ → O O O O R O O O

58-80% Ref [193]

$C_6F_6$ + $h\nu$ → F F F F F F

35% Ref [194]

$C_6F_6$ + $h\nu$ → F F F F F F + F F F F F F

47% 20% Ref [195]

(121 continued)

There is still considerable uncertainty about the mechanisms involved in photocycloaddition reactions to benzenes. An extended analysis has been made [177] in terms of correlation diagrams which leads to the assignment of *allowed* or *forbidden* labels to various possible concerted mechanisms (e.g. excited $S_1$ aromatic + ground state alkene or ground state aromatic + excited state alkene). However, other modes of reaction are feasible, particularly those that involve charge-transfer interaction between the two components or those that involve prior formation of a biradical or zwitterionic intermediates from the aromatic component (e.g. the prefulvene species **66**). It is likely that different mechanisms operate for different pairs of substrate types. For example, the formation of 1,3-cycloadducts occurs by way of $S_1$ benzene and seems either to be fully concerted or to proceed by way of concerted addition of the alkene to a pre-formed intermediate. In contrast the reaction of benzene with acrylonitrile to give a 1,2-cycloadduct (**122**)

CN + hν → CN

exo : endo ratio 5:1

(122)

takes place by way of initial excitation of the alkene [182, 196] and the reaction of benzene with maleimide is similar [197] except that the initial adduct undergoes Diels–Alder reaction with a second molecule of maleimide or can be trapped by a powerful dienophile such as tetracyanoethylene. The reaction of benzene with maleic anhydride is formally related to that of benzene with maleimide, but the mechanism begins with excitation of a ground-state charge-transfer complex. No 1 : 1-cycloadduct can be detected or intercepted by dienophiles and it is proposed [198] that a zwitterionic species is involved that reacts with a second molecule of maleic anhydride or that is diverted in acid solution to give phenylsuccinic anhydride (**123**).

O O O + hν → O O $O^-$ + $CF_3CO_2H$ → O O O

(123)

The photoaddition of alkynes to benzenes generally gives a 1,2-cycloadduct as the primary photoproduct, and in these examples the alkyne is usually the species that is excited initially, as in the case of benzene and acrylonitrile or maleimide.

Despite the large number of examples studied, there has not been widespread success in applying these reaction types to synthetic procedures. In part this is due to the low quantum yields encountered in some systems and to the difficulty in separating the desired product from other isomeric compounds. The 1,2-cycloadduct of benzene with 1,1-dimethoxyethylene is a useful precursor to cyclo-octatrienone (**124**) [199], and the major 1,3-

MeO OMe + hν OMe OMe H⁺ O

(124)

cycloadduct from anisole and cyclopentene leads to a carbonyl-bridged perhydroazulene (**125**) [200].

OMe + hν OMe H⁺ O 80% KOBu$^t$ $CO_2H$

(125)

The tricyclic compounds obtained from benzenes and alkenes by 1,3-photocycloaddition are generally susceptible to cleavage leading to bicyclo[3.3.0]octane systems (**126**) [201]. The reaction between indane and

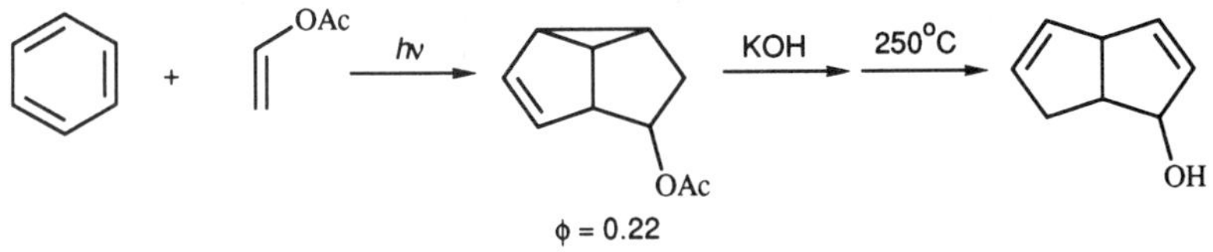

$\phi = 0.22$

(126)

vinyl acetate (**127**) has been used in a relatively short synthesis of *modhephene*

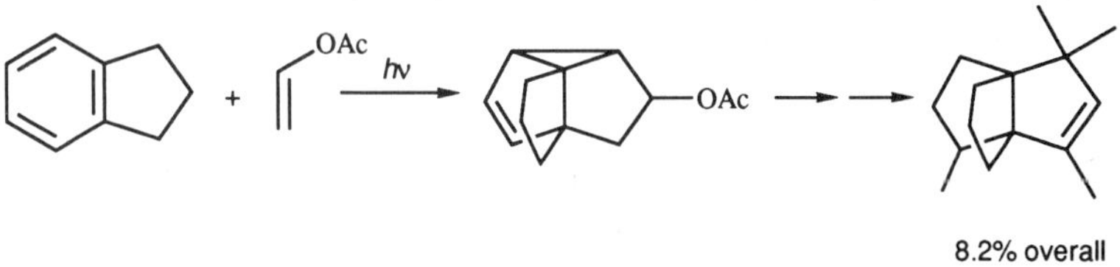

8.2% overall

(127)

[202]. Intramolecular examples of the reaction have been successfully incorporated in elegant syntheses of terpenoids such as *isocomene* [203] or *hirsutene* (**128**) [204].

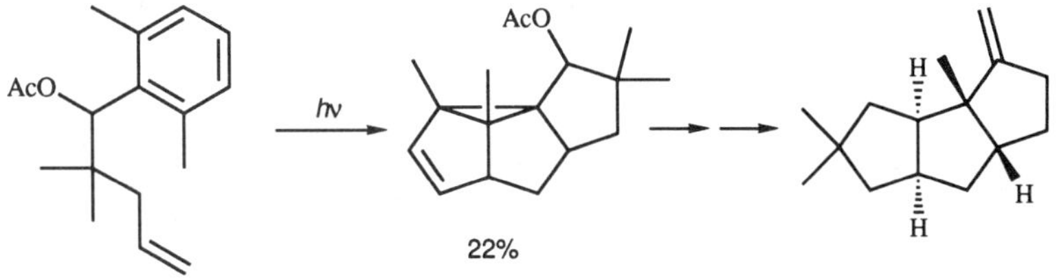

22%

(128)

The factors controlling the orientation and mode of cycloaddition in the intramolecular photochemistry of model bichromophoric phenyl-alkene systems have been investigated [205]. The length and nature of the bridging unit play a major role, as do the electronic effects of substituents.

With mono-alkenes benzenes normally give only small amounts of 1,4-adducts although an acyclic 1,4-adduct is sometimes formed by way of an *ene* process (**129**) [197]. 1,2-Dienes lead to 1,4-cycloadducts as major

$h\nu$ (254 nm)

(129)

products on irradiation with benzene (**114**) and 1,3-dienes give 1,4-cycloadducts as well as the expected 1,3-adducts. However, the 1,4-adducts obtained with conjugated dienes are those derived from (4 + 4) rather than (4 + 2) cycloaddition. Benzene and buta-1,3-diene give both *cis*- and *trans*-isomers of the 1,4-adducts (**130**) [206]. The *trans*-compound is usually isolated as a

$h\nu$

dimer

(130)

dimer and the *cis*-product can be prepared in reasonable yield by irradiation in the presence of iodine [207]. Similar 1,4-cycloadducts are obtained with other dienes (e.g. **131**) [208], or with substituted benzenes (e.g. **132**) [209].

$h\nu$

5%

(131)

$h\nu$

Me

(132)

There are a number of (4 + 2)-photocycloaddition reactions in which a styrene unit provides the $4\pi$-system as in the dimerization of 2-phenyl propene (**133**) [210]. These reactions are promoted by electron-acceptor initiators, of

hv
pyromellitic dianhydride
Ph
>60%

(133)

which the most widely used is *p*-dicyanobenzene, and it seems likely that electron transfer plays a part in the mechanism of these processes. A formally similar reaction occurs when singlet oxygen reacts with styrenes, and this process, together with other cycloadditions of singlet oxygen to aromatic systems, is considered with the photo-oxidation of conjugated compounds (see section 2.4.3).

Naphthalene and substituted naphthalenes normally react photochemically with alkenes to give (2 + 2)-cycloadducts where the addition has taken place across the 1,2-positions (**134**). Yields can be high [211]. There is evidence

CN
+
hv
NC
38%

(134)

for wavelength dependency as shown in (**135**) [212]. Irradiation at 313 nm

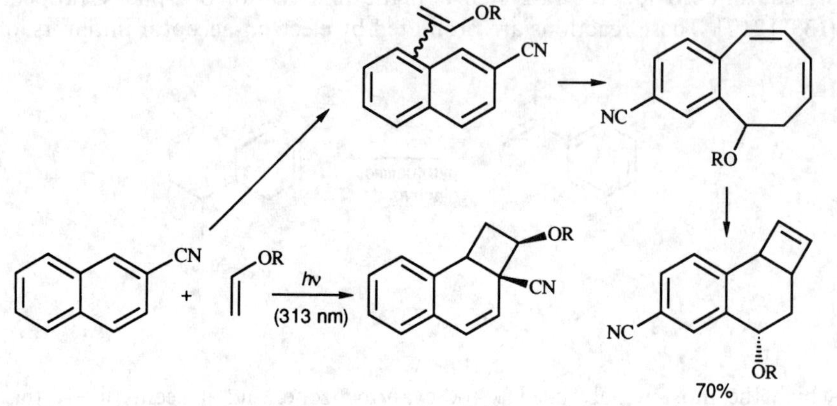

(135)

yields the 1,2-adduct, whereas using Pyrex-filtered light ($\lambda > 280$ nm) affords an isomer. This latter product is formed from the exciplex illustrated, leading to a cyclo-octadiene derivative which subsequently ring-closes to the observed product. The use of naphthol with acrylonitrile (**136**) [213], or of cyanonaph-

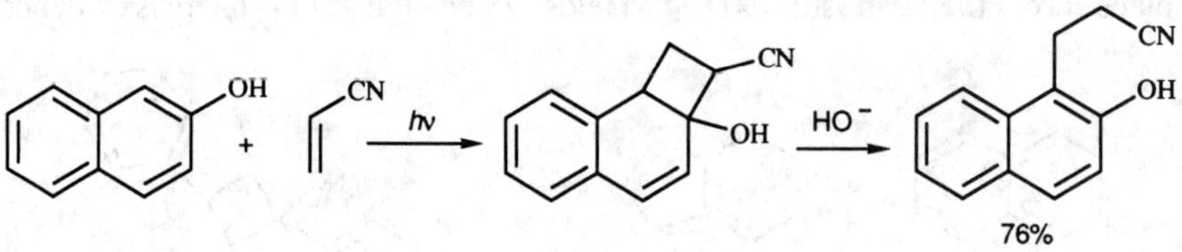

(136)

thalene with a protected enol (**137**) [214], offers a route to the synthesis of

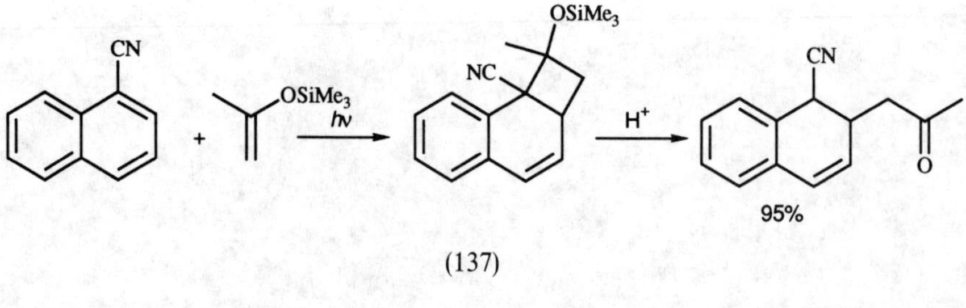

(137)

compounds that are products formally of a Michael addition reaction.

Intramolecular examples of naphthalene–alkene (2 + 2)-photocycloaddition reactions have been described [215], as have intermolecular reactions with *trans*-cyclo-octene that lead to 1,3- and 1,4-cycloadducts rather than the normal 1,2-adducts [216]. Conjugated dienes that are rigidly held in a *cisoid* conformation give rise to (4 + 4)-products (e.g. **138**) [217].

85%

(138)

Phenanthrene behaves like naphthalene in giving (2 + 2)-cycloadducts with certain alkenes (**139**) [218], but anthracenes provide a wider range of product

(139)

types. With mono-alkenes anthracene gives (4 + 2)-cycloaddition products on irradiation (**140**) [219]. Conjugated dienes, however, give both (4 + 2)-

$\phi = 0.02$

(140)

and (4 + 4)-adducts, including products that arise by reaction at a terminal rather than the central ring of the anthracene system (**141**) [220]. The major

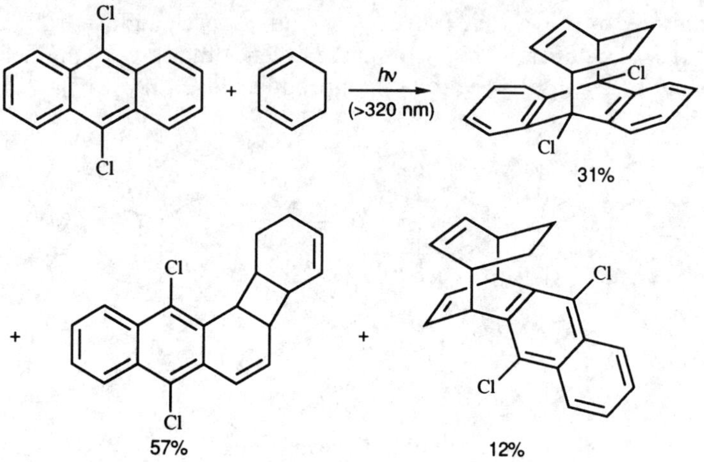

(141)

products, however, are usually those derived by reaction at the 9,10-positions, and selective formation of one or other type of adduct can be achieved by the choice of suitable reaction conditions, since the initially formed (4 + 4)-adduct at least in some systems is a precursor to the (4 + 2)-adduct (**142**) [221].

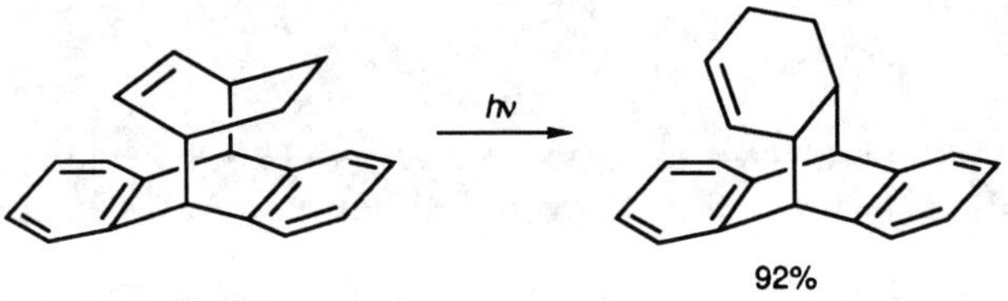

(142)

The photodimerization of anthracenes has been studied fairly intensively, partly with a view to finding systems suitable for the chemical storage of solar energy. The basic reaction (**143**) involves the formation of a (4 + 4)-

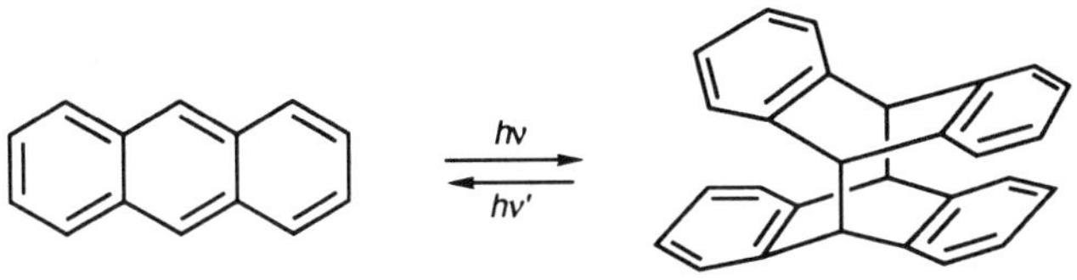

(143)

cycloadduct across the 9,10-positions of both molecules, and the reaction can be reversed if shorter-wavelength radiation is used [222]. Intramolecular examples have been reported (e.g. **144**) [223], and by choosing suitable

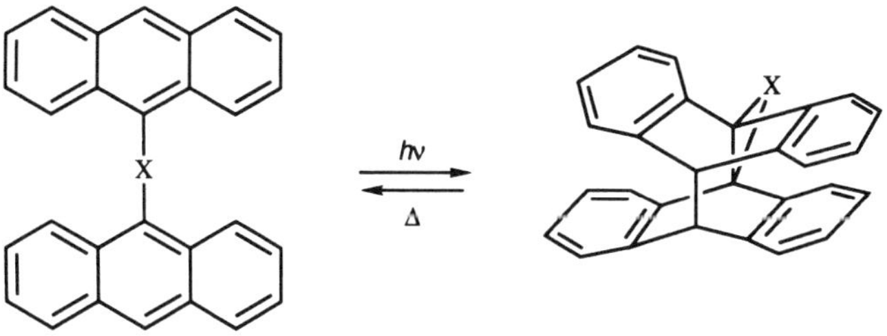

(144)

substrate molecules it is possible to design syntheses of polycyclic systems (**145**) arising by way of other modes of cycloaddition, such as the (4 + 4)-

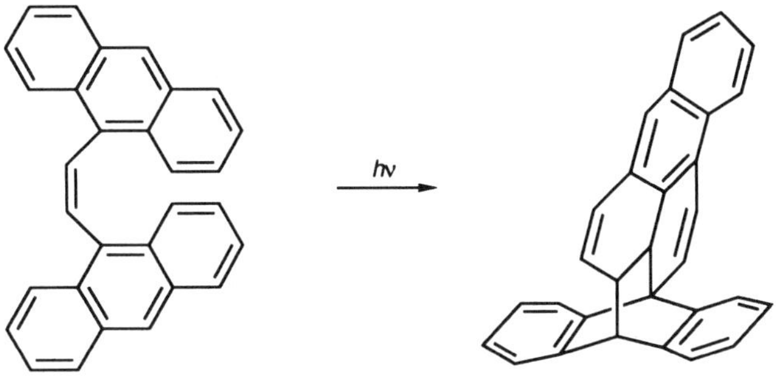

50%

(145)

addition path involving a terminal ring of one of the anthracenes or (4 + 2)-addition involving the 1,2-positions of one anthracene [224].

Analogous (4 + 4)-intramolecular dimers have been reported for a few naphthalenes (e.g. **146**) [225], but only in one intramolecular example (in a

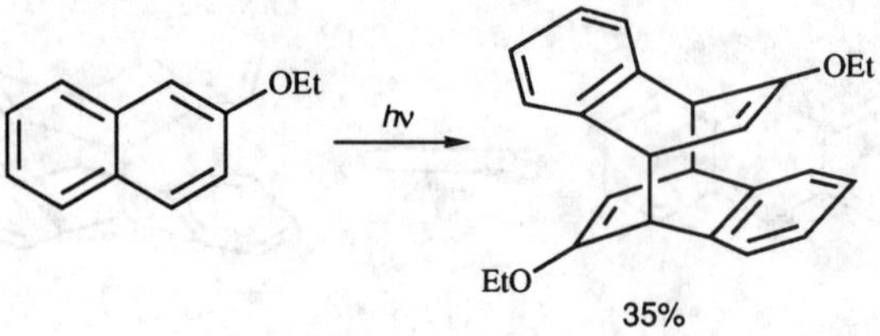

35%

(146)

cyclophane) are benzene ring dimers produced [226]. Photocycloadditions involving five-membered heteroaromatic compounds are quite common (e.g. **147**) [227], and in many respects these aromatic compounds behave like

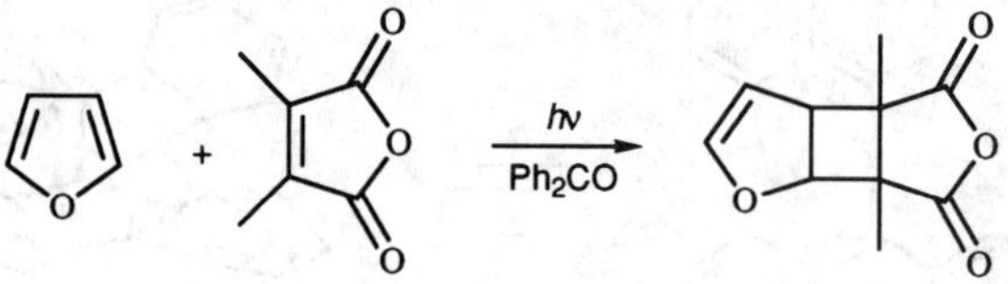

(147)

simple alkenes in such reactions. Cycloadditions to six-membered heteroaromatics are much less widespread. Pyridines substituted with electron-withdrawing groups are reported to give 1 : 2 adducts with alkenes (**148**) by

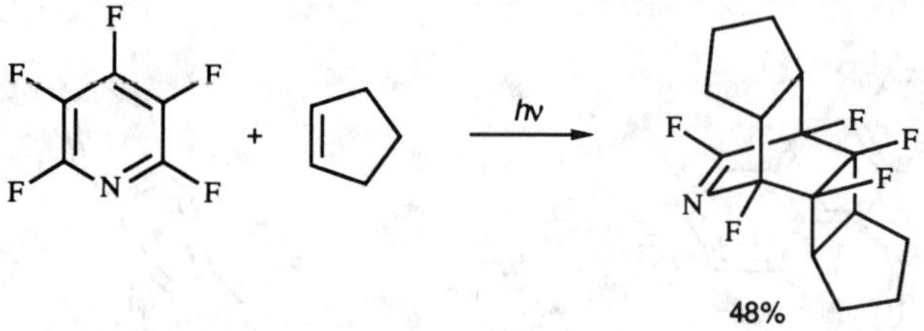

48%

(148)

way of an initial 1,2-cycloadduct [228]. In some cases the presumed biradical intermediate is diverted to give non-cyclic adducts (**149**) [229] or products

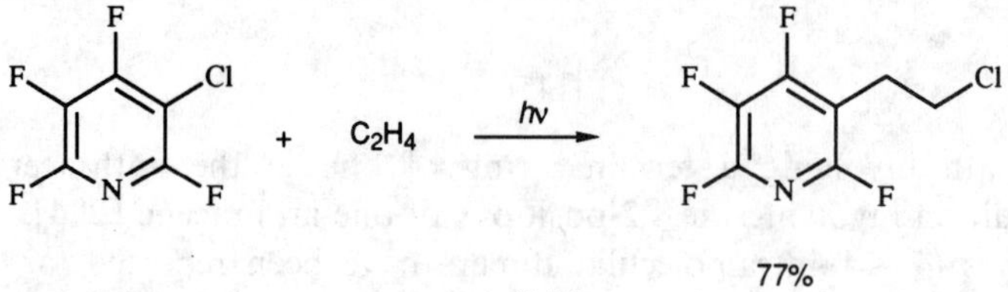

77%

(149)

derived by reaction at a cyano-substituent (**150**) [230]. 2-Pyridone behaves

(i) $h\nu$ (ii) $H_2O$

36%

(150)

like a conjugated carbonyl compound and gives (4 + 4)-cyclodimers on irradiation (see Chapter 3).

### 2.2.7 Photocyclization

Many photocyclization reactions involving an aromatic ring are related, at least formally, to the reaction in which *cis*-stilbene gives 4a,4b-dihydrophenanthrene (**151**). The *cis*-isomer of stilbene can be generated photo-

$h\nu$ $h\nu$ $h\nu$ $h\nu$ [O]

(151)

chemically from the more stable *trans*-compound and the photocyclization represents both a side-reaction to the more quantum-efficient geometrical isomerization and a chemically efficient route to phenanthrene by oxidation of the dihydrophenanthrene.

The reaction occurs from the lowest excited singlet state of *cis*-stilbene and the 4a,4b-dihydrophenanthrene (which is coloured) is formed stereospecifically in the *trans*-configuration [231]. This configuration is the result of a conrotatory six-electron electrocyclic ring-closure. The dihydrophenanthrene undergoes photochemical ring-opening to give *cis*-stilbene, and in this and related cyclizations the quantum yield of phenanthrene formation can be low if oxidizers are rigorously excluded. The chemical yield can also be low, since phenanthrene formation then requires stilbene to act as the oxidizing agent. However, when air (or oxygen) is present, or when iodine is added, chemical and quantum yields are much higher. The cyclization is also enhanced by added electron acceptors such as tetracyanoethylene [232].

Photocyclization works well for a wide range of stilbenes substituted on the aromatic rings or on the alkene double bond (e.g. **152**) [233] and provides

Ph Ph Ph Ph $\xrightarrow[I_2]{h\nu}$ 65%

(152)

a method for making many substituted phenanthrenes [234]. Yields of 60–85% can be expected in most cases although some stilbenes do not react (e.g. nitro- or dimethylamino-stilbenes). A number of procedures have been used for predicting such lack of reactivity and for predicting or rationalizing the orientation of reaction in unsymmetrical compounds. One of the easiest methods to apply involves the criterion that the sum of the free valence indices for the first excited state at the carbon atoms forming the new bond must be greater than unity. With appropriate substitution patterns the simple stilbene reaction has been adapted to the synthesis of aporphine alkaloids (**153**) [235] and to the preparation of a dibenzophosphonin (**154**) [236].

MeO MeO N $CO_2Et$ $\xrightarrow[I_2]{h\nu}$ MeO MeO N $CO_2Et$ 35%

(153)

hν, $I_2$, $O_2$ — 56% — $O_3$ — 58%

(154)

Dihydrophenanthrenes can be isolated in reasonable yield in certain systems. With α-phenyl cinnamates and others with electron-withdrawing groups [237], the initial 4a,4b-dihydro compound isomerizes in a protic solvent to the 4a,9-dihydro isomer and then the 9,10-dihydro compound is produced as the final product (**155**). In the presence of added primary amine

$CO_2Me$ — hν — $\phi = 0.03$

(155)

[238], stilbenes undergo photocyclization to give 1,4-dihydro-phenanthrenes (**156**).

Ph — hν (254 nm), $Pr^iNH_2$ — 70%, $\phi = 0.085$

(156)

The photocyclization of stilbenes to give phenanthrenes is overall an oxidative process, but non-oxidative reactions are possible either if there is an *ortho*-substituent capable of being eliminated together with a hydrogen atom from the other ring-junction position of the dihydro intermediate or two leaving groups which can result in the aromatization of the dihydrophenanthrene intermediate. Thus *o*-methoxystilbene gives largely the unsubstituted phenanthrene (**157**) on irradiation. The alternative approach

OMe hν 58%

(157)

has been observed in the synthesis of substituted phenanthrenes and, for example, the irradiation of the adduct (**158**) [239] in propan-2-ol under

Ph Ph O $Cl_4$ Ph O Me O hν $Pr^iOH/N_2$ Ph Me O 49%

(158)

nitrogen affords the cyclopentaphenanthrenone in 49 % yield while the adduct (**159**) [240] follows the same path yielding a phenanthrofuran in 79 % yield.

Ph Ph O $Cl_4$ Ph O O Ph hν $Pr^iOH/N_2$ Ph Ph O 79%

(159)

However, the balance between oxidative and non-oxidative cyclization can be influenced by other substituents, and 2,5-dimethoxystilbene gives a dimethoxyphenanthrene (**160**), that can be converted subsequently to 1,4-

MeO
OMe
$h\nu$
$I_2$
OMe
MeO
71%

(160)

phenanthraquinone [241]. *ortho*-Methyl groups can also be eliminated, but this can be minimized by carrying out the photocyclization in the presence of a primary amine (see reaction **156**) and oxidizing the dihydrophenanthrene so produced [242].

Reactions related to the stilbene cyclization have been used to make fused polycyclic aromatic systems. One approach is to start with a 1,2-diarylethylene such as styrylnaphthalene which provides a useful synthesis of the very potent carcinogen 5-methylchrysene or of other 5-substituted chrysenes (**161**)

X
$h\nu$
X
29% (X = Me)
76% (X = CN)

(161)

[243]. This approach is also the basis for routes to many helicenes [244], and a study of dinaphthylethylenes [245] has highlighted the importance of conformational factors in governing the direction of cyclization. As with the stilbene reactions, a dihydro intermediate can be isolated (e.g. **162**) in the

$h\nu$
H
H

(162)

absence of an added oxidizing agent [246]. Optically pure [8]-, [9]-, [10]-, [11]-, and [13]-helicenes have been produced starting with alkenes made from optically pure 2-formyl [6]-helicene [247].

An alternative approach is to use a distyryl arene as substrate. Thus *p*-distyrylbenzene on irradiation gives benzo[*ghi*]perylene (**163**) [248], and

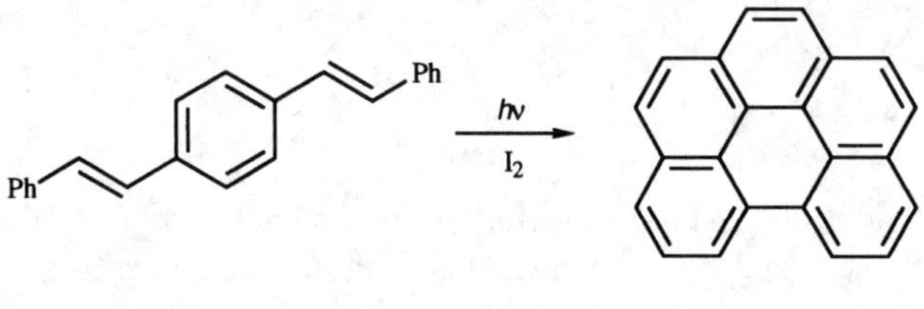

(163)

2,7-distyrylnaphthalene can be used as a precursor to 5,6-dihydro[6]helicene [249]. A third approach employs compounds in which the central double bond of the stilbene unit is itself part of an aromatic ring. *o*-Terphenyl readily gives triphenylene on oxidative photocyclization (**164**) [250]. Unexpected

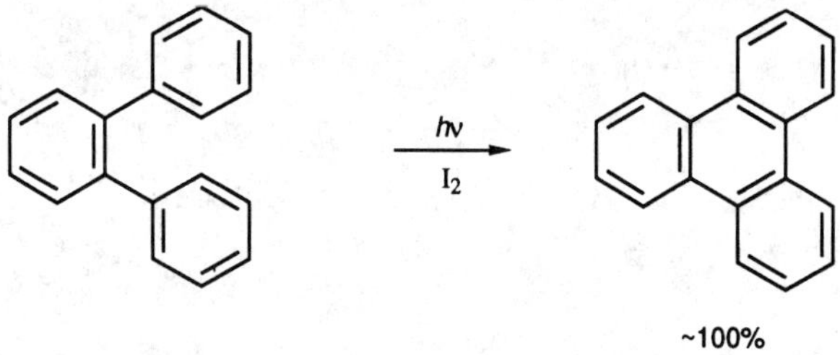

(164)

substituent shifts have been reported such as the migration of fluorine in the cyclization to a chlorofluoro [5]-helicene (**165**) [251].

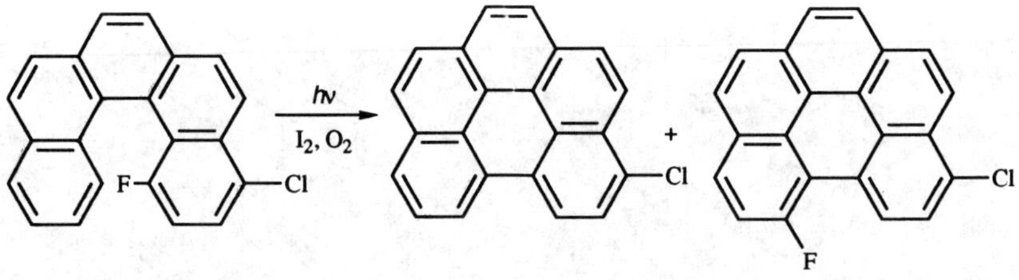

(165)

Many examples have been reported in which 1,2-diphenyl heterocyclic systems undergo photocyclization to give a fused phenanthrene. Illustrations of this in (**166**) [252, 253], (**167**) [254–256], (**168**) [257, 258] and (**169**) [259]

85% Ref [252]

95% Ref [253]

(166)

50% Ref [254]

(167) 74% Ref [255]

Ph, Ph, N, O, Ph; $h\nu$, $I_2$ or $O_2$; N, O, Ph

50% Ref [256]

(167 continued)

Ph, S; $h\nu$, $I_2$, air; S

40% Ref [257]

Ph, N, N, $CO_2Me$, $CO_2H$; $h\nu$, $I_2$; N, N, $CO_2H$, $CO_2Me$

55% Ref [258]

(168)

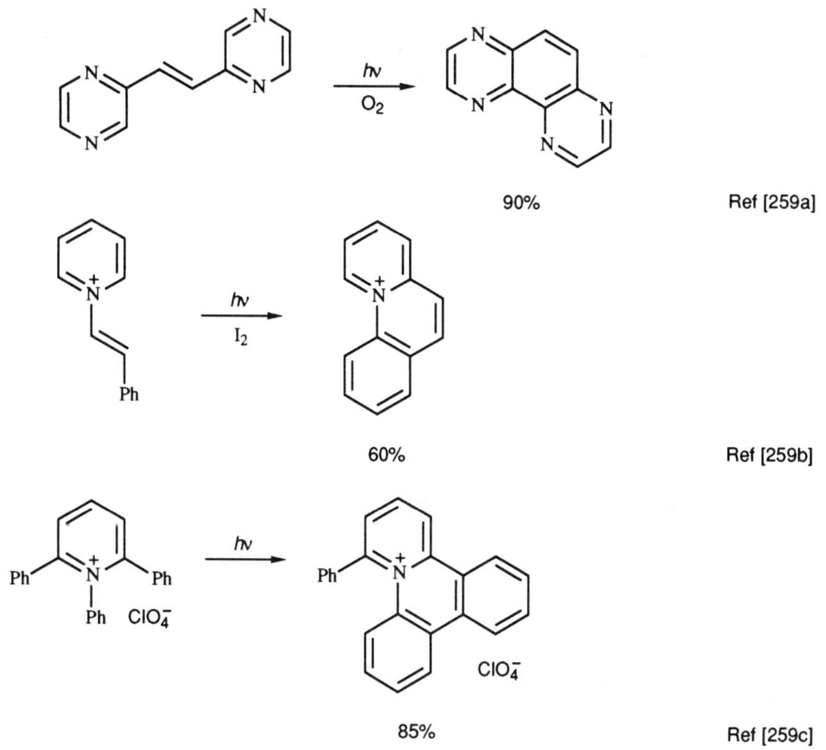

90% Ref [259a]

60% Ref [259b]

85% Ref [259c]

(169)

show some of these together with examples of the cyclization of systems in which one or both of the aryl groups in a 1,2-diarylethylene is heteroaromatic.

The basic $6\pi$-electron system for the photocyclization process can be made up of any combination of aromatic and alkene double bonds. *o*-Vinylbiphenyls undergo a reaction on irradiation that leads [260] to the formation of a 9,10-dihydrophenanthrene (**170**), by way of a 1,5-sigmatropic shift of

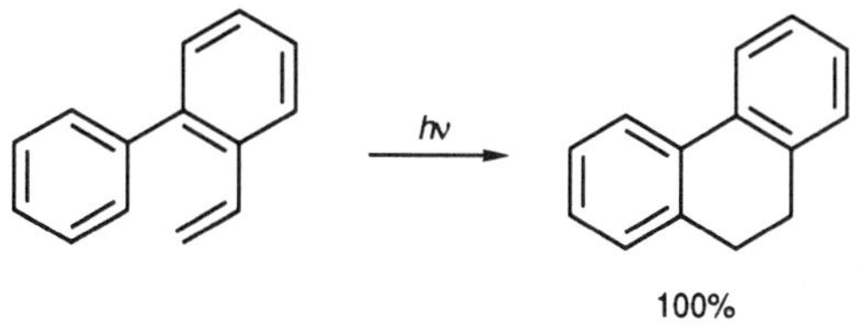

100%

(170)

hydrogen in the initial 8a,9-dihydro compound. This has been employed [261] in a synthetic route to the cytotoxic phenol *juncusol* (**171**). An attempt

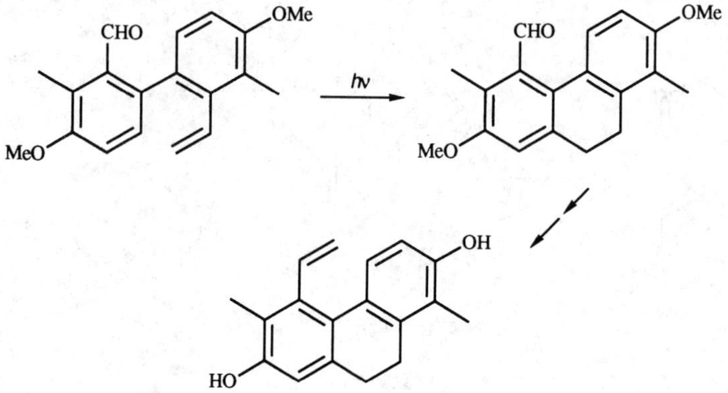

(171)

to use a 2,2′-divinylbiphenyl to provide a shorter synthetic route was unsuccessful since such systems give largely the tetrahydropyrene (**172**) [262].

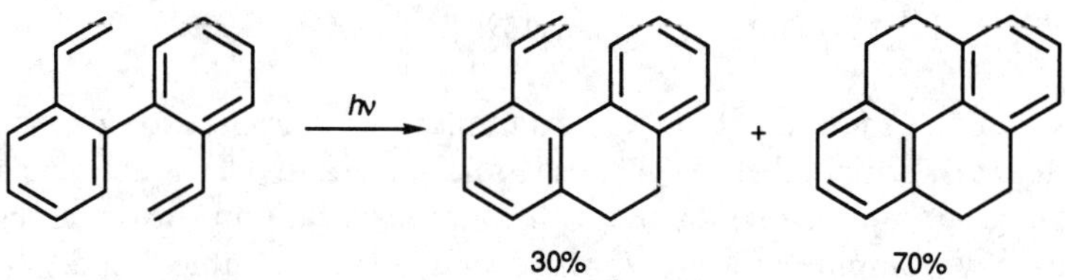

(172)

9-Fluorophenanthrene can be formed directly in 60% yield when *o*-(2,2-difluorovinyl)biphenyl is irradiated [263].

In a similar way 1-phenylbuta-1,3-diene systems give rise to dihydronaphthalenes on irradiation or naphthalenes if oxidative conditions are employed. Many of the reported examples involve a higher degree of ring fusion, such as the reaction (**173**) leading to a hexahydrophenanthrene. If oxidative

MeO CO2Me hv N2 MeO CO2Me OMe OMe 86%

(173)

conditions are used in this instance the corresponding tetrahydrophenanthrene is obtained in 65% yield [264]. Similar reactions have been used to make benzophenanthridiones (**174**) [265], benzocarbazoles [266] or ben-

Ph hv (254 nm) $I_2$ N H O N H O 84%

(174)

zoxanthenones [267]. A more complex example (**175**) forms a part of a

$EtO_2C$ H N $CO_2Me$ hv $(PhSe)_2$ N $CO_2Me$ $EtO_2C$ H N $CO_2Me$ N $CO_2Me$

(175)

synthetic route to *methoxatin*, a coenzyme in *methylotroph bacteria* [268].

Photocyclization of 2,3-bis(arylmethylene)butyrolactones provides a route to the apolignan system (**176**) [269], and related bis(alkylidene)succinic

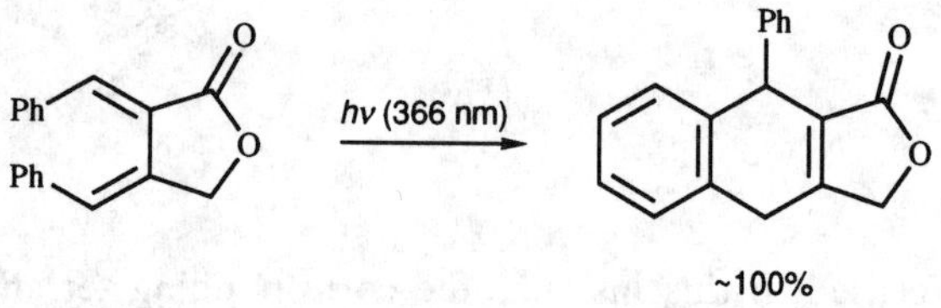

(176)

anhydrides are useful photochromic compounds. The furan derivative in reaction (**177**) has proved to be a simple and good actinometric compound

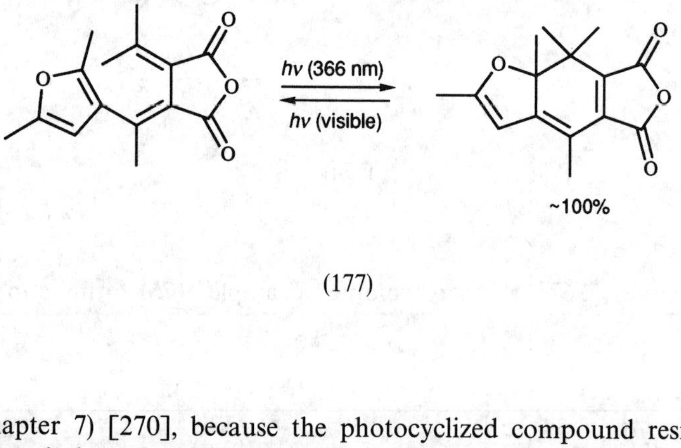

(177)

(see Chapter 7) [270], because the photocyclized compound resists both thermal and photochemical degradation. Thus the colour does not fade and the compound can be recycled many times. Naphthalenes can be formed directly from phenylbutadienes without the need of subsequent oxidation if a suitable substituent can be eliminated. This happens in one-pot, two-stage photochemical preparation of benzanthraquinones from 2-bromo-3-methoxy-1,4-naphthoquinones and aryl alkenes (**178**) (see also Chapter 3)

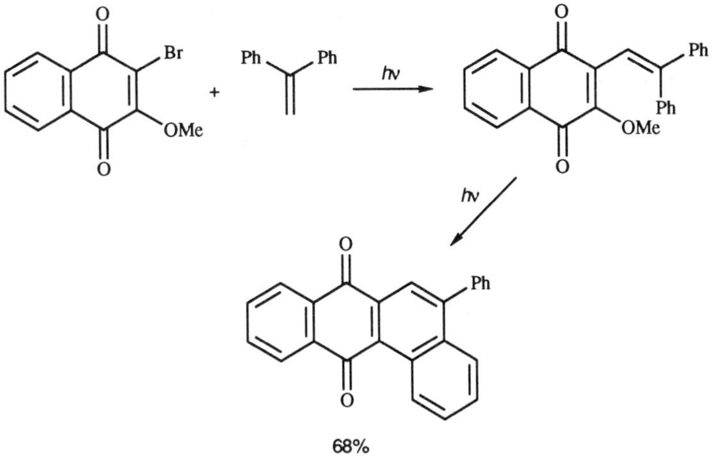

68%

(178)

[271].

1,4-Diarylbutenynes undergo non-oxidative photocyclization to give 1-arylnaphthalenes directly (**179**) [272]. Although formation of an intermediate

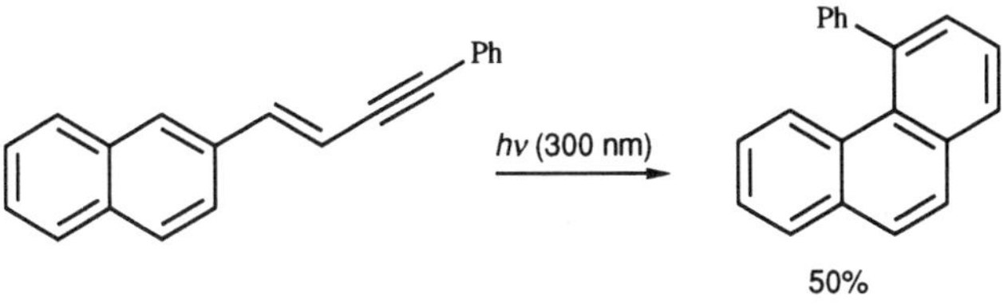

*hv* (300 nm)

50%

(179)

cyclic allene is a feasible path, a zwitterionic or biradical intermediate is probably a better representation. 1-Vinylnaphthalenes behave like *o*-vinylbiphenyls, giving high yields of acenaphthenes on irradiation under reducing conditions (**180**). In the presence of oxygen, acenaphthylenes are formed [273].

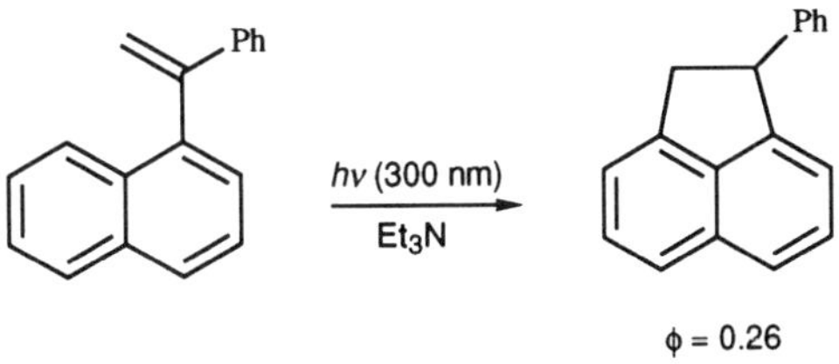

*hv* (300 nm)
$Et_3N$

$\phi = 0.26$

(180)

In the 6$\pi$-electron system that undergoes photocyclization it is not essential for all the atoms involved to be carbon, and some examples have nitrogen atoms at the point of attachment of the styryl group to the second aromatic unit. *N*-Arylimines of aromatic ketones or aldehydes (see Chapter 5) give phenanthridines on irradiation (**181**) [274], and *cis*-azobenzene (or more

OMe hν OMe O O N O O N 56%

(181)

likely the protonated species) gives benzo[*c*]cinnoline (**182**) in the presence

N N hν $FeCl_3$, $H^+$ N N

(182)

of iron(III) chloride as the oxidizing agent [275].

Anilinodivinylboranes lead to cyclized products on irradiation (**183**) [276]

Ph N B hν N B 90%

(183)

in a process that bears a formal similarity to the amide cyclizations discussed in Chapter 5.

Yet another group of photocyclizations involving a six-electron cyclic process are those arising from diarylamines or aryl vinyl ethers or sulphides [277a]. The basic reaction (**184**) leads to an ylide that can undergo a proton

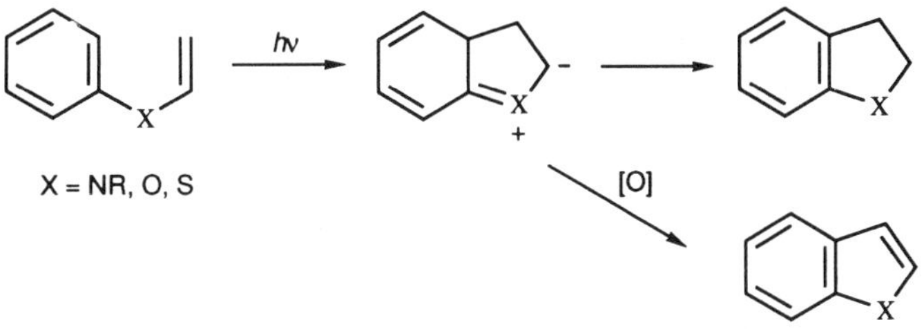

X = NR, O, S

hν

[O]

(184)

shift (common in the aryl vinyl systems) or can be oxidized (common in the diaryl systems). The reaction of *N*-methyldiphenylamine gives initially a 4a,4b-dihydrocarbazole which, in the absence of oxygen, disproportionates to yield *N*-methyltetrahydrocarbazole and *N*-methylcarbazole [277b]. The illustrations (**185**) and (**186**) [277a, 278–284] give a range of representative

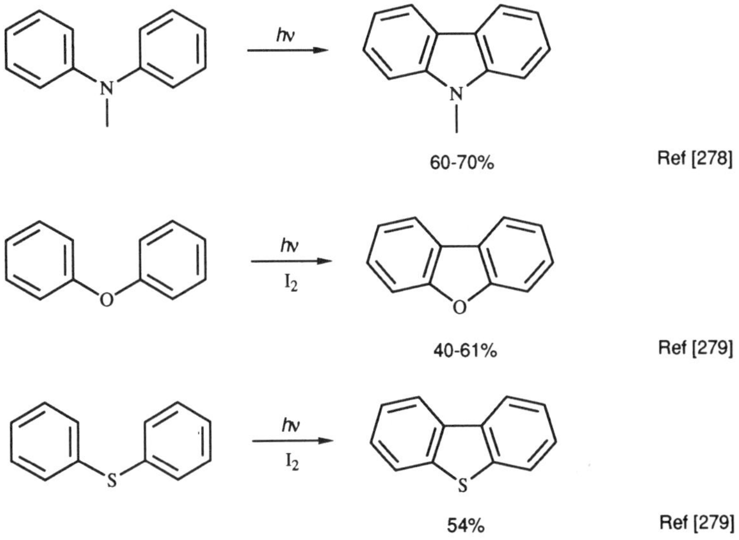

hν

60-70% Ref [278]

hν
$I_2$

40-61% Ref [279]

hν
$I_2$

54% Ref [279]

(185)

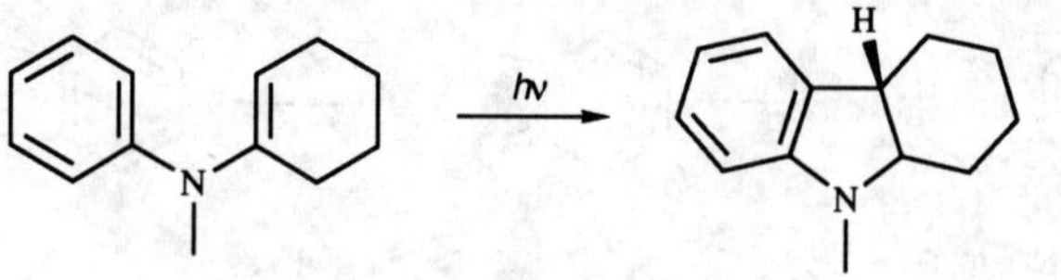

hν

H

N

N

71% Ref [280]

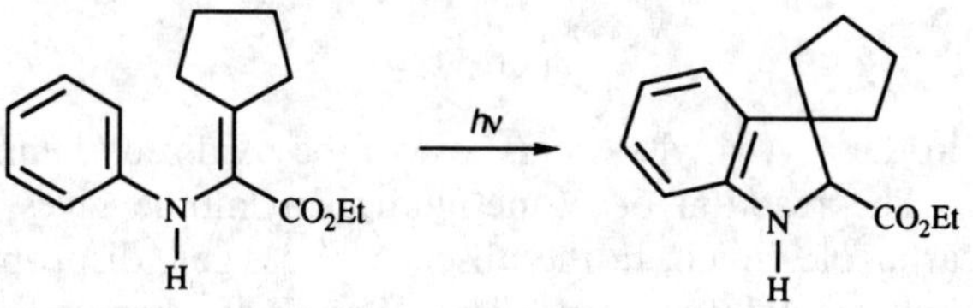

hν

N $CO_2Et$ H

N $CO_2Et$ H

70% Ref [281]

(185 continued)

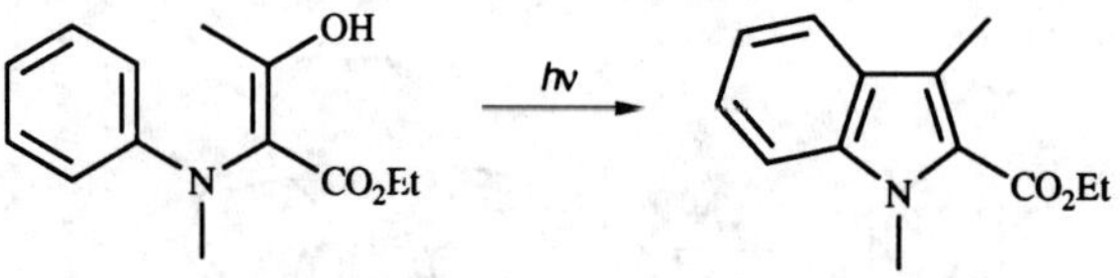

OH

N $CO_2Et$

hν

N $CO_2Et$

92% Ref [282]

S

hν

S

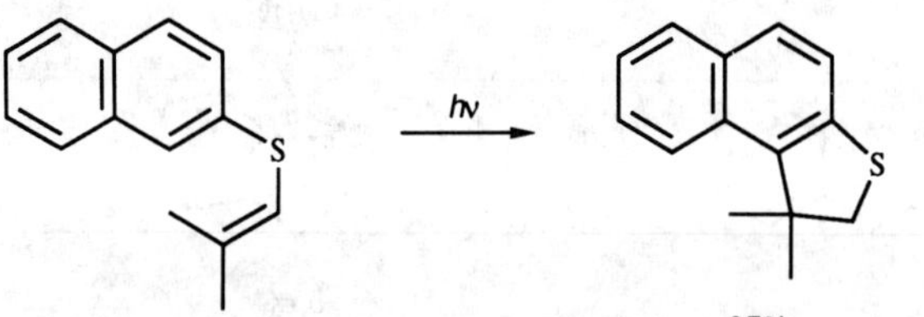

95% Ref [283]

(186)

88% Ref [277a]

95% Ref [284]

(186 continued)

reactions, including those in which an oxidative step is not needed because of the elimination of a hydroxy substituent derived from the enol group in the substrate. There are some limitations to the scope of the reaction. Thus, for example, *N*-arylnaphthylamines do not undergo photocyclization and an indirect route (**187**) has to be taken to obtain the corresponding benzocar-

65%

chloranil

41%

(187)

bazoles [285]. It should be noted that non-photochemical routes can sometimes be effective, as with the palladium(II) acetate-promoted oxidative cyclization of diphenylamines

A similarity to the arylenamine systems can be seen in a photocyclization (**188**) that has been used in the synthesis of a gibberellin precursor [286].

hν

$CO_2H$

80-85%

(188)

However, a number of miscellaneous cyclizations cannot operate by a related mechanism because there is no conjugation between the alkene and aryl components. Such examples include the reactions of 4-arylbut-1-enes [287], *N*-benzylenamines (e.g. **189**) [288], and *N*-allylpyridinium salts (**190**) [289].

hν

MeO

80-85%

(189)

hν MeOH

$H_2/Pt$

$ClO_4^-$

OMe

60%

(190)

**2.2.8 Lateral–nuclear photorearrangements**

The general lateral–nuclear rearrangement process (**191**, X = O, NH, S etc.),

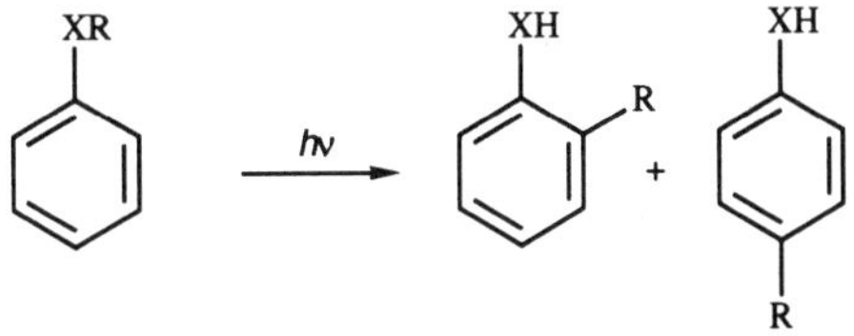

(191)

particularly the Fries reaction of phenol esters and related reactions of *N*-substituted anilines, is well-established in thermal chemistry. Many of the thermal processes have their photochemical counterparts and the photo-Fries reaction is the most extensively investigated [290]. This reaction occurs by way of homolytic cleavage to give acyl and phenoxy radicals which combine within the solvent cage to form cyclohexadienones that tautomerize to the substituted phenols. The phenoxy radicals and the dienones have been detected by Raman spectroscopy [291]. The acyl radicals have been trapped by added 2-methyl-2-nitrosopropane [292]. Phenoxy radicals that escape from the solvent cage can produce the parent phenol, and when phenyl acetate is irradiated in the gas phase, phenol is the only aromatic product formed [293].

The photo-Fries reaction has been used with esters of thymol (**192**) [294],

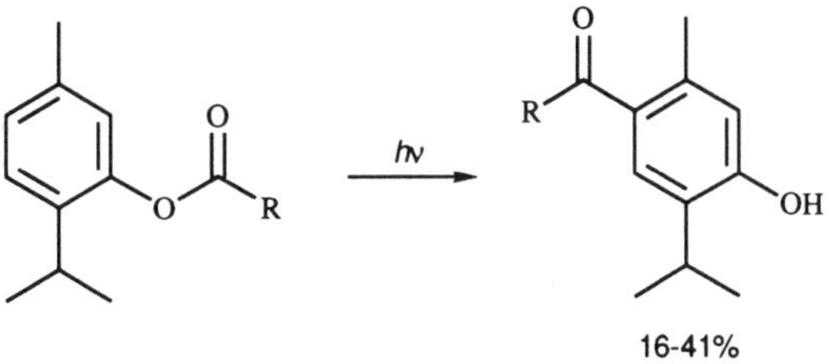

16-41%

(192)

and with esters of 1-naphthol (**193**) [295] to provide rearrangement products

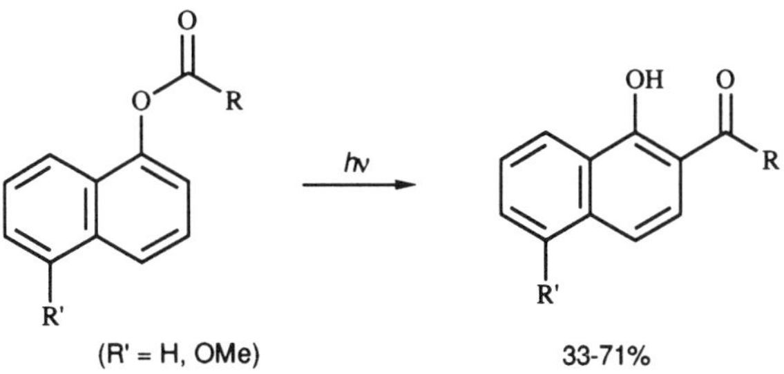

(R' = H, OMe)

33-71%

(193)

more cleanly than in the corresponding thermal processes. In the latter case the aim was to make analogues of adriamycinone, and a more heavily substituted naphthyl acetate has been employed [296] in a synthesis of spirochrome A. Carbonate esters are also reactive and afford dihydroxybenzophenone derivatives via a double photo-Fries process (**194**) [297].

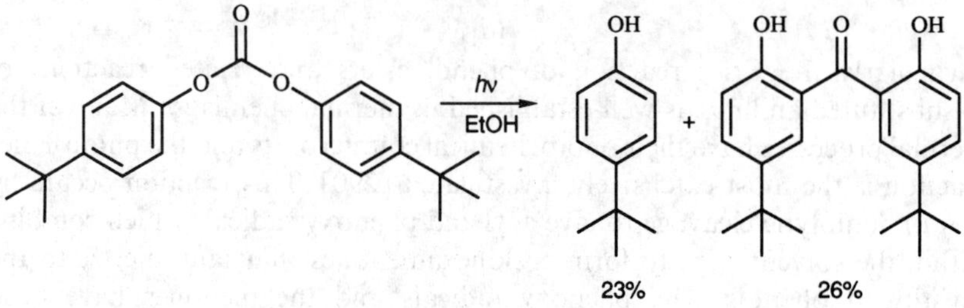

hν EtOH

23% 26%

(194)

The succinate ester of *p*-dihydroxybenzene can be used successfully [298] in a photochemical, but not thermal, Fries reaction to provide a route to aryldihydrofuranones (**195**). Cyclization involving the new phenol group

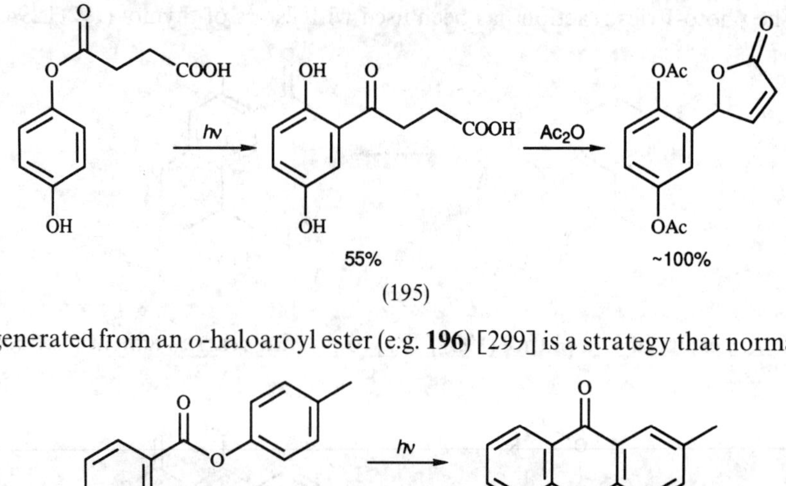

hν 55% $Ac_2O$ ~100%

(195)

generated from an *o*-haloaroyl ester (e.g. **196**) [299] is a strategy that normally

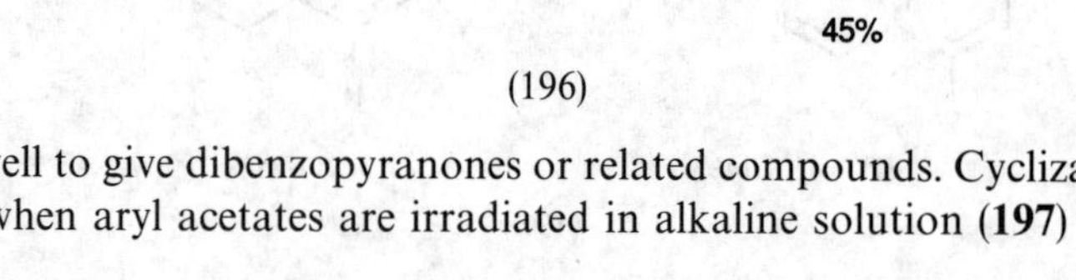

hν 45%

(196)

works well to give dibenzopyranones or related compounds. Cyclization also occurs when aryl acetates are irradiated in alkaline solution (**197**) [300].

hν
$C_6H_6$,
aq. NaOH

91% (R = H)
82% (R = Me)

(197)

Although the Fries rearrangement normally gives *ortho*- and *para*-substituted acyl phenols, a *meta* product can be obtained in systems where the alternative positions of attack are impossible. An example of this (**198**) has

hν

~100%

(198)

been employed in synthesis directed towards mitomycin analogues [301].

Other lateral–nuclear photorearrangements have been described and in (**199**) [302–305] and (**200**) [306–308] there are examples involving allyl aryl

hν

$\phi$ = 0.061 $\phi$ = 0.055 $\phi$ = 0.058 Ref [302]

hν

(n = 5-11) 80-87% Ref [303]

(199)

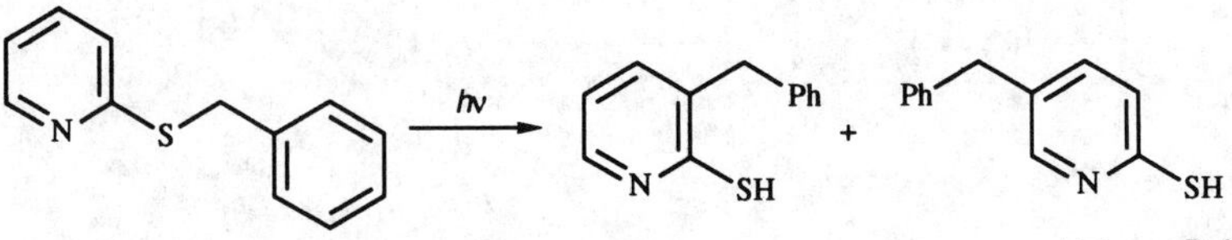

$h\nu$

Ph Ph

N S N SH N SH

26% 14% Ref [304]

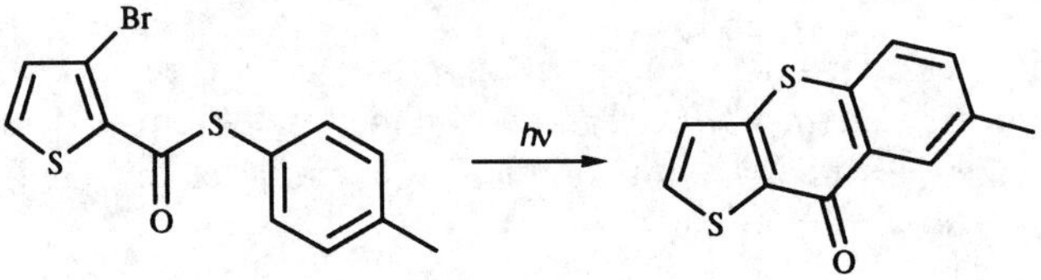

Br

S S $h\nu$ S

O S O

17% Ref [305]

(199 continued)

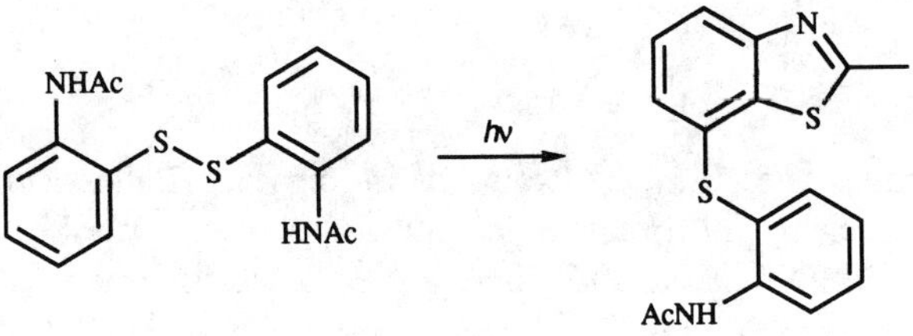

NHAc N

S S $h\nu$ S

HNAc S

AcNH

62% Ref [306]

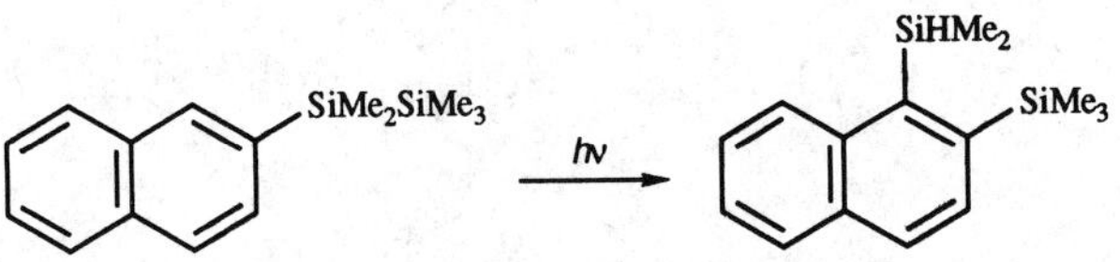

$SiHMe_2$

$SiMe_2SiMe_3$ $SiMe_3$

$h\nu$

40% Ref [307]

$NH{-}CO_2Et$ $NH_2$ $NH_2$ $NH_2$

$CO_2Et$

$h\nu$ + +

$CO_2Et$

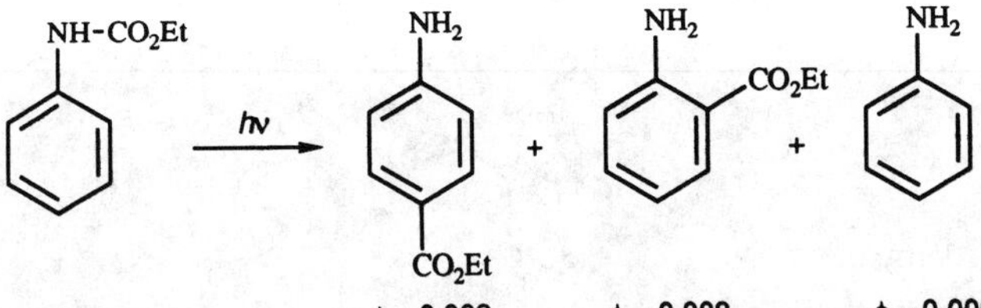

$\phi = 0.003$ $\phi = 0.008$ $\phi = 0.005$ Ref [308]

(200)

ethers, *N*-acylanilines, *N*-arylcarbamates, aryl benzyl sulphides, thiol esters or aroyl acids, diaryl sulphides and aryldisilanes. Formally similar reactions are found with diaryl ethers (**201**) [309], and with arylcyclopropenes (**202**)

OH

OH

O

hν

+

20% 13%

(201)

Ph

Ph

R

hν

Ph

Ph

Ph

R

Ph

64% (R = $CH_2Ph$)

(202)

[310], although the mechanisms are different in these systems.

## 2.3 ALKANES

### 2.3.1 Excited states and spectra

As far as the organic photochemist is concerned the alkanes are of least interest owing to the fact that they have no chromophores in the readily accessible region of the spectrum. Indeed they are essentially transparent to near ultraviolet irradiation of wavelengths longer than 180 nm. The bonding electrons in alkanes are found in $\sigma$-orbitals and this photochemical excitation of these compounds will involve $\sigma \rightarrow \sigma^*$ transitions. These will be of high energy, and decomposition within the molecule is likely to result in the cleavage of a $\sigma$-bond.

The presence of only $\sigma$-bonding electrons with the accompanying $\sigma\sigma^*$ excited states results in spectra with strong $\sigma \rightarrow \sigma^*$ absorptions in the vacuum ultraviolet region. As with the alkenes, increased substitution moves the absorption maxima towards longer wavelengths; some examples are shown in Table 2.2 [311].

**Table 2.2.** Alkane absorption maxima

| Alkane | $\lambda_{max}$/nm |
|---|---|
| Methane | 122 |
| Ethane | 135 |
| 2,4-Dimethylpentane | 154 |

The lifetime of the resultant $\sigma\sigma^*$ excited state is very short, although these states are probably not totally repulsive states since there are reports of weak fluorescence from some linear alkanes [312a] and also from cyclic alkanes such as cyclohexane [312b].

### 2.3.2 Photochemistry of alkanes

The reaction paths open to the excited states of alkanes have high rate constants, and this often leads to molecular rather than radical fission processes.

A reaction typical of this is the direct irradiation of methane in the gas phase at 124 nm. Here there is a preference for the formation of methylene and hydrogen by the so-called molecular detachment process [313], rather than a simple hydrogen-carbon bond fission to yield a hydrogen and a methyl radical [314]. Subsequent reaction of methylene with methane affords ethane, the principal product of the reaction. The study of the influence of wavelength changes in the 106–142 nm range on the outcome of the reaction confirms the involvement of these reaction paths (**203**) [315]. Ethane again shows

$$CH_4 \xrightarrow{h\nu} :CH_2 + H_2$$

$$:CH_2 \xrightarrow{CH_4} CH_3CH_3$$

$$CH_4 \xrightarrow{h\nu} H^{\bullet} + {}^{\bullet}CH_3$$

(203)

a preference for molecular rather than radical pathways when irradiated at or near its absorption maximum of 135 nm. Again hydrogen is the major product formed from the reaction as shown in **204** [316]. The influence

$$CH_3CH_3 \xrightarrow{h\nu} H_2 + CH_3\ddot{C}H$$

$$CH_3CH_3 \xrightarrow{h\nu} H_2 + H_2C{=}CH_2$$

(204)

of using shorter wavelength irradiation on both methane and ethane results in an increase in homolytic reactions. With ethane, using 110-nm light, C—H bond fission accounts for 41% of the primary processes, while the C—C cleavage accounts for 31%. Molecular processes correspondingly decrease in importance and amount to only 26% [317]. Propane behaves somewhat differently in that the formation of molecular hydrogen is less important. Carbon-carbon bond fission appears to dominate, with the formation of ethane arising by an intramolecular process [318]. The formation of propene and ethene are also important reaction paths. Other straight-chain alkanes such as pentane react to form molecular hydrogen ($\Phi = 0.42$) on irradiation in the gas phase at 147 nm [319]. The influence of substitution on the outcome of such reactions is illustrated by the gas-phase decomposition of 2,2,3,3-tetramethylbutane (**205**) at 147 nm, where the

$$(H_3C)_3C{-}C(CH_3)_3 \xrightarrow[147\ \text{nm}]{h\nu} 2\ (H_3C)_3C\cdot$$

(205)

principal reaction is the fission of the central C—C bond to afford *t*-butyl radicals with a quantum yield of 0.78 [320].

With cyclic systems the reaction can often follow a variety of paths, as in the irradiation of *trans*- and *cis*-1,2-dimethylcyclopropane, which affords many products on irradiation at 124 and 147 nm, although the principal reaction path with the highest quantum yield is that shown in **206** [321]. If,

$MeCH_2CH{=}CHCH_2Me$ ← hν, 185 nm — (Et, Et cyclopropane) — hν, 185 nm → $MeCH_2CH{=}CH_2$ + $MeCH{=}CH_2$

via (Et, Et biradical)

Ref [321a]

(Me, Me cyclopropane) — hν, 147 or 124 nm → (Me, Me biradical) → Me alkene

Ref [321b]

(206)

however, the substituents on the cyclopropane ring are phenyl groups that are light-absorbing themselves, then a benzenoid $\pi\pi^*$ state will be involved [322], and a variety of usefully synthetic reactions can be found [323a]. A typical example of this is shown in **207** for the solution-phase irradiation in

hν, t-BuOH → 38% (via Ph, Ph carbene); Ph, Ph, $Bu^t$, O

(207)

*t*-butanol of a 1,1-diphenyl-substituted cyclopropane [323b]. The presence of the 1,1-diphenyl substitution assists in a two-bond rupture process to afford a carbene and an alkene. The presence of a single phenyl substituent also exerts control during the irradiation of **208** where C—C one-bond fission

$\phi = 0.045$ $\phi = 0.022$

(208)

yields the best biradical possible from the molecule. The reactions of this either bring about *trans–cis* isomerization or the formation of an alkene by 1,2-hydrogen migration or cyclization to yield 1-phenylindane [324]. Hydrogen migration is a fairly common reaction within the biradical generated on C—C bond fission in cyclopropanes. Other examples of hydrogen migration within biradicals formed on photo-ring fission of cyclopropane derivatives in the singlet excited state have been reported, and examples are shown in **209** [325]. Isomerization is also reported for the cyclopropane **210**

Ref [325a]

| Ar | % | % |
|---|---|---|
| $p$-$CF_3C_6H_4$ | 67 | 8 |
| $m$-$CF_3C_6H_4$ | 61 | 7 |
| $p$-$AcC_6H_4$ | 74 | 10 |
| $m$-$MeC_6H_4$ | 53 | 14 |
| Ph | 65 | 16 |
| $m$-$MeOC_6H_4$ | 50 | 5 |
| $p$-$MeOC_6H_4$ | 48 | 36 |

Ref [325b]

(209)

and these products are accompanied by alkenes [326]. An ionic mechanism is suspected in this reaction to account for the striking substituent effect, a

| Ar | % | % |
|---|---|---|
| Ph | 13 | 6 |
| *p*-MeOC$_6$H$_4$ | 24 | 10 |
| *p*-MeC$_6$H$_4$ | 25 | 18 |
| *m*-MeC$_6$H$_4$ | 18 | 2 |
| *m*-CF$_3$C$_6$H$_4$ | 21 | 13 |

(210)

result of an aryl ring-cyclopropane charge-transfer process. The subject of charge-transfer interactions in the photochemistry of small ring compounds has been reviewed [327]. Ring opening can also be brought about by irradiation on ZnS or CdS surfaces [328]. Other examples of cyclopropane ring opening are shown in **211** [326, 329].

31% Ref [326]

| R | % |
|---|---|
| Ph | 50 |
| *p*-ClC$_6$H$_4$ | 24 |

Ref [329]

(211)

## 2.4 PHOTO-OXIDATION OF ALKENES AND DIENES

Unsaturated compounds such as alkenes and dienes all undergo reaction with singlet oxygen [330]. The formation of singlet oxygen can be carried out either thermally or photochemically. The photochemical path utilizes visible radiation and the transfer of energy from sensitizers such as methylene blue ($E_T = 168$ kJ mol$^{-1}$) or Rose Bengal ($E_T = 176$ kJ mol$^{-1}$). The energy-transfer mechanism, shown in **212**, was substantiated by Foote [331] in support of a much earlier proposal [332]. Foote showed that chemical methods for the production of singlet oxygen, e.g. alkaline decomposition of hydrogen peroxide, yielded identical results and exhibited the same stereoselectivity of products as those produced by the photo-oxidation

$$\text{Sensitizer} \xrightarrow{h\nu} [\text{sens.}]^{T_1}$$

$$[\text{sens.}]^{T_1} + {}^{3}O_2 \longrightarrow \text{sens.} + {}^{1}O_2^{*}$$

$$A + {}^{1}O_2^{*} \longrightarrow AO_2$$

(212)

procedure. The photochemical formation of the singlet excited state of oxygen ($E_S = 94.6$ kJ mol$^{-1}$) from ground-state triplet oxygen by transfer of triplet energy from a triplet sensitizer is a spin-allowed process.

### 2.4.1 Reactions with alkenes

In the reaction of singlet oxygen with alkyl-substituted alkenes the major process involves allylic hydrogen abstraction followed by recombination to give an allylic hydroperoxide. Normally these are not isolated but are reduced *in situ* to the allylic alcohol. A classical example of this is the photo-oxidation of (+)-limonene (**213**) [332] or of the alkene (**214**) [333]. The mechanism by

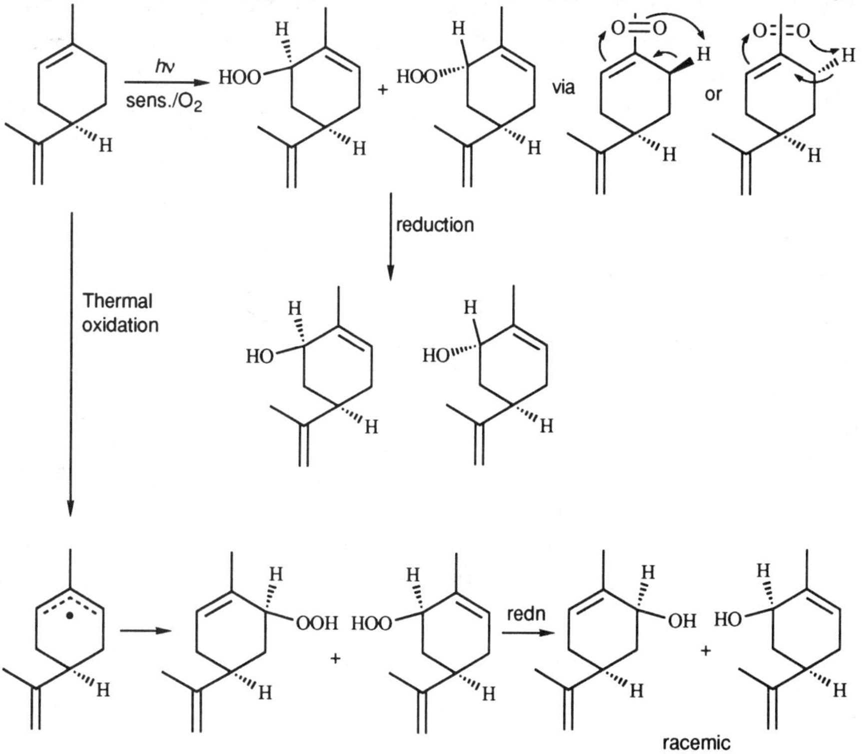


(213)

hν
sens./$O_2$
OH
+
HO
48%
52%
after reduction

(214)

which the addition occurs is as shown, with the weight of evidence in favour of a concerted process similar to an 'ene' reaction. However, other possibilities cannot be ruled out. The fact that free radicals are not involved is supported by the results obtained from the limonene (**213**) reaction, which yields only the *cis*- and the *trans*-alcohols as shown. Furthermore thermal oxidation yields only a racemate. Oxygenation occurs faster with more heavily substituted alkenes. Within one molecule, if there is a choice between a secondary or a tertiary hydroperoxide, then the ratio of products is close to 1 : 1, as seen in **214**. Slight variations encountered in the reactions are often due to stereochemical rather than electronic effects. Such stereochemical effects result from reaction of only the allylic hydrogens which are *cis* to the attacking oxygen. Such an effect can be seen in the example using a steroid molecule (**215**) [334]. Other examples of the usefulness of the process are

$C_8H_{17}$
HO
H
D
hν
sens./$O_2$
HO
OH
Ref [334a]
via
HO
H
O
O
D

(215)

$C_8H_{17}$ — $h\nu$, sens./$O_2$ — 40-70% — Ref [334b]

via

(215 continued)

illustrated in **216** [335].

$h\nu$, sens./$O_2$, -50°C — OOH, OMe — 55% + OMe, O—O — 45% — Ref [335a]

1. $h\nu$/sens./$O_2$ 2. reduction — OH + OH — ratio 70:30 — Ref [335b]

(216)

## 2.4.2 1,2-Addition

Some substituted alkenes, usually those classed as electron-rich, react by a different path with singlet oxygen and yield dioxetanes as shown in **216** [335a] and **217** [336]. Some dioxetanes [337] can show remarkable stability

Ph, Ph — $h\nu$, sens./$O_2$ — Ph, Ph, O—O — 80%

(217)

even to relatively high temperatures, whereas others decompose readily to yield a 1,2-dicarbonyl compound. The addition to the electron-rich alkene appears to be a 1,2-cycloaddition process and is stereospecific. This is seen to advantage in the cycloaddition shown in **218** where the dioxetane products retain the geometry of the original alkene [338].

EtO, EtO — $h\nu$, sens./$O_2$ → EtO, EtO, O–O; quant.

(218)

**2.4.3 Reaction with dienes**

Oxygen in its excited singlet state is electronically similar to ethene in its ground state. Such a similarity has been inferred from the behaviour of singlet oxygen in the 'ene' type reaction with alkenes. A further similarity occurs in the reactions of singlet oxygen with dienes where a Diels–Alder type process takes place [339]. The addition reaction of isoprene and other dienes (**219**)

$h\nu$, sens./$O_2$ → O–O; 50% Ref [340a]

$h\nu$, sens./$O_2$ → O–O; Ref [340b]

(219)

with oxygen is typical of this and affords cyclic peroxides [340]. Similar reactivity is observed with cyclic dienes (**220**) [341]. Other systems are also

hν sens./$O_2$ HN=NH

30% Ref [341a]

$R^1$ $R^2$ hν sens./$O_2$

$R^1 = R^2 = OEt$, 66%
$R^1 = N_3$, $R^2 = H$, 28% Ref [341b]

hν sens./$O_2$ HN=NH

R = Me, 92%
R = Et, 80% Ref [341c]

(220)

reactive, as in the addition to eneynes (**221**) [342] or vinylnaphthalenes (**222**)

hν sens./$O_2$ → $R^1$—CHO + $R^2$—C(=O)—C≡CH

a: $R^1 = R^2 = Ph$ a: 80% 84%
b: $R^1 = p\text{-}MeOC_6H_4$, $R^2 = Ph$ b: 83% 85%

(221)

OMe
OMe
$h\nu$
sens./$O_2$
OMe
O
O
OMe
98%

(222)

[343]. Some aromatic compounds also react with singlet oxygen to yield stable cyclic peroxides referred to as *endoperoxides*, the photochemical reactions of which are discussed in Chapter 3. The process affording these is, however, restricted to polycyclic aromatic compounds such as anthracene and naphthacene, as illustrated in **223** [344]. Oxidation of furan derivatives is

Ph
Ph
$h\nu$
sens./$O_2$
Ph
O
O
Ph
quant.

OMe
OMe
$h\nu$
sens./$O_2$
OMe
O
O
OMe quant.

(223)

also well known but the endoperoxides are unstable and usually break down to a dicarbonyl compound as shown in **224** [345].

| $R^1$ | $R^2$ | $R^3$ | % |
| --- | --- | --- | --- |
| Ph | H | H | 70 |
| Ph | Me | H | 90 |
| Me | $CO_2Me$ | H | 63 |

(224)

## REFERENCES


[1] A. J. Mercer and R. S. Mulliken, *Chem. Rev.*, 1969, **69**, 639.

[2] G. J. Collin, *Adv. Photochem.*, 1988, **14**, 135.

[3] P. J. Kropp, C. Ouannes and R. Beugelmans, *Elements de Photochemie Avancée.* ed. P. Courtot, Hermann, Paris, 1972, p. 231.

[4] J. B. Lambert, H. F. Shurvell, L. Verbit, R. G. Cooks and G. H. Stout, *Organic Structural Analysis*, Macmillan, New York, 1976, p. 345.

[5] J. N. Murrell, *The Theory of the Electronic Spectra of Organic Molecules*, Methuen, London, 1963, p. 138.

[6] M. Itoh and R. S. Mulliken, *J. Phys. Chem.*, 1969, **73**, 4332.

[7] G. S. Hammond, J. Saltiel, A. A. Lamola, N. J. Turro, J. S. Bradshaw, D. O. Cowan, R. C. Counsell, V. Vogt and C. Dalton, *J. Am. Chem. Soc.*, 1964, **86**, 3197.

[8] A. Gilbert, in *Spec. Publ.-R. Soc. Chem.*, 1986, **57** (Photochem. Org. Synth.), 141.

[9] G. S. Hammond, *The Spectrum*, 1990, **3**, 1.

[10] H. Gusten and L. Klasinc, *Tetrahedron Lett.*, 1968, 3097.

[11] D. Gegiou, K. A. Muszkat and E. Fischer, *J. Am. Chem. Soc.*, 1968, **90**, 3907.

[12] M. G. Rockley and K. Salisbury, *J. Chem. Soc., Perkin Trans. 2*, 1973, 1582.

[13] J. Bergman, K. Osaki, G. M. J. Schmidt and F. I. Sonntag, *J. Chem. Soc.*, 1964, 2021.

[14] D. Lenoir, J. E. Gano, and J. McTague, *Tetrahedron Lett.*, 1986, **27**, 5339.

[15] F. P. Tise and P. J. Kropp, *J. Am. Chem. Soc.*, 1981, **103**, 7293; W. Adam, *Z. Naturforsch. Tiel B.* 1981, **36**, 658.
[16] R. P. Johnson, *Org. Photochem.*, 1985, **7**, 75.
[17] O. Rodriguez and H. Morrison, *Chem. Commun.*, 1971, 679.
[18] T. J. Steirman and R. P. Johnson, *J. Am. Chem. Soc.*, 1985, **107**, 3971.
[19] C. S. Drucker, V. G. Toscano and R. G. Weiss, *J. Am. Chem. Soc.*, 1973, **95**, 6482.
[20] F. H. Watson, jun., A. T. Armstrong and S. P. McGlynn, *Theoret. Chim. Acta.*, 1970, **16**, 75: F. H. Watson, jun. and S. P. McGlynn, *Theoret. Chim. Acta.*, 1971, **21**, 309.
[21] P. J. Kropp, Ę. J. Reardon, jun., Z. L. F. Gaibel, K. F. Williard and J. H. Hattaway, *J. Am. Chem. Soc.*, 1973, **95**, 7058.
[22] P. J. Kropp, *Org. Photochem.*, 1979, **4**, 1.
[23] H. G. Fravel, jun. and P. J. Kropp, *J. Org. Chem.*, 1975, **40**, 2434.
[24] T. R. Fields and P. J. Kropp, *J. Am. Chem. Soc.*, 1974, **96**, 7559.
[25] P. J. Kropp, T. R. Fields and H. G. Fravel, jun., unpublished results, reported in [22].
[26] Y. Inoue, S. Takamuku and H. Sakurai, *J. Chem. Soc., Perkin Trans. 2*, 1977, 1635; R. Srinivasan and K. H. Brown, *J. Am. Chem. Soc.*, 1978, **100**, 4602.
[27] D. R. Arnold and V. Y. Abraitys, *J. Chem. Soc., Chem, Commun.*, 1967, 1053.
[28] H. Yamazaki and R. Cvetanovic. *J. Am. Chem. Soc.*, 1969. **91**, 520; H. Yamazaki, R. Cvetanovic and R. S. Irwin, *J. Am. Chem. Soc.*, 1976, **98**, 2198.
[29] P. J. Kropp, J. J. Snyder, P. C. Rawlings and H. J. Fravel, jun., unpublished results reported in [27].
[30] D. R. Arnold and V. Y. Abraitys, *Mol. Photochem.*, 1970, **2**, 27.
[31] (a) H. H. Stechl, *Angew. Chem. Int. Ed. Engl.*, 1963, **2**, 743.
(b) R. R. Kostikov, M. Yu. Kiselev, A. P. Molchanov and A. De Maeyre, *Zh. Org. Khim.*, 1986, **22**, 2464.
[32] (a) R. Srinivasan and K. A. Hill, *J. Am. Chem. Soc.*, 1966, **88**, 3765.
(b) J. H. Penn, L. Gan, T. A. Eaton, E. Y. Chan and Z. Lin, *J. Org. Chem.*, 1988, **53**, 1519; J. H. Penn, L. Gan, E. Y. Chan and P. D. Loesel, *J. Org. Chem.*, 1989, **54**, 601.
[33] H.-D. Scharf and F. Korte, *Chem. Ber.*, 1964, **97**, 2425.
[34] M. Tada, T. Kokubo and T. Sato, *Bull. Chem. Soc. Jpn.*, 1970, **43**, 2161; R. G. Salomon, K Folting, W. E. Streib and J. K. Kochi, *J. Am. Chem. Soc.*, 1974, **96**, 1145.
[35] P. J. Kropp, *J. Am. Chem. Soc.*, 1969, **91**, 5783; H.-D. Scharf and F.

Korte, *Tetrahedron Lett.*, 1963, 821; D. R. Arnold, D. J. Trecker and E. B. Whipple, *J. Am. Chem. Soc.*, 1965, **87**, 2596; D. R. Arnold, *Adv. Photochem.*, 1968, **6**, 301; S. H. Schroeter, *Mol. Photochem.*, 1972, **4**, 473.

[36] H.-D. Scharf, *Tetrahedron*, 1967, **23**, 3057; G. O. Schenck and R. Steinmetz, *Chem. Ber.*, 1963, **96**, 520.

[37] T. C. Chou, and J. H. Chiou, *J. Chin. Chem. Soc. (Taipei)*, 1986, **33**, 227 (*Chem. Abstr.*, 1987, **107**, 133 945).

[38] L. A. Paquette and C.-C. Shen, *Tetrahedron Lett.*, 1988, **29**, 4069.

[39] D. J. Trecker, J. P. Henry and J. E. McKeon, *J. Am. Chem. Soc.*, 1965, **87**, 3261; D. J. Trecker and R. S. Foote, *Org. Photochem. Syn.*, 1971, **1**, 81.

[40] R. G. Salomon and J. K. Kochi, *J. Am. Chem. Soc.*, 1974, **96**, 1137.

[41] R. G. Salomon and S. Ghosh, *Org. Synth.*, 1984, **62**, 125 (*Chem. Abstr.*, 1985, **102**, 166 345).

[42] R. G. Salomon, S. Ghosh, S. R. Raychaudhuri, and T. S. Miranti, *Tetrahedron Lett.*, 1984, **25**, 3167.

[43] K. Lal, E. A. Zarate, W. J. Youngs and R. G. Salomon, *J. Am. Chem. Soc.*, 1986, **108**, 1311.

[44] H. J. T. Bos, V. H. M. Elferink and D. Van der Ploeg, *Recl: J. R. Neth. Chem. Soc.*, 1984, **103**, 301 (*Chem. Abstr.*, 1985, **102**, 23 585).

[45] I. Sauers, L. A. Grezzo, S. W. Staley and J. H. Moore, jun., *J. Am. Chem. Soc.*, 1976, **98**, 4218.

[46] W. G. Herkstroeter and G. S. Hammond, *J. Am. Chem. Soc.*, 1966, **88**, 4769.

[47] A. H. Davidson and I. H. Wallace, *J. Chem. Soc. Chem. Commun.*, 1986, 1759.

[48] Y. Cao, W. Wu, Y. Ming and B. Zhang, *Ganguang Kexue Yu Kuang Huaxue*, 1985, 35 (*Chem. Abstr.*, 1986, **104**, 148 046).

[49] M. Comtet, *J. Am. Chem. Soc.*, 1969, **91**, 7761.

[50] E. J. Corey, F. A. Carey, and R. A. E. Winter, *J. Am. Chem. Soc.*, 1965, **87**, 934; E. J. Corey and J. I. Shulman, *Tetrahedron Lett*, 1968, 3655.

[51] J. A. Marshall and A. R. Hochstetler, *J. Chem. Soc., Chem. Commun.*, 1968, 296.

[52] P. J. Kropp, *J. Am. Chem. Soc.*, 1969, **91**, 5783.

[53] P. J. Kropp, *J. Am. Chem. Soc.*, 1966, **88**, 4091; P. J. Kropp and H. J. Krauss, *J. Am. Chem. Soc.*, 1967, **89**, 5199.

[54] J. A. Marshall and R. D. Carroll, *J. Am. Chem. Soc.*, 1966, **88**, 4092.

[55] T.-Y. Leong, T. Imagawa, K. Kimoto and M. Kawanisi, *Bull. Chem.*

*Soc. Jpn.*, 1973, **46**, 596.
[56] R. Srinivasan, *Adv. Photochem.*, 1966, **4**, 113.
[57] J. Saltiel, L. Metts, A. Sykes and M. Wrighton, *J. Am. Chem. Soc.*, 1971, **93**, 5302.
[58] R. Srinivasan, *J. Am. Chem. Soc.*, 1968, **90**, 4498.
[59] R. Srinivasan, *J. Am. Chem. Soc.*, 1962, **84**, 4141.
[60] R. B. Woodward and R. W. Hoffmann, *The Conservation of Orbital Symmetry*, Verlag Chemie, Weinheim, 1970; T. L. Gilchrist and R. C. Storr, *Organic Reactions and Orbital Symmetry*, Cambridge University Press, London, 2nd Edition 1979
[61] P. Datta, T. D. Goldfarb and R. S. Boiken, *J. Am. Chem. Soc.*, 1971, **93**, 5189.
[62] E. E. van Tamelen, T. L. Burkoth and R. H. Greeley, *J. Am. Chem. Soc.*, 1971, **93**, 6120.
[63] E. E. van Tamelen, S. P. Pappas and K. Kirk, *J. Am. Chem. Soc.*, 1971, **93**, 6092.
[64] W. H. Laarhoven, *Org. Photochem.*, 1987, **9**, 129.
[65] J. S. Swenton and G. L. Smyser, *J. Org. Chem.*, 1978, **43**, 165.
[66] J. W. F. Keana and R. H. Morse, *Tetrahedron Lett.*, 1976, 213.
[67] (a) W. G. Dauben and R. M. Coates, *J. Org. Chem.*, 1964, **29**, 2761.
(b) S. Abramson, J. Zizuashvili and B. Fuchs, *Tetrahedron Lett.*, 1982, **23**, 1377.
(c) R. Gleiter and U. Steuerle, *Tetrahedron Lett.*, 1987, **28**, 6159.
[68] P. J. Kropp, *Pure Appl. Chem.*, 1970, **24**, 585.
[69] (a) W. G. Dauben and J. S. Ritschers, *J. Am. Chem. Soc.*, 1970, **92**, 2925.
(b) P. G. Gassman and W. E. Hymans, *Tetrahedron*, 1968, **24**, 4437.
(c) J. A. Barltrop and H. E. Browning, *J. Chem. Soc., Chem. Commun.*, 1968, 1481.
[70] H. E. Zimmerman and A. C. Pratt, *J. Am. Chem. Soc.*, 1970, **92**, 6267.
[71] E. Block and H. W. Orf, *J. Am. Chem. Soc.*, 1972, **94**, 438.
[72] H. E. Zimmerman, P. Baeckstrom, T. Johnson and D. W. Kurtz, *J. Am. Chem. Soc.*, 1972, **94**, 5504.
[73] H. E. Zimmerman, D. W. Kurtz and L. M. Tolbert, *J. Am. Chem. Soc.*, 1973, **95**, 8210.
[74] (a) H. E. Zimmerman and D. N. Schissel, *J. Org. Chem.*, 1986, **51**, 196.
(b) H. E. Zimmerman, D. Armesto, M. G. Amezua, T. P. Gannett and R. P. Johnson, *J. Am. Chem. Soc.*, 1979, **101**, 6367.
(c) H. E. Zimmerman and L. M. Tolbert, *J. Am. Chem. Soc.*, 1975, **97**, 5497.

(d) H. E. Zimmerman and R. T. Klun. *Tetrahedron*, 1978, **34**, 1775; H. E. Zimmerman, W. T. Gruenbaum, R. T. Klun, M. G. Steinmetz and T. R. Welter, *J. Chem. Soc., Chem. Commun.*, 1978, 228.

[75] D. C. Lankin. D. M. Chihal, G. W. Griffin and N. S. Bhacca, *Tetrahedron Lett.*, 1973, 4009.

[76] G. W. Griffin, D. M. Chihal, J. Perreten and N. S. Bhacca, *J. Org. Chem.*, 1976, **41**, 3931.

[77] J. J. Brophy, *Austral. J. Chem.*, 1976, **29**, 2445.

[78] (a) H. E. Zimmerman, M. G. Steinmetz and C. L. Kreil, *J. Am. Chem. Soc.*, 1978, **100**, 4146.
(b) H. E. Zimmerman, T. P. Gannet and G. E. Keck, *J. Org. Chem.*, 1979, **44**, 1982.

[79] H. E. Zimmerman and G. L. Grunewald, *J. Am. Chem. Soc.*, 1966, **88**, 183.

[80] (a) R. G. Paddick, K. E. Richards and G. J. Wright, *Austral. J. Chem.*, 1976, **29**, 1005.
(b) C. O. Bender and J. Wilson, *Helv. Chim. Acta*, 1976, **59**, 1469.

[81] J. R. Scheffer, J. Trotter, M. Garcia-Garibay and F. Wireko, *Cryst. Liq. Cryst.*, 1988, **156** (Pt. A), 63; P. R. Pokkuluri, J. R. Scheffer and J. Trotter, *Tetrahedron Lett.*, 1989, **30**, 1607.

[82] (a) L. A. Paquette and E. Bay, *J. Am. Chem. Soc.*, 1984, **106**, 6693.
(b) L. A. Paquette, A. Varadarajan, and E. Bay, *J. Am. Chem. Soc.*, 1984, **106**, 6702.
(c) L. A. Paquette, M. J. Coughlan, C. E. Cottrell, T. Irie and H. Tanida, *J. Org. Chem.*, 1986, **51**, 696.

[83] (a) M. Kuzuya, M. Ishikawa, T. Okuda and H. Hart, *Tetrahedron Lett.*, 1979, 523; M. Kuzuya, E. Mano, M. Ishikawa, T. Okuda and H. Hart, *Tetrahedron Lett.*, 1981, **22**, 1613.
(b) M. Kuzuya, M. Adachi, A. Noguchi and T. Okuda, *Tetrahedron Lett.*, 1983, **24**, 2271.
(c) D. P. Kjell and R. S. Sheridan, *J. Am. Chem. Soc.*, 1986, **108**, 4111.

[84] H. Prinzbach, *Pure Appl. Chem*, 1968, **16**, 17.

[85] (a) H. Tamiaki and K. Maruyama, *Chem. Lett.*, 1988, 875.
(b) L. A. Paquette and U. S. Racherla, *J. Org. Chem.*, 1987, **52**, 3250.
(c) H. Prinzbach, G. Kaupp, R. Fuchs, M. Jiyeux, R. Kitzing and J. Markert, *Chem. Ber.*, 1973, **106**, 3824.
(d) H. Prinzbach and J. Rivier, *Helv. Chim. Acta*, 1970, **53**, 2201.

[86] H. Hayakawa, H. Taoda, T. Yumoto, H. Yamakita, M. Tazawa and K. Kawase, *Jp. Kokai, Tokkyo Koho*, JP 61 53232; H. Taoda, T. Yumoto, K. Hayakawa, K. Kawase, H. Yamakita and M. Tazawa,

*Nagoya Kogyo Gijutsu Shikensho Hokoku*, 1986, **35**, 1.

[87] K. Maruyama and H. Tamiaki, *Chem. Lett.*, 1987, 683.

[88] W. L. Dilling, *J. Am. Chem. Soc.*, 1967, **89**, 2742.

[89] J. H. Tumlinson, R. C. Gueldner, D. D. Hardee, A. C. Thompson, P. A. Hedin and J. P. Minyard. *J. Org. Chem.*, 1971, **36**, 2616.

[90] R. A. Whitney, *Can. J. Chem.*, 1983, **61**, 1158.

[91] W. Reid, O. Bellinger and J. W. Bats, *Chem. Ber.*, 1983, **116**, 3794.

[92] T. Hiyama, S. Fujita and H. Nozaki, *Bull. Chem. Soc. Jpn*, 1971, **44**, 3222.

[93] A. C. Pratt, *J. Chem. Soc., Chem. Commun.*, 1974, 183; J. M. Hornback, *Tetrahedron Lett.*, 1976, 3389.

[94] R. B. Cundall, D. A. Robinson and L. C. Pereira, *Adv. Photochem.*, 1977, **10**, 147.

[95] A. J. Rest, K. Salisbury and J. R. Sodeau, *J. Chem. Soc., Faraday Trans. 2*, 1977, 1396.

[96] T. S. Godfrey and G. Porter, *Trans, Faraday Soc.*, 1966, **62**, 7.

[97] F. Hirayama, T. A. Gregory and S. Lipsky, *J. Chem. Phys.*, 1973, **58**, 4696.

[98] C. E. Easterly, L. G. Christophorou and J. G. Carter, *J. Chem. Soc., Faraday Trans. 2*, 1973, 471.

[99] I. Yamazaki and H. Baba, *J. Chem. Phys.*, 1977, **66**, 5826.

[100] D. Bryce-Smith and A. Gilbert, in *Rearrangements in Ground and Excited States*, ed P. de Mayo, Academic Press, 1980, vol 3, p. 349.

[101] S. A. Lee, J. M. White and W. A. Noyes, *J. Chem. Phys.*, 1976, **65**, 2805.

[102] T. J. Katz, E. J. Wang and N. Acton, *J. Am. Chem. Soc.*, 1971, **93**, 3782.

[103] D. Bryce-Smith and A. Gilbert, *Tetrahedron*, 1976, **32**, 1309.

[104] J. J. C. Mulder, *J. Am. Chem. Soc.*, 1977, **99**, 5177.

[105] E. Farenhorst, *Tetrahedron Lett.*, 1966, 6465.

[106] (a) I. Haller, *J. Am. Chem. Soc.*, 1966, **88**, 2070.
(b) M. G. Barlow, R. N. Haszeldine and R. Hubbard, *J. Chem. Soc. (C)*, 1970, 1232.

[107] (a) K. E. Wilzbach and L. Kaplan, *J. Am. Chem. Soc.*, 1965, **87**, 4004.
(b) R. West, M. Furue and V. N. M. Rao, *Tetrahedron Lett.*, 1973, 911.
(c) Y. Tobe, K. Kakiuchi, Y. Odaira, T Hosakai, Y. Kai and N. Kasai, *J. Am. Chem. Soc.*, 1983, **105**, 1376.

[108] Y. Kobayashi and I. Kumadaki, *Acc. Chem. Res.*, 1981, **14**, 76.

[109] G. Maier and K.-A. Schneider, *Angew. Chem. Int. Ed. Engl.*, 1980, **19**, 1022.

[110] H. Wiebe, S. Braslovsky and J. Heicklen, *Can. J. Chem.*, 1972, **50**, 2721; Y. Kobayashi, I. Kumadaki, A. Ohsawa and Y. Sekine,

*Heterocycles*, 1977, **6**, 1587; Y. Kobayashi, K. Kawada, A. Ando, and I. Kumadaki, *Heterocycles*, 1983, **12**, 174.
[111] R. W. Franck, W. L. Mandella, K. J. Falci, *J. Org. Chem.*, 1975, **40**, 327.
[112] Z. Yoshida, F. Kawamoto, H. Miyoshi and H. Ikikoshi, *Jpn Kokai Tokkyo Koho*, 79, 138 549 (*Chem. Abstr.*, 1980, **92**, 163 763).
[113] H. Gusten, M. Mintas and L. Klasinc, *J. Am. Chem. Soc.*, 1980, **102**, 7936.
[114] H. Hart and B. Ruge, *Tetrahedron Lett.*, 1977, 3143.
[115] L. Kaplan, D. J. Rausch and K. E. Wilzbach, *J. Am. Chem. Soc.*, 1972, **94**, 8638.
[116] J. A. Berson and N. M. Hasty, *J. Am. Chem. Soc.*, 1971, **93**, 1549.
[117] J. W. Pavlik, T. E. Snead and J. R. Tata, *J. Heterocyclic Chem.*, 1981, **18**, 1481.
[118] R. F. Childs, G. S. Shaw and A. Varadarajan, *Synthesis*, 1982, 198.
[119] R. B. Cundall, D. A. Robinson and A. J. R. Voss, *J. Photochem.*, 1974, **2**, 239.
[120] A. W. Burgstahler and P.-L. Chen, *J. Am. Chem. Soc.*, 1964, **86**, 2940.
[121] L. Kaplan, K. E. Wilzbach, W. G. Brown and S. S. Yang, *J. Am. Chem. Soc.*, 1965, **87**, 675.
[122] J. A. Barltrop and A. C. Day, *Chem Commun.*, 1975, 177.
[123] J. A. Barltrop, R. Carder, A. C. Day, J. R. Harding and C. Samuel, *Chem. Commun.*, 1975, 729.
[124] J. A. Barltrop, A. C. Day and C. J. Samuel, *Chem. Commun.*, 1976, 823.
[125] J. A. Barltrop, A. W. Baxter, A. C. Day and E. Irving, *Chem. Commun.*, 1980, 606.
[126] A. Padwa, *Rearrangements in Ground and Excited States*, ed P. de Mayo, Academic Press, 1980, vol 3, p. 501; Y. Kobayashi and I. Kumadaki, *Adv. Heterocyclic Chem.*, 1982, **31**, 169.
[127] S. Boue and R. Srinivasan, *J. Am. Chem. Soc.*, 1970, **92**, 1824.
[128] E. E. van Tamelen and T. Whitesides, *J. Am. Chem. Soc.*, 1971, **93**, 6129.
[129] R. D. Chambers, A. A. Lindley, and H. C. Fielding, *J. Fluorine Chem.*, 1978, **12**, 337.
[130] A. Couture, A. Delevallee, A. Lablache-Combier and C. Parhanyi, *Tetrahedron*, 1975, **31**, 785.
[131] H. Wynberg, *Acc. Chem. Res.*, 1971, **4**, 65.
[132] H. Wynberg, R. M. Kellogg, H. van Drielm and G. E. Beekhuis, *J. Am. Chem. Soc.*, 1966, **88**, 1966.
[133] J. A. Barltrop, A. C. Day and E. Irving, *J. Chem. Soc., Chem. Commun.*, 1979, 966.

[134] Y. Kobayashi, A. Ando, K. Kawada and I. Kumadaki, *J. Org. Chem.*, 1980, **45**, 2968.
[135] A. Melhorn, F. Fratev and V. Monev, *Tetrahedron*, 1981, **37**, 2627.
[136] J. Becher, L. Finsen, I. Winckelmann, R. R. Koganty and O. Buchardt, *Tetrahedron*, 1981, **37**, 789.
[137] C. Guyon, P. Boule and J. Lemaire, *Tetrahedron Lett.*, 1982, **23**, 1581.
[138] M. Maerky, H. Meier, A. Wunderli, H. Heimgartner, H. Schmid and H. J. Hansen, *Helv. Chim. Acta*, 1978, **61**, 1477; H. Gotthardt and F. Reiter, *Chem. Ber.*, 1979, **112**, 1477; K.-H. Pfoertner and J. Foricher, *Helv. Chim. Acta*, 1980, **63**, 653.
[139] A. Holm and N. H. Toubro, *J. Chem. Soc., Perkin Trans. 1*, 1978, 1445.
[140] A. Krantz and J. Laureni, *J. Am. Chem. Soc.*, 1981, **103**, 486.
[141] J. Font, M. Torres, H. E. Gunning, and O. P. Strausz, *J. Org. Chem.*, 1978, **43**, 2487.
[142] O. Zimmer and H. Meier, *J. Chem. Soc., Chem. Commun.*, 1982, 481.
[143] G. Lodder and E. Havinga, *Tetrahedron*, 1972, **28**, 5583.
[144] W. J. Spillane, in *Isotopes in Organic Chemistry*, Eds E. Buncel and C. C. Lee, Elsevier, 1978, vol 4, p. 51.
[145] W. P. Neumann and H. Hillgartner, *Synthesis*, 1971, 537.
[146] J. P. H. Muller, H. Parlar and K. Korte, *Synthesis*, 1976, 524.
[147] T. Takaki, *Bull. Chem. Soc. Jpn*, 1978, **51**, 1145.
[148] J. Cornelisse, G. Lodder and E. Havinga, *Rev. Chem. Intermed.*, 1979, **2**, 231; C. Parkanyi, *Pure Appl. Chem.*, 1983, **55**, 331.
[149] R. L. Letsinger and J. H. McCain, *J. Am. Chem. Soc.*, 1966, **88**, 2884.
[150] A. van Vliet, M. E. Kronenberg, J. Cornelisse and E. Havinga, *Tetrahedron*, 1970, **26**, 1061.
[151] M. Sawaura and T. Mukai, *Bull. Chem. Soc. Jpn*, 1981, **54**, 3213.
[152] C. A. G. O. Varma, J. J. Tamminga and J. Cornelisse, *J. Chem. Soc., Faraday Trans. 2*, 1981, 265.
[153] G. Grabner, W. Rauscher, J. Zechner and N. Getoff, *J. Chem. Soc., Chem. Commun.*, 1980, 222.
[154] J. A. Barltrop, N. J. Bunce and A. Thomson. *J. Chem. Soc. (C)*, 1967, 1142.
[155] S. Nilsson, *Acta Chem. Scand.*, 1973, **27**, 329.
[156] K Fujiki, T. Nishio, and Y. Omote, *Bull. Chem. Soc. Jpn*, 1979, **52**, 614.
[157] G. C. Miller, M. J. Miile, D. G. Crosby, S. Sontum and R. G. Zepp, *Tetrahedron*, 1979, **35**, 1797.
[158] O. V. Kul'bitskaya, A. N. Frolov and A. V. Ed'tsov. *Zh. Org. Khim.*, 1979, **15**, 440 (*Chem. Abstr.*, 1979, **91**, 4996).
[159] G. G. Wubbels, A. M. Halverson and J. D. Oxman, *J. Am. Chem.*

*Soc.*, 1980, **102**, 4848.
[160] K. Mizuno, C. Pac and H. Sakurai, *J. Chem. Soc., Perkin Trans. 2*, 1983, 1269.
[161] N. Suzuki, K. Shimazu, T. Ito and Y. Izawa, *J. Chem. Soc., Chem. Commun.*, 1980, 1253.
[162] C. Parkanyi, *Bull. Soc. Chim. Belg.*, 1981, **90**, 599.
[163] M. B. Groen and E. Havinga, *Mol. Photochem.*, 1974, **6**, 9.
[164] M. Bellas, D. Bryce-Smith, M. T. Clarke, A. Gilbert, G. Klunkin, S. Krestonosich, C. Manning and S. Wilson, *J. Chem. Soc., Perkin Trans. 1*, 1977, 2571.
[165] A. Gilbert, S. Krestonosich and D. L. Westover, *J. Chem. Soc., Perkin Trans. 1*, 1981, 295.
[166] D. Bryce-Smith, A. Gilbert and G. Klunkin, *J. Chem. Soc., Chem. Commun.*, 1973, 330.
[167] J. A. Barltrop and R. Owers, *J. Chem. Soc., Chem. Commun.*, 1970, 1462
[168] A. Sugimoto, R. Sumida, N. Tamai, H. Inoue and Y. Otsuji, *Bull. Chem. Soc. Jpn*, 1981, **54**, 3500.
[169] M. Yasuda, C. Pac and H. Sakurai, *Bull. Chem. Soc. Jpn*, 1981, **54**, 2352.
[170] S. Tazuke and H. Ozawa, *J. Chem. Soc., Chem. Commun.*, 1975, 237.
[171] J. A. Barltrop, *Pure Appl. Chem.*, 1973, **33**, 179.
[172] M. Yasuda, C. Pac and H. Sakurai, *J. Org. Chem.*, 1981, **46**, 788.
[173] Y. Kanaoka, Y. Hatanaka, E. N. Deusler, I. L. Karle and B. Witkop, *Chem. Pharm. Bull.*, 1982, **30**, 3028.
[174] H. J. F. Angus and D. Bryce-Smith, *J. Chem. Soc. (C)*, 1960, 4791.
[175] J. Cornelisse and R. Srinivasan, *Chem. Phys. Letters*, 1973, **20**, 278.
[176] D. Bryce-Smith, B. E. Foulger and A. Gilbert, *J. Chem. Soc., Chem. Commun.*, 1972, 664.
[177] D. Bryce-Smith and A. Gilbert, *Tetrahedron*, 1977, **33**, 2459.
[178] M. F. Mirbach, M. J. Mirbach and A. Saus, *Tetrahedron Lett.*, 1977, 959.
[179] D. Bryce-Smith, A. Gilbert, B. H. Orger and P. J. Twitchett, *J. Chem. Soc., Perkin Trans. 1*, 1978, 232.
[180] Y. Katsuhara, T. Nakamura, A. Shimizu, Y. Shigemitsu and Y. Odaira, *Chem. Lett.*, 1972, 1215.
[181] J. Mattay, J. Runsink, H. Leismann and H.-D. Scharf, *Tetrahedron Lett.*, 1982, **23**, 4919; D. Bryce-Smith, B. Foulger, J. Forrester, A. Gilbert, B. H. Orger and H. M. Tyrell, *J. Chem. Soc., Perkin Trans. 1*, 1980, 55.

[182] A. Gilbert, G. N. Taylor and A. Collins, *J. Chem. Soc., Perkin Trans. 1*, 1980, 1218.
[183] A. Gilbert and P. Yianni, *Tetrahedron*, 1981, **37**, 3275.
[184] M. G. B. Drew, A. Gilbert, P. Heath, A. J. Mitchell and P. W. Rodwell, *J. Chem. Soc., Chem. Commun.*, 1983, 750.
[185] V. Y. Merritt, J. Cornelisse and R. Srinivasan, *J. Am. Chem. Soc.*, 1973, **95**, 8250.
[186] D. Bryce-Smith, B. E. Foulger, A. Gilbert and P. J. Twitchett, *J. Chem. Soc., Chem. Commun.*, 1971, 794.
[187] A. Gilbert and G. Taylor, *Tetrahedron Lett.*, 1977, 469.
[188] J. Mattay, H. Leismann, and H.-D. Scharf, *Chem. Ber.*, 1979, **112**, 577.
[189] R. J. Atkins, G. I. Fray, A. Gilbert and M. W. bin Samsudin, *Tetrahedron Lett.*, 1977, 3579.
[190] A. W. H. Jans, J. J. Van Dijk-Knepper and J. Cornelisse, *Recl.: J. R. Neth. Chem. Soc.*, 1982, **101**, 275.
[191] R. Sheridan, *Tetrahedron Lett.*, 1982, **23**, 267.
[192] M. Ohashi, Y. Tanaka and S. Yamada, *J. Chem. Soc., Chem. Commun.*, 1976, 800; A. Gilbert and P. Yianni, *Tetrahedron Lett.*, 1982, **23**, 255.
[193] B. A. Zhubanov, O. A. Almabekov and Zh. M. Ismailova, *Zh. Org. Khim.*, 1981, **17**, 996 (*Chem. Abstr.*, 1981, **95**, 114 883).
[194] B. Sket and M. Zupan, *J. Chem. Soc., Chem. Commun.*, 1976, 1053.
[195] B. Sket and M. Zupan, *Tetrahedron Lett.*, 1977, 2811.
[196] A. Gilbert and P. Yianni, *Tetrahedron Lett.*, 1982, **23**, 4611.
[197] D. Bryce-Smith, *Pure Appl. Chem.*, 1968, **16**, 47.
[198] D. Bryce-Smith, R. Deshpande, A. Gilbert and J. Groznka, *J. Chem. Soc., Chem. Commun.*, 1970, 561.
[199] A. Gilbert, G. N. Taylor and M. W. bin Samsudin, *J. Chem. Soc., Perkin Trans. 1*, 1980, 869.
[200] J. A. Ors and R. Srinivasan, *J. Org. Chem.*, 1977, **42**, 1321.
[201] A. Gilbert and M. W. bin Samsudin, *J. Chem. Soc., Perkin Trans. 1*, 1980, 1118.
[202] P. A. Wender and G. B. Dreyer, *J. Am. Chem. Soc.*, 1982, **104**, 5805.
[203] P. A. Wender and G. B. Dreyer, *Tetrahedron*, 1981, **37**, 4445.
[204] P. A. Wender and J. J. Howbert, *Tetrahedron Lett.*, 1982, **23**, 3983.
[205] A. Gilbert and G. N. Taylor, *J. Chem. Soc., Perkin Trans. 1*, 1980, 1761.
[206] K. Kraft and G. Koltzenburg, *Tetrahedron Lett.*, 1967, 4357, 4723
[207] H. P. Loeffler, *Tetrahedron Lett.*, 1974, 787.
[208] N. C. Yang, C. V. Neywicke and K. S. Srinivasachar, *Tetrahedron*

*Lett.*, 1975, 4313.

[209] T. S. Cantrell, *J. Org. Chem.*, 1981, **46**, 2674.

[210] Y. Yamamoto, M. Irie, Y. Yamamoto and K. Hayashi, *J. Chem. Soc., Perkin Trans. 1*, 1979, 1517.

[211] M. Yasuda, C., Pac and H. Sakurai, *Bull. Chem. Soc., Jpn,* 1980, 502.

[212] K. Mizumo, C. Pac and H. Sakurai, *J. Chem. Soc., Perkin Trans. 1*, 1975, 2221.

[213] I. A. Akhtar and J. J. McCullough, *J. Org. Chem.*, 1981, **46**, 1447.

[214] C. Pac, K. Mizumo, H. Okamoto and H. Sakurai, *Synthesis* 1978, 589.

[215] J. J. McCullough, W. K. MacInnis, C. J. L. Lock and R. Faggiani, *J. Am. Chem. Soc.*, 1982, **104**, 4644.

[216] Y. Inoue, K. Nishida, K. Ishibe, T. Hakushi and N. J. Turro, *Chem. Lett.*, 1982, 471.

[217] K. T. Mak, J. Srinivasachar and N. C. Yang, *J. Chem. Soc., Chem. Commun.*, 1979, 1038.

[218] D. A. Holden and J. E. Guillet, *J. Polym. Sci., Polym. Letters Ed.*, 1979, **17**, 15.

[219] N. Selvarajam and V. Ramakrishnan, *Indian J. Chem., Sect. B*, 1979, **18B**, 331.

[220] W. K. Smothers, M. C. Meyer and J. Saltiel, *J. Am. Chem. Soc.*, 1983, **105**, 545; N. C. Yang, R. L. Yates, J. Masnovi, D. M. Shold and W. Chiang, *Pure Appl. Chem.*, 1979, **51**, 173.

[221] T. Wang, J. Ni, J. Masnovi and N. C. Yang, *Tetrahedron Lett.*, 1982, **23**, 1231; G. Kaupp, H. W. Grueter and E. Teufel, *Chem. Ber.*, 1983, **116**, 630.

[222] D. O. Cowan and R. L. Drisko, *Elements of Organic Photochemistry*, Plenum Press, N. Y., 1976, p. 37.

[223] W. R. Bergmark, G. Jones, T. E. Reinhardt and A. M. Halpern, *J. Am. Chem. Soc.*, 1978, **100**, 6665.

[224] H.-D. Becker, K. Sandros and K. Anderson, *Angew. Chem. Int. Ed. Engl.*, 1983, **22**, 495.

[225] T. Teitei, D. Wells, T. H. Spurling and W. H. F. Sasse, *Aust, J. Chem.*, 1978, **31**, 85.

[226] H. Higuchi, K. Takatsu, T. Otsubo, Y. Sakata and S. Misumi, *Tetrahedron Lett.*, 1982, **23**, 671.

[227] G. O. Schenck, W. Hartmann, S.-P. Mannsfeld, W. Metzner and C. H. Krauch, *Chem. Ber.*, 1962, **95**, 1642.

[228] M. G. Barlow, R. N. Haszeldine and J. R. Langridge, *J. Chem. Soc., Perkin Trans. 1*, 1980, 129; B. Sket, N. Zupancic and M. Zupan, *J.*

*Org. Chem.*, 1982, **47**, 4462.
[229] M. G. Barlow, R. N. Haszeldine and J. R. Langridge, *J. Chem. Soc., Chem. Commun.*, 1979, 608.
[230] I. Saito, K. Kanehira, K. Shimozono and T. Matsuura, *Tetrahedron Lett.*, 1980, **21**, 2737.
[231] K. A. Muszkat, *Topics in Current Chemistry*, 1980, **88**, 89.
[232] J. Bendig, M. Beyermann and D. Kreysig, *Tetrahedron Lett.*, 1977, 3659.
[233] D. Billen, N. Brens and F. C. De Schryver, *J. Chem. Res.*, 1979, 79; M. V. Sargent and C. J. Timmons, *J. Chem. Soc.*, 1964, 5544.
[234] A. J. Floyd, S. F. Dyke and S. E. Ward, *Chem. Rev.*, 1979, **76**, 509; F. R. Stermitz, *Org. Photochem.*, 1967, **1**, 247.
[235] T. Mametani and K. Fukomoto, *Acc. Chem. Res.*, 1972, **5**, 212.
[236] E. D. Middlemas and L. D. Quin, *J. Am. Chem. Soc.*, 1980, **102**, 4838.
[237] P. H. G. op het Veld and W. H. Laarhoven, *J. Am. Chem. Soc.*, 1977, **99**, 7221.
[238] A. Buquet, A. Couture and A. Lablache-Combier, *J. Org. Chem.*, 1979, **44**, 2300.
[239] W. M. Horspool *J. Photochem.*, 1984, **27**, 122.
[240] W. M. Horspool and D. T. Anderson, *J. Photochem.*, 1984, **27**, 124.
[241] B. I. Rosen and W. P. Weber, *J. Org. Chem.*, 1977, **42**, 3463.
[242] R. Lapouyade, A. Veyres, N. Hanafi, A. Couture and A. Lablache-Combier, *J. Org. Chem.*, 1982, **47**, 1361.
[243] C. E. Brown, T. K. Dobbs, S. S. Hecht and E. J. Eisenbraun, *J. Org. Chem.*, 1978, **43**, 1656; P. H. Gore and F. S. Kamonah, *Synthesis*, 1978, 773.
[244] R. H. Martin, *Angew. Chem. Int. Ed. Engl.*, 1974, **13**, 649.
[245] T. Wismonski-Knittel and E. Fischer, *J. Chem. Soc., Perkin Trans. 2*, 1979, 449.
[246] W. H. Laarhoven, Th. J. H. M. Cuppen and H. H. K. Brinkhof, *Tetrahedron*, 1982, **38**, 3179.
[247] R. H. Martin and V. Libert, *J. Chem. Research (S)*, 1980, 130.
[248] T. S. Skorokhdova, G. N. Ivanov, V. I. Luk'yanov, Yu. G. Yu'rev, V. F. Kam'yanov and E. B. Merkushev, *Neftekhimiya*, 1979, **19**, 839 (*Chem. Abstr.*, 1980, **92**, 215 131).
[249] J. H. Borkent, J. W. Diesveld and W. H. Laarhoven, *Rec. Trav. Chim. Pays Bas,* 1981, **100**, 114.
[250] T. Sato, Y. Goto and K Hata, *Bull. Chem. Soc. Jpn*, 1967, **40**, 1994.
[251] F. D. Mallory and C. W. Mallory, *J. Org. Chem.*, 1983, **48**, 526.
[252] A. Buquet, A. Couture, A. Lablache-Combier and A. Pollet, *Tet-*

*rahedron*, 1981, **37**, 75.

[253] J. Glinka, *Zesz. Nauk. Politech. Slask., Chem.*, 1978, **75** (*Chem. Abstr.*, 1980, **92**, 110 814).

[254] I. Lantos, *Tetrahedron Lett.*, 1978, 2761.

[255] O. Tsuge, K. Oe and H. Inoue, *Heterocycles*, 1979, **12**, 217.

[256] V. N. R. Pillai and M. Ravindran, *Indian J. Chem.*, 1977, **15B**, 1043.

[257] Y. Tominaga, M. L. Lee and R. N. Castle, *J. Heterocyclic Chem.*, 1981, **18**, 977.

[258] G. Cooper and W. J. Irwin, *J. Chem. Soc., Perkin Trans. 1*, 1976, 75.

[259] (a) S. C. Shim and S. K. Lee, *Synthesis*, 1980, 116.
(b) R. E. Doolittle and C. K. Bradsher, *J. Org. Chem.*, 1966, **31**, 2616.
(c) A. R. Katritzky, Z. Zakaria and E. Lunt, *J. Chem. Soc., Perkin Trans. 1*, 1980, 1979.

[260] S. W. Horgan, D. D. Morgan and M. Orchin, *J. Org. Chem.*, 1973, **38**, 3801; P. H. G. op het Veld and W. H. Laarhoven, *J. Chem. Soc., Perkin Trans. 2*, 1978, 915.

[261] A. S. Kende and D. P. Curran, *J. Am. Chem. Soc.*, 1979, **101**, 1857.

[262] A. Padwa, C. Doubleday and A. Mazzu, *J. Org. Chem.*, 1977, **42**, 3271.

[263] R. Lapouyade, N. Hanafi and J.-P. Morand, *Angew. Chem. Int. Ed. Engl.*, 1982, **21**, 766.

[264] R. Srinivasan, V. Y. Merritt, J. N. C. Hsu, P. H. G. op het Veld and W. H. Laarhoven, *J. Org. Chem.*, 1978, **43**, 980.

[265] V. Arisvaran, M. Ramesh, S. P. Rajendran and P. Shanmugan, *Synthesis*, 1981, 821; K. Veeramani, K. Paramasivan, S. Ramakrishnansubramanian and P. Shanmugan, *Synthesis*, 1978, 855.

[266] C. Minot, P. Roland-Gosselin and C. Thal, *Tetrahedron*, 1980, **36**, 1209.

[267] I. Yokoe, K. Higuchi, Y. Shirataki and M. Komatusu, *Chem. Pharm. Bull.*, 1981, **29**, 2670.

[268] J. B. Hendrickson and J. G. de Vries, *J. Org. Chem.*, 1982, **47**, 1148.

[269] H. G. Heller and P. J. Strydom, *J. Chem. Soc., Chem. Commun.*, 1976, 50.

[270] H. G. Heller and J. R. Lang, *J. Chem. Soc., Perkin Trans. 2*, 1981, 341.

[271] K. Maruyama, T. Otsuki and K. Mitsui, *J. Org. Chem.*, 1980, **45**, 1424.

[272] A. H. Tinnemans and W. H. Laarhoven, *J. Chem. Soc., Perkin Trans. 2*, 1976, 1111, 1115; *Org. Photochem. Synth.*, 1976, **2**, 93.

[273] R. Lapouyade, R. Koussini, and H. Bouas-Laurent, *J. Am. Chem. Soc.*, 1977, **99** 7374.

[274] T. Onaka, Y. Kanda and M. Natsume, *Tetrahedron Lett.*, 1974, 1179.

[275] G. R. Badger, R. J. Drewer and G. E. Lewis, *Aust. J. Chem.*, 1964,

**17**, 1036.
[276] T. J. Sobieralski and K. G. Hancock, *J. Am. Chem. Soc.*, 1982, **104**, 7533.
[277] (a) A. G. Schultz and L. Motyka, *Org. Photochem.*, 1983, **6**, p. 1.
(b) K. H. Grellmann, W. Kuhnle, H. Weller and T. Wolff, *J. Am. Chem. Soc.*, 1981, **103**, 6889.
[278] W. Carruthers, *J. Chem. Soc., Chem. Commun.*, 1966, 202.
[279] K.-P. Zeller and H. Petersen, *Synthesis*, 1975, 532.
[280] O. L. Chapman, G. L. Eian, A. Bloom, and J. Clardy, *J. Am. Chem. Soc.*, 1971, **93**, 2981.
[281] A. G. Schultz and C.-K. Sha, *Tetrahedron*, 1980, **36**, 1757.
[282] A. G. Schultz and W. Hagmann, *J. Chem. Soc., Chem. Commun.*, 1976, 726; *idem.*, *J. Org. Chem.*, 1978, **43**, 4231.
[283] A. G. Schultz and M. B. DeTar. *J. Am. Chem. Soc.*, 1976, **98**, 3564.
[284] A. G. Schultz, R. D. Lucci, W. Y. Fu, M. H. Berger, J. Erhardt and W. K. Hagmann, *J. Am. Chem. Soc.*, 1978, **100**, 2150.
[285] R. J. Olsen and O. W. Cummings, *J. Heterocyclic Chem.*, 1981, **18**, 439.
[286] S. C. Roy and U. R. Ghatak, *J. Chem. Research (S)*, 1983, 138.
[287] R. C. Cookson, D. E. Sadler and K. Salisbury, *J. Chem. Soc., Perkin Trans. 2*, 1981, 774.
[288] A. U. Rahman and M. Ghazala, *Heterocycles*, 1981, **16**, 261.
[289] U. C. Yoon, S. L. Quillen, P. S. Mariano, R. Swanson, J. L. Stavinoha and E. Bay, *J. Am. Chem. Soc.*, 1983, **105**, 1204; P. S. Mariano and J. L. Stavinoha, in *Synthetic Organic Photochemistry*, ed. W. M. Horspool, Plenum Press, New York, 1984, p. 145.
[290] D. Bellus, *Adv. Photochem.*, 1971, **8**, 109.
[291] S. M. Beck and L. E. Brus, *J. Am. Chem. Soc.*, 1982, **104**, 1805.
[292] I. Rosenthal, M. M. Mossoba and P. Riesz, *Can. J. Chem.*, 1982, **60**, 1486.
[293] J. W. Meyer and G. S. Hammond, *J. Am. Chem. Soc.*, 1972, **94**, 2219.
[294] V. K. Pathak and R. N. Khanna, *Synthesis*, 1981, 882.
[295] D. J. Crouse, S. L. Hurlbut and D. M. S. Wheeler, *J. Org. Chem.*, 1981, **46**, 374.
[296] F. Farina, R. Martinez-Utrilla and M. C. Paredes, *Tetrahedron*, 1982, **38**, 1531.
[297] W. M. Horspool and P. L. Pauson, *J. Chem. Soc.*, 1965, 5162.
[298] R. Martinez-Utrilla and M. A. Miranda, *Tetrahedron Lett.*, 1980, **21**, 2281.
[299] K. Beelitz, and K. Praefcke, *Liebiegs Ann. Chim.*, 1979, 1081.
[300] J. Primo, R. Tormo and M. A. Miranda, *Heterocycles*, 1982, **19**, 1819.

[301] D. R. Crump, R. W. Franck, R. Gruska, A. A. Ozorio, M. Pagnotta, G. J. Siuta and J. G. White, *J. Org. Chem.*, 1977, **42**, 105.
[302] N. Shimamura and A. Sugimori, *Bull. Chem. Soc. Jpn*, 1971, **44**, 281.
[303] M. Fischer, *Chem. Ber.*, 1969, **102**, 342.
[304] J. L. Fourrey and P. Jouin, *J. Org. Chem.*, 1979, **44**, 1892.
[305] K. Beelitz, G. Buchholz and K. Praefcke, *Liebigs Ann. Chem.*, 1979, 2043.
[306] Y. Maki and M. Sako, *J. Chem. Soc., Perkin Trans. 1*, 1979, 1478.
[307] M. Ishikawa, M. Oda, M. Miyoshi, L. Fabry, M. Kumada, T. Yamada, K. Akagi and K. Fukui, *J. Am. Chem. Soc.*, 1979, **101**, 4612.
[308] J. E. Herweh and C. E. Hoyle, *J. Org. Chem.*, 1980. **45**, 2195.
[309] Y. Ogata, K. Takagi and I. Ishino, *Tetrahedron*, 1970, **26**, 2703.
[310] J. O. Stoffer and J. T. Bohanon, *J. Chem. Soc., Perkin Trans. 2*, 1978, 692; A. Padwa, T. J. Blacklock, R. Loza and R. Polniaszek. *J. Org. Chem.*, 1980, **45**, 2181.
[311] S. F. Mason, *Quart. Rev.*, 1961, **15**, 313.
[312] (a) S. Dellonte, L. Flamigni, F. Bariqelletti, L. Wojnarovits and G. Orlandi, *J. Phys. Chem.*, 1984, **88**, 58; L. Flamigni and G. Orlandi, *J. Photochem. Photobiol. A*, 1988, **88**, 241.
(b) P. Ausloos, *Mol. Photochem.*, 1972, **4**, 39.
[313] H. Okabe, in *Photochemistry of Small Molecules*, Wiley-Interscience, New York, 1978, p. 67.
[314] R. E. Rebbert, S. G. Lias, and P. Ausloos, *Chem. Phys. Letters*, 1971, **12**, 323; B. H. Mahan and R. Mandal, *J. Chem. Phys.*, 1962, **37**, 207.
[315] L. C. Lee and C. C. Chiang, *J. Chem. Phys.*, 1983, **78**, 688; T. G. Slanger and G. Black, *J. Chem. Phys.*, 1982, **77**, 2432.
[316] B. C. Roquitte, *J. Phys. Chem.*, 1970, **74**, 1204; H. Okabe and J. R. McNesby, *J. Chem. Phys.*, 1961, **34**, 668.
[317] S. G. Lias, G. J. Collin, R. E. Rebbert and P. Ausloos, *J. Chem. Phys.*, 1970, **52**, 1841.
[318] J. H. Vorachek and R. D. Koob, *Can. J. Chem.*, 1973, **51**, 344.
[319] R. A. Holroyd. *J. Am. Chem. Soc.*, 1969, **91**, 2208.
[320] S. K. Tokach and R. D. Koob, *J. Phys. Chem.*, 1980, **84**, 6.
[321] (a) R. Srinivasan and J. A. Ors, *J. Org. Chem.*, 1979, **44**, 3426.
(b) T. S. Pendleton, M. Kaplan and R. D. Doepker, *J. Phys. Chem.*, 1980, **84**, 472.
[322] R. S. Becker, L. Edwards, R. Bost, M. Elam and G. Griffin, *J. Am. Chem. Soc.*, 1972, **94**, 6584.
[323] (a) G. W. Griffin, *Angew. Chem. Int. Ed. Engl.*, 1971, **10**, 537.
(b) H. E. Zimmerman and A. C. Pratt, *J. Am. Chem. Soc.*, 1970, **92**,

6259.
[324] E. W. Valyocsik and P. Sigal, *J. Org. Chem.*, 1971, **36**, 66.
[325] (a) P. H. Mazzocchi, R. S. Lustig and G. W. Craig, *J. Am. Chem. Soc.*, 1970, **92**, 2169.
(b) S. S. Hixson and C. R. Gallucci, *J. Org. Chem.*, 1988, **53**, 2711.
[326] S. S. Hixson, L. A. Franke, J. A. Gere and Y. Xing, *J. Am. Chem. Soc.*, 1988, **110**, 3601.
[327] T. Miyashi and M. Kamata, *Yuki Gosei Kagaku Kyokaishi*, 1986, **44**, 986 (*Chem. Abstr.*, 1987, **106**, 213127).
[328] P. A. Carson and P. de Mayo, *Can. J. Chem.*, 1987, **65**, 976.
[329] J. W. Blunt, J. M. Coxon, W. T. Robinson and H. A. Schuyt, *Aust. J. Chem.*, 1983, **36**, 565.
[330] R. W. Denny and A. Nickon, *Org. Reactions* 1973, **20**, 133.
[331] C. S. Foote, *Acc. Chem. Res.*, 1968. **1**, 104; *idem, Pure Appl. Chem.*, 1971, **27**, 635.
[332] G. O. Schenck, K. Gollnick, G. Buchwald, S. Schroeter and G. Ohloff, *Ann. Chem.*, 1964, **674**, 93.
[333] M. Orfanopoulos, S. M. B. Grdina and L. M. Stephenson, *J. Am. Chem. Soc.*, 1979, **101**, 275.
[334] (a) A. Nickon and J. F. Bagli, *J. Am. Chem. Soc.*, 1959, **81**, 6330; 1961, **83**, 1498.
(b) N. Furutachi, Y. Nakadaira and K. Nakanishi, *J. Chem. Soc., Chem. Commun.*, 1968, 1625.
[335] (a) G. Rousseau, A. Lechevallier, F. Huet and J. M. Conia, *Tetrahedron Lett.*, 1978, 3287.
(b) K. Sato, H. Adachi, T. Iwaki and M. Ohashi, *J. Chem. Soc., Perkin Trans. 1*, 1979, 1806.
[336] K. A. Zaklika, A. L. Thayer and A. P. Schaap, *J. Am. Chem. Soc.*, 1978, **100**, 4916.
[337] S. Mazur and C. S. Foote, *J. Am. Chem. Soc.*, 1970, **92**, 3225.
[338] P. D. Bartlett and A. P. Schaap, *J. Am. Chem. Soc.*, 1970, **92**, 3223.
[339] K. Gollnick and G. O. Schenck, in *1,4-Cycloaddition Reactions*, ed J. Hamer, Academic Press, New York, 1957, p. 255.
[340] (a) K. Kondo and M. Matsumoto, *J. Chem. Soc., Chem. Commun.*, 1972, 1332.
(b) J. P. Hagenbuch, J. L. Birbaum, J. L. Metral and P. Vogel, *Helv. Chim. Acta*, 1982, **65**, 887.
[341] (a) W. Adam and H. J. Eggelte, *J. Org. Chem.*, 1977, **42**, 3987.
(b) W. Adam, N. Gretzke. L. Hasemann, G. Klug, E. M. Peters, K. Peters, H. G. von Schnering and B. Will, *Chem. Ber.*, 1985, **118**, 3357.

(c) W. Adam, M. Balci, B. Pietrzak and H. Rebollo, *Synthesis*, 1980, 820; W. Adam and M. Balci, *Angew. Chem. Int. ed. Engl.*, 1978, **17**, 954; W. Adams and H. Rebollo, *Tetrahedron Lett.*, 1982, **23**, 4907.

[342] E. Lee-Ruff, M. Maleki, P. Duperrouzel, M. H. Lien and A. C. Hopkinson, *J. Chem. Soc., Chem. Commun.*, 1983, 346.

[343] M. Matsumoto, K. Kuroda and Y. Suzuki, *Tetrahedron Lett.*, 1981, **22**, 3253.

[344] J. Rigaudy, *Pure Appl. Chem.*, 1968, **16**, 169.

[345] M. L. Graziano, M. R. Iesce, B. Carli and R. Scarpati, *Synthesis*, 1982, 736.

# 3

# Oxygen-containing compounds

This chapter deals with the reaction of the group of organic molecules containing at least one oxygen atom. As with the previous chapter the saturated and unsaturated examples are discussed and the material includes some gas phase processes which, although of little synthetic value, allow for a broad discussion of the area in general.

## 3.1 ABSORPTION SPECTRA OF ALCOHOLS, ETHERS AND PEROXIDES

### 3.1.1 Alcohols

The absorbance of saturated alcohols in the ultraviolet starts around 190 nm, and the first maximum appears as an unstructured band around 180–185 nm [1]. This moderately intense band arises from the excitation of an electron from the non-bonding pairs on the oxygen and leads to an excited state by $n \rightarrow \sigma^*$ [2, 3] excitation although others [4] have assigned this transition to a Rydberg-type involving $n \rightarrow 3s$ excitation. Typical examples of the intensity ($\varepsilon$ in $dm^3\ mol^{-1}\ cm^{-1}$) of the absorptions at 185 nm shown by some simple alcohols as neat liquids are as follows: MeOH 6 [5, 6], *i*-PrOH 32 [7], and *t*-BuOH 90 [7]. The absorbances are more intense in the gas phase owing to the absence of hydrogen bonding, and some examples of values of $\varepsilon$ for these are: MeOH 150 [8, 9], *i*-PrOH 240 [8, 9] and *t*-BuOH 1150 [9].

### 3.1.2 Ethers

With ethers the first band in the ultraviolet for cyclic compounds has moved towards longer wavelength and starts around 200 nm [1]. Open-chain ethers or the more strained oxiranes also absorb in this area but at slightly shorter wavelengths. The transitions observed for these species are the same as for alcohols [3,10]. The absorptions shown by the ethers are generally more intense than those of alcohols, and typical values for the intensity ($\varepsilon$ in $dm^3$ $mol^{-1}$ $cm^{-1}$) of the absorptions (at 185 nm) are: diethyl ether 1500 [11], diisopropyl ether 500 [12], di-*t*-butyl ether 2200 [13], oxirane 100 [14], oxetane 1200 [14], tetrahydrofuran 700 [15], and *p*-dioxane 4500 [15].

### 3.1.3 Peroxides

Peroxides also absorb in the ultraviolet region [16, 3]. In the main the absorptions for peroxides are structureless and simply exhibit a continuous absorption curve. Hydrogen peroxide [17], dialkyl peroxides [18–20] and alkyl hydroperoxides [21] have absorptions with an onset around 300 nm increasing in intensity towards 200 nm. Diacyl peroxides [22] show similar behaviour but the absorption bands start at the slightly shorter wavelength of 280 nm. Only the diaroyl peroxides have some structure, with bands peaking at 230 and 273 nm reminiscent of benzene derivatives [18, 23]. Again excitation involves an $n\sigma^*$ transition. In this state the predominant mode of reaction is dissociative resulting in O—O bond cleavage to yield vibrationally excited alkoxy radicals [24, 25].

## 3.2 PHOTOCHEMICAL REACTIONS

### 3.2.1 Alcohols

Early work readily established that irradiation of simple alcohols [26] such as methanol, ethanol, and propan-2-ol was brought about using the 185 nm line from a low-pressure Hg lamp. Excitation at wavelengths shorter than 185 nm does not bring about a marked change in the outcome of the reaction [26]. The reaction of the simpler alcohols showed that hydrogen was the principal product and the initial step was the fission of the O—H bond to afford an alkoxy radical [27]. Studies also showed that molecular elimination of hydrogen could take place as well as C—H bond fission as shown for methanol (**1**). This illustration also highlights the fact that in both the gas

$$\mathrm{CH_3OH} \xrightarrow[\text{185 nm, neat liquid or gas phase}]{h\nu} [\mathrm{CH_3OH}]^* \longrightarrow \mathrm{CH_3O}^{\bullet} + \mathrm{H}^{\bullet} \quad \text{88\% liquid}$$

$$[\mathrm{CH_3OH}]^* \longrightarrow \dot{\mathrm{C}}\mathrm{H_2OH} + \mathrm{H}^{\bullet} \quad \text{79\% gas}$$

(1)

phase and the neat liquid the principal reaction is O—H fission. As substitution on the alcohol is increased so there is a change in the products obtained and the molecular path to hydrogen formation becomes more important. A detailed analysis [28] has shown that the formation of hydrogen and the corresponding aldehyde or ketone by the molecular elimination path increases from methanol (6%), to ethanol (47%) and propan-2-ol (60%). Irradiated *t*-butanol in the liquid differs drastically in its behaviour from the previously discussed compounds, and C—C bond fission is dominant in the neat liquid, as shown by Yang and his coworkers [26]. The study of the photochemistry of *t*-butanol (**2**) has shown that hydrogen, methane, and

$$\mathrm{(CH_3)_3COH} \xrightarrow[\text{185 nm, neat liquid}]{h\nu} [\mathrm{(CH_3)_3COH}]^* \longrightarrow \dot{\mathrm{C}}\mathrm{H_3} + \mathrm{CH_3}\dot{\mathrm{C}}(\mathrm{OH})\mathrm{CH_3}$$

$$[\mathrm{(CH_3)_3COH}]^* \longrightarrow \mathrm{CH_4} + \mathrm{CH_3COCH_3} \quad \text{(67\%)}$$

$$[\mathrm{(CH_3)_3COH}]^* \longrightarrow \mathrm{H_2} + \text{2,2-dimethyloxirane } \mathrm{(CH_3)_2C\!-\!CH_2O} \quad \text{17\%}$$

(2)

acetone are the main products. The formation of hydrogen in this example does not occur by a hydrogen atom route but instead is produced by either an intramolecular or an intermolecular path [29].

In more heavily substituted alcohols other processes can occur such as retro-Aldol reactions as with the phenyl ethanols (**3**) [30]. Other examples

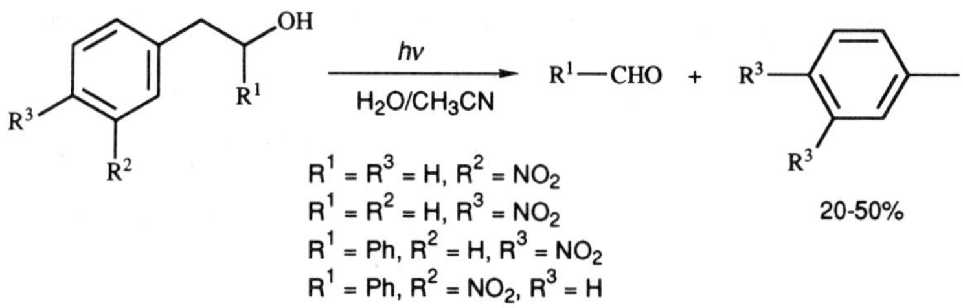


(3)

of these processes have also been reported [31]. Electron-transfer reactivity has also been observed, as in the fragmentation of the amino alcohols shown in (**4**) [32, 33].

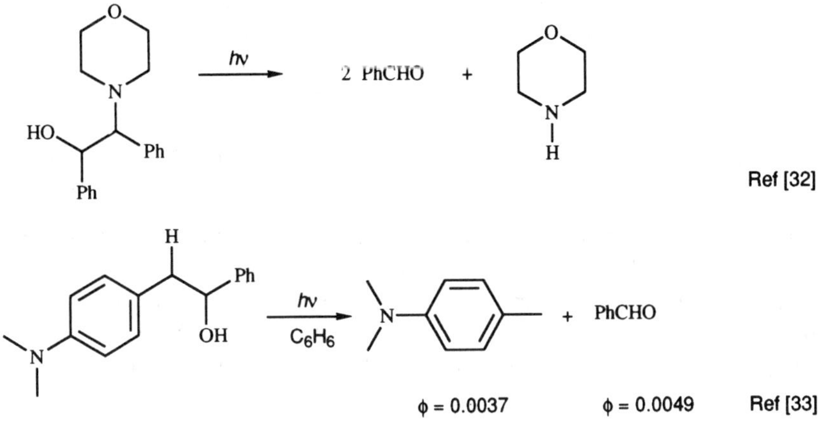


(4)

## 3.2.2 Ethers

### *3.2.2.1 Acyclic ethers*

Like alcohols the photochemical reactivity of ethers is dominated by the $n\sigma^*$ transition of the non-bonding pairs on oxygen and fission of a C—O bond results [1, 2, 34–36]. Irradiation of simple open-chain ethers in the liquid phase yields alcohols as the most important products. These are produced mainly by homolytic fission of the C—O bond, although molecular processes can compete successfully. This latter reaction path is indistinguishable from radical disproportionation processes [1]. The alkoxy radicals, once formed, abstract hydrogen to produce the alcohol, among other products. These processes are summarized for the irradiation of diethyl ether (**5**) in the liquid

$$C_2H_5OC_2H_5 \xrightarrow[185\text{ nm}]{h\nu} [C_2H_5OC_2H_5]^* \longrightarrow \begin{cases} C_2H_5^{\bullet} + C_2H_5O^{\bullet} & 70\% \\ CH_2{=}CH\text{-}O\text{-}CH_2CH_3 + H_2 & 11\% \\ C_2H_6 + CH_3CHO & 10\% \\ CH_2{=}CH_2 + CH_3CH_2OH & 8\% \end{cases}$$

(5)

phase at 185 nm [34] and for *t*-butyl methyl ether (**6**) [13]. As can be seen

$$(CH_3)_3C\text{—}OCH_3 \xrightarrow[185\text{ nm}]{h\nu} [(CH_3)_3C\text{—}OCH_3]^* \longrightarrow (CH_3)_3C^{\bullet} + CH_3O^{\bullet} \quad 52\%$$

$$[(CH_3)_3C\text{—}OCH_3]^* \longrightarrow (CH_3)_3C\text{—}O^{\bullet} + {}^{\bullet}CH_3 \quad 30\%$$

(6)

in this reaction sequence, the asymmetrically substituted ethers can undergo fission of either of the C—O bonds and they do so with different probabilities. By using quantum yield measurements for the formation of the two possible alcohols, an approximate measure of the primary processes can be obtained (**7**) [1, 11, 13, 34, 35]. Di-*t*-butyl ether is the exception to the above

$$C_2H_5OC_2H_5 \xrightarrow[185\text{ nm}]{h\nu} C_2H_5OH \ (\phi = 0.46)$$

$$C_2H_5OCH_2CH_2CH_3 \xrightarrow[185\text{ nm}]{h\nu} C_2H_5OH \ (\phi = 0.31) + CH_3CH_2CH_2OH \ (\phi = 0.28)$$

$$CH_3OCH_2CH_2CH_3 \xrightarrow[185\text{ nm}]{h\nu} CH_3OH \ (\phi = 0.16) + CH_3CH_2CH_2OH \ (\phi = 0.70)$$

$$CH_3O\text{—}C(CH_3)_2\text{—}CH_3 \xrightarrow[185\text{ nm}]{h\nu} CH_3OH \ (\phi = 0.41) + CH_3\text{—}C(CH_3)_2\text{—}OH \ (\phi = 0.20)$$

(7)

generalizations. In this case the molecular path is followed 80% of the time, yielding *t*-butanol ($\Phi = 0.84$) and 2-methylpropene [13]. Other studies have examined the gas phase decomposition of acyclic ethers such as dimethyl ether [37] and diethyl ether [38], where it has been shown that the mechanisms encountered are similar to those in the liquid phase.

### *3.2.2.2 Cyclic ethers*

The situation regarding the cyclic ethers is more complicated, but in general it can be assumed that the principal photochemical event is C—O bond fission analogous to that encountered in the acyclic examples. However, there are often variations dependent on the ring size. Thus, oxetane (**8**) is reported

$$\text{oxetane} \xrightarrow{h\nu} CH_2O + \underset{>95\%}{CH_2{=}CH_2}$$

(8)

to yield methanal and ethene as the exclusive products [14, 39], while the 1,2-dimethyl derivative (**9**) follows two paths to yield acetone, methanal,

$$\xrightarrow{h\nu} CH_2{=}CH_2 + (CH_3)_2C{=}CH_2 + CH_2O + (CH_3)_2CO$$

(9)

ethene and 2-methylpropene [14, 39].

Tetrahydrofuran (**10**) also undergoes reaction on irradiation at 185 nm in

$$\xrightarrow{h\nu} \underset{\phi = 0.06}{\text{H, O}} + \underset{\phi = 0.02}{\text{HO}} + \underbrace{\triangle + CH_2O}_{\phi = 0.19}$$

(10)

the liquid phase affording a miscellany of products [1]. The predominant reaction mode appears to be C—O bond fission, and convincing evidence for this reaction mode comes from the fact that the 2,5-dimethyl derivative (**11**) undergoes interconversion from *cis* to *trans* and vice versa via a biradical

$$\underset{h\nu}{\rightleftharpoons} \text{O}\bullet \underset{h\nu}{\rightleftharpoons}$$

(11)

intermediate. Reactivity in the gas phase is dominated by fragmentation processes [40, 41]. Similar photochemical reactivity to that observed for tetrahydrofuran is found on irradiation of tetrahydropyran (**12**) in the liquid

$\phi = 0.13$ $\phi = 0.4$ $CH_2O$ $\phi = 0.002$

(12)

phase at 185 nm [42].

The above examples illustrate the types of reaction which cyclic ethers undergo. However, there are not many synthetic uses to which these systems can be put. An exception to this is the photochemical behaviour of oxiranes. Gas phase studies on the photochemical behaviour of ethylene oxide (**13**)

$$\xrightarrow{h\nu} CO + H_2 + CH_3CH_3 + CH_4 + CH_3CHO$$

(13)

have shown that it decomposes at wavelengths above 170 nm to yield carbon monoxide, hydrogen, ethane, methane and acetaldehyde [39, 43, 44]. In solution phase the photochemical investigation of oxiranes has been focussed on the more substituted derivatives where the ultraviolet absorptions are in a range more accessible to solution–phase photochemists, and several reviews have been published [45–50]. The photochemical behaviour of oxiranes is quite varied with reactions involving rearrangement, homolysis by C—O or C—C bond fission, and the formation of carbenes.

In the C—O cleavage reaction of oxiranes, evidence has been obtained which suggests that the species produced by this has ionic character. Thus the irradiation of styrene oxide (**14**) in methanol affords the hydroxy ether

$h\nu$ MeOH Ph HO OMe Ph 99%

(14)

where addition of the alcohol has occurred at the phenyl-substituted site, which is better able to stabilize a positive charge [51]. Other examples (**15**)

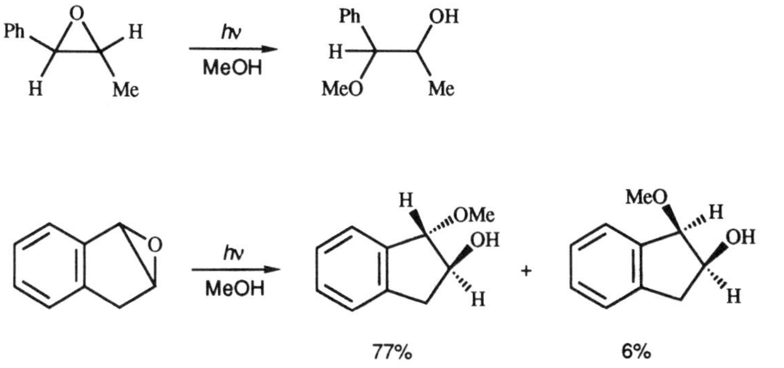
Ph
O
H
H
Me
hν
MeOH
Ph
OH
H
MeO
Me
O
hν
MeOH
H
OMe
OH
H
+
MeO
H
OH
H
77%
6%

(15)

have been observed to follow this path [50]. The outcome of these reactions may be due to the well-known fact that the irradiation of methanol solutions often gives rise to acidic material and the reactions observed could be influenced by the presence of acid.

As well as C—O ring-opening processes involving ionic species, there is evidence that oxiranes can undergo C—O bond fission in a homolytic fashion, as shown in (**16**) [52] where fission of the C—O bonds of the oxirane ring

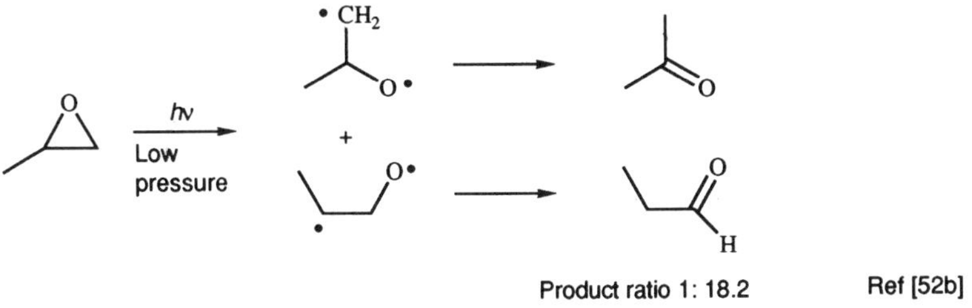
O
hν
Low
pressure
• $CH_2$
O•
+
O•
O
O
H
Product ratio 1: 18.2
Ref [52b]

(16)

occurs to afford two biradical intermediates and subsequently two products. Biradical intermediates are also formed on irradiation of the spiro-derivative

(**17**), which affords the products shown [53].

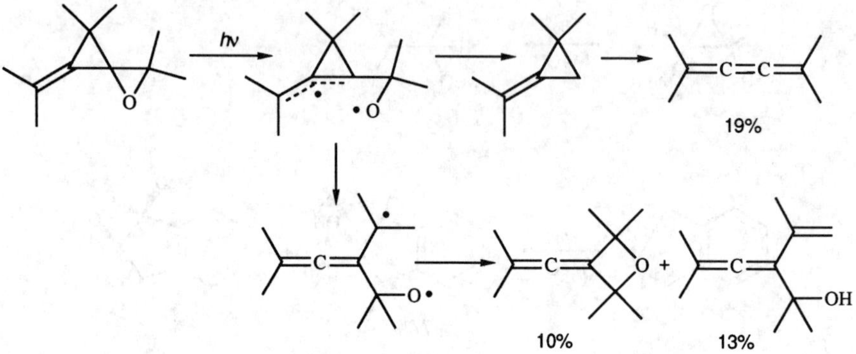

(17)

An alternative ring-opening mode is by fission of the C—C bond of the ring. An example of this is the irradiation of the norbornadiene derivative (**18**), resulting in a (2 + 2)-addition where any intermediate formed is

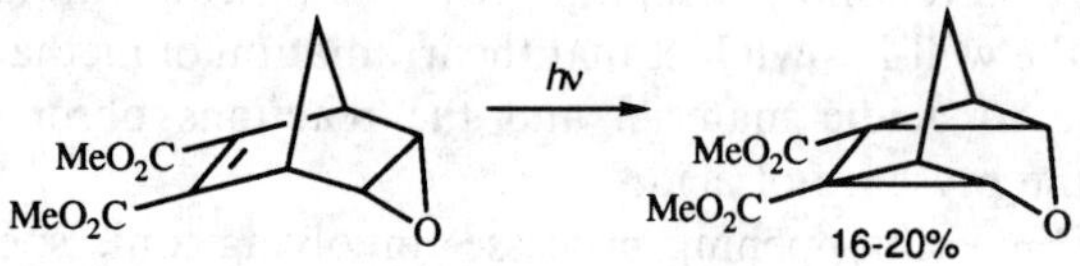

(18)

intramolecularly trapped [54]. With other substituents, such as aryl groups or electron-withdrawing groups, the fission of the C—C bond of the oxirane ring occurs by a heterolytic path to afford a carbonyl ylide. The formation of the ylides can be brought about either by direct irradiation or under sensitized conditions, thus implicating a triplet-state reaction. Fission of this intermediate ylide yields a carbene. An examination of this reaction has shown that the carbonyl ylide is formed by disrotatory ring opening, and the presence of the ylide can be detected at low temperatures. Thus the irradiation of (**19**) at 254 nm affords the intermediate, which subsequently

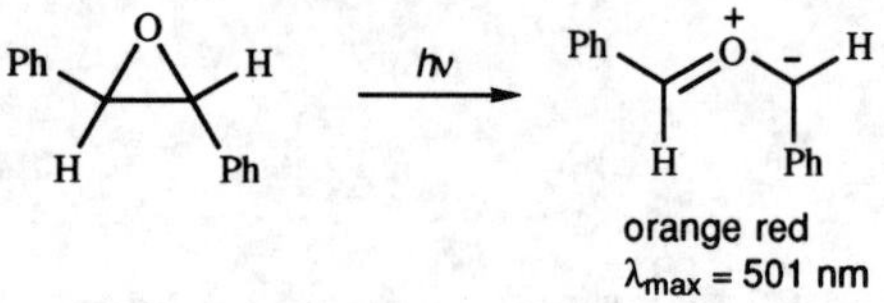

orange red
$\lambda_{max}$ = 501 nm

(19)

undergoes fission to the aldehyde and the carbene on irradiation with visible light [55]. Additional evidence for the generation of the ylide is obtained from the irradiation of oxiranes in a matrix at 77 K, when a coloured species is developed. Warming the matrix brings about bleaching with the reformation of the oxirane [56]. Further substantiation of the ylide route was obtained from the photolysis and thermolysis of the strained oxirane (**20**) which

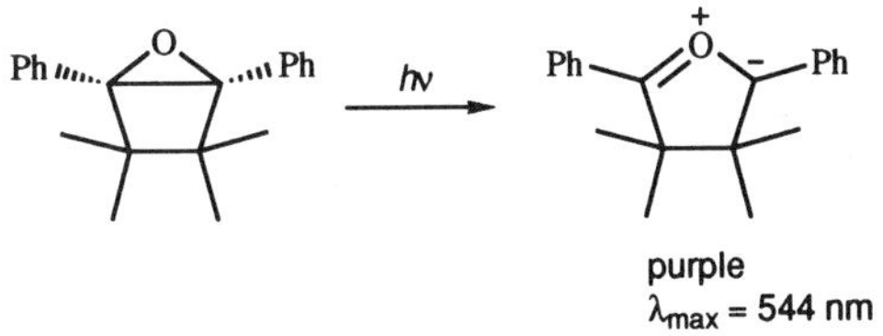

purple
$\lambda_{max}$ = 544 nm

(20)

generates the relatively stable and highly coloured ylide [57]. Photochromism of this type has been demonstrated for a variety of arylcyclopentenones (**21**) [58]. The fission of the carbonyl ylide to a carbene has been

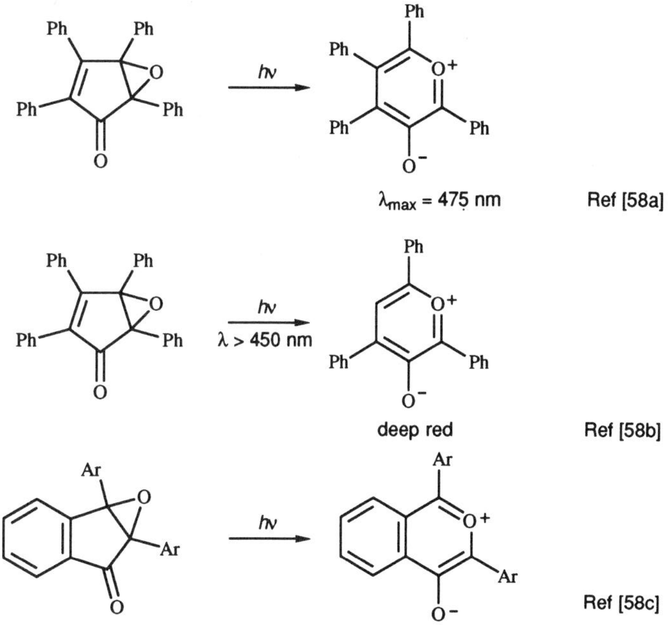

$\lambda_{max}$ = 475 nm Ref [58a]

$\lambda$ > 450 nm

deep red Ref [58b]

Ref [58c]

(21)

demonstrated for a variety of systems, as shown in **22** [59, 60] and the carbene

Ph Ph O Ph Ph —hν→ Ph—C̈—Ph 88-100%

Ph CN O Ph Ph —hν→ Ph—C̈—CN 71% + Ph—C̈—Ph 5%

(22)

produced from such a fission process can be trapped by alkenes to afford a synthetic route to cyclopropanes [61]. The ylide intermediates produced by the ring-opening of oxiranes have synthetic potential and can be used in cycloaddition reactions to afford heterocyclic compounds, as shown in **23**

Ph CN O H Ph —hν, sens./acetone, $CO_2Me$→ H Ph O Ph CN $CO_2Me$ 23% + H CN O Ph Ph $CO_2Me$ 77% Ref [62a]

—$MeO_2C$, $CO_2Me$→ Ph CN O H Ph $MeO_2C$ $CO_2Me$ 42% Ref [62b]

(23)

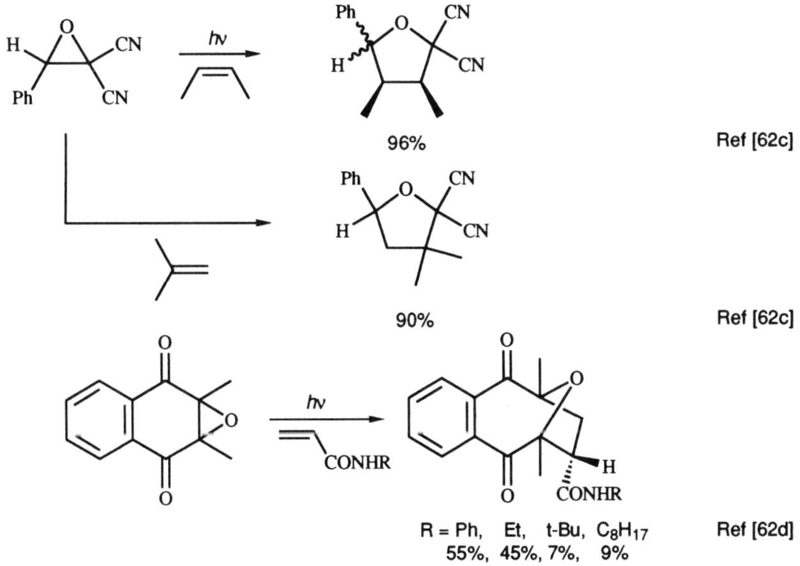

96% Ref [62c]

90% Ref [62c]

R = Ph, Et, t-Bu, $C_8H_{17}$
55%, 45%, 7%, 9% Ref [62d]

(23 continued)

[62]. The formation of ylides is now an accepted, although sometimes minor, reaction path for the photochemical behaviour of oxiranes, and some examples are shown in **24** [63].

hν
254 nm
$CH_3CN$

51% 9% 11% 8% Ref [63a]

hν
254 nm

4% 6%

1% 1% 1%

45% Ref [63b]

(24)

Ring opening of oxiranes can also be brought about by electron-transfer reactions such as that of the oxiranes (**25**), which behave as an electron donor

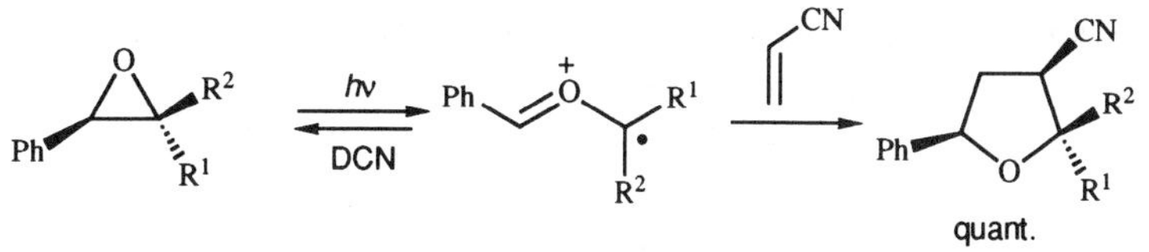

(25)

or irradiation in the presence of 1,4-dicyanonaphthalene. Under these conditions the oxirane undergoes *cis–trans*-isomerization in non-nucleophilic solvents such as acetonitrile [64]. The isomerization process involves the formation of the radical cation followed by ring opening to the carbonyl ylide. If a dipolarophile such as acrylonitrile is added, the products isolated are tetrahydrofuran derivatives.

Keto-oxiranes were introduced above with the arylcyclopentenone oxides (**21**). The reactivity in these arises solely by the C—C bond fission path to the ylides. Simpler keto-oxiranes, however, can undergo other reactions, such as the formation of 1,3-diketones, as shown in **26** [65]. Many examples of

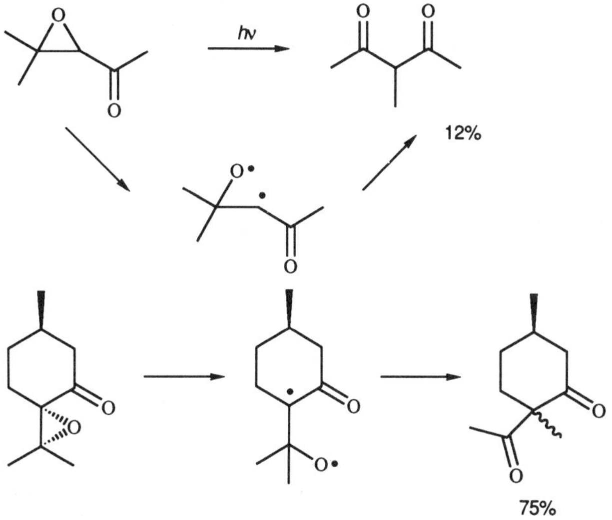

(26)

this sort have been reported and some of these are illustrated in **27** [66]. Mechanistically this reaction arises from a singlet $n\pi^*$ state and results in the fission of the C—O bond. The migratory aptitudes shown within such compounds is best explained via the involvement of a biradical species formed by C—O bond fission. As can be seen in **27** the migrations can occur

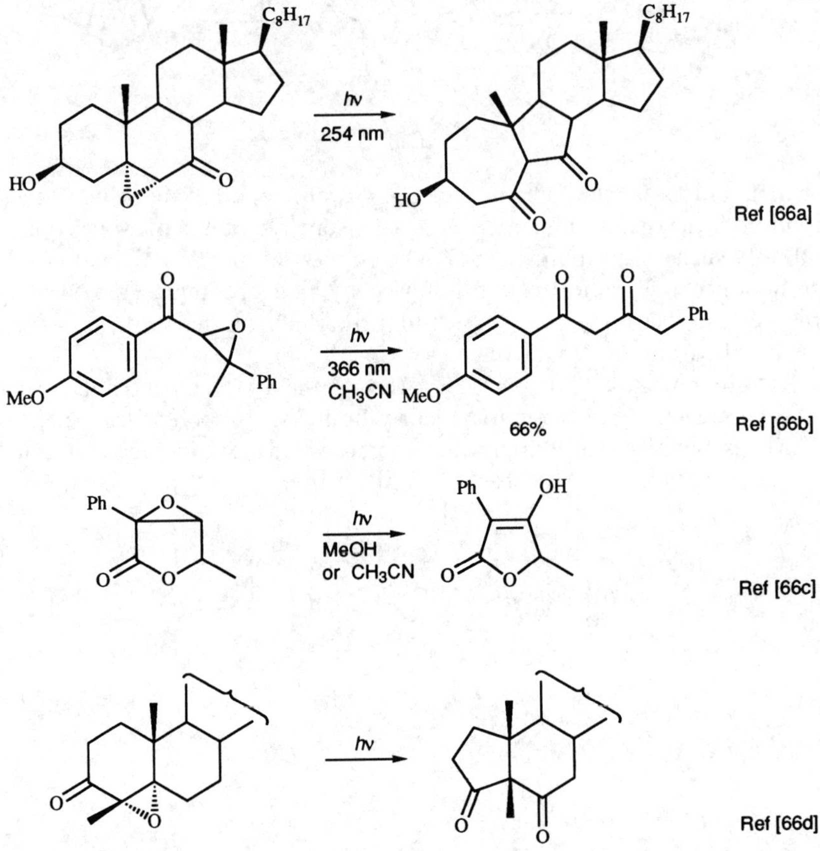

(27)

stereospecifically [66d].

### 3.2.3 Peroxides

As mentioned in section 3.1.3 excitation of peroxides involves an $n\sigma^*$ transition. In this state, the predominant mode of reaction is dissociative, resulting in O—O bond cleavage to yield vibrationally excited alkoxy radicals [24, 25]. A process typical of this is shown in **28**, where excitation of *t*-butyl

$$\text{t-BuOOH} \xrightarrow{h\nu} \text{t-BuO}^{\bullet} + \text{HO}^{\bullet}$$

(28)

hydroperoxide at 254 nm yields a hydroxy and a *t*-butoxy radical which have been identified by electron spin resonance (e.s.r.) spectroscopy [67]. The outcome of this reaction is formation of another peroxide by combination of the alkoxy radicals. Radicals of this type can be used for the oxidation of alcohols to ketones [68]. Dialkyl peroxides behave in a similar manner and, for example, the irradiation of di-*t*-butyl peroxide (**29**) yields *t*-butoxy radicals

$$(CH_3)_3C{-}O{-}O{-}C(CH_3)_3 \xrightarrow{h\nu} (CH_3)_3C{-}O^{\bullet} \rightarrow CH_3{-}CO{-}CH_3 + {}^{\bullet}CH_3 \quad \text{or} \quad (CH_3)_3C{-}OH$$

(29)

[69]. The fate of such radicals can be decomposition to acetone and a methyl radical [70] or else abstraction of hydrogen from a solvent molecule [71]. Cyclic peroxides also decompose on irradiation to afford alkoxy radicals as shown in **30** [72a] where chemiluminescence can be observed [72b], and in

hν, 310-350 nm; 2

(30)

**31** [73]. In this latter example the products are formed via a 1,3-biradical

hν, 350 nm; 70%

(31)

followed by either ring closure to give the epoxide or rearrangement to a ketone [74]. In larger ring systems such as **32** the initial O—O cleavage

hν, 310-350 nm

99% 90%

(32)

reaction to the biradical is followed by fission to yield acetone and ethene [75]. Diacyl peroxides such as benzoyl peroxide (**33**) also readily undergo

$$\mathrm{Ph{-}C({=}O){-}O{-}O{-}C({=}O){-}Ph} \xrightarrow[C_6H_6]{h\nu} 2\,\mathrm{Ph{-}C({=}O){-}O^{\bullet}} \longrightarrow \underset{13\%}{PhCO_2Ph} + \underset{30\%}{PhCO_2H}$$

$$\mathrm{Ph{-}C({=}O){-}O^{\bullet}} \xrightarrow{C_6H_6} \underset{13\%}{PhCO_2Ph}$$

$$\mathrm{Ph{-}C({=}O){-}O^{\bullet}} \xrightarrow{\text{H - abstraction}} \underset{30\%}{PhCO_2H}$$

$$\mathrm{Ph^{\bullet}} + CO_2;\quad \mathrm{Ph^{\bullet}} \xrightarrow{C_6H_6} \underset{33\%}{\mathrm{Ph{-}Ph}}$$

(33)

O—O bond fission from the singlet state on excitation in solution. As can be seen by the reactions which the acyl radicals undergo the process provides a useful route to aryl radicals [76]. Phthaloyl peroxide (**34**) is an interesting

hν, ambient → + $CO_2$

hν/matrix, 8 K → + CO +

(34)

example of a diaroyl peroxide, since irradiation at room temperature yields benzyne by a double decarboxylation [77], whereas in a matrix at 8 K benzopropiolactone and a ketoketene are formed [78]. These products are prone to interconversion which can be brought about on further irradiation [79].

In Chapter 2 the synthesis of endoperoxides by the photochemical addition of singlet oxygen to dienes was discussed. A theoretical treatment of the photochemical reactivity of these peroxides has concluded that an $n\sigma^*$ excited state, either triplet or singlet, is generated [80]. Irradiation of such a compound in the long-wavelength region results in fission of the O—O bond and production of a biradical. Evidence obtained from the use of triplet sensitizers has indicated that a triplet state is the reactive species. The photoconversion of the peroxide (**35**) into the bisoxirane and the keto-oxirane

366 nm

27%

OH

22%

(35)

is in support of this proposal, and hydrogen migration to the keto-oxirane occurs in competition with bond formation to yield the bisoxirane [81]. Other studies with *ascaridole* (**36**) have shown that the outcome of the irradiation is wavelength-dependent and that irradiation at 185 nm leads to

185 nm

+ $O_2$

366 nm

(36)

deoxygenated products as well as the bisoxirane, whereas irradiation at 366 nm affords only the bisoxirane [82]. These compounds have remarkable synthetic utility and reviews have highlighted their value [83]. Reactions of this type have been described over the years and some are illustrated in **37** [81, 84].

OMe MeO O

$h\nu$
435 nm

OMe MeO O

Ref [84]

$h\nu$
366 nm

$CO_2Me$ $CO_2Me$ + $CO_2Me$

31% 10% Ref [81]

(37)

## 3.3 CARBONYL COMPOUNDS

### 3.3.1 Spectra and excited states

There is little doubt that the photochemistry of carbonyl compounds is one of the most thoroughly and extensively investigated areas. These substrates are particularly attractive since their absorptions are very accessible and in the early days of photochemistry such compounds could be irradiated with sunlight or with the poorly dependable lamps available at the turn of the century [85].

Saturated aliphatic aldehydes and ketones exhibit four main bands at around 280, 195, 170 and 155 nm in the ultraviolet and vacuum ultraviolet regions [86]. For the solution-phase chemist the band at 280 nm is the one of interest. The band is weak ($\varepsilon = 10\text{–}30\ \mathrm{dm^3\ mol^{-1}\ cm^{-1}}$) and corresponds to the symmetry-forbidden $n\pi^*$ transition arising by the promotion of a non-bonding electron on oxygen from an orbital localized in the plane of the carbonyl group into a higher-energy antibonding $\pi^*$ orbital delocalized over the carbonyl group. The lowest excited states for such compounds are the $n\pi^*$ singlet and triplet states, and it is these states which are primarily responsible for the observed photochemical behaviour of simple aldehydes and ketones. The exact position of these weak absorptions at 280 nm is influenced by substitution. Thus, increasing alkyl substitution on the carbon adjacent to the carbonyl group causes the $n\pi^*$ to move to longer wavelength, e.g. for acetone $\lambda_{max} = 274$ nm and for 4-methylpentan-2-one $\lambda_{max} = 283$ nm [87].

The exact energies of the singlet and triplet states of the aliphatic ketones have not been accurately determined in most cases. This is due to the broad structureless bands which these compounds exhibit in both absorption and emission spectroscopy. Acetone is one exception and, for example, exhibits weak fluorescence ($\Phi_F = 0.0009$, $\tau_F = 2.1 \times 10^{-9}$ s, $E_S = 377$ kJ mol$^{-1}$) in solution at room temperature. Owing to a high intersystem crossing efficiency ($\Phi_{ISC} = 0.98$) the phosphorescence of acetone is stronger ($\Phi_P = 0.043$, $\tau_P = 0.33$ ms, $E_T = 310$ kJ mol$^{-1}$) [88].

The situation for aromatic ketones is more clear-cut. Aromatic ketones do not exhibit fluorescence because of high intersystem crossing efficiency. Typically the intersystem crossing efficiency of benzophenone, acetophenone, and substituted derivatives is unity. Benzophenone phosphoresces strongly ($\Phi_P = 0.90$, $E_T = 285$ kJ mol$^{-1}$) and can be observed at room temperature in completely inert solvents [89]. Unsaturated ketones fall between these two extremes, and both fluorescence and phosphorescence are weak, e.g. acrolein shows $\Phi_F = 0.07$, $\Phi_P = 0.00004$ [90]. The u.v. spectra of these compounds usually show discrete $n\pi^*$ and $\pi\pi^*$ transitions, although occasionally the weaker $n\pi^*$ is hidden by the strong $\pi\pi^*$ transition.

Photochemical excitation of a simple carbonyl compound like formaldehyde [91], which has a weak ($\varepsilon$ *c.* 10–30 dm$^3$ mol$^{-1}$ cm$^{-1}$) band in the 280–320 nm region, involves the promotion of an electron from a lower orbital into a higher one. The electron undergoing excitation comes from the non-bonding $n$-orbital and ends up in the anti-bonding $\pi^*$-orbital. This is referred to as an $n\pi^*$ transition. In the excited state the electrons which were spin-paired (antiparallel spin) in the ground state ($S_0$) can have either antiparallel spin, a singlet ($S_1$) state, or parallel spin, a triplet ($T_1$) state. The singlet and triplet states are designated $S_1$ and $T_1$ since Kasha's rule [92] permits the assumption that only vibrationally equilibrated lowest excited states will be involved in primary photochemical processes of organic molecules in solution. The population of the triplet state on direct irradiation is dependent on the intersystem crossing efficiency, as mentioned above.

The reactivity of the $n\pi^*$ state becomes obvious when we examine a simple model for the excited state. The outcome of the electronic excitation changes the spatial electronic distribution around the carbonyl group. Thus there is a stereo-electronic factor intrinsic in the $n\pi^*$ state of such a group. This is illustrated in **38**, where it can be seen that the excitation yields a half-filled electrophilic $n$-orbital in the plane of the molecular structure and a nucleophilic $\pi^*$-orbital perpendicular to the molecular plane. These changes

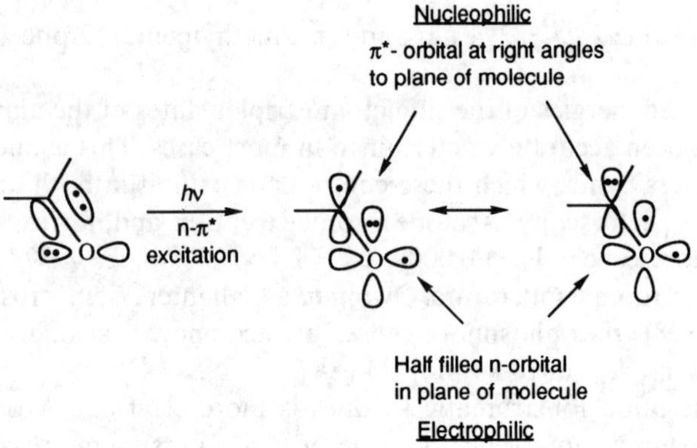

Nucleophilic
$\pi^*$- orbital at right angles
to plane of molecule

$h\nu$
n-$\pi^*$
excitation

Half filled n-orbital
in plane of molecule
Electrophilic

(38)

make nucleophilic attack at the carbon of the carbonyl group an unlikely event in the excited state, and the outstanding feature is the unpaired electron in a *p*-type orbital on oxygen. The similarity between this and an alkoxy radical, which also has an odd electron on oxygen, should be obvious [93]. As a result of this it is not surprising that many of the reactions of a carbonyl group in its $n\pi^*$ state will be similar to reactions of alkoxy radicals. Indeed this simple model gives a reasonable account of all the processes which arise from the carbonyl $n\pi^*$ excited state such as inter- and intra-molecular hydrogen abstraction, photoaddition, and photofragmentation. The situation is slightly different in $\alpha,\beta$-unsaturated ketones since there is mixing of the two chromophores, the carbonyl $\pi$-orbital with the olefinic $\pi$-orbital, producing two $\pi$-orbitals and two $\pi^*$-orbitals of different energies.

## 3.3.2 Norrish Type I fragmentation reactions

### *3.3.2.1 Acyclic ketones and aldehydes*

Norrish Type I reactions or Type I cleavage is a term used to denote cleavage of a carbon-carbon bond adjacent to the carbonyl group. This process arises from the $n\pi^*$ state of the carbonyl compound. The bond cleavage in the $n\pi^*$ excited carbonyl compound will only take place if the excitation energy is converted into vibrational energy localized in the bond undergoing fission. In the gas-phase photochemistry of saturated carbonyl compounds, localization of vibrational energy usually occurs in the weaker of the bonds attached to the carbonyl group. Fission results in the formation of an acyl and an alkyl radical as a result of the primary photochemical process. Typical of this is the fission of acetone into a methyl radical and an acetyl radical [94].

Excitation at 313 or 254 nm affords carbon monoxide with a quantum yield of $\Phi = 1$ and methane and ethane as the hydrocarbon products. These compounds are formed by hydrogen abstraction and dimerization. The decarbonylation process occurs from both the singlet and the triplet state, with the triplet state more efficient [95]. Both of the $\sigma$-bonds in acetone are of comparable strength (297 kJ mol$^{-1}$). However, this is not always the case and usually there is a preference for the fission of the bond which yields the 'better' radical pair, as is shown in the photolysis of acetyl chloride where excitation affords only the acetyl radical and a chlorine atom as the primary process [96]. With butan-2-one (**39**) excitation to the $n\pi^*$ state permits fission

$$\xrightarrow[n\pi^*]{h\nu} [\ ]^* \longrightarrow CH_3\dot{C}H_2 + CH_3\dot{C}O$$
$$\longrightarrow \dot{C}H_3 + CH_3CH_2\dot{C}O$$

(39)

by the two possible routes in a ratio of 40:1 using 313-nm excitation [97]. This partition of reaction paths is wavelength-dependent and falls to 2.4:1 when 254-nm excitation is employed. This example shows that the cleavage reaction is relatively slow from the zero vibrational level and also reflects the enhanced stability of ethyl radicals compared with methyl radicals. However, the rate constant for bond fission increases as higher energy irradiation or temperature of reaction increases the population of the higher vibrational levels. Such an effect explains the influence of wavelength and temperature changes on the outcome of either singlet or triplet reactions and also why in the gas phase the singlet contribution to reaction increases as a factor of temperature. Di-*t*-butylketone is another example of facile decarbonylation ($\Phi = 1$) in the gas phase at elevated temperatures [98]. Here again the primary photochemical event is fission into a pivaloyl radical and a *t*-butyl radical.

In solution phase the $\alpha$-fission process is less common for acyclic ketones if alternative reaction paths are available. However, di-*t*-butyl ketone is also reactive in solution, and both the singlet and triplet states are involved in the Norrish Type I fission [99]. On irradiation in pentane, fission results in the production of the pivaloyl radical and *t*-butyl radical which, by subsequent dark reactions, yields the products, carbon monoxide (90%), 2-methylpropane, 2-methylpropene, and 2,2,3,3-tetramethylbutane. Similar behaviour is observed for the irradiation of dibenzylketone (**40**) in benzene solution.

$$PhCH_2COCH_2Ph \xrightarrow[C_6H_6]{h\nu} Ph\dot{C}H_2 + PhCH_2\dot{C}O \longrightarrow PhCH_2CH_2Ph + CO$$

PhCH₂–O–N(2,2,6,6-tetramethylpiperidine) + $PhCH_2COO$–N(2,2,6,6-tetramethylpiperidine)

(40)

This triplet state reaction yields carbon monoxide efficiently ($\Phi = 0.7$). The reaction is still two-step, involving $\alpha$-fission affording a phenylacetyl and a benzyl radical. A two-step rather than a concerted mechanism was substantiated by trapping experiments where the two radical species were intercepted by 2,2,6,6-tetramethylpiperidine-1-oxyl to yield the two products shown [100a]. Alternatively photo-CIDNP (photo-chemically induced dynamic nuclear polarization) can be used to demonstrate the involvement of a two-step radical reaction path [100b]. Several examples of such Norrish Type I processes in solution phase are recorded in (**41**) [101].

$h\nu$ 44% Ref [101a]

$h\nu$, $C_6H_6$

R = Et, $C_3H_7$, $C_6H_{13}$ or $C_8H_{15}$ 38-53% Ref [101b]

(41)

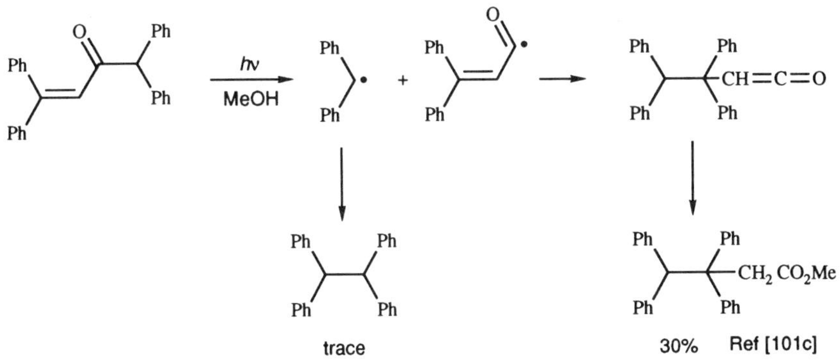

Ph, O, Ph, Ph, Ph — $h\nu$ / MeOH → Ph₂C• + Ph₂C=CH–C•=O → Ph₂CH–CPh₂–CH=C=O

↓ Ph₂CH–CHPh₂ (trace); ↓ Ph₂CH–CPh₂–$CH_2CO_2Me$ (30%) Ref [101c]

(41 continued)

Aryl ketones such as benzophenone and acetophenone are not reported to undergo the Norrish Type I process, owing, no doubt, to the lower energy available in excited aryl ketones which is insufficient to bring about the fission of the C—C bond. The balance can, however, be tipped in favour of α-fission if a reasonably stabilized radical can be formed. Such is the case with the benzoin ether (**42**) [102], which undergoes fission to afford benzoyl

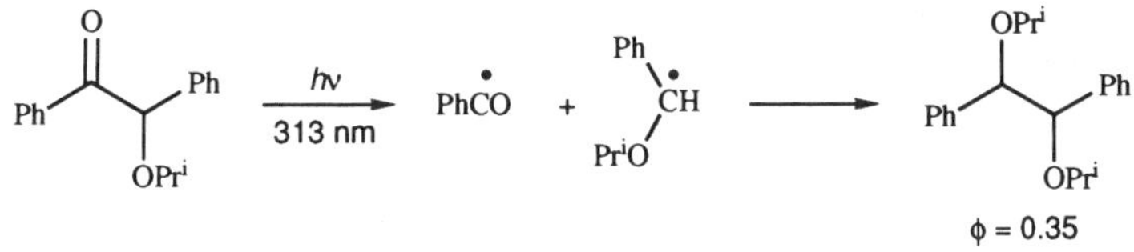

PhCO–CH(Ph)OPr$^i$ —$h\nu$, 313 nm→ PhĊO + Ph–ĊH–OPr$^i$ → PhCH(OPr$^i$)–CH(OPr$^i$)Ph

$\phi = 0.35$

(42)

and benzyl radicals. Subsequent dark reactions afford the product shown. Similar behaviour is exhibited for other deoxybenzoin derivatives (**43** [103]

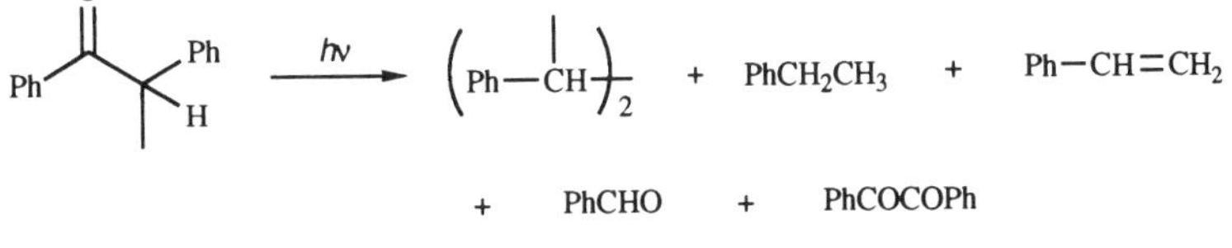

PhCO–CH(Ph)CH$_3$ —$h\nu$→ (Ph—CH(CH$_3$)–)$_2$ + $PhCH_2CH_3$ + Ph—CH=$CH_2$

+ PhCHO + PhCOCOPh

(43)

and **44** [104]). Lactones (**45**) have also been shown to undergo the α-fission

$$\mathrm{Ph{-}C(=O){-}CH(OAc){-}Ph} \longrightarrow \mathrm{Ph{-}\overset{O}{\overset{\|}{C}}\cdot} + \mathrm{\cdot CH(OAc){-}Ph} \longrightarrow \mathrm{PhCOCOPh} +$$

AcO OAc Ph Ph + 15%

(44)

R = *cis*-Ph
R = *trans*-Ph

$\phi$ = 0.3-0.18

(45)

process, affording carbon dioxide and the mixture of 1,2-diphenylcyclopropanes shown [105]. The influence of surfaces and micelles on the outcome of photochemical reactions of ketones such as dibenzyl ketone, as an approach to the measurement of rate processes of free radicals, has also been studied in considerable detail [106].

Aldehydes, such as acetaldehyde, undergo the photochemical α-fission process in the gas phase. In this case irradiation at 334 nm yields an acetyl radical and a hydrogen atom while 254 nm excitation affords methane and carbon monoxide directly by what is considered to be a concerted or nearly concerted reaction [107]. Again, as with the acyclic ketones, the α-fission of aldehydes is not very common. Indeed studies have shown that acetaldehyde in the triplet excited state in solution abstracts the aldehydo-hydrogen from a ground-state molecule to afford the radicals (**46**) [108]. Similar behaviour

$$2\ \mathrm{CH_3CHO} \xrightarrow[\text{gas phase}]{h\nu} \mathrm{CH_3{-}\dot{C}{=}O} + \mathrm{CH_3{-}\dot{C}H{-}OH}$$

(46)

is reported for triplet-state benzaldehyde [109] and for the cyclopropylaldehyde (**47**) [110]. Pivalaldehyde [111] and propionaldehyde (**48**) [112], also

CHO

$h\nu$ / 310 nm → $CH_3CH_2CH_2CHO$ + O O + O O HO + O HO OH O

Product ratio 4:3:5:2

(47)

CHO $\xrightarrow{h\nu}$ $CH_3CH_2\dot{C}HOH$ + $R\dot{C}O$

→ $CH_3CH_3$ + CO

→ $CH_3\dot{C}H_2$ + $\dot{C}HO$

(48)

undergo this type of reaction, but in addition α-fission in the triplet excited state is also observed, yielding the radicals shown. Pentaacetylglucose (**49**)

CHO, OAc, AcO, OAc, OAc, OAc $\xrightarrow[C_6H_6]{h\nu}$ OAc, AcO, OAc, OAc, OAc + Ph, OAc, AcO, OAc, OAc, OAc + 2, OAc, AcO, OAc, OAc, OAc

16% 2.5% 1%

(49)

is also reactive in this mode and undergoes α-fission to afford the products shown [113]. Further examples of the decarbonylation reactions of aldehydes are shown in **50** [114].

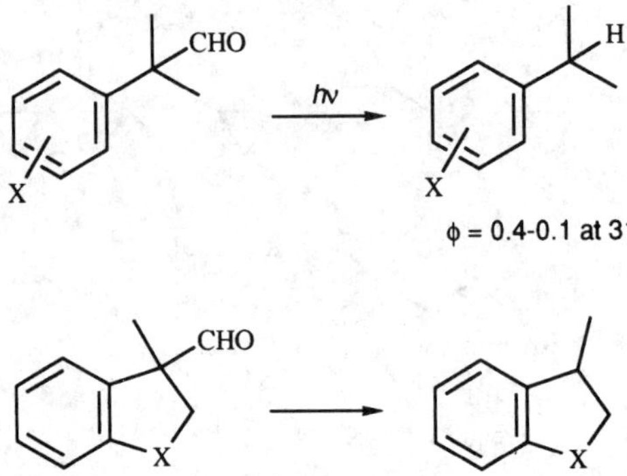

$\phi$ = 0.4-0.1 at 313 nm Ref [114a]

X = O; 95%
X = S; 93% Ref [114b]

(50)

### *3.3.2.2 Cyclic ketones*

Cyclic ketones, in contrast to the acyclic variety, show a greater tendency to undergo an $\alpha$-fission process to furnish acylalkyl biradicals. After the cleavage reaction has occurred, the radicals can undergo several dark reactions, yielding a variety of products. However, the outcome of the reaction can be affected by structural features such as changes in ring size. Thus photoextrusion of carbon monoxide from cyclopropanones and cyclobutanones, is quite facile but this reaction becomes less likely with cyclopentanones and above. Thus with a cyclic ketone four possible reaction types can be identified:

(1) Decarbonylation and bond formation to yield ring-contracted products.
(2) $\beta$-Fission within the biradical.
(3) Intramolecular hydrogen abstraction yielding a ketene and/or an unsaturated aldehyde.
(4) Ring expansion to afford a carbene subsequently trapped by solvent.

#### *3.3.2.2.1 Decarbonylation and bond formation*

As pointed out above, the decarbonylation process is dependent to some extent on the ring size of the ketone. Cyclopropanones are usually thermally unstable and few photoreactions have been described. However, the stable cyclopropanone (**51**) does undergo facile photodecarbonylation to afford an quantitative Ref [115a]

(51)

alkene [115a] whereas aziridones (**52**) decarbonylate to yield imines [115b].

97% Ref [115b]

(52)

The more stable cyclopropenones (**53**) provide other examples of decarbonylation of three-membered ring ketones [116].

$h\nu$, $C_6H_6$ → Ar—C≡C—Ar + CO

Ar = Ph or 4-OH-3,5-$Bu^tC_6H_2$
quantitative

(53)

In the gas phase, cyclobutanone on direct excitation populates the singlet excited state in which decarbonylation and $\beta$-fission are predominant (**54**)

$h\nu$, 313 nm, gas phase

CO + ... $\phi = 0.35$

$\beta$-cleavage

$C_2H_4$ + $CH_2{=}C{=}O$ $\phi = 0.51$

(54)

[117]. In solution cyclobutanone also follows these paths and yields ethylene, cyclopropane and propylene. The outcome of the reaction appears to be solvent-independent, as shown in **55** [118]. In some cases the cyclopropane-

hν 313 nm solution → CO + ethylene + propene + cyclopropane

| Solvent | φ (ethylene) | φ (cyclopropane) |
| --- | --- | --- |
| n-heptane | φ = 0.18 | φ = 0.14 |
| t-butanol | φ = 0.17 | φ = 0.10 |
| $CH_3CN$ | φ = 0.18 | φ = 0.085 |

(55)

forming process is predominant, as with the hexafluoro derivative (**56**) [119].

hν → hexafluorocyclopropane + CO

quantitative

(56)

Generally, however, in these systems decarbonylation and cleavage of the $\beta$-bond, and formation of ketene, are the principal reactions, owing no doubt to the inherent strain in the cyclobutanone. Other examples of the fission process are shown in **57** [120]. Fission processes such as illustrated in **57** do

hν 325 nm → 2-butene + $CH_2{=}C{=}O$

φ = 0.13

Ref [120a]

hν MeOH

$R^1$ = MeO, $R^2$ = H or viceversa

26%

7%

30%

Ref [120b]

(57)

occur in methanol, but an alternative reaction path, that of ring expansion to a carbene, can also be observed (**58**) [121] and has been the subject of

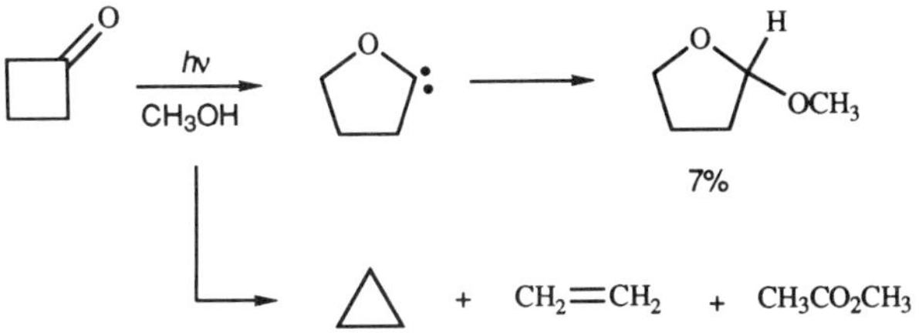

(58)

extensive reviews [112]. The ring-expansion path is in conflict with the evidence obtained in explanation of the decarbonylation and $\beta$-fission route. This latter process involves a biradical which subsequently undergoes fission. This was substantiated by the results from the irradiation of cyclobutanone at $-78\,^{\circ}C$ in the presence of buta-1,3-diene when the cyclohexanone was isolated (**59**) [123]. However, the demonstration that the ketosteroids (**60**)

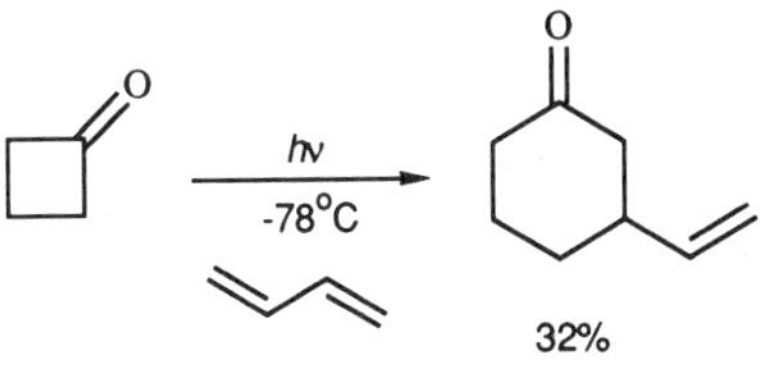

(59)

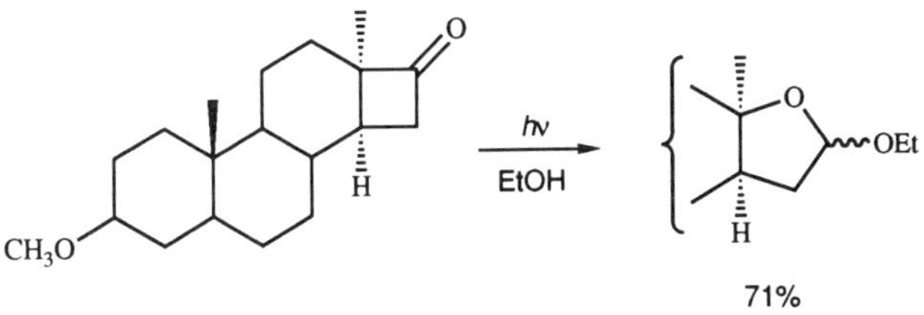

(60)

[124a] and the simpler cyclobutanones (**61**) [124b] rearrange with the

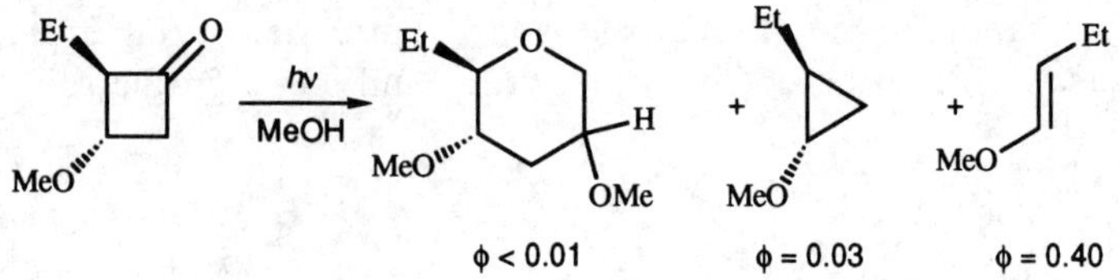

$\phi < 0.01$ $\phi = 0.03$ $\phi = 0.40$

(61)

retention of the bridgehead geometry shows that a freely rotating acyl-alkyl biradical is not involved and that the reaction is likely to be concerted. Some examples of the processes are illustrated [125] and applications to the synthesis of natural products are shown (**62**) [126].

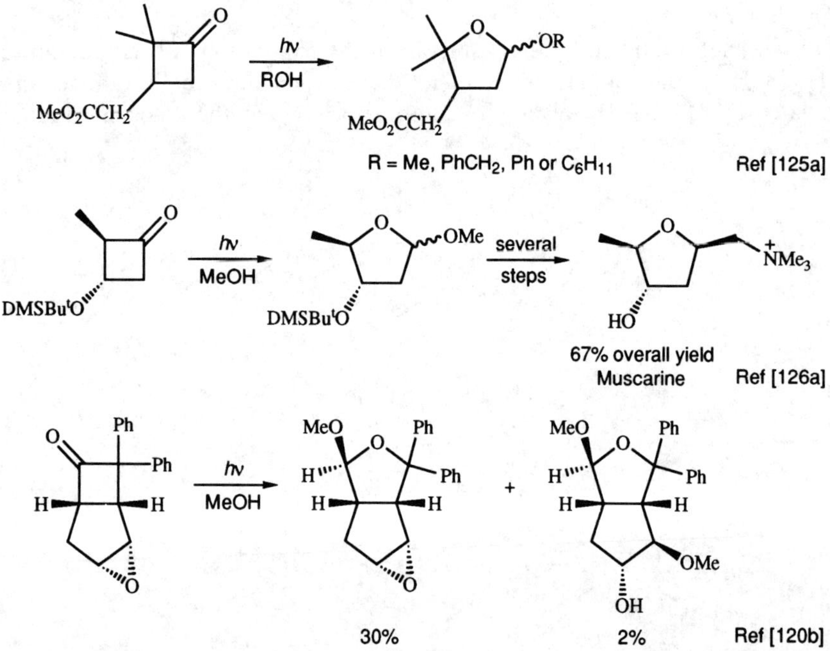

R = Me, $PhCH_2$, Ph or $C_6H_{11}$ Ref [125a]

67% overall yield
Muscarine Ref [126a]

30% 2% Ref [120b]

(62)

hν MeCN/$H_2O$

43%

one step

PG $F_{2α}$

Ref [126b]

(62 continued)

While the carbene ring-expansion route is common in cyclobutanone photochemistry, few examples of ring expansion in larger ring compounds have been reported. However, an example is the ring expansion of the furanone (**63**) to yield a dioxane [127], or the formation of **64** in 27% yield

hν MeOH

R = H or Me

(63)

hν MeOH

27%

(64)

from the irradiation of the ketone at $-60°C$ in methanol–diethyl ether [128]. Another is the ring expansion of the silaketone (**65**), which affords the

Ph Ph Si O hν MeOH Ph Ph Si O H OMe >95%

(65)

compounds shown [129]. Ring expansion is also found in bicyclic systems where strain influences the outcome of the reaction (**66**) [130].

$MeO_2C$ O hν MeOH $MeO_2C$ O OMe $\phi = 0.03$

(66)

Cyclopentanone also decarbonylates on irradiation in the gas phase at 147 nm [131]. Again a two-step process is involved, affording a biradical which decarbonylates to another biradical which either fragments to ethene or undergoes bond formation to cyclobutane. Fission to ethene is much more efficient with $\Phi = 0.76$ as opposed to cyclobutane formation where $\Phi = 0.02$. In solution the loss of carbon monoxide from a cyclopentanone is a major path only when the radical centres formed are stabilized by alkyl substitution, double bonds or cyclopropyl rings (**67**) [132]. The non-concerted nature of

O hν $\lambda > 200$ nm 45% + H 55%

O hν 70%

(67)

the decarbonylation in cyclopentanones such as **68** is substantiated by the

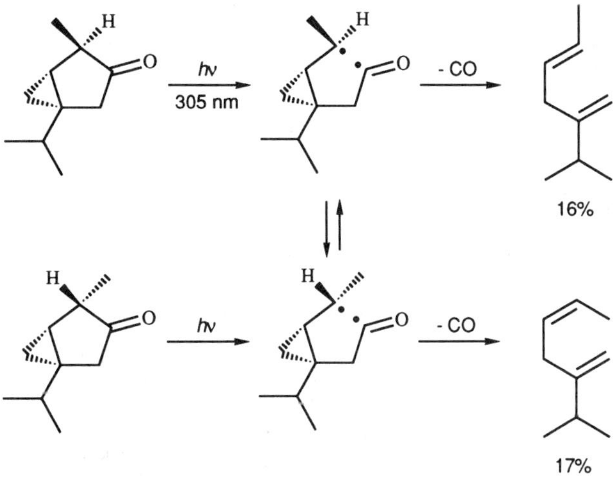

hν
305 nm
- CO
16%
hν
- CO
17%

(68)

fact that irradiation of either of the isomeric ketones yields the same mixture of hydrocarbons [133]. A similar effect is seen with cyclohexanone derivatives where the tetramethyl derivative (**69**) decarbonylates efficiently (*c*. 51%) [132].

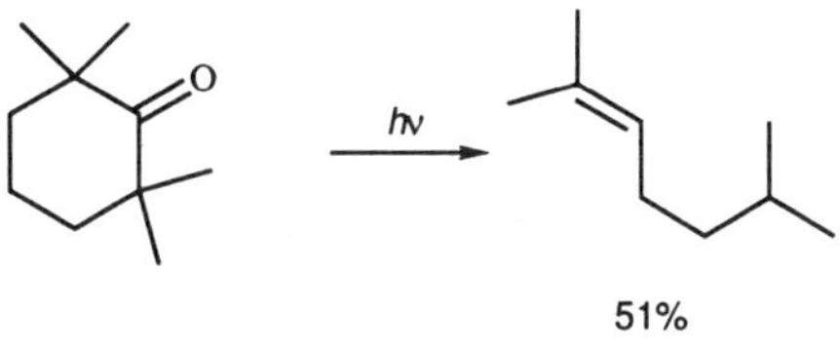

hν
51%

(69)

The biradical formed in this reaction undergoes hydrogen abstraction to yield the observed product. Decarbonylation reactions are used extensively in the synthesis of a variety of molecules, either where dienes are the target or where bond formation following the decarbonylation affords bicyclic compounds as shown (**70**) [134].

$MeO_2C$ — $h\nu$, 313 nm — $CO_2Me$ — $CO_2Me$

10% Ref [134a]

Ph, Ph, $CO_2Me$, $CO_2Me$ — $h\nu$ — Ph, Ph, $CO_2Me$, $CO_2Me$

Ph, Ph + $CO_2Me$, $CO_2Me$

Ref [134b]

Ar, Ar, R, R — $h\nu$ — Ar, Ar, R, R

R = H, Ar = Ph, p-$ClC_6H_4$ 1:1 cis:trans 34-57% Ref [134c]

R = Me, Ar = Ph 89%

$h\nu$

90% Ref [134d]

$R^1$, $R^2$ — $h\nu$ — $R^1$, $R^2$

$R^1 = CO_2Me$, $R^2$ = H Ref [134e]

or viceversa

(70)

*3.3.2.2.2 Intramolecular hydrogen abstraction*

The principal reaction observed with cyclopentanones and cyclohexanones, when they undergo Norrish Type I cleavage producing a biradical intermediate, is intramolecular hydrogen abstraction. This abstraction can occur in two ways to afford a mixture of ketenes and aldehydes (**71**) [135]. Clearly

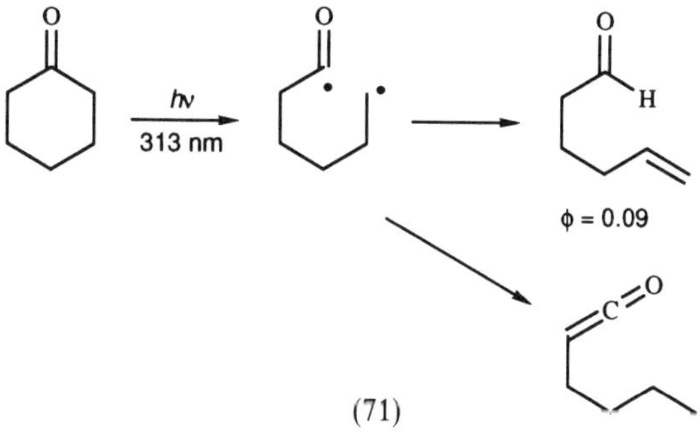

hν
313 nm
$\phi = 0.09$

(71)

in the absence of a nucleophilic solvent the ketene will not be trapped. A mechanistic study of this has shown that the triplet state is primarily involved and that all of the products can be quenched on the addition of a triplet quencher. The scope of the reaction has been studied in detail and it is clear that α-substitution leads to a more efficient reaction owing to the ability of the alkyl group in stabilizing the alkyl radical [136]. Thus fragmentation of 2-methylcyclohexanone follows the pathway which generates the biradical incorporating the secondary alkyl radical in preference to the path leading to the less stable biradical. Two isomeric aldehydes are formed in this reaction and possible intermediates for their formation are shown in **72**. β-

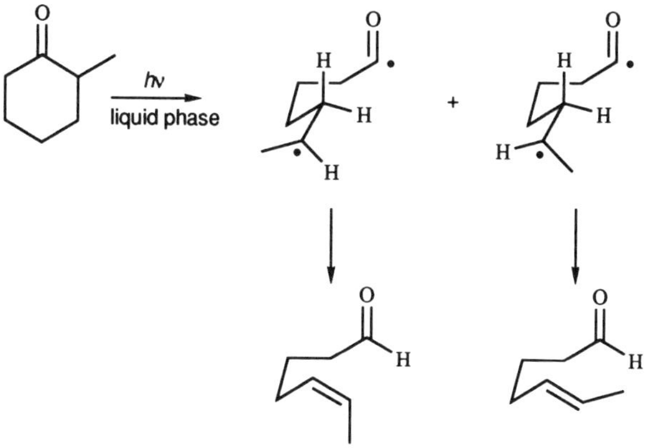

hν
liquid phase

(72)

Substitution also has an adverse effect on the cleavage reaction and this may be due to several factors, such as decreased availability of β-hydrogens for abstraction, the influence of β-substituents on recoupling and the reformation of starting material, or the presence of rotational barriers impeding hydrogen transfer (**73**) [135, 137]. The influence of a γ-substituent has also been

$\phi = 0.033$

(73)

evaluated and these results show that increasing size of the γ-substituent adversely influences the formation of alkenal products (**74**) [138]. It is

| $R^1$ | $R^2$ | $R^3$ | Ratio |
|---|---|---|---|
| H | H | H | 1.6:1 |
| Me | H | H | 2.5:1 |
| H | Me | H | 0.4:1 |
| H | H | $Bu^t$ | 0.18:1 |

(74)

interesting to note that cyclopentanone and some derivatives undergo more efficient ring opening than the corresponding cyclohexanones, indicating that relief of strain can be an important factor in establishing the rate at which cyclic ketones undergo Norrish Type I fission [137, 139].

Irradiation of cycloalkanones in methanol illustrates the operation of the alternative reaction path of ketene formation. In this reaction mode, α-substitution increases the proportion of aldehyde formed while β- or

$\gamma$-substitution increases the amount of ester formed. Results from a series are shown in **74** [138]. It is clear that in most cases of Norrish Type I reactivity, the reaction involves a biradical intermediate, although in some cases a concerted mechanism might be involved. Evidence for the biradical process has already been given in the acyclic case and for cyclic ketones evidence for a biradical comes from the irradiation of the ketones (**75**) where the same

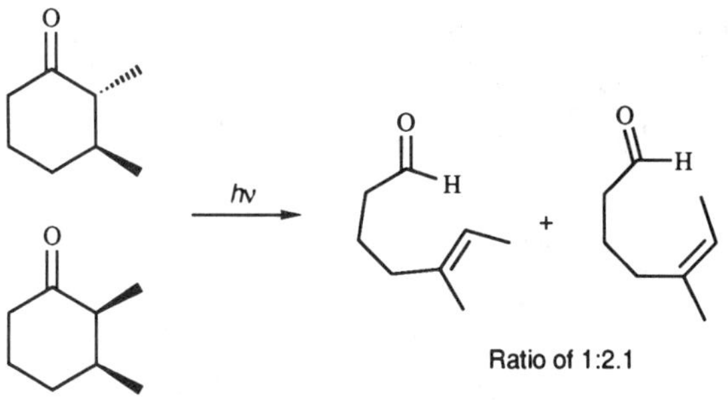

Ratio of 1:2.1

(75)

mixture of alkenes is obtained from either starting material [140a]. In this case no evidence for concertedness was obtained. Further proof comes from a study of the back-recombination process in the hydroindanones (**76**) where

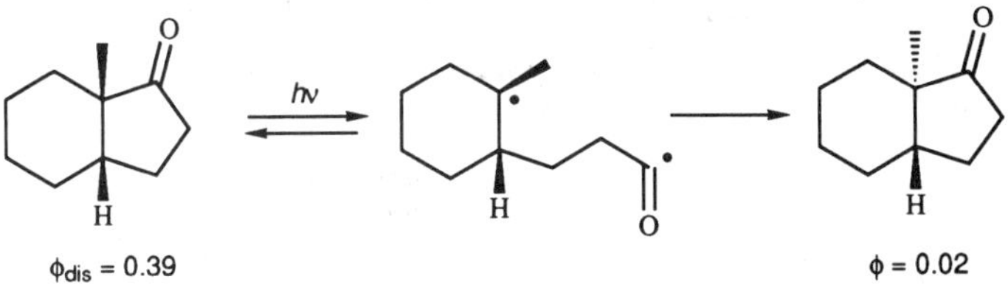

$\phi_{dis}$ = 0.39

$\phi$ = 0.02

(76)

irradiation of one can result in its conversion into the other [140b]. Again a biradical intermediate has to be involved. This reaction mode has been shown to be of synthetic value, and many examples appear in the literature, some of which are shown in **77** and **78** [141].

CHO

$h\nu$

$\phi$ = 0.28

Ref [141a]

O

$h\nu$

300 nm

29%

44%

Ref [141b]

$h\nu$

45%

20%

Ref [141c]

(77)

HO

$h\nu$

Ratio 96:4
yield 55%

$h\nu$

EtOH

Ratio 97:3
yield 33%

(78)

Ratio 97:3
yield 41%

Ref [141d]

*hν*
MeOH

(78 continued)

Ref [141e]

### 3.3.3 α-Fission of β,γ-unsaturated compounds

β,γ-Unsaturated ketones exhibit strong absorptions in the u.v. around 290 nm. Intensification of these absorptions arises if certain geometrical and electronic requirements are fulfilled [142, 143]. Thus the greatest intensification occurs when the chromophores, the carbonyl group and the alkene, lie in such a way that interaction can result. A detailed discussion of these features has been made in a review [144]. β,γ-Unsaturated ketones are also capable of undergoing Norrish Type I fission from the singlet state. Direct irradiation of enones (**79**) results in a 1,3-migration of the acyl group [145,

*hν*
pentane

thermal

X = $CH_2$, 10%
X = O, 12%

X = $CH_2$, 42%
X = O, 57%

(79)

146]. Although a biradical intermediate was thought to be involved in this transformation, results obtained from the irradiation of the enone (**80**) have

$h\nu$

$CH_2CO_2H$

($\alpha_D$ = -130°) ($\alpha_D$ = +147°) Ref [148a]

(80)

demonstrated that the reaction is concerted and stereospecific and that there is no evidence for the racemization of starting material [147, 148a]. The reactions encountered in such systems often vary from one system to another as illustrated (**81**) [148]. Photochemical 1,3-migrations also arise in enol

$h\nu$

$\phi$ = 0.054 Ref [148a]

HO C≡CH

$h\nu$ / $C_6H_6$

$H_2C$

36% Ref [148b]

$Bu^n$

$h\nu$ / $C_6H_6$

several steps → (±) ptilocaulin

64% Ref [148c]

(81)

esters and detailed reviews of this should be consulted [149–151]. A typical example of the reactions encountered is found in the irradiation of the enol ester (**82**) by excitation of the $\pi\pi^*$ state [152] that affords a 1,3-diketone. If

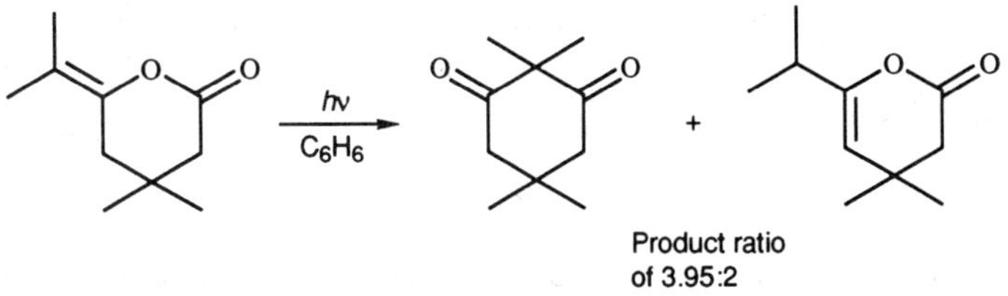

hν
$C_6H_6$
+
Product ratio
of 3.95:2

(82)

the diketone cannot enolize then reasonable conversions can be obtained. If, however, the product 1,3-diketone can enolize, this compound behaves as an efficient light filter and only low yields of product are obtained [153]. Exceptions to the usual low conversions can be found, as in the irradiation of the ketone (**83**) which undergoes 1,3-migration, affording the product in

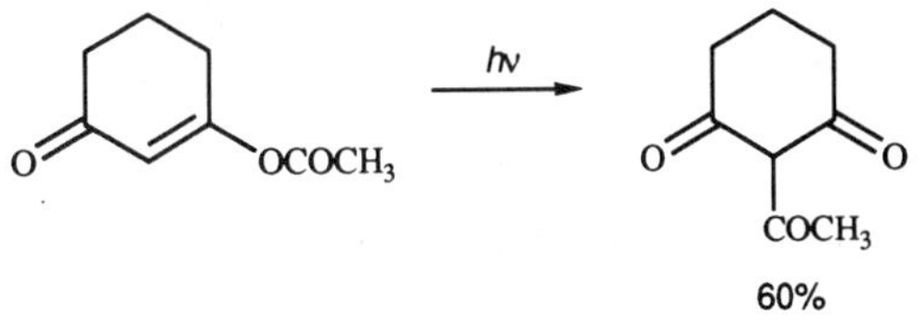

$OCOCH_3$
hν
$COCH_3$
60%

(83)

60% yield [154]. In some instances both a 1,3- and a 1,5-migration can arise, as in irradiation of **84**, affording the products shown [155]. The steroidal

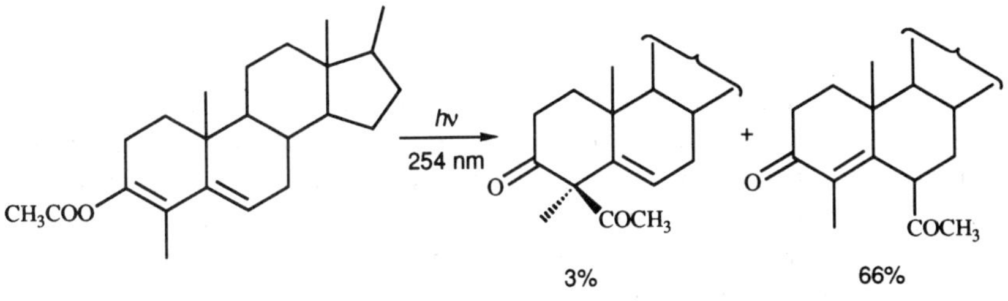

$CH_3COO$
hν
254 nm
+
$COCH_3$
$COCH_3$
3%
66%

(84)

lactone (**85**) is also photochemically reactive. Fission to a biradical followed

OAc hν 3% 40%

(85)

by 1,3-migration affords the cyclobutanone derivative, whereas decarbonylation of this product and cyclization yields the cyclopropane [156].

### 3.3.4 The photo-Fries reaction

The aryl ester analogue of the foregoing 1,3-migration process is known as the photo-Fries reaction, some examples of which are discussed in Chapter 2 [149–151]. Generally this lateral rearrangement process of aryl compounds (**86**) occurs in compounds in which the bond between X and R is readily

X—R hν XH R

(86)

cleaved homolytically. The aryl ester reaction (**87**) was first described by

$OCOCH_3$ OH hν OH OH $COCH_3$ + OH OH $CH_3CO$ + OH OH

22% 18% 46%

(87)

Anderson and Reese [157] and affords products of 1,3- and 1,5-migration as well as the corresponding phenol. Several studies of the mechanism have been carried out [158–165] and the consensus is that the reaction arises from the singlet state with the formation of a radical pair within a solvent cage. Recombination within the cage affording the migration product is faster than escape from the cage yielding the phenol. The scope for the reaction is considerable and some examples (**88**) are listed [166]. The process is quite

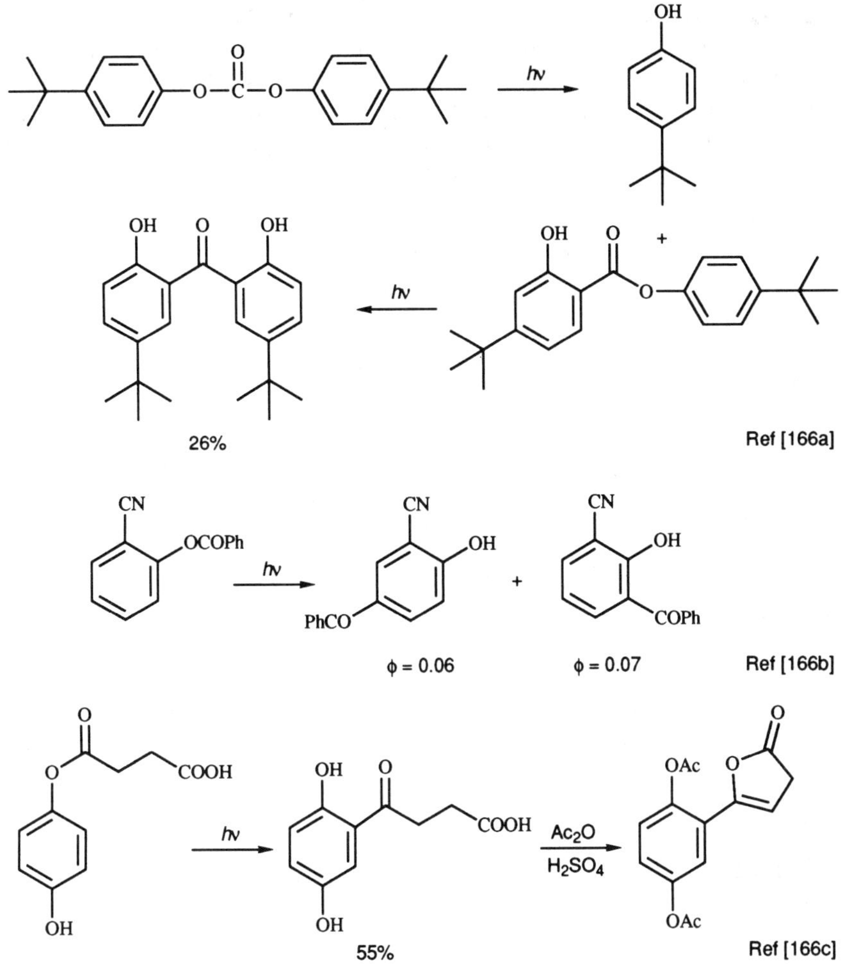

hν

26%

Ref [166a]

hν

$\phi = 0.06$ $\phi = 0.07$

Ref [166b]

hν $Ac_2O$ $H_2SO_4$

55%

Ref [166c]

(88)

general and can be carried out with *N*-arylamides [167], *N*-arylcarbamates [168], and urethanes [169]. Sulphur [170], and selenium analogues [171], and derivatives containing silicon [172] are also reactive.

### 3.3.5 1,2-Migration in *β*,*γ*-unsaturated ketones

In a previous section (3.3.3) the singlet-state reactivity of the enones (e.g. **80**) was discussed. When sensitization is employed or when intersystem crossing is efficient compared with other processes, an alternative reaction mode is operative. This process affords cyclopropane derivatives by what has been referred to as an oxa-di-$\pi$-methane rearrangement and is analogous to the di-$\pi$-methane rearrangement discussed in Chapter 2. Mechanistic studies [144] on this reaction have shown that the most likely excited state involved is a low-lying $\pi\pi^*$ triplet state. The process formally involves the 1,2-migration of the acyl group shown for the acyclic enone (**89**) where this reaction mode

$* = {}^{14}C$

hv

acetic acid

(89)

was substantiated by labelling studies [173]. The reaction does not involve $\alpha$-fission, but a stepwise process is involved where the acyl group migrates to yield an intermediate biradical. Re-formation of the carbonyl group followed by bonding in the resultant biradical affords the isolated products. A stepwise mechanism is supported by results from Schaffner and his coworkers [174] who observed that the rearrangement of the steroidal enone (**90**) yields a mixture of cyclopropane derivatives where the methyl and the

hv 254 nm

19% 4%

$R^1 = CD_3, R^2 = Me$

$R^1 = Me, R^2 = CD_3$

(90)

trideuteromethyl groups are scrambled. The reaction has considerable synthetic utility for the construction of a variety of tricyclic compounds with high enantioselectivity or as key steps in the synthesis of natural products (**91**) [175, 176].

hν sens. acetone 84% Ref [175a]

hν $\phi = 0.93$ Ref [175b]

OMEM hν acetone 75% Ref [175c]

(91)

Interestingly aldehydes (**92**), studied in detail by Schaffner and his coworkers [177–184], do not undergo this 1,2-migration reaction but decarbonylate instead. The reaction shown in **92**, a decarbonylation, is

CHO hν 70% 30%

(92)

followed, presumably because excitation involves the aldehydo group which results in C—C bond fission rather than bridging as shown in **89**. In the

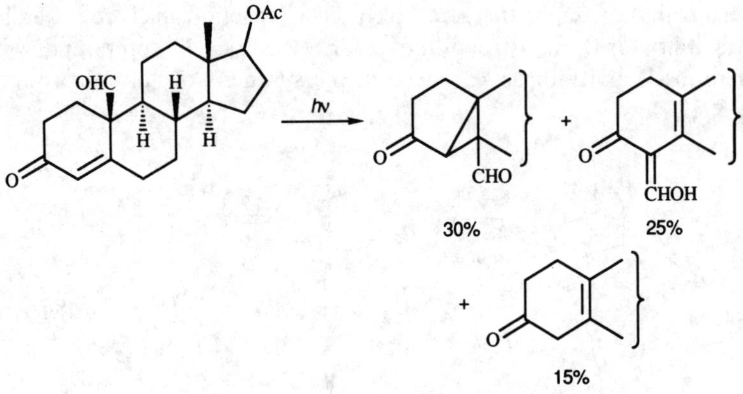

30%

25%

15%

(93)

bichromophoric compound **93** decarbonylation occurs but this is accompanied by a 1,2-migration [183]. This is presumably formed by an oxa-di-$\pi$-methane process and in fact this is the only example involving a $\beta,\gamma$-unsaturated aldehyde. A further example of the decarbonylation of aldehydes is shown in **94** [185].

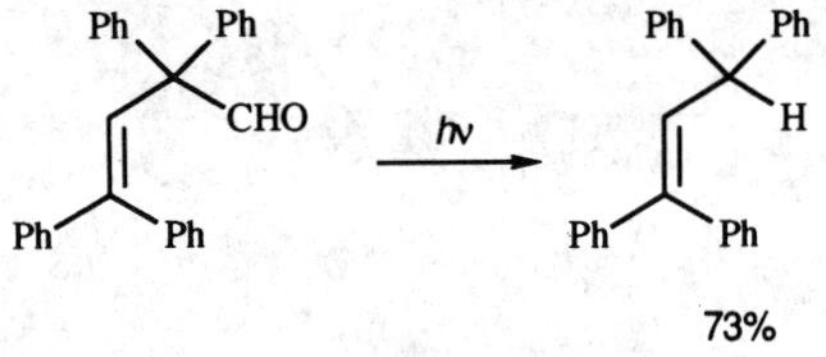

73%

(94)

The failure of aldehydes to undergo the oxa-di-$\pi$-methane rearrangement can be overcome by conversion of the aldehyde to an imine. Thus when aldehyde **95** is converted to the imine and irradiated under sensitized

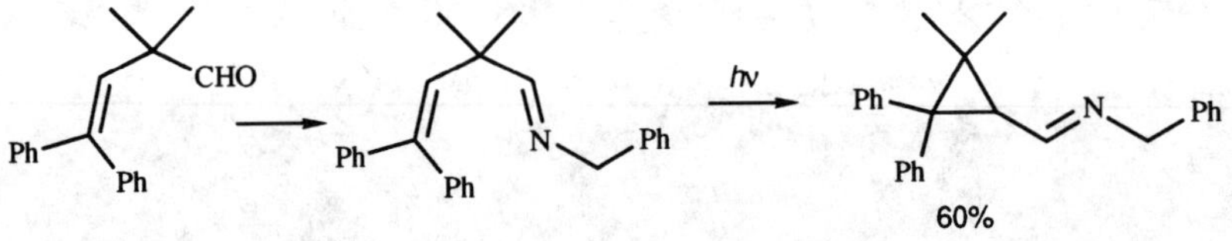

60%

(95)

conditions, the normal cyclopropane-forming process takes place [186]. The formation of the cyclopropane by this path has been demonstrated to be an example of an aza-di-$\pi$-methane process. This reaction path is followed in this case since the triplet energy is now absorbed by the 1,1-diarylalkenyl moiety rather than by the imine moiety. Bridging is preferred to $\alpha$-fission in this excited state and so the di-$\pi$-methane path is followed. A similar path has also been shown to be involved with the oxime acetate (**96**), again affording

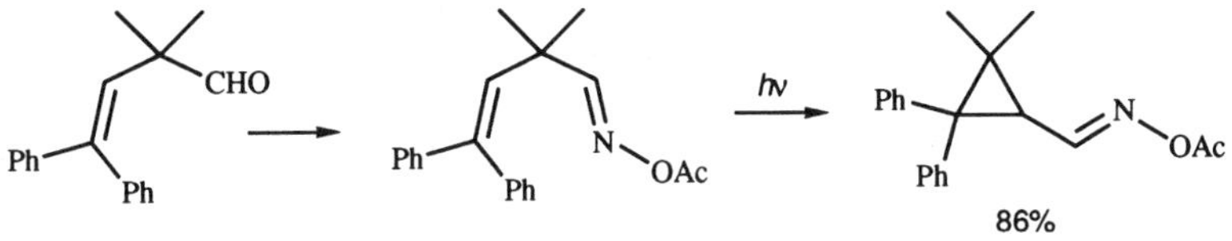

(96)

the cyclopropane product [187]. Details of this reaction mode, referred to as the aza-di-$\pi$-methane rearrangement, are included in Chapter 5.

### 3.3.6 $\beta$-Cleavage reactions

The fission of bonds other than the $\alpha$-bond sometimes occurs on photochemical excitation of carbonyl compounds. The first report was made as early as 1954, when it was observed that irradiation of the cyclopropyl ketone (**97**) brought about conversion to a butenone [188]. Reactions of this type

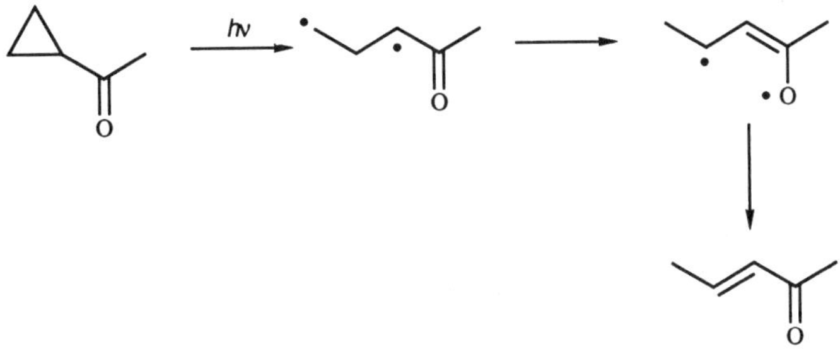

(97)

are distant relations of the Norrish Type I process but involve the fission of the weaker C—C bond in preference to the fission of the bond adjacent to the carbonyl group. Indeed in some systems the two fission processes are often in competition, as shown in **98** [189]. As with $\alpha$-fission, the reaction is

20% 19%

(98)

susceptible to changes in substitution as outlined in **99** [190, 191]. The

(99)

mechanism of the reaction has been shown to involve the formation of a biradical intermediate, and no evidence for the participation of electron-poor (carbocation) or electron-rich (carbanion) species was obtained from a study of the isomerization of the ketone (**100**) [192]. $\beta$-Cleavage is also an important

Ar = $p$-MeOC$_6$H$_4$ Ratio 3.4:1

(100)

reaction in the photochemical conversion of compounds such as lumisantonin (**101**) [193] and the cyclopropyl ketone (**102**) [194]. Other examples

(101)

hν
low -temperature
matrix

(102)

have already been referred to in an earlier section involving the C—C fission of keto-oxiranes [195].

### 3.3.7 Hydrogen abstraction reactions

Hydrogen abstraction is one of the oldest and most extensively studied photochemical reactions of the carbonyl group [196]. It can occur both intra- and inter-molecularly and can lead either to reduction of the carbonyl function or to the synthesis of new compounds by cyclization or fragmentation.

#### *3.3.7.1 Intermolecular hydrogen abstraction reactions of ketones*

The irradiation of a carbonyl compound such as acetone in a solvent can result in the abstraction of hydrogen from the solvent. This reaction can occur with a variety of substrates, such as alkenes, alkanes, alcohols, and ethers. The result of this is the production of free radicals which can undergo disproportionation or coupling. The irradiation of acetone (**103**) in *n*-hexane,

hν, 313 nm
hexane
RH
OH + R•
H—OH

(103)

for example, yields propan-2-ol efficiently [197]. The formation of the product is quenched by added diene and thus a triplet state is implicated. Indeed the reduction arises solely from the triplet state. This ties in with the fact that the intersystem crossing efficiency of acetone is almost unity. Other alkyl ketones such as hexan-2-one also undergo reduction from the triplet state but do so less efficiently than acetone. The intersystem crossing efficiency of hexan-2-one is only 0.27 [198]. Cyclic ketones fit into this general pattern but, having a low efficiency of intersystem crossing, they undergo reduction inefficiently from the triplet state [199]. Aryl ketones also undergo reduction.

Benzophenone is particularly unique and has been studied almost for as long as modern photochemistry has excited interest [196]. The reduction reaction of benzophenone is highly efficient and in a hydrocarbon solvent such as toluene affords the pinacol with a quantum yield of 0.39. The solvent in which the reaction is carried out often plays a large part in determining the overall efficiency of the process. Thus in propan-2-ol (**104**) the quantum yield

$$Ph_2CO \xrightarrow[313\ nm]{h\nu} [Ph_2CO]^{T_1} \xrightarrow{(CH_3)_2CHOH} Ph_2\dot{C}OH + \cdot C(CH_3)_2OH$$

$$Ph_2\dot{C}OH \longrightarrow Ph_2C(OH)\text{—}C(OH)Ph_2$$

(104)

for the disappearance of benzophenone has a limiting value of 2. The excitation of the benzophenone yields the triplet state of the carbonyl group which abstracts a hydrogen from C-2 of the alcohol yielding the two ketyl radicals shown. The latter of these can transfer hydrogen to another molecule of benzophenone, thus giving a molecule of acetone and another partially reduced benzophenone. Substantiation of this mechanism has been obtained from an e.s.r. study which identified both ketyl radicals. The reaction described above is only efficient when low light intensities are employed. With high light intensity, highly coloured compounds such as those illustrated in **105** are

(105)

produced and act as filters, thus preventing photoreduction [200]. As was pointed out earlier, other substrates, such as thiols, sulphides, ethers and amines, can act as donors of hydrogen to the excited-state carbonyl group. The involvement of electron-transfer steps has been shown to be important in photoreduction in the presence of amines [201]. This is outlined for pinacol formation from benzophenone and secondary amines (**106**). Because of the

$$Ph_2CO \xrightarrow[Et_2NH]{h\nu} Ph_2\dot{C}\bar{O} + H\dot{\overset{+}{N}}Et_2 \longrightarrow Ph_2\dot{C}{-}OH + H{-}N(CH_2CH_3)(\dot{C}HCH_3)$$

$$\longrightarrow Ph{-}C(Ph)(OH){-}C(Ph)(OH){-}Ph + CH_3CH{=}N{-}CH_2CH_3$$

(106)

intervention of the electron-transfer step, the reduction of the ketone cannot be quenched by triplet quenchers even though the reaction does involve the triplet state of the carbonyl group. Sulphides and arylamines behave in a like manner.

Many aryl ketones and aldehydes fail to undergo photoreduction in alcoholic media. The ability of diaryl ketones and aldehydes to abstract hydrogens from a substrate depends to a great extent on the substitution pattern on the molecule. Substituents on the molecule can change the nature of the lowest triplet state from $n\pi^*$ to $\pi\pi^*$. In the $n\pi^*$ state the excitation energy is localized on the carbonyl group while in the $\pi\pi^*$ the excitation is associated with the whole system of the aromatic molecule and so tends to increase the electron density on oxygen thus diminishing its electrophilicity. Thus compounds which fail to photoreduce are those where the $\pi\pi^*$ triplet state is lower in energy than the $n\pi^*$ state, e.g. *o*-hydroxybenzo. *p*-methoxy, and *o*-methyl acetophenone derivatives and *p*-hydroxy, *p*-amino benzophenones. The last compound mentioned above, *p*-aminobenzophenone (**107**), exhibits photochemical properties which depend on the polarity of the

$$C_6H_5{-}CO{-}C_6H_4{-}NH_2$$

(107)

solvent in which the reaction is being carried out. In propan-2-ol the hydrogen abstraction process is very inefficient owing to the presence of a charge-transfer state which is stabilized relative to the $n\pi^*$ state. In non-polar solvents the charge-transfer state does not occur and the $n\pi^*$ triplet state is then lower in energy. Thus in cyclohexane, photoreduction does take place, yielding the corresponding pinacol. The formation of the charge-transfer state can be suppressed in propan-2-ol by the addition of HCl. This protonates the nitrogen and so prevents the occurrence of the charge transfer. Under these conditions the photoreduction to afford pinacol does take place. Analogous behaviour is reported for dimethylaminobenzophenone [202]. It is possible to bring about the photoreduction of ketones with low-lying $\pi\pi^*$ states using a single electron transfer from an amine as illustrated for fluorenone (**108**) [203].

hν
$Et_3N/Pr^iOH$
+ $Et_3\dot{N}^{+}$
+ $CH_3\dot{C}HNEt_2$
H OH
OH
O
O−

(108)

### *3.3.7.2 1,2-Dicarbonyl compounds*

1,2-Diketones and *o*-benzoquinones (3.3.12.2.) have long-wavelength absorptions, are often highly coloured, and also participate in hydrogen abstraction reactions. Thus pinacols and other coupling products can arise from such irradiations. This is shown in general for 1,2-diketones (**109**) for the formation

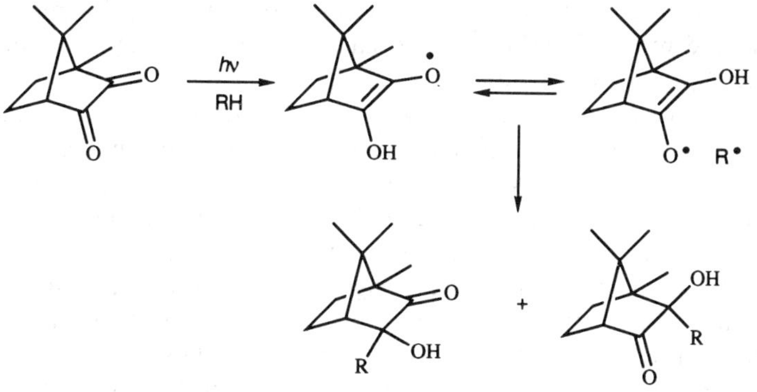

(109)

of coupling products. The existence of a semidione radical intermediate has been proven by flash photolysis and e.s.r. studies [204].

### *3.3.7.3 Intramolecular hydrogen abstraction reactions*

#### *3.3.7.3.1 γ-Hydrogen abstraction*

*3.3.7.3.1.1 Fragmentation reactions.* In the model for the excited state of carbonyl compounds (**38**) it was pointed out that the half-filled orbital on oxygen was electron-deficient. Thus on excitation many carbonyl compounds with a γ-hydrogen undergo a characteristic 1,5-hydrogen transfer from carbon to oxygen to yield a 1,4-biradical. This biradical can either react by cleavage or by cyclization or can undergo back-hydrogen transfer to yield the starting material. The first two processes are outlined in **110**. This is a well-known

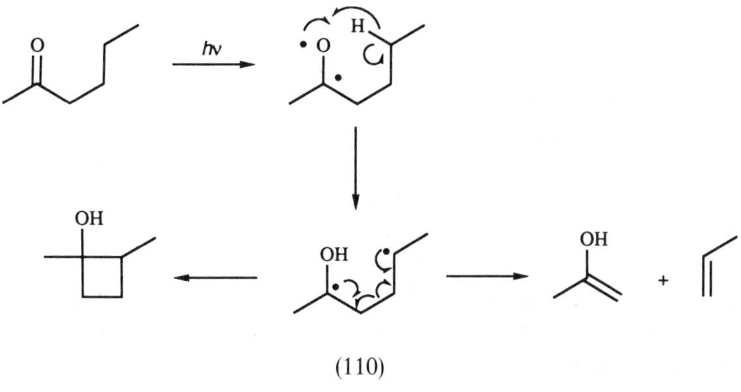

(110)

reaction referred to as a Norrish Type II process [205]. The fragmentation step to alkene and enol outlined in **110** has its analogy in mass spectrometry where the McLafferty rearrangement [206] brings about a similar process involving the abstraction of the $\gamma$-hydrogen. The hydrogen abstraction in the $n\pi^*$ state will occur to the half-filled electron-deficient (electrophilic) $n$-orbital on the oxygen. Thus the hydrogen to be transferred has to be capable of close approach to this centre. Wagner [207] has studied this problem in considerable detail for both the acyclic and the cyclic situations. For a conventional open chain system involving an excited triplet-state $\gamma$-hydrogen abstraction, the 1,5-hydrogen transfer is the most rapid. Available data shows that such transfers are 20 times faster than 1,5-transfers in cyclic systems. Thus the molecule, because of its flexibility, adopts the most strain-free transition state available, a six-membered system, from all the possible transition states open to it. The situation in a cyclic system is quite different and there are only a few conformations available to the molecule within which a hydrogen transfer can take place. This is illustrated for the 2-*n*-propyl-4-*t*-butyl cyclohexanones (**111**) where it can be seen that there

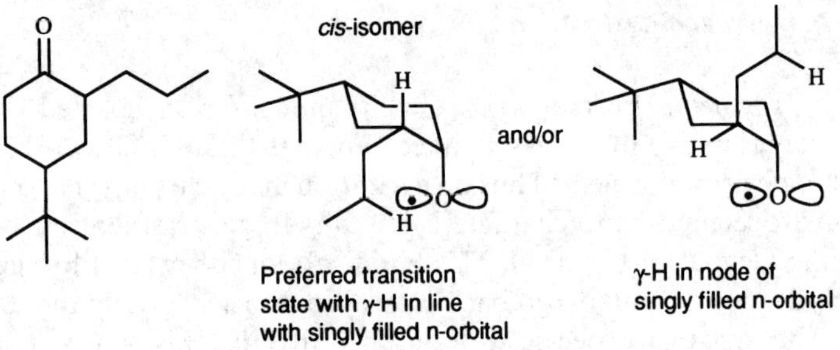

*cis*-isomer

and/or

Preferred transition state with γ-H in line with singly filled n-orbital

γ-H in node of singly filled n-orbital

(111)

is a stereo-electronic requirement for hydrogen abstraction. In the *cis*-isomer, the conformation shown has the $\gamma$-hydrogen in line with the singly filled orbital on oxygen and so hydrogen abstraction involving a six-membered transition state is easily achieved. With the *trans*-isomer, however, the more stable conformation has the hydrogen to be abstracted lying in the nodal plane of the singly filled orbital and so abstraction does not occur [208]. Aliphatic ketones undergo Norrish Type II hydrogen abstraction from both the singlet and the triplet $n\pi^*$ states. The evidence for this is that only part of the reaction can be quenched by the addition to the reaction mixture of known triplet-state quenchers. The percentage of reaction arising from the singlet is dependent on the strength of the $\gamma$-hydrogen bond. For example, pentan-2-one with a $\gamma$-H bond strength of 410 kJ mol$^{-1}$ reacts almost completely from the triplet state, 5-methylhexan-2-one, with a $\gamma$-H bond strength of 380 kJ mol$^{-1}$, reacts equally from the singlet and the triplet, and hexan-2-one, with a $\gamma$-H bond strength of 395 kJ mol$^{-1}$, reacts from the singlet and the triplet in a ratio of 1 : 2 [209].

Aryl ketones, because of their unit intersystem crossing efficiency, always react from the triplet state. There is a dependency on substitution of the rate at which the $\gamma$-hydrogen atom is abstracted. In the series of ketones (**112**) it

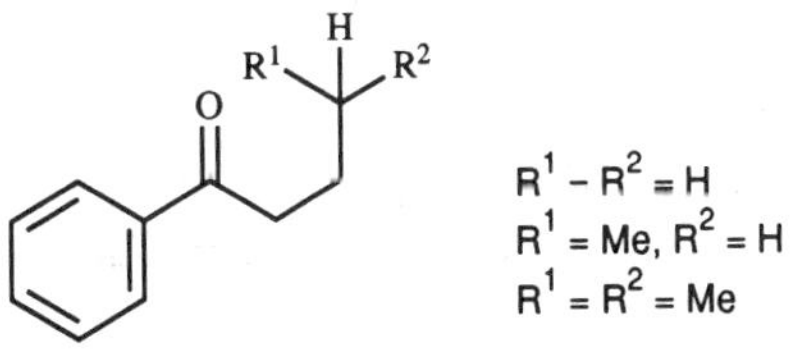

H
R$^1$ R$^2$
O

$R^1 - R^2 = H$
$R^1 = Me, R^2 = H$
$R^1 = R^2 = Me$

(112)

was demonstrated that the tertiary $\gamma$-hydrogens were 165 times more reactive than primary $\gamma$-hydrogens. There is also a rate dependence on the nature of the substituent on the $\gamma$-carbon. Thus the abstraction of a $\gamma$-hydrogen is dependent upon the substituents, and radical stability must play a part [209a, 209b, 210]. The abstraction of the $\gamma$-hydrogen yields a biradical which implies, as a result of the concepts of conservation of spin, that a singlet excited carbonyl group will give rise to a singlet biradical whereas a triplet-state carbonyl group will yield a triplet biradical. This influences the lifetime of the biradical, and the disappearance of a singlet biradical will be considerably faster than that of a triplet owing to the need for spin inversion in the latter case. Substituents have an influence on the lifetime of a triplet biradical, as does the solvent in which the reaction is carried out. Scaiano [211] has demonstrated that the triplet lifetime of the biradical is enhanced in polar solvents. Apparently the solvent effect shows itself as a redistribution of the biradical conformational populations. However, Wagner [212] was the first to observe that alcohol solvents could be used to enhance the fragmentation in ketones (e.g. **113**) and that racemization, by a back H-transfer, could be

$h\nu$, 313 nm

| | | | |
|---|---|---|---|
| $C_6H_6$ | $\phi = 0.23$ | $\phi = 0.78$ | $\phi = 0.03$ |
| $Bu^tOH$ | $\phi = 0.94$ | $\phi = 0$ | $\phi = 0.06$ |

(113)

completely suppressed in such solvents. These two effects can be rationalized in terms of H-bonding by solvent. Thus the H-bonded biradical is not free to undergo reverse H-transfer on intersystem crossing to the singlet state and so fragmentation can compete favourably. The existence of biradical intermediates in such processes is without doubt and their presence has been substantiated by spectroscopic techniques and by trapping using a variety of reagents such as di-*t*-butylselenoketone, alkenes, and oxygen as shown in **114** [213].

(114)

A typical example of the outcome of $\gamma$-H abstraction is shown in **113**. In some ways this has become the archetypal reaction and illustrates all the possible events which can happen to the biradical: racemization, fragmentation and cyclization [214]. The fragmentation reaction does have considerable synthetic potential. One of the synthetic uses is shown in **115** where a

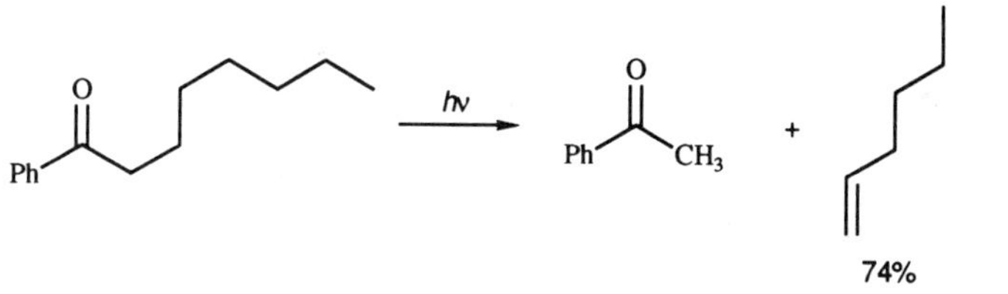

(115)

pure alkene is formed in high yield [215]. The use of the fragmentation reaction in the photoremoval of a side chain, a protecting group, from some carbohydrate systems is illustrated in **116** [216]. This scheme (**116**) also

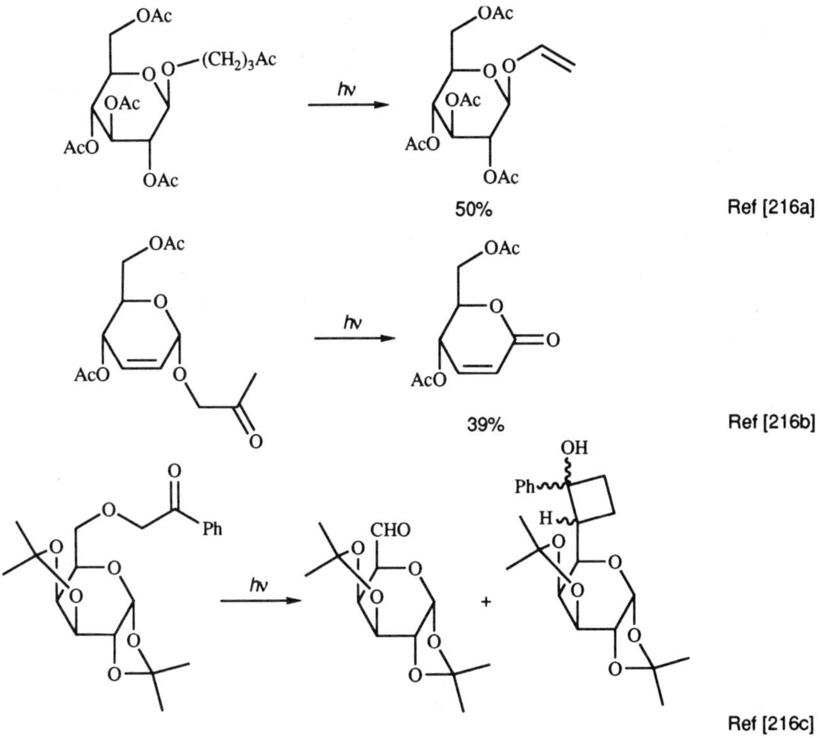

(116)

$h\nu$
EtOH
$\lambda > 280$ nm
H D
Ph S
D
quant.
+ Ph—C(=O)—SH
Ref [217a]

O
OH
$h\nu$
hexane
H
OH
H
23%
Ref [217b]

(116 continued)

illustrates the photofragmentation of a thiobenzoate in the selective *syn*-elimination in the steroid derivative [217a] and elimination in a route to the synthesis of optically active *drimenol* derivatives [217b]. Another application of the Norrish Type II fragmentation has been to chain-breaking reactions in suitably substituted polymers [218].

*3.3.7.3.1.2 Cyclization reactions.* The 1,4-biradical resulting from $\gamma$-H abstraction has also the possibility of undergoing cyclization. Turro and his coworkers [219] have demonstrated that steric factors can be important. These factors manifest themselves in the stereo-electronic requirements necessary for good overlap. In the absence of overlap, cyclization occurs in preference to fragmentation. This is seen in **117** where hydrogen abstraction

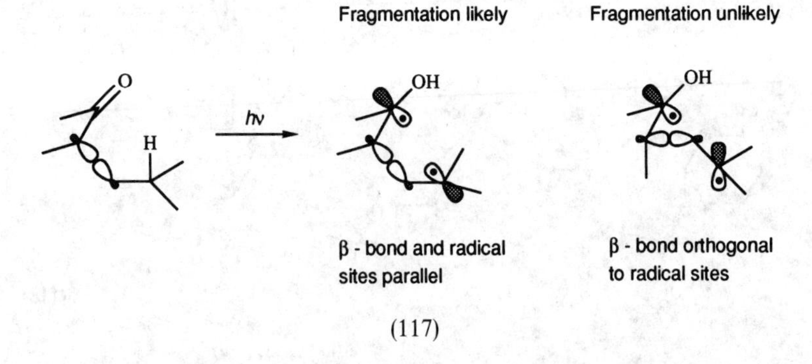


(117)

hν
$C_6H_5$
Vycor filter
β - bond orthogonal to radical centre
$R^1$ = Me, $R^2$ = OH
$R^1$ = OH, $R^2$ = Me

(117 continued)

yields a 1,4-biradical which has the β-bond, i.e. the bond which usually fragments, orthogonal to the orbitals of the biradical. Thus cyclization occurs in preference to fission and yields the cyclobutanols.

Cyclobutanol synthesis by this route is well-exploited. Much of the effort has been directed towards the synthesis of strained molecules, the formation of which would be tedious or impossible by ground-state paths. Several examples are outlined in **118** [220].

hν
quantitative
Ref [220a]
hν
80%
Ref [220b]

(118)

hν

26%

Ref [220c]

hν

27%

Ref [220d]

(118 continued)

*3.3.7.3.2 Hydrogen abstraction from other sites*

While the normal reaction path of the Norrish Type II process involves a six-membered transition state, alternative reaction paths are fairly common. In these either larger or smaller transition states provide biradicals which cannot fragment. Often the route followed in the hydrogen abstraction can be controlled by the absence of $\gamma$-hydrogens or the presence of hetero atoms which help to stabilize the resultant biradicals. The following examples will illustrate the use that can be made of the reaction for the synthesis of a variety of cyclic compounds.

*3.3.7.3.2.1 $\beta$-Hydrogen abstraction.* This class of hydrogen abstraction can be used for the synthesis of cyclopropanols. Usually this hydrogen abstraction path arises when there are no abstractable $\gamma$-hydrogens. If this condition is satisfied, excitation affords the triplet excited carbonyl group which abstracts a $\beta$-hydrogen affording a 1,3-biradical that undergoes ring closure. The yields of product can be variable from such reactions and some examples are illustrated in **119** [221]. In other examples (**120**) [222, 223] the yields of

$R^1 = R^2 = H$ or Me; $R^3$ = Me, Et, $CH_2Ph$, $CH_2CH_2Ph$
$R^1$ = Me or $CHMe_2$; $R^2 = H$

72%

(119)

R = $CH_2Ph$

Ratio 3:1

and

(120)

cyclopropane are lowered because of competition between alternative reaction paths. Thus irradiation leads to two biradicals, one from the $\beta$-abstraction path yielding the cyclopropanol while the five-membered ring is produced by a path involving $\delta$-hydrogen abstraction. In the former example isomeric cyclopropanols are formed in a ratio of 3:1.

*3.3.7.3.2.2 $\delta$-Hydrogen abstraction.* As illustrated above, $\delta$-hydrogen abstraction is also fairly common and again arises in situations where $\gamma$-hydrogens are not available. Some examples of this type of behaviour have been reported [224, 225] for carbohydrate derivatives (**121**). Excitation again

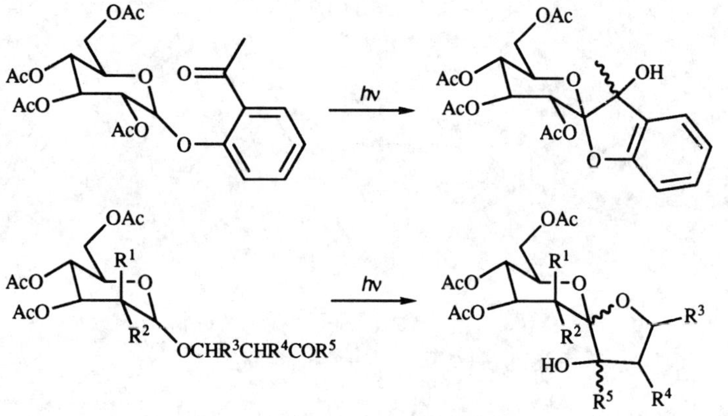

| $R^1$ | $R^2$ | $R^3$ | $R^4$ | $R^5$ |
|---|---|---|---|---|
| H | HO, H or OAc | H | H | β-Me |
| OAc | H | H | H | α-Me |
| H | OAc | Ph | Ph | H or β-Me |
| H | HO | H | H | α-Me |

(121)

results in hydrogen abstraction and cyclization within the resultant 1,5-biradical yielding the cyclized products shown. Other studies [226] have also demonstrated that cyclizations of this type occur in simpler molecules such as the substituted aryl ketones (**122**). Irradiation of these brings about cyclization

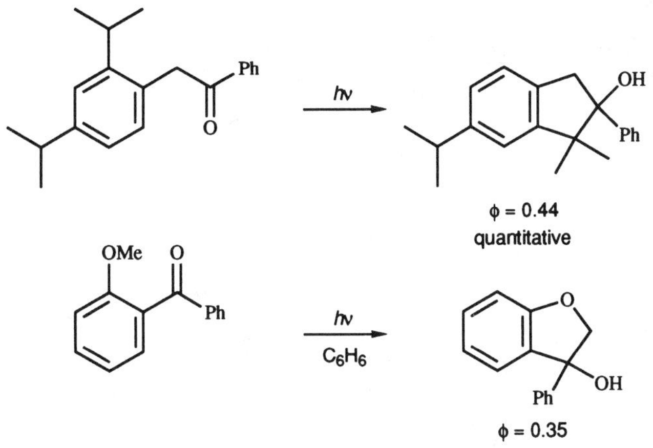

$\phi = 0.44$
quantitative

$\phi = 0.35$

(122)

to afford indanol derivatives. Another example of this type of cyclization is also illustrated in **122** and involves abstraction from an *o*-methoxy group on benzophenone, yielding a dihydro-benzofuranol [227]. Analogous reactivity is also seen [226] in the irradiation of *o*-*t*-butylbenzophenone **123**. For

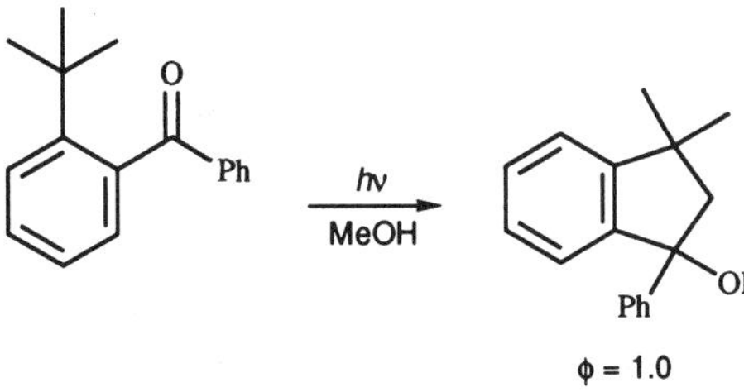

$\phi = 1.0$

(123)

many years it had been thought that *t*-butyl groups were reluctant donors of hydrogen in radical reactions, but this has been proven to be wrong. The H-abstraction from a methyl of a *t*-butyl group leads to the formation of a primary radical which undergoes cyclization to yield an indanol. The reaction is efficient with a quantum yield of unity in methanol.

Another area where $\delta$-H abstraction is common is in phthalimide photochemistry. Some examples of this methodology (**124**) are used in the formation of five-membered rings. Most of these examples involve the directional influence of a heteroatom in the side chain. The presence of the heteroatom

hν

R = H, 65%
R = Me, 82%

hν

R = Me; X = S, 3%
R = Ph; X = S, 27%
R = Ph; X = NR, 8-12%
R = H, Me or Ph; X = O, 12-65%

(124)

can complicate the interpretation of the mechanism of the process since electron-transfer reactions could be involved [228].

*3.3.7.3.2.3 ε-Hydrogen abstraction.* Hydrogen abstractions in this classification are not as well known. Two examples of the process are shown (**125**).

hν
$C_6H_6$

$R^1$ = H, $R^2$ = Ph
$R^1$ = Ph, $R^2$ = H
$\phi = 0.045$
Ref [229a]

(125)

61%, ratio 1:1
$R^1$ = vinyl, $R^2$ = H
$R^1$ = H, $R^2$ = vinyl

Ref [228b]

(125 continued)

The first of these [229] illustrates the formation of a dihydropyran from the cyclization of the aryl ketone derivative. The second example in **125** [229] shows the formation of a tetrahydropyran. The reaction path involves excitation of the ketone followed by hydrogen abstraction from the ε-position. The attack at this site is controlled to some extent by the presence of the oxygen atom. Thus the ε-hydrogen abstraction affords a stable allyl radical additionally stabilized by the heteroatom that is to be preferred over the secondary radical which could have arisen by β-hydrogen abstraction.

*3.3.7.3.2.4 Hydrogen abstraction from distant sites* Considerable interest has been shown in reactions within this classification. The principal area in which they are seen is in the study of phthalimide derivatives which are known to undergo hydrogen abstraction and cyclization, as illustrated previously in **124**. Moderate- to large-ring aza systems can be prepared by the irradiation of long-chain substituted phthalimides as shown (**126**). The

n = 5-12 25-78% 0-10%

(126)

yields are variable, although it is possible to achieve high-yield cyclizations producing large rings, as illustrated in **127** [228].

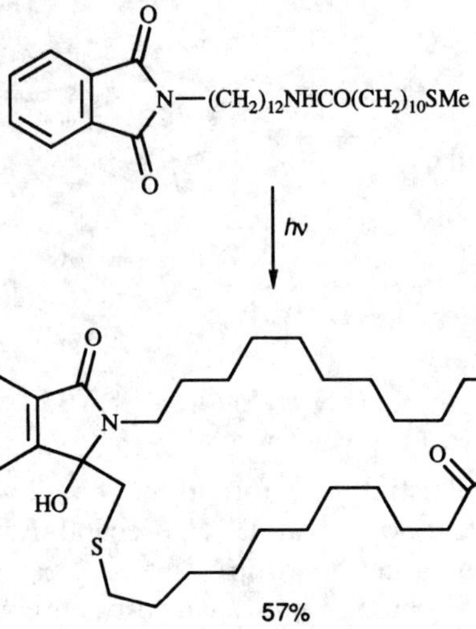

N—$(CH_2)_{12}NHCO(CH_2)_{10}SMe$

$h\nu$

57%

(127)

*3.3.7.3.2.5 Biomimetic hydrogen abstraction* Interest has been shown in hydrogen-abstraction processes which are designed to mimic those which occur in biological or biochemical environments. The research of Breslow and his coworkers [230] is particularly notable; these studies involve hydrogen abstraction from sites far removed from the carbonyl function undergoing the excitation. The attachment of a benzophenone unit to a long alkyl chain as in **128** is one method of approach. Irradiation affords a complex mixture of lactones corresponding to hydrogen abstraction from a variety of sites along the chain. A nine-carbon chain was found to be the minimum and sample results are recorded [231]. Others [232] have also observed that the rate constant for the hydrogen abstraction from remote sites is very small in esters similar to those in **128** when the chain length is less than nine carbons. The rigidity of the backbone of the molecule was also studied using benzophenone attached to a steroidal molecule. In this way the selectivity of the photochemical hydrogen abstraction from various points on the steroid by the photo-excited carbonyl oxygen was assessed. In the example shown

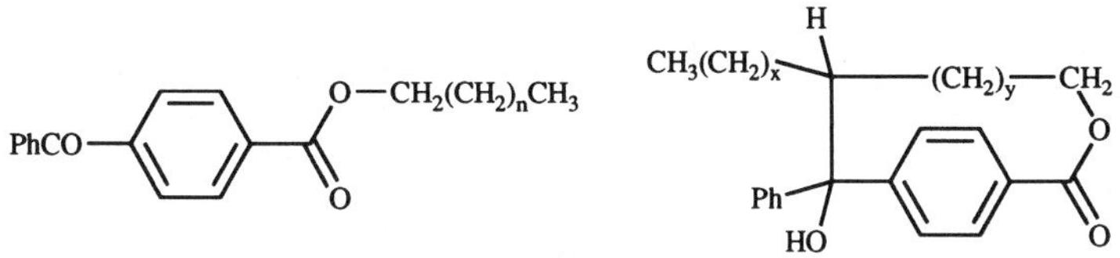

Remote oxidation of alkyl side-chains (% yields)

| Oxidation site | Number of carbon atoms in side-chain | | | |
|---|---|---|---|---|
| | n = 12 | n = 14 | n=16 | n = 18 |
| C-9 | 1.4 | 1.1 | 0.1 | 2 |
| C-10 | 3 | 7.8 | 8 | 5 |
| C-11 | 11 | 12 | 17 | 15 |
| C-12 | 49 | 13 | 21 | 20 |
| C-13 | 22 | 10 | 18 | 19 |
| C-14 | | 56 | 12 | 19 |
| C-15 | | 7 | 5 | 13 |
| C-16 | | | 13 | 8 |
| C-17 | | | 6 | 0.7 |

(128)

**(129)** the chain holding the two units together permits the carbonyl oxygen

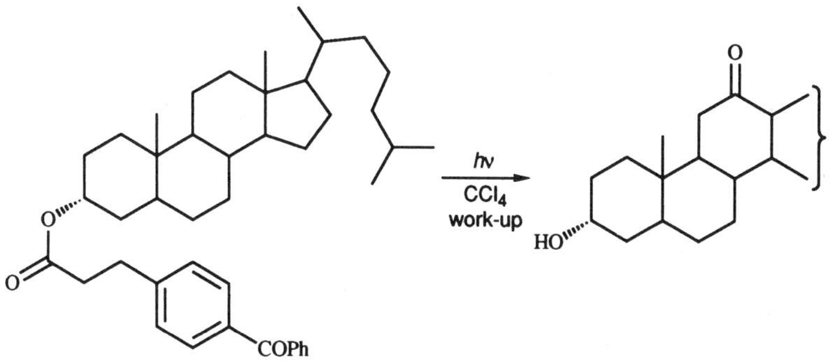

(129)

to abstract hydrogen from the C-12 carbon which, after further chemical reaction, yields the ketocholestanol in 16% yield [233]. The influence of longer chains was also studied **(130)**. The extended carbon chain gives the

(130)

photoexcited carbonyl group a longer reach and in this example the C-17 hydrogen comes under attack. However, the chain is still flexible, and attack closer to the point of attachment is also encountered, i.e. at C-14 [234].

*3.3.7.3.2.6 Formation of photoenols* The formation of photoenols arises by a Norrish Type II process in *ortho*-substituted aryl ketones and are produced by the path shown in **131**. The enols are formed from *o*-alkyl aryl carbonyl



(131)

compounds and involve a complicated series of excited states [235]. Detailed studies have identified the states and conformations involved. Thus the *syn*-triplet state shown in **131** is the only conformation in which H-abstraction can occur. The biradical so formed is also a triplet which can be quenched by oxygen, but in the absence of quencher the excited state decays to the *syn* and *anti* ground-state enols (**131**).

It should be obvious, because of the ease of formation of photoenols, that their trapping is of use in synthesis. An example [236] is shown in **132** where

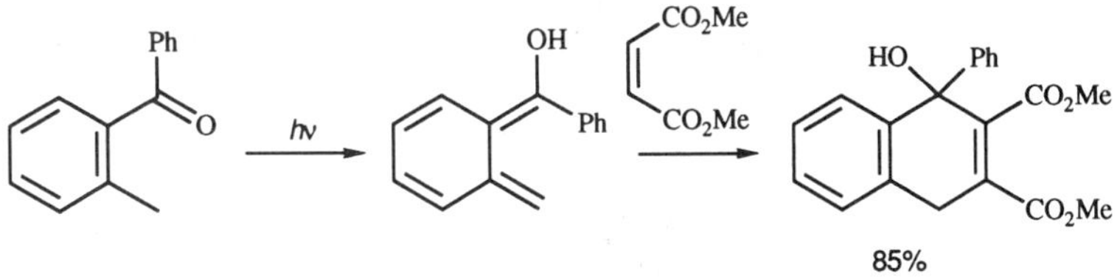

(132)

the photoenol is trapped by the addition of a dienophile to afford an adduct. Another example is illustrated in **133** [237]. Some elegant studies have

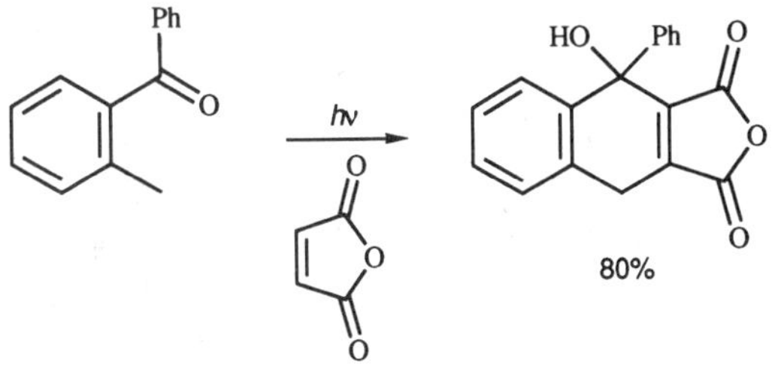

(133)

been made on the use of this reaction process and the work of Quinkert and his co-workers [238] is especially notable among the many experiments where trapping has been employed. This example (**134**) is an approach to the synthesis of oestrone. The initial irradiation affords the two isomeric photoenols shown. The major product is obtained from the least strained transition state, the *Z*/*exo*/-conformation. The possible transition state leading to the minor isomer is less easy to rationalize and it could equally be formed from the two transition states, *Z*/*endo*/- or *E*/*exo*/-, represented in **134**. Subsequent development of this approach has sought to synthesize enantiomerically pure oestrone.

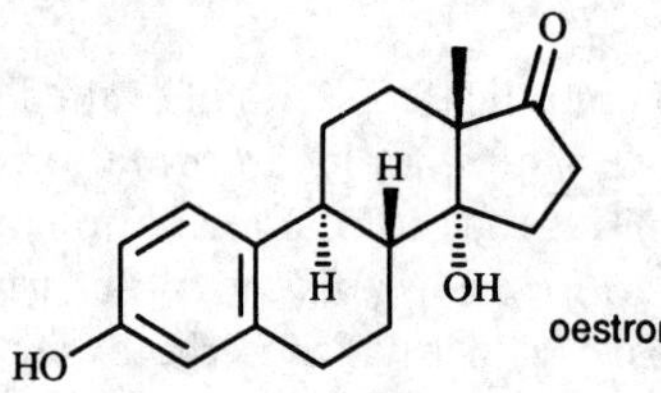

oestrone

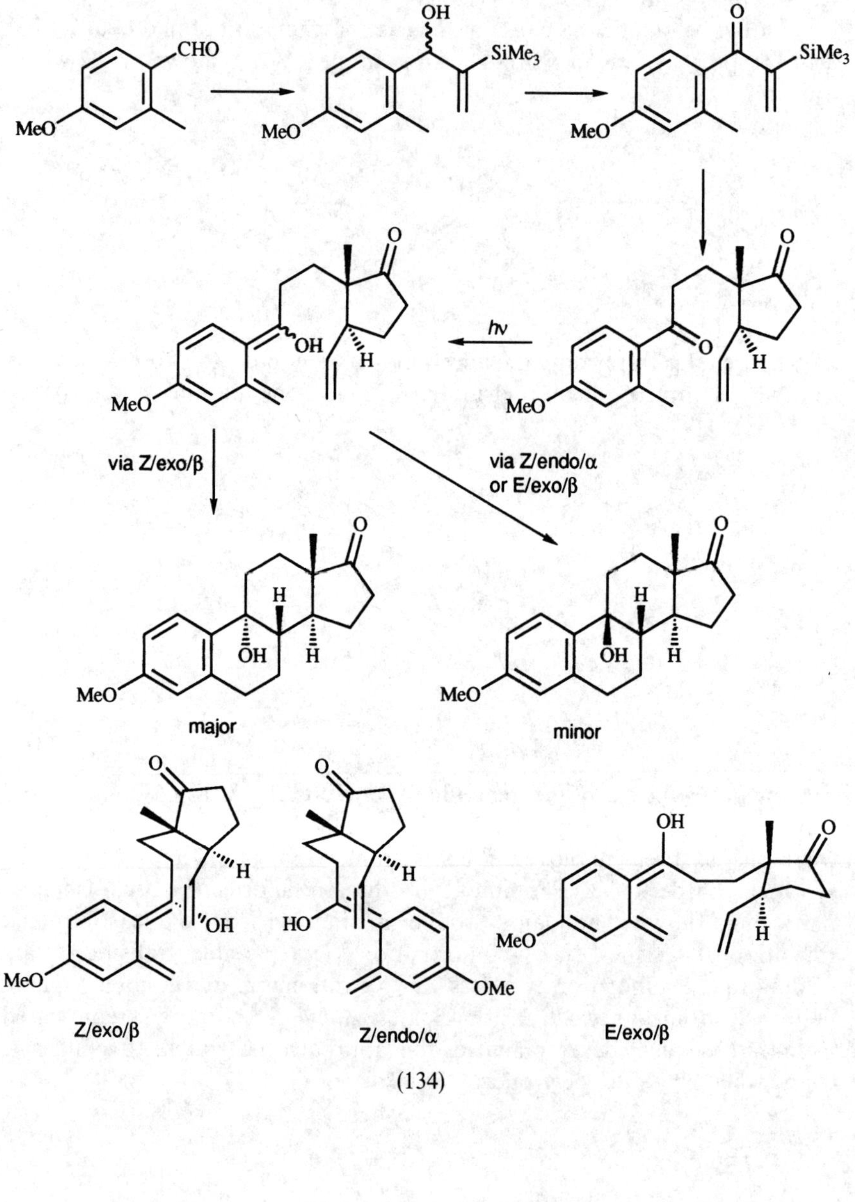

*hν*

via Z/exo/β

via Z/endo/α
or E/exo/β

major

minor

Z/exo/β

Z/endo/α

E/exo/β

(134)

## 3.3.8 Cycloaddition reactions of carbonyl compounds

### *3.3.8.1 Additions to electron-rich systems*

The photoaddition of a carbonyl compound such as an aldehyde or a ketone to an alkene results in the formation of a four-membered heterocycle known as an oxetane [239]. The excited state involved in such processes is dependent upon the initial carbonyl compound and alkyl ketones or aldehydes will react from either the triplet or the singlet state whereas an aryl ketone or aryl aldehyde will react only from the triplet state. The excited state of the alkene has seldom been implicated in such an addition [240]. The mechanism for the addition involves a 1,4-biradical intermediate, and spectroscopic evidence has been obtained in substantiation of such a species [241]. Other studies have identified the involvement of an exciplex between the carbonyl compound and the alkene. In these cases the alkene behaves as an electron donor and the excited-state carbonyl compound as the electron acceptor. Proof of these proposals comes from studies of alkene isomerization, deuterium isotope effects, and temperature effects on fluorescence quenching. The exciplex mechanism also gives a reasonable explanation for the low-to-moderate quantum yields for oxetane formation by the provision of an energy-wasting step in the collapse of the exciplex back to starting material [242].

For oxetane formation to be successful two important rules have to be satisfied: (1) only carbonyl compounds with a low-lying $n\pi^*$ state will react and (2) the energy of the carbonyl excited state must be lower than that of the alkene (or diene), thus eliminating the possibility of energy transfer from the carbonyl excited state to the alkene. The first example of such an addition was reported by Paterno and Chieffi (**135**) [243]. A later reinvestigation of

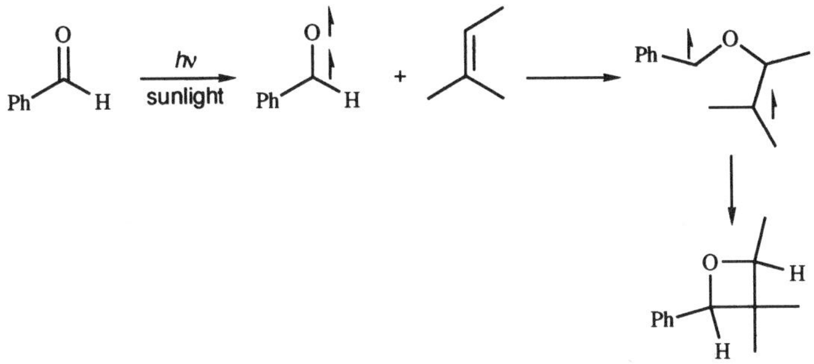

(135)

the addition reaction by Buchi in the 1950s has led to the reaction being named the Paterno–Buchi reaction and since then many examples of the reaction have been described. Buchi *et al.* [244] were the first to describe the process as involving the addition of the electrophilic oxygen atom of the excited carbonyl group to an electron-rich alkene affording the biradical shown in **135**. The specificity of the reaction can, therefore, be rationalized in terms of the more stable biradical, which leads in this case to the formation of the oxetane. Further substantiation of a biradical comes from the observation that photoaddition of acetone to *cis*-but-2-ene (**136**) affords two

O + hν O + O

Ratio 1:1.6

(136)

adducts. The loss of stereochemistry of the alkene component occurs within the 1,4-biradical, which exists long enough for bond rotations prior to bond closure [245].

### *3.3.8.2 Reactions with alkenes*

In general the reactions of ketones and aldehydes with alkenes is efficient. The addition of benzophenone to a variety of methyl-substituted alkenes such as 2-methylpropene yields oxetanes in the 58%–93% range [246]. Providing that substitution in the ketone does not change the ordering of the excited state and that the $n\pi^*$ is still the lower state, then the efficient addition of substituted benzophenones will also occur. Many examples of such additions have been reported over the years and an encyclopaedic coverage cannot be given here. Examples are included in **137** [246, 247] and

Ref [247a]

R* = (-)8-phenylmenthyl 99% Ref [247b]

Ratio 9:1
Total yield 73%
Ref [246]

MeCHO + MeCH=CHMe $\xrightarrow{h\nu}$

$\phi$ = 0.13-0.16 Ref [247c]

(137)

**138** where the addition of benzaldehyde to the bicyclic alkene affords an

$\xrightarrow[PhCHO]{h\nu}$ 40%

(138)

adduct (**138**) that has been used as a starting material for the synthesis of a prostaglandin analogue [248]. Other examples (**139**) have focused on

MeCO N O + Ph O O O R* $\xrightarrow{h\nu}$ Ph O O H H N Ac O OR* 53%

R* = (-) 8-phenylmenthyl

MeO HO Ph O N Ac O OR*

(139)

asymmetric induction in the addition reactions and are reported to be useful in the synthesis of carbohydrates [249].

Intramolecular addition is also of synthetic value, as in the formation of the two products from irradiation of the ketoalkene (**140**) [250]. An approach

$\xrightarrow[\text{pentane}]{h\nu}$ +

Total yield 56%, ratio 2:5

(140)

to the synthesis of bicyclic structures also uses the intramolecular cycloaddition in the ketoalkene (**141**), affording the oxetanes in a ratio of 1:0.9 [251, 252].

Ratio 1:0.9

(141)

Additions have also been reported in the norbornene compounds (**142**)

R = H or Me

57%

(142)

which afford the strained oxetanes shown [253]. The influence of the presence of aryl groups (**143**) has also been investigated [254].

$h\nu$

λ>300 nm

R =

yield 88% 58%

(143)

### *3.3.8.3 Reactions with dienes and trienes*

If the triplet energy of a diene is lower than that of a ketone or aldehyde, then energy transfer to the diene results in the formation of dimers as discussed in Chapter 2. Regardless of this caveat there have been reports of the addition of carbonyl compounds to such species providing that a high concentration of the diene is present. In general the addition of an aliphatic ketone or aldehyde to a diene involves the singlet state [255]. Thus, for example, acetone (**144**) adds to buta-1,3-diene to afford two oxetanes in low yield [255c] and

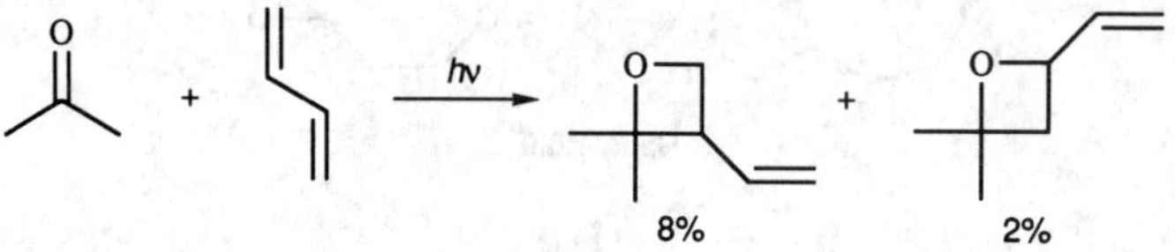

hν
8%
2%

(144)

cyclobutanone adds to the same diene at $-78\,^{\circ}\mathrm{C}$ to yield another oxetane in 20% yield [256]. Other dienes such as penta-1,3-diene (**145**) are also

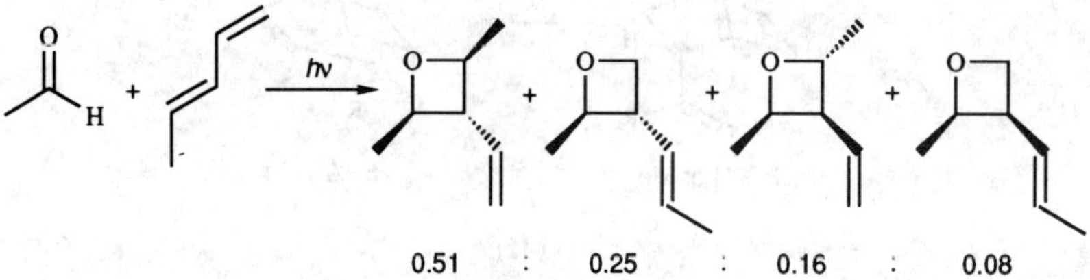

hν
0.51 : 0.25 : 0.16 : 0.08

(145)

reactive, affording a mixture of oxetanes with acetaldehyde as the carbonyl component [257]. Addition of acetaldehyde (**146**) to methylene cyclohexene

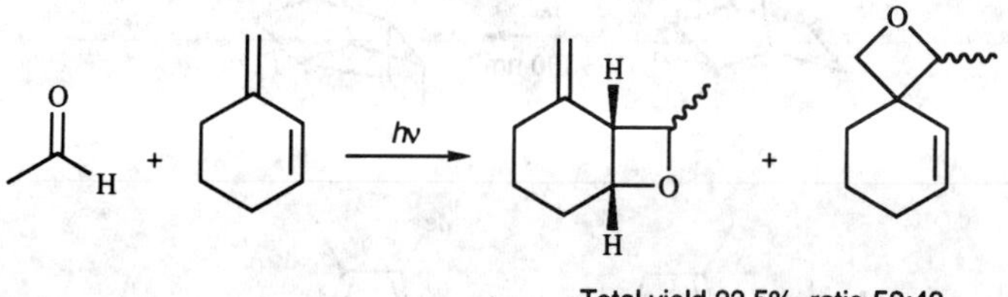

hν
Total yield 23.5%, ratio 52:48

(146)

also takes place, affording two oxetanes [258] while the addition of propanal (**147**) to cyclohexa-1,3-diene affords bicyclic oxetanes [259]. Aromatic ketones

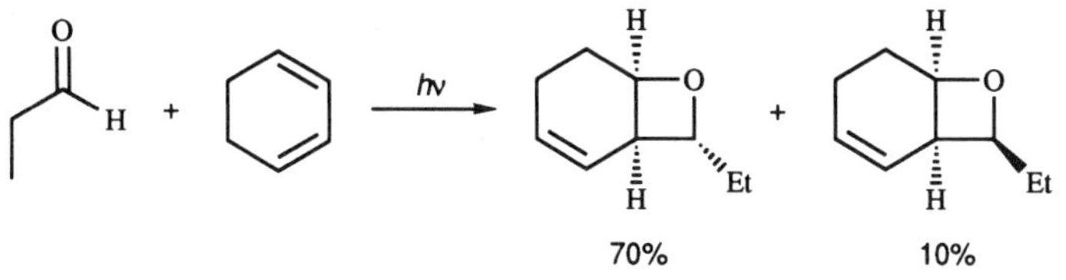

70% 10%

(147)

do add to dienes in a few cases. In these instances the addition is more efficient than the energy-transfer path [260]. Thus benzophenone adds to buta-1,3-diene, 2-methylbuta-1,3-diene, 2,5-dimethylhexa-2,4-diene, and to 2,3-dimethylbuta-1,3-diene [260]. Addition of aromatic aldehydes (**148**) to

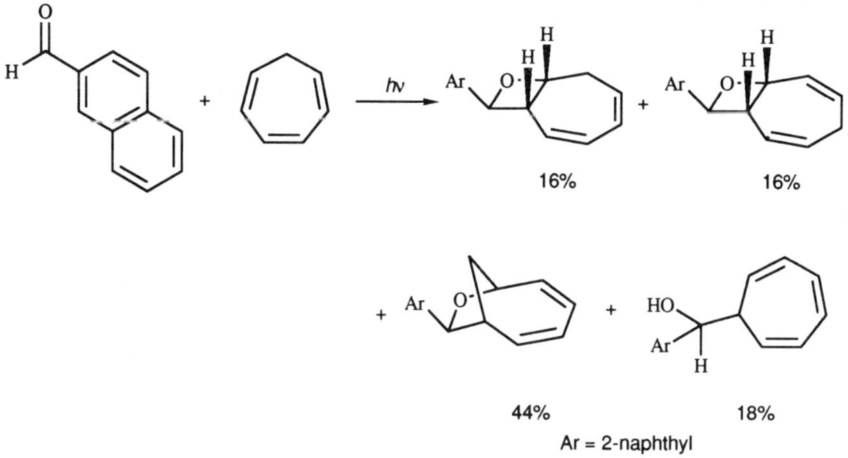

16% 16%

44% 18%

Ar = 2-naphthyl

(148)

cycloheptatriene has also been reported, and this affords a mixture of products arising from (2 + 2)-addition, (2 + 6)-addition or hydrogen abstraction [261].

Norbornadiene, a homo-conjugated diene, is a special case because of the triplet-sensitized formation of quadricyclane (see Chapter 2) which can be brought about by aliphatic and aromatic carbonyl compounds [262]. When benzophenone (**149**), for example, is irradiated in the presence of norbornadiene the products shown are obtained. These in fact arise from the attack of benzophenone on quadricyclane and not by attack on norbornadiene [263].

Benzophenone also adds to furan and substituted derivatives by the conventional triplet-state ketone addition to ground-state furan, affording 1,4-biradicals. The addition occurs with high stereoselectivity to afford the

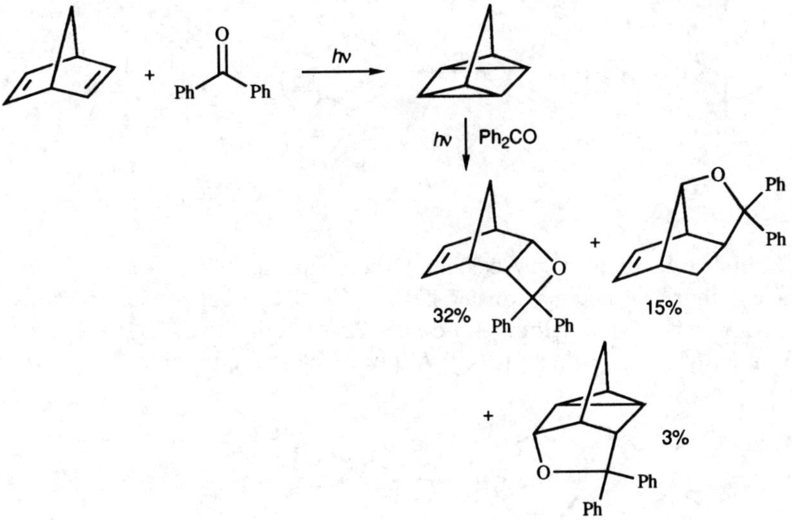

+ Ph(CO)Ph —$h\nu$→

$h\nu$ ↓ $Ph_2CO$

32% Ph Ph + O Ph Ph 15% + O Ph Ph 3%

(149)

1:1 adduct (**150**) [264]. A second addition can also take place, affording a

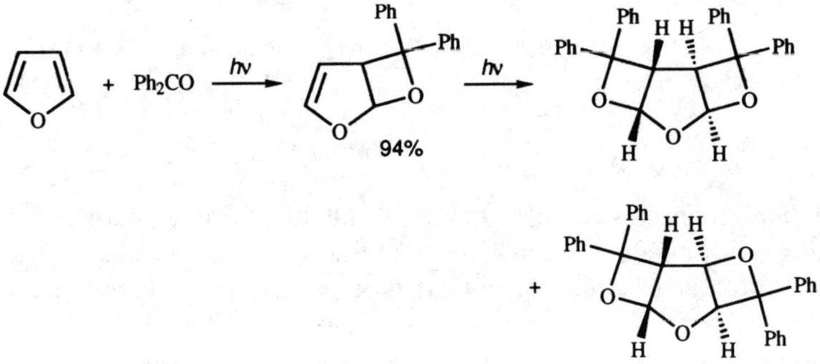

O + $Ph_2CO$ —$h\nu$→ Ph Ph O O 94% —$h\nu$→ Ph Ph H H Ph Ph O O O H H

+ Ph Ph H H O Ph O O H H Ph

(150)

bis-adduct [265]. Addition occurs between aldehydes and furan, affording analogous adducts [266]. Such reactions can provide useful starting materials for elaboration into naturally occurring compounds (**151**) [267].

Bu$_3$Sn + BuCO$_2$CHO $\xrightarrow{h\nu}$ H H CO$_2$Bu O Bu$_3$Sn O H 35%

Ref [267a]

H O OCH$_2$Ph + $\xrightarrow{h\nu}$ OCH$_2$Ph O O H 63%

Ref [267b]

O + O $\xrightarrow{h\nu}$ H O O H

Ref [267c]

O + H O $\xrightarrow{h\nu}$ H C$_8$H$_{17}$ O O H

quantitative Ref [267d]

(151)

### *3.3.8.4 Reaction with allenes and ketenimines*

The photoaddition of aryl ketones to allenes (**152**) to afford 1:1 adducts is

O Ph Ph + C $\xrightarrow{h\nu}$ O Ph Ph + O O Ph Ph Ph Ph + Ph Ph O O Ph Ph

37% 28% 15%

(152)

regiospecific, and the formation of the adduct in 37% yield arises by attack of the oxygen atom of the benzophenone triplet state at the central carbon of the allene. Further addition to this oxetane affords two 2:1 adducts in 28% and 15%, but the addition is obviously less specific [268]. Addition of benzophenone (**153**) to allene itself fails to yield a 1:1 adduct and only yields

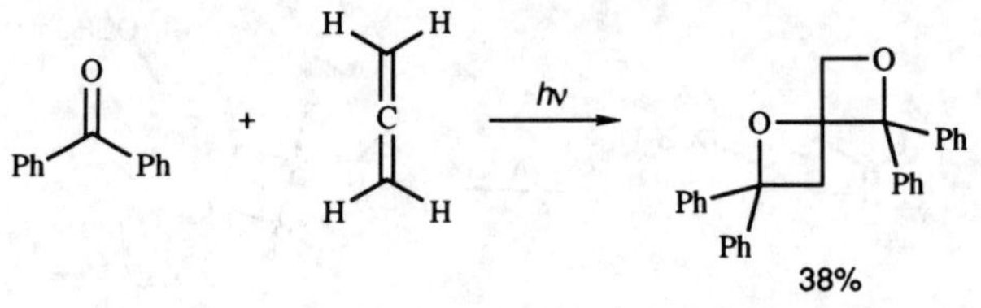

38%

(153)

the bis-adduct in 38% yield [269]. Benzophenone also adds photochemically to ketenimines, again via the triplet state and, dependent on the substitution pattern on the ketenimine, affords both α- and β-iminooxetanes. Thus addition to ketenimines (**154**) affords only the α-isomers, while addition to

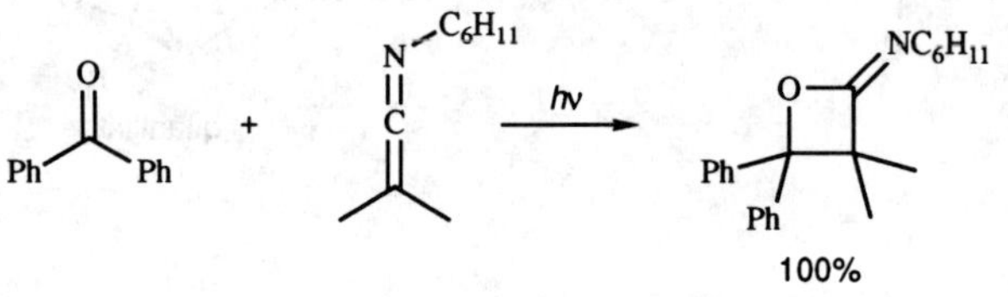

100%

(154)

ketenimine (**155**) gives the two adducts shown in a ratio of 2:3 with

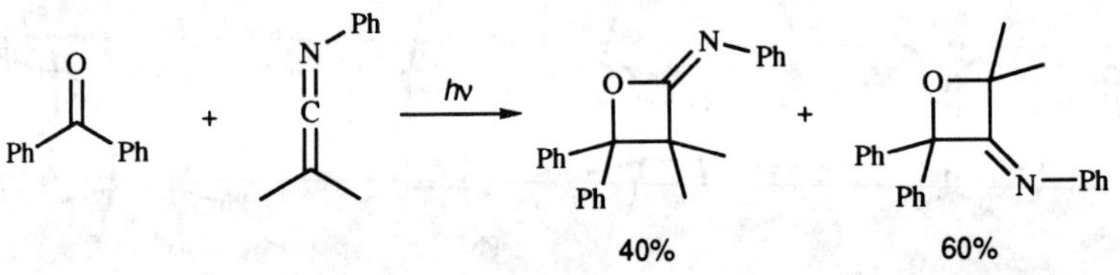

40% 60%

(155)

preferential formation of the β-iminooxetane [270]. Acetone, in its excited singlet state, is less regiospecific in its addition to ketenimines (**156**) and yields

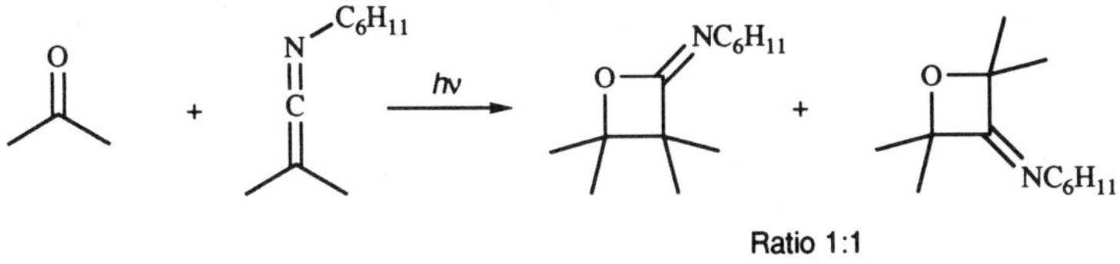

Ratio 1:1

(156)

both $\alpha$- and $\beta$-isomers. It is of interest to note that the $\alpha$-oxetane is thermally unstable and undergoes ring opening on chromatography [271].

### *3.3.8.5 Reaction with vinyl ethers*

Photoaddition of acetone to ethoxyethene (**157**) yields two adducts via the

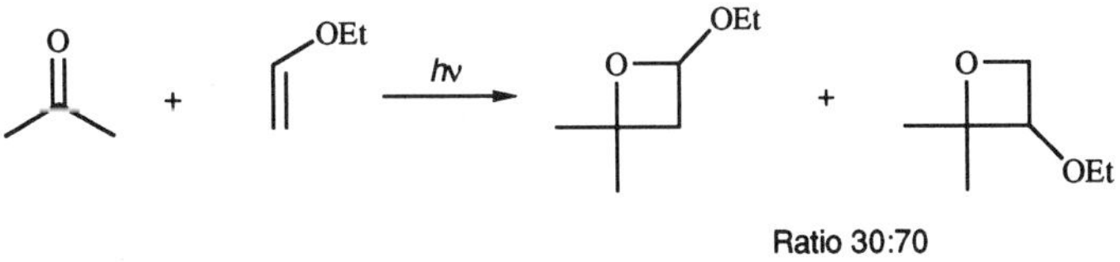

Ratio 30:70

(157)

two possible addition modes in a ratio of 3:7 [272]. Aldehydes are also reactive with the same alkene and yield analogous adducts but with higher selectivity (1:2) [272]. Detailed study of the photochemical addition of acetone to 1-methoxybut-1-ene (**158**) has shown that both the triplet and the singlet

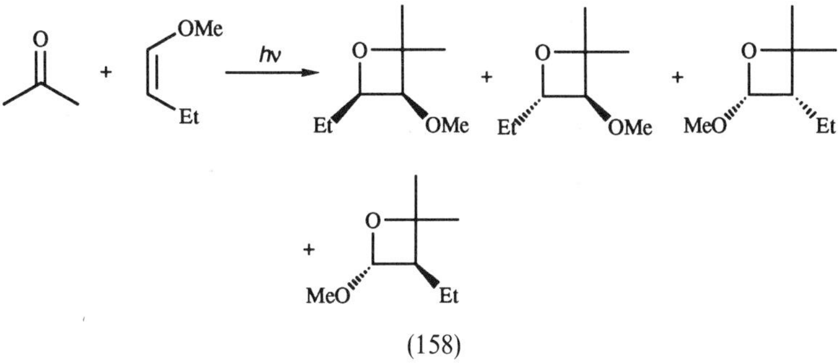

(158)

states are involved [273]. The initial stereochemistry of the alkene is partially retained when the singlet state of acetone undergoes addition, but there is no stereoselectivity in the addition of triplet acetone. Two types of oxetane are produced in this reaction but there is little selectivity with a ratio of 1.53:1 being obtained. Use has been made of silyl ethers in photoadditions for the synthesis of natural products. Thus the oxetane (**159**) can be obtained

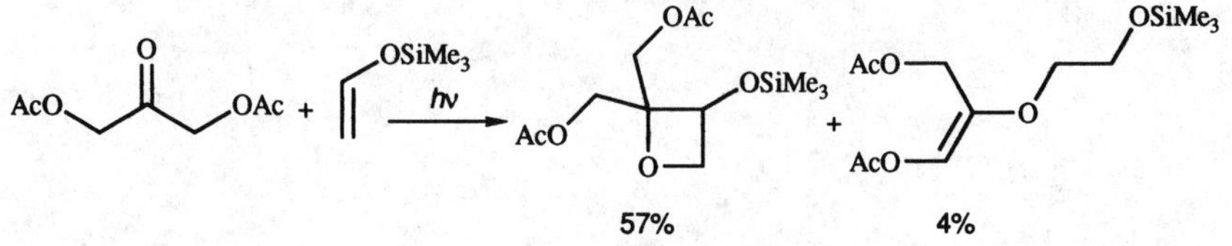

57% 4%

(159)

in high yield from the addition of the 1,3-diacetoxypropanone to the vinyl silyl ether [274]. The second product formed in this reaction is the result of hydrogen abstraction within the initially formed 1,4-biradical. Addition of aliphatic carbonyl compounds also occurs to cyclic ethers (**160**) with high

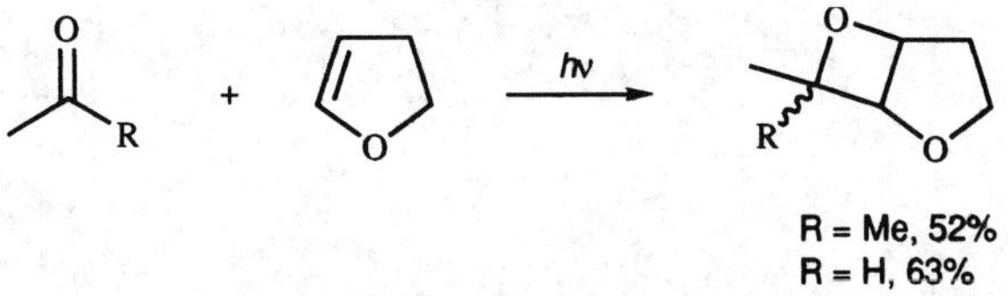

R = Me, 52%
R = H, 63%

(160)

efficiency and selectivity to yield the adducts shown [275]. Other reactions with cyclic ethers are also of synthetic value, such as the reactions of acetone (**161**) [276] and benzophenone (**162**) [277] yielding the oxetanes shown.

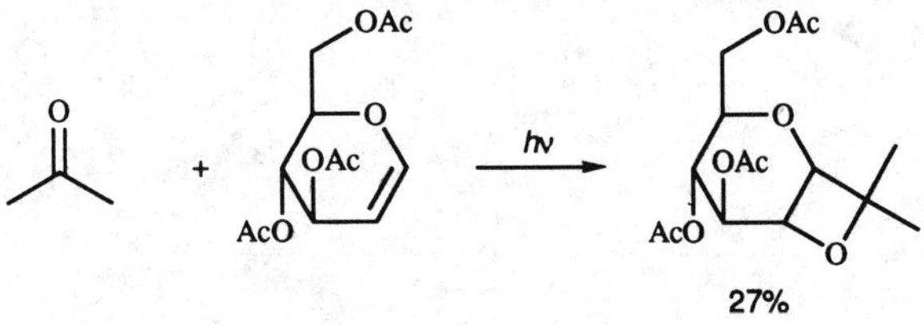

27%

(161)

82%

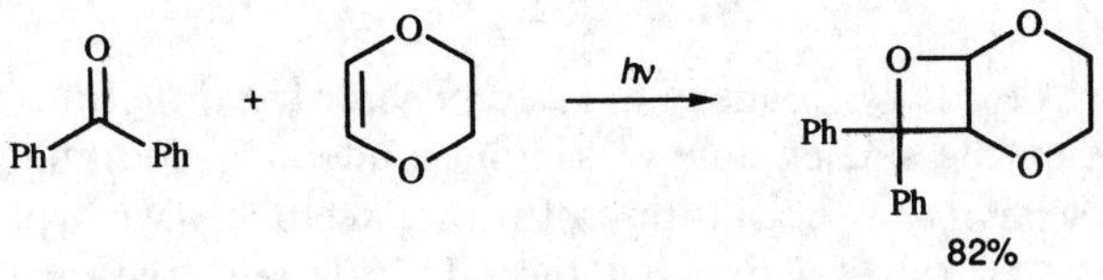

(162)

### *3.3.8.6 Addition to electron-deficient alkenes*

Only alkyl and cycloalkyl ketones are effective in additions to electron-deficient alkenes since there is clear evidence that the singlet state of the ketone is implicated. In the reactions of 1,2-dicyanoethylene (**163**) there is

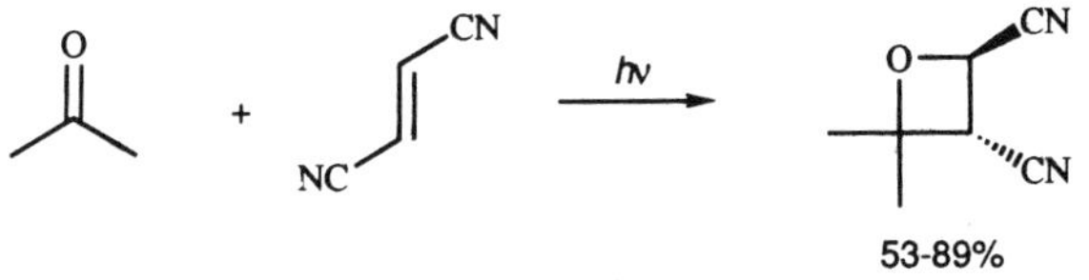

53-89%

(163)

evidence that the addition is very efficient chemically but does not involve a biradical. Indeed it is likely that an excited-state complex is involved which would account for the low quantum efficiency of the additions [278]. The products formed in this reaction mode are stereospecific as illustrated (**163**), retaining the original geometry of the alkene. With unsymmetrical alkenes (**164**) the addition is shown to be regiospecific and only one oxetane is formed

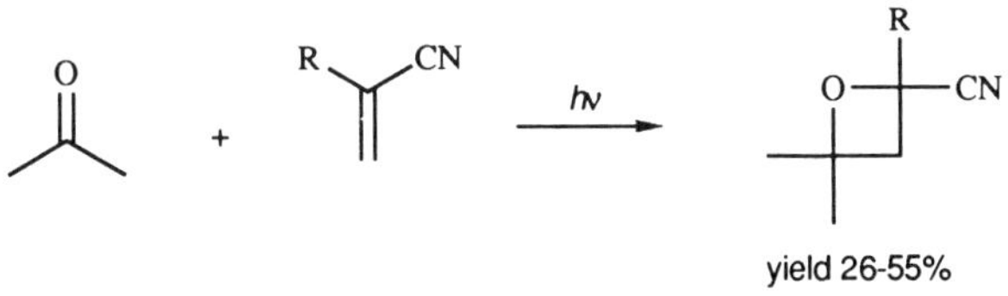

yield 26-55%

(164)

on reaction of acetone with acrylonitrile. This mode of addition can be rationalized by the application of a simple MO treatment [279]. Thus all of the additions are highly regioselective, as in the addition of acetone in its singlet state to fumaronitrile (**163**), which affords the single adduct. The triplet state of acetone is also populated during this reaction, and energy transfer to the alkene competes with the addition process [280]. Similar specificity is seen in the addition of norbornanone (**165**) to fumaronit-rile [281].

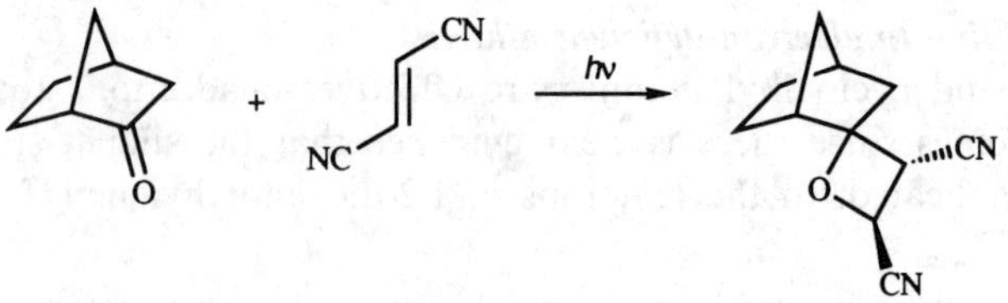

(165)

### 3.3.9 Enone and dienone rearrangements

Conjugated enones have intense absorptions in the 210–250 nm region for the $\pi\pi^*$ transitions overlapping the weaker $n\pi^*$ transitions in the 310–330 nm region [282]. Because of these intense absorptions the compounds have been popular candidates for photochemical study. There are many types of reaction which these compounds undergo, but this section will deal exclusively with rearrangement processes found in the cyclic enones and dienones. Several reviews have highlighted various aspects of the chemistry of these compounds such as the mechanistic interpretation [283] and cycloaddition reactions [284].

#### *3.3.9.1 Cyclic enone rearrangements*

Irradiation of cyclohexenones in *t*-butanol yields products of the bicyclo[3.1.0]hexane type, usually to the exclusion of competing processes. The route by which these are formed is dependent upon the type of substitution at the C-4 carbon and this is illustrated for 4,4-dimethylcyclohex-2-enone (**166**). When C-4 is unsubstituted or is singly substituted, rearrangement to

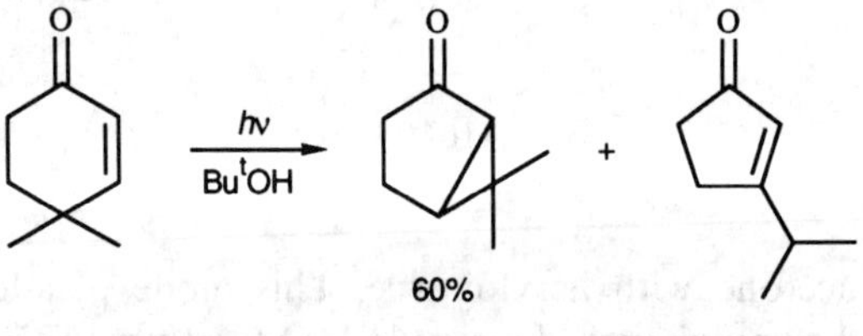

(166)

the bicyclic compound does not occur. However, when the C-4 site is dialkyl substituted, the sensitized irradiation affords a product by the migration of the C-4 ring carbon. This rearrangement involves a $\pi\pi^*$ excited state. The rearrangement product is accompanied by a cyclopentenone formed by ring-opening of the initially formed bicyclic ketone [285]. The reaction, affording what is called the lumiketone, is referred to as a Type A process and proceeds with a low quantum yield [284a]. The scope of the rearrangement of 4,4-dialkyl-substituted enones has been studied by Dauben and his coworkers [286]. More detailed study with the optically active compound **167**

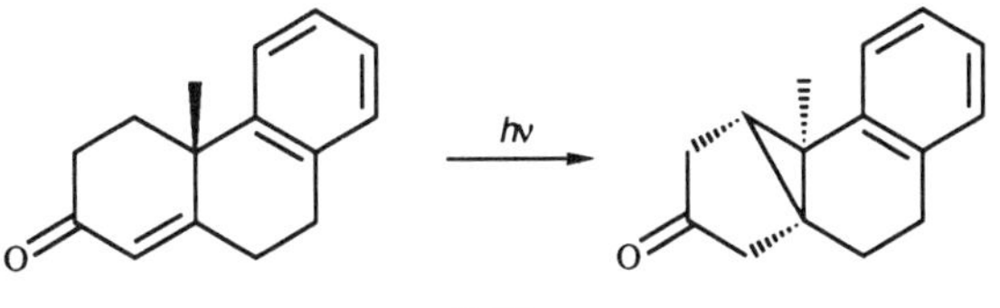

(167)

has shown that the stereochemical outcome of the reaction involves inversion of the optical centre at C-10, affording the product shown with 95% retention of optical purity [287]. Such a result could imply that the reaction was concerted, but arguments have been put forward, based on consideration of molecular models, that bond formation could only take place from the bottom face of the molecule because of the presence of the angular methyl group [288]. Thus a variety of paths, such as zwitterionic, biradical, or concerted, have been considered for the transformations. Zwitterionic intermediates are considered to be operative in irradiations carried out in acidic media when the enone **168** is converted into the corresponding acetate and the lumiketone

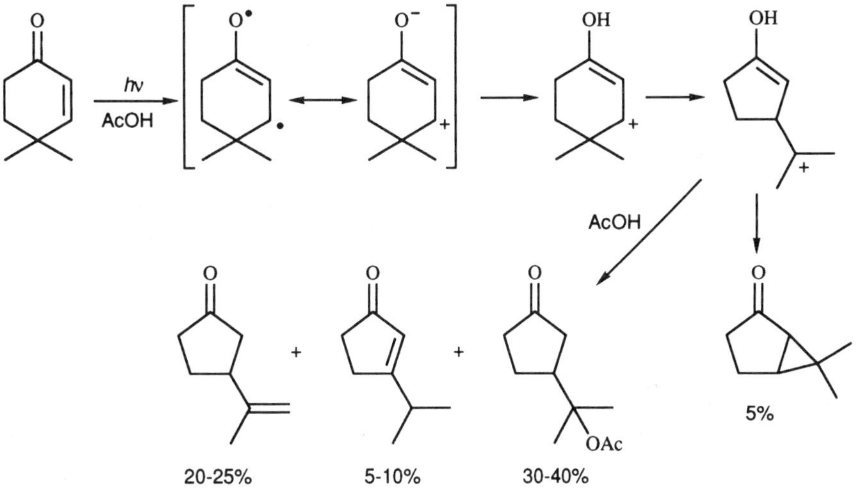

20-25% 5-10% 30-40% 5%

(168)

[285]. Steroidal enones are also reactive under such conditions and one example is shown in **169** [289].

$H_2SO_4$ HO + $h\nu$ O H 24% + O H 29%

(169)

Cyclohexenones with aryl substitution at C-4 (**170**) behave differently on

O• Ph Ph O• Ph Ph O Ph Ph $h\nu$ O H H Ph Ph + O H H Ph Ph + O Ph Ph

$\phi = 0.043$ $\phi = 0.0003$ $\phi = 0.0002$

(170)

sensitized irradiation in benzene. The excitation generates the $n\pi^*$ excited state and rearrangement yields the bicyclic product by migration of an aryl group prior to the formation of the three-membered ring [290]. The mechanism favoured for this rearrangement is as shown. A study of the influence of substituents on the aryl groups has shown that the excited state is not electron-deficient at the $\gamma$-carbon, again substantiating a biradical rather than a zwitterionic intermediate [291]. The enone **171** also undergoes

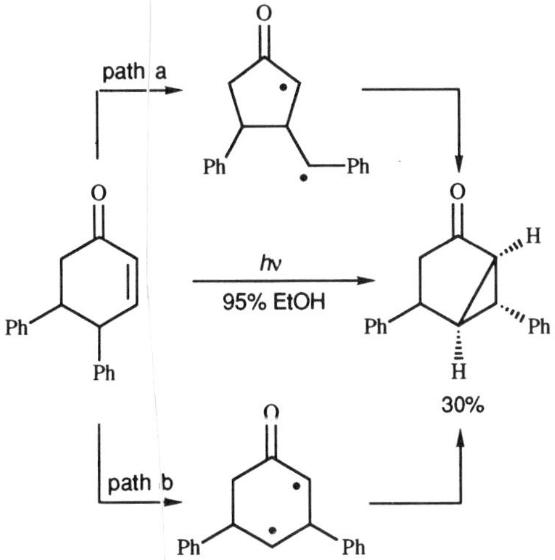

(171)

photorearrangement, affording the Type A product (98.6%) with the aryl migration path (path b) accounting for only 1.4%. The actual details of this were obtained by the use of $^{14}C$ labelling. The preference for the Type A path in this case could be due to the absence of aryl stabilization of the intermediate biradical. This is apparently an important feature in the aryl migration reaction, as is relief of strain and the need for at least one aryl group to be in an axial conformation to permit migration [292]. Some other examples of the rearrangements of enones of this type are illustrated (**172**) [293].

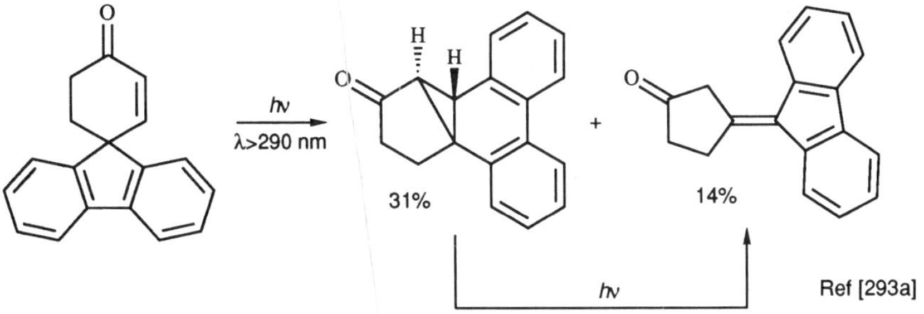

(172)

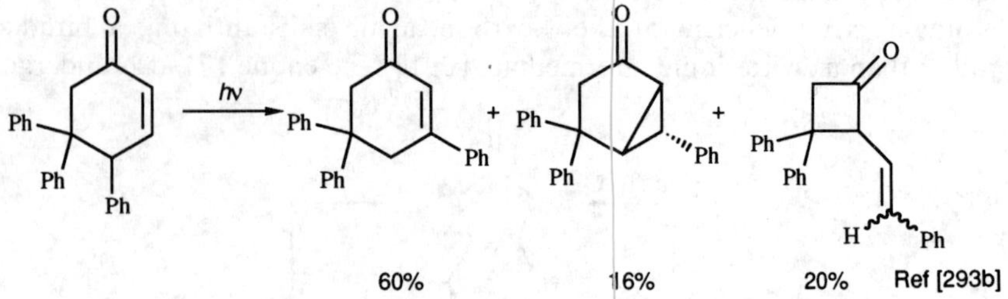

60% 16% 20% Ref [293b]

(172 continued)

In larger ring enones (**173**), the photoreactions are dominated by *cis*–

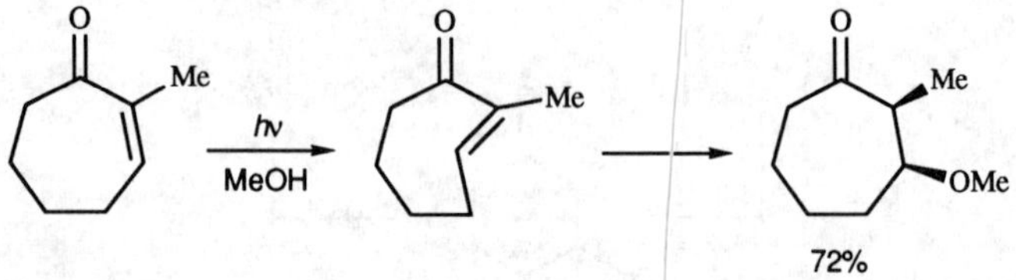

72%

(173)

*trans*-isomerization of the double bond. If the reactions are carried out in protic media then *syn*-addition of the solvent affords the addition product [294].

### 3.3.9.2 *Hydrogen abstraction reactions*

Enones are also capable of undergoing hydrogen abstraction reactions such as those shown in **174**. Here *trans*–*cis*-isomerization, by irradiation at 300 nm,

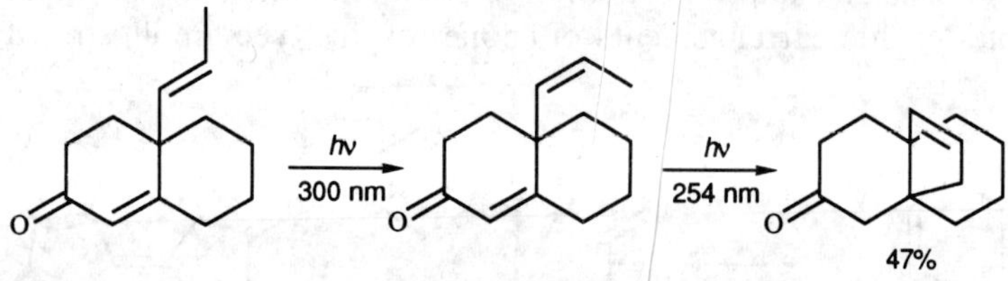

47%

(174)

is followed by hydrogen abstraction by the α-carbon of the enone (irradiation at 254 nm) followed by cyclization of the biradical [295]. Hydrogen abstraction by the β-carbon of an enone can also take place as in the intramolecular cyclization of enone (**175**). This process affords a biradical which subsequently

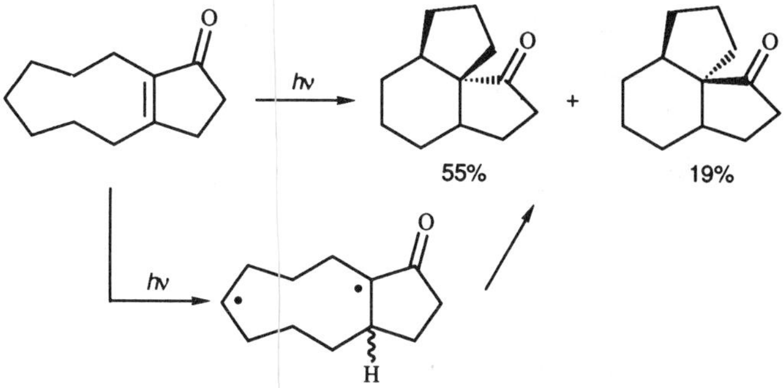

55% 19%

(175)

cyclizes to afford the isomeric tricyclic ketones [296].

One of the areas which still attracts considerable attention and which falls within the hydrogen abstraction path of enones is the deconjugation process. This reaction of $\alpha,\beta$-unsaturated enones yielding the isomeric $\beta,\gamma$-unsaturated systems has been developed over the years. The deconjugation reaction, fundamentally similar to the Norrish Type II reaction, is fairly common for acyclic enones with an alkyl group on the $\gamma$-carbon and involves a 1,5-hydrogen transfer which, in the examples illustrated, affords a dienol (**176**)

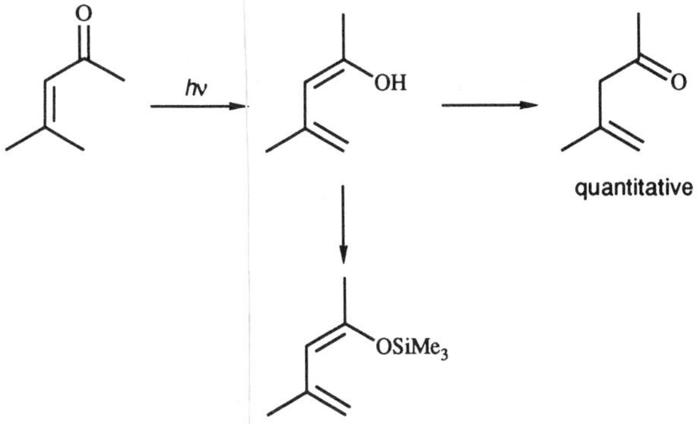

OH quantitative

$OSiMe_3$

(176)

[297]. The existence of this dienol has been proven in some cases by trapping as the silyl derivative. In the absence of a trap, re-ketonization can afford either the starting material or the $\beta,\gamma$-isomer. Studies have shown that, when the irradiations are carried out in the presence of weak bases, the yield of the deconjugated product is greatly enhanced [298]. In more recent work attempts have been made to obtain enantioselective reactions with some

success, as shown in **177** where irradiation in the presence of, for example,

(177)

(1*R*, 1*S*)-1-phenyl-2-*i*-propylaminopropanol affords the deconjugated product with an enantiomeric excess of 70% [299].

### *3.3.9.3 (2 + 2)-Cycloaddition reactions*

Perhaps the most useful reaction of α,β-unsaturated enones is the (2 + 2)-photocycloaddition reaction with alkenes which affords cyclobutane derivatives. This reaction has been utilized with great success in the construction of key intermediates in the syntheses of naturally occurring compounds. A simplified mechanistic interpretation [184a, 300, 301] which accounts for the reactions which enones undergo in the presence of alkenes is shown in **178**.

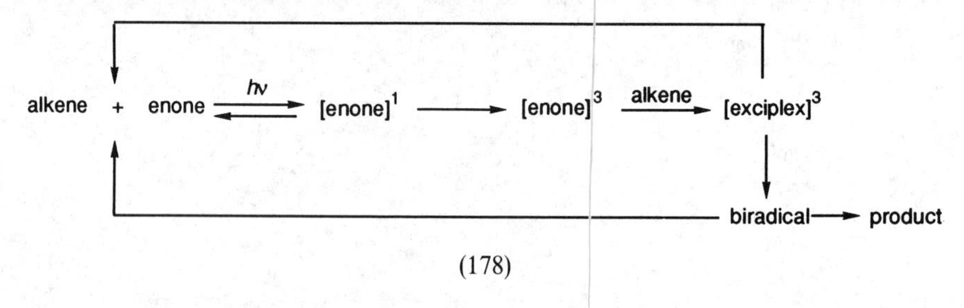


(178)

In this scheme excitation of the enone affords the singlet excited state which, in open-chain enones or large ring enones, can give rise to *cis–trans*-isomerization or decay back to the ground state. In systems where there is less flexibility, intersystem crossing to the triplet state competes efficiently [302]. In the presence of an alkene an exciplex is formed whose fate can be either decay to the ground state or, more frequently, formation of a biradical. The biradical can be formed by bonding at the α or the β carbon of the enone to the alkene, although which site is preferred has not been full resolved. The results from the addition of 2-methylpropene to cyclohexenone (**179**) is typical of this, with bonding at the α and the β carbon yielding cyclobutanes by bond formation or products of hydrogen abstraction [303]. Another

O H H 6.5% O H H 26%

α-bond

O + $h\nu$ O + 14%

β-bond

O H H 6% O 8%

(179)

feature of the mechanism of the process relates to regiochemistry. With an unsymmetrical alkene there are two possible orientations for the addition; thus the addition of ethoxyethene to 3-methylcyclohex-2-enone affords the *head to tail* adduct shown in **180a** whereas the addition of acrylonitrile to the same enone yields a product by *head to head* addition (**180b**) [154]. The

δ– δ+ OEt O H Me OEt 91%

δ– O δ+ $h\nu$ δ+ O δ– (180a)

(180b) δ– CN δ+ O H CN Me 46%

(180)

control exercised appears to be electronic and the outcome of the reaction can be rationalized in terms of the orientation of the dipole of the excited enone with the dipole of the ground-state alkene within the exciplex. Other subtle effects, such as change in solvent, can also be used to effect in the control of the regiochemistry of cycloaddition [304]. Recent results have questioned the authenticity of the mechanism presented in **178**. This study [300b], using laser-flash techniques, has cast doubt on the involvement of an exciplex in some enone–alkene systems. Recent results have questioned the authenticity of the mechanism presented in **178**. This study [300b], using laser-flash techniques, has cast doubt on the involvement of an exciplex in enone–alkene cycloaddition reactions. Other studies have also addressed this problem and suggested that, while the route to products shown by (**178**) is useful for predicting the outcome of the cycloaddition, the stability of the biradicals involved in the cycloaddition is more important in determining the eventual outcome of the process [300c, d].

The stereochemistry of the products formed in the cycloaddition is also important, and some general statements can be made based on the large number of examples reported in the literature [284]. Clearly the involvement of a biradical intermediate can lead to two products with either a *cis* or a *trans* junction between the rings. However, there are constraints on this, dependent on the enone and the alkene components used. Thus the ring junction between the cyclobutane and the cycloalkenone ring is always *cis* with a cyclopentenone or when alkynes, allenes, and cyclobutenes undergo addition to a cyclohexenone. With more flexible alkenes, additions to cyclohexenones tend to give a mixture of *cis* and *trans* ring fusion, but usually there is a preponderance of the *cis*-adduct. Furthermore, with additions of cyclic alkenes to cyclic enones the rings are usually *anti* to each other. Some examples which illustrate these points are illustrated (**181**) [305].

hν
pentane
Ratio 72:28
Ref [304a]

$CO_2Me$
hν
dichloromethane
(181)
Ratio 29:71
Ref [304 b]

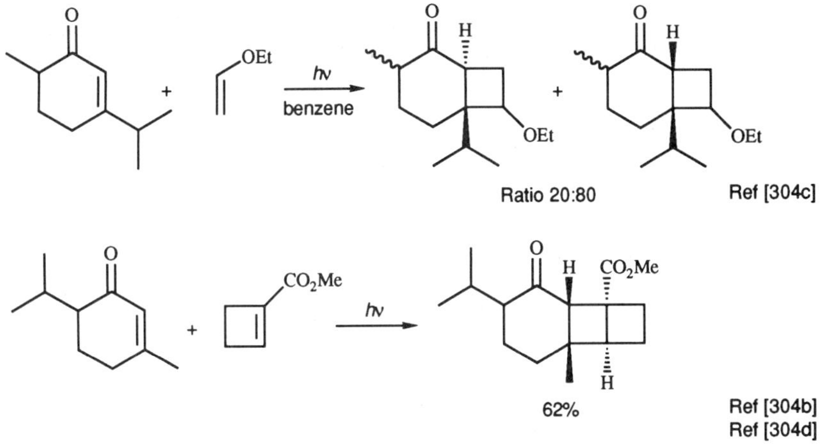

Ratio 20:80 Ref [304c]

62% Ref [304b]
Ref [304d]

(181 continued)

*3.3.9.3.1 Intermolecular addition*

Cycloadditions provide clearly a facile route to natural product molecules. Some examples from the early literature are the synthesis of $\alpha$ and $\beta$-*bourbonenes* (**182**) from the photoaddition of cyclopentenone to 1-methyl-

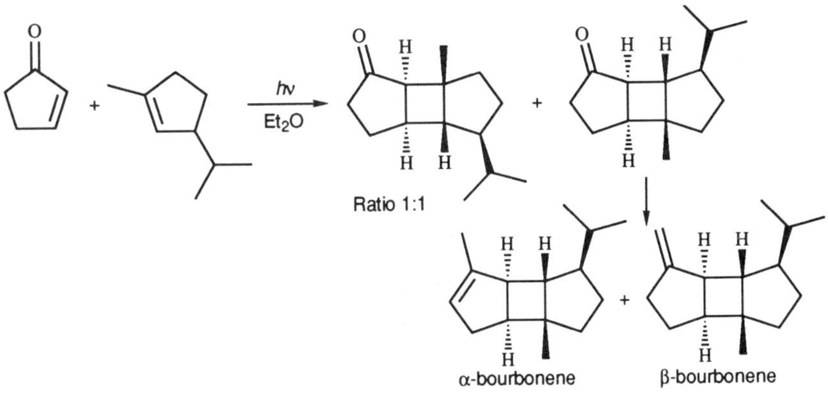

Ratio 1:1

α-bourbonene β-bourbonene

(182)

3-*i*-propylcyclopentene [306], the synthesis of *Grandisol* from the addition of ethylene to a cyclic lactone (**183**) [307], or the elegant synthesis of

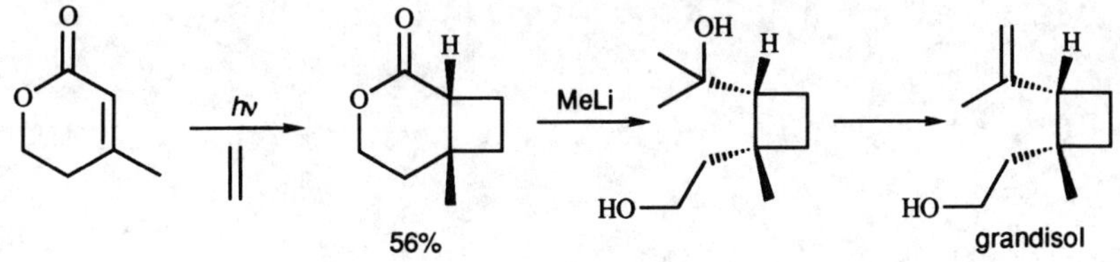

hν
MeLi
56%
grandisol

(183)

α-caryophyllene alcohol from the addition of 4,4-dimethylcyclopentene to 3-methylcyclohexenone (**184**) [308]. Other examples illustrating the power of

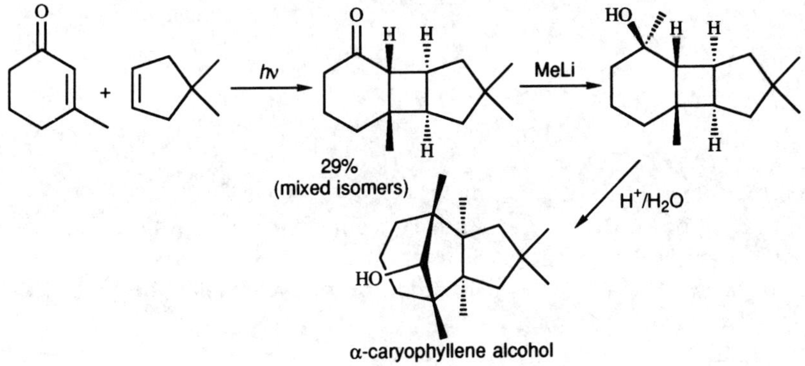

hν
MeLi
29%
(mixed isomers)
$H^+/H_2O$
α-caryophyllene alcohol

(184)

the method are shown in **185** [309]. Other substrates, such as allenes and

$MeO_2C$ H $PhSO_2$ O + $h\nu$ $MeO_2C$ H $PhSO_2$ O H 71% + $MeO_2C$ H $PhSO_2$ O H 23%

sterpuric acid

+ OTMS OTMS $h\nu$ 360 nm OTMS H OTMS O 24%

OH H OH balanitol

several steps

Δ

Ref [309c]

O O

daucene

Ref [309b]

(185)

alkynes, also undergo addition (**186**) [310].

hν
allene
several
steps
75%
Ref [310a]

hν
Ratio 6:1

hν
n-Bu−C≡CH
Buⁿ
Ratio 69:31 (35% yield)
Ref [310b]

AcO
$R^1$ = H or MeO
hν/polarized light
$R^2$−C≡CH
$R^2$ = H or Buⁿ
Ref [310c]

(186)

Among the many enones used as substrates for the photochemical (2 + 2)-addition, the reaction of enolizable 1,3-diketones and 1,2-diketones has provided a reaction of synthetic utility. The process was originally described by de Mayo [311] and his coworkers although in the initial examples the cyclobutane was not isolated. The adduct obtained from this simple reaction undergoes a retro-Aldol process to yield 1,5-diketones (**187**).

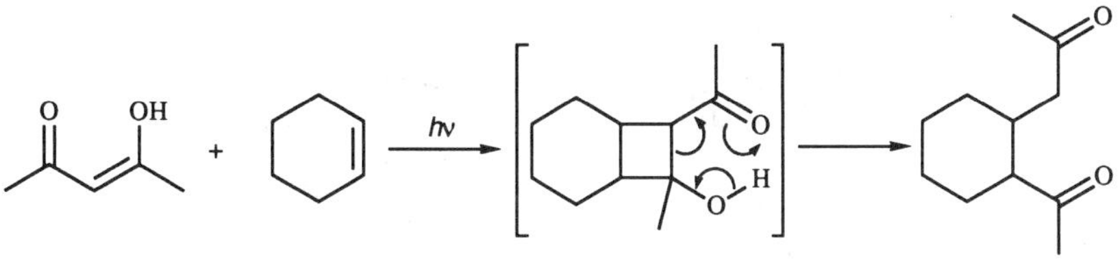

(187)

Another reported addition (**188**) affords a quantitative yield of the 1,5-

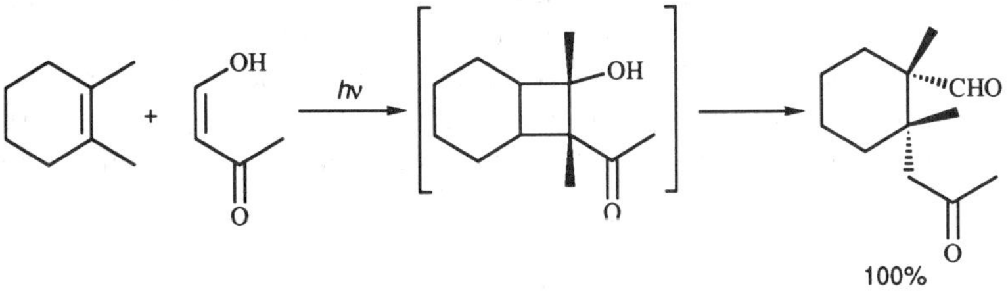

100%

(188)

dicarbonyl compound which was subsequently converted to valerane and isovalerane [312]. More recently a short route to hirsutene has been described using cycloaddition of 5,5-dimethylcyclohexa-1,3-dione (**189**) [313].

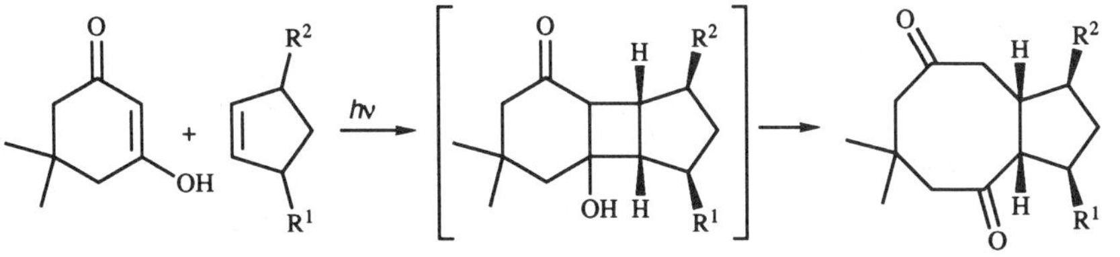

$R^1 = R^2 = H$, >90%
$R^1 = H$, $R^2 = OH$, 38%
$R^1 = OH$, $R^2 = H$, 49%

(189)

*3.3.9.3.2 Intramolecular addition*

Intramolecular cycloadditions also provide a valuable route in synthesis. The first intramolecular enone addition was the carvone to camphor conversion (**190**) [314]. The regiochemistry observed in such intramolecular photoadditions is very predictable and, as in the case shown in **190**, when the bonds

hν

9%

(190)

are 1,5 to each other, i.e. part of a hexa-1,5-diene, then the initial cyclization affords a five-membered ring by what is referred to as the *Rule of Five*. This rule is also obeyed in the majority of cases when the diene is part of a hexa-1,6-diene. In this latter case there are examples where a six-membered ring is formed. Examples of these cycloadditions are shown in **191**, where it

hν

35-42% Ref [315a]

hν

87% Ref [315b]

hν

70% + 15% Ref [315c]

(191)

hν
λ>350 nm
87%
Ref [315d]

hν
hexane
Pyrex filter
70%
Ref [315e]

(191 continued)

can be seen that complex molecules can be obtained in high yield [315].

### 3.3.10 Cross-conjugated dienone rearrangements

The photorearrangement of the cross-conjugated dienone α-santonin (**192**),

(192)

was first described in 1830 [316a] although the identification of the product(s) was not made for many years [316b]. Since then, many examples of the rearrangement of cross-conjugated dienones have been described and the mechanism has been studied in detail [317]. The rearrangement of the simpler dienones (**193**) arises from the triplet $n\pi^*$ excited state, and in aqueous solution

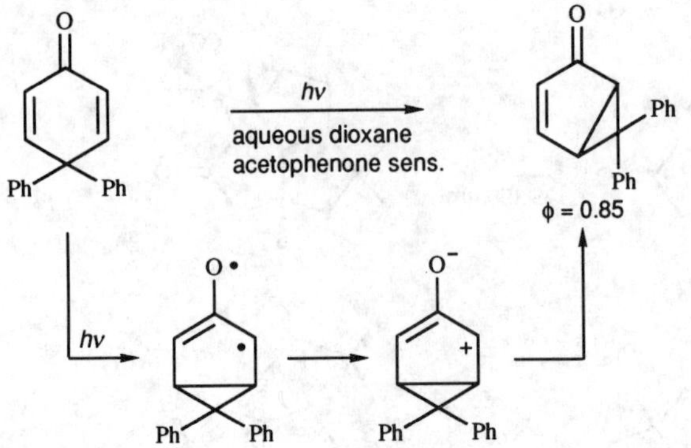

(193)

dioxane affords the bicyclic product, referred to as the *lumiketone*, with high efficiency [318]. The mechanism favoured for this reaction involves zwitterionic intermediates, as illustrated in **193**, although there has been some debate over the matter. It would appear that both biradical and zwitterionic mechanisms can be operative. Thus the gas-phase irradiation of the dienone (**194**) affords the bicyclic ketone as the primary photoproduct. However, this

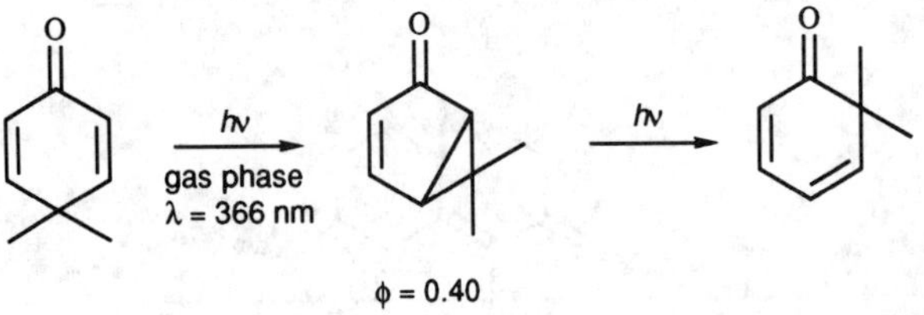

(194)

product is subject to secondary photochemical conversion into the final product shown [319]. Both of these products are obtained on irradiation in aprotic solvent, but in a protic medium the enone (**195**) and two phenols are obtained, suggesting the operation of a zwitterionic process. Both biradical and zwitterionic mechanism are operative in the photoreactions of the dienone

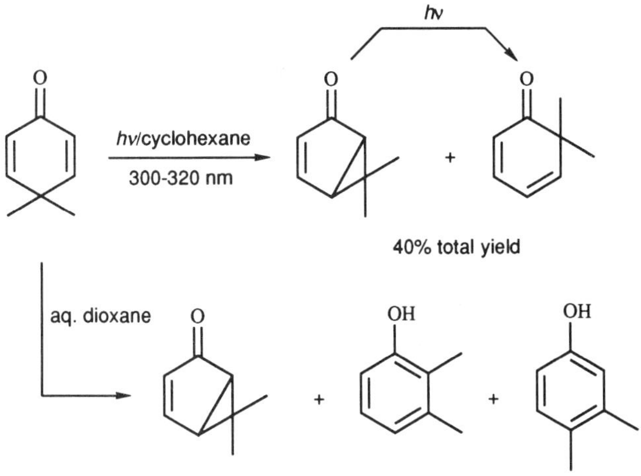

(195)

(**196**) [320]. The isolation of the oxetane from the irradiation of this dienone

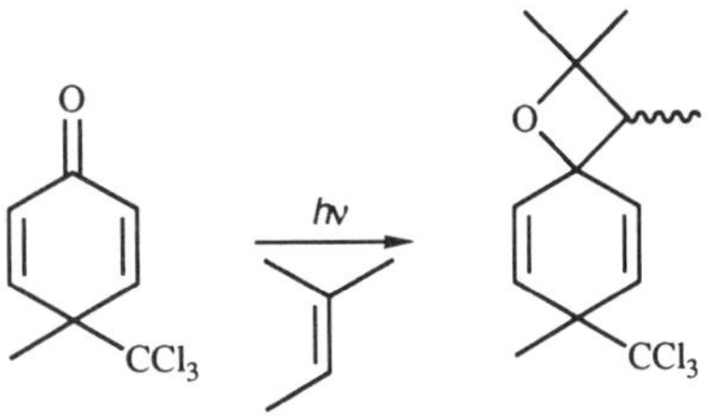

(196)

(**196**) in 2-methylbut-2-ene demonstrates that an $n\pi^*$ excited state is involved in the reaction. A biradical mechanism is, however, involved in the rearrangement of the dienones (**197**) [321] and (**198**) [322] where opening of the

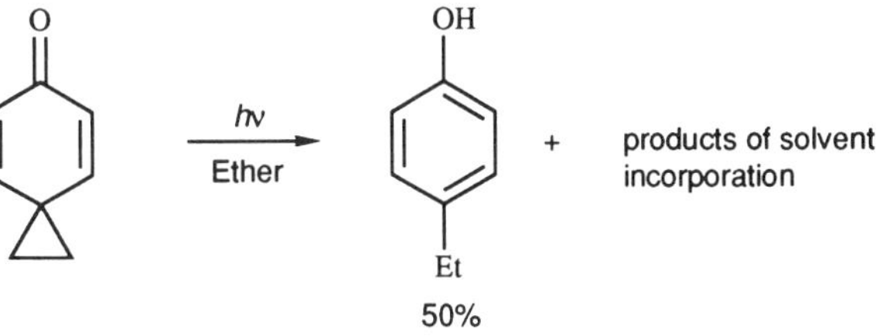

(197)

hν
254 nm
+
H Pr$^i$
Et
Me migration
H migration

(198)

cyclopropane ring dominates the reactions.

Notwithstanding the involvement of a biradical reaction path in some systems, by far the commoner reaction is that involving zwitterions. Support for zwitterionic intermediates comes from trapping experiments such as irradiation of the dienone (**199**) [323] in methanol when the bicyclic adduct

hν
HCl/MeOH

(199)

is obtained or on irradiation of the dienones (**200**) in methanol, which affords

hν
MeOH
R = alkyl
R = Me, 67%
R = Et, 74%
R = Bu$^t$, 37%

(200)

the cyclopentenone by ring opening and trapping by solvent [324]. As far as the synthetic potential of the rearrangement is concerned, the principal feature is in the control which can be exercised on the rearrangement path. α-Santonin, for example, rearranges in aprotic media into *lumi*santonin, which on prolonged irradiation is converted to the linearly conjugated dienone. This compound is again prone to secondary photolysis and in water ring–opens to a ketene which is trapped as the acid (**201**). An alternative

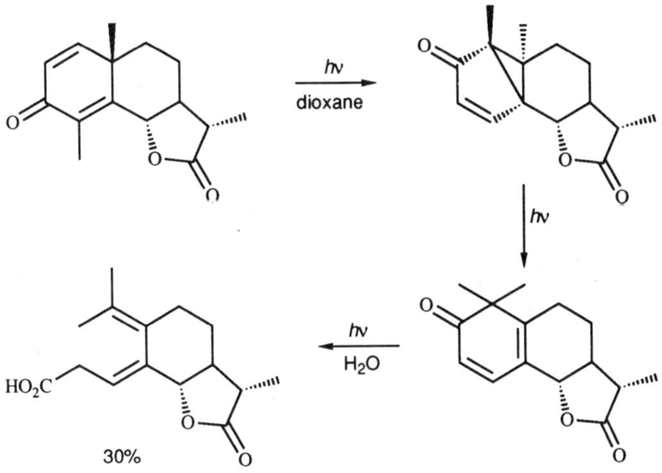

(201)

reaction path is followed if α-santonin is irradiated in aqueous acid, when the hydroazulene skeleton is formed (**202**). This latter reaction mode can be

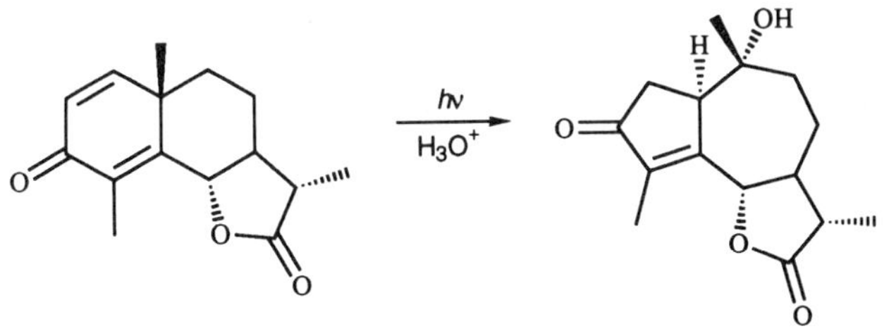

(202)

readily understood in terms of protonation of the zwitterion followed by rearrangement within the resultant carbocation. Thus when cyclohexadienones, unsubstituted at C-4, are irradiated in an acid medium, both hydroazulene and spiroketone products are obtained. The route to both of these and the control which substitution can exercise on the outcome of the reaction is shown in **203**. In this scheme the stability of the cation is important

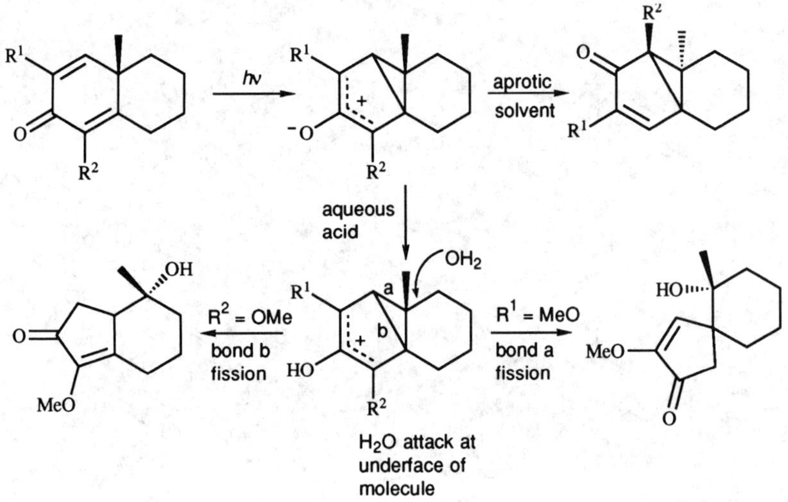

(203)

and when a group at C-4 capable of stabilizing is present then a hydroazulene skeleton is formed. Conversely when a group at C-2 stabilizes the cation then a spiro skeleton is produced. Examples of the use to which this control can be put are shown in **204** [325].

$h\nu$, $CH_3COOH/H_2O$ — 50% — Ref [325a]

$h\nu$, $CH_3COOH$ — 91% — Ref [325b]

$h\nu$, $CH_3COOH$ — 89% — Ref [325c]

(204)

The *lumi*-ketones are also photochemically reactive and can be converted into a linearly conjugated cyclohexadienone as shown in the irradiation of α-santonin (**201**). The reaction path involves the generation of a zwitterion within which methyl group migration yields the observed product. The zwitterion generated on the irradiation of *lumi*-ketones can be trapped intramolecularly before group migration can compete. This reaction path is illustrated in (**205**) [326].

R = Me, $CH_2OH$, CHO
20%, 71%, 10%

Ref [326a]

quantitative

Ref [326b]

**(205)**

### 3.3.11 Linearly conjugated dienone rearrangements

Linearly conjugated dienones have been referred to above and can be obtained by the photochemical ring opening of *lumi*-ketones **(194)**. Dienones of this type are prone to undergo ring opening on excitation to the $n\pi^*$ state involving an $\alpha$-fission process and affording a *cis*-diene ketene, as illustrated for the rearrangement to photosantonic acid **(201)** [327]. Ketenes so produced can be trapped intermolecularly to provide a route to a variety of esters or amides [328, 329]. Some examples of the process are shown in **206** and **207** [330]. The reaction has some synthetic potential, as illustrated by the synthesis of the saffron pigment *dimethyl crocetin*, from the reaction of the dicyclo-

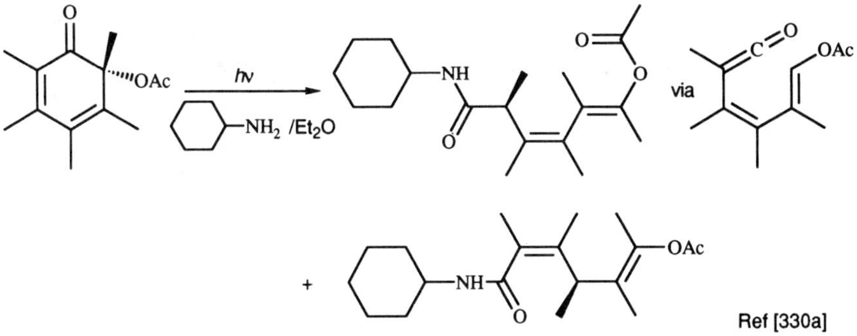

(206)

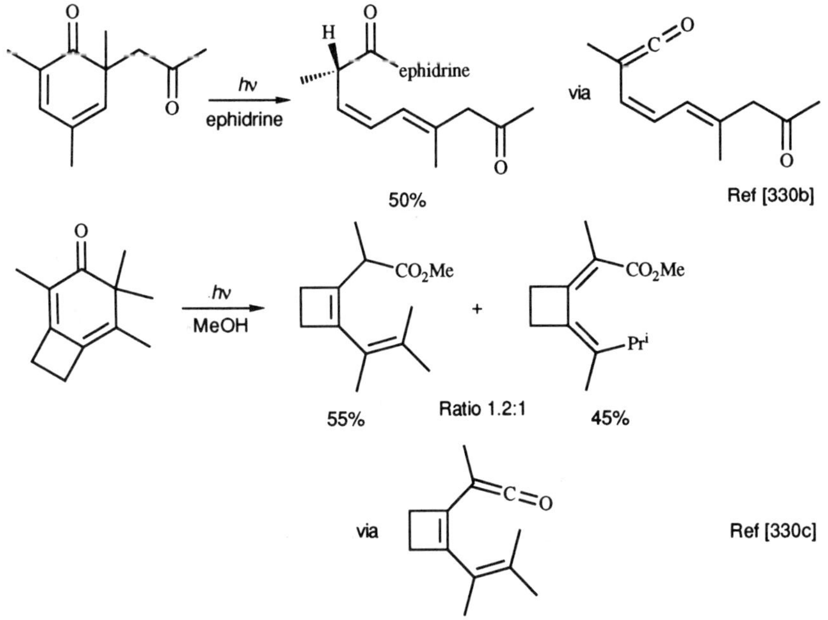

(207)

hexadienone (**208**) [331]. An elegant synthesis of macrolides (**209**) has also been developed and involves both inter- and intramolecular trapping [332].

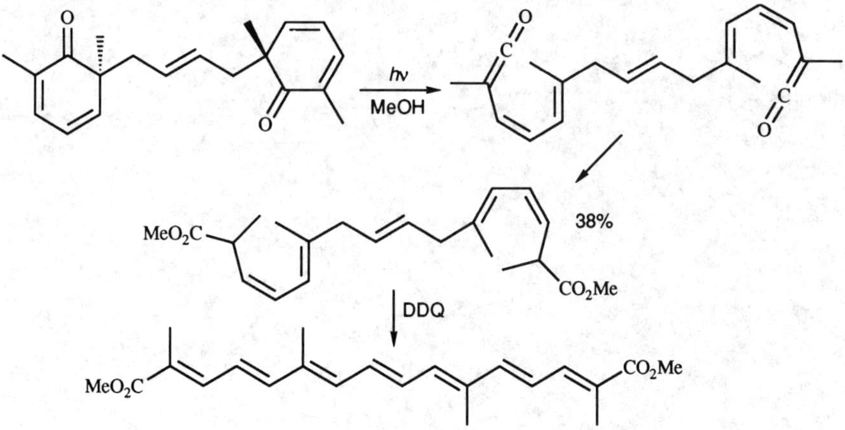

$h\nu$ MeOH

38%

$MeO_2C$ $CO_2Me$

DDQ

$MeO_2C$ $CO_2Me$

(208)

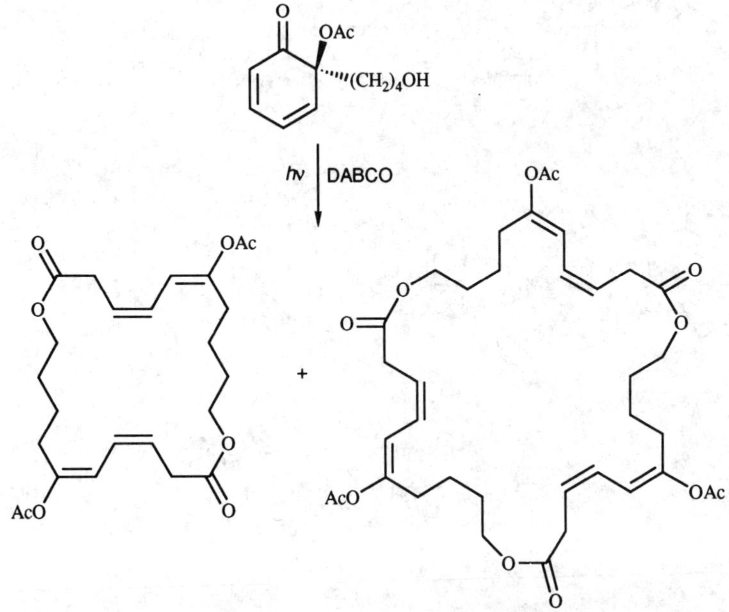

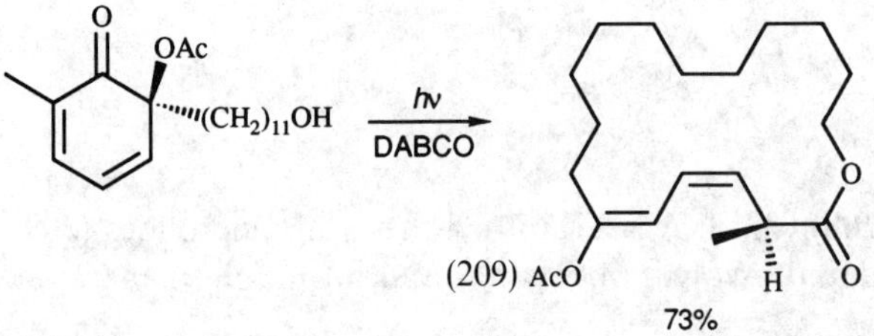

OAc $(CH_2)_4OH$

$h\nu$ DABCO

OAc AcO OAc OAc AcO

+

OAc $(CH_2)_{11}OH$

$h\nu$ DABCO

(209) AcO H

73%

This work has led to a synthetic approach to the naturally occurring macrolide (+)-*aspicilin* (**210**) [333].

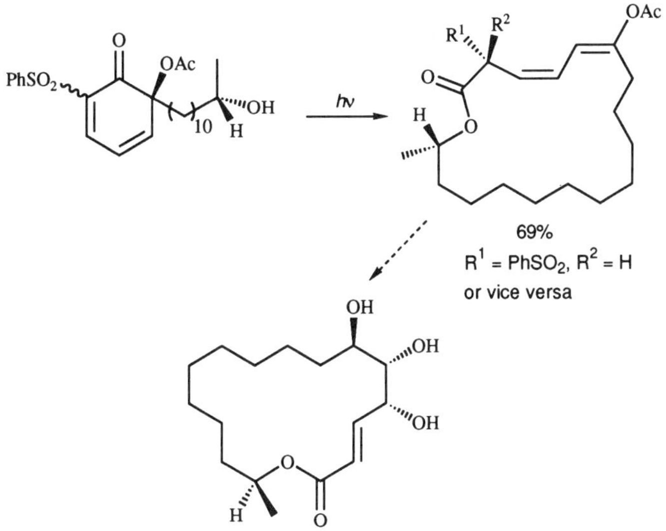

69%
$R^1$ = $PhSO_2$, $R^2$ = H
or vice versa

(210)

There are some examples, such as with the dienone (**211**), where irradiation

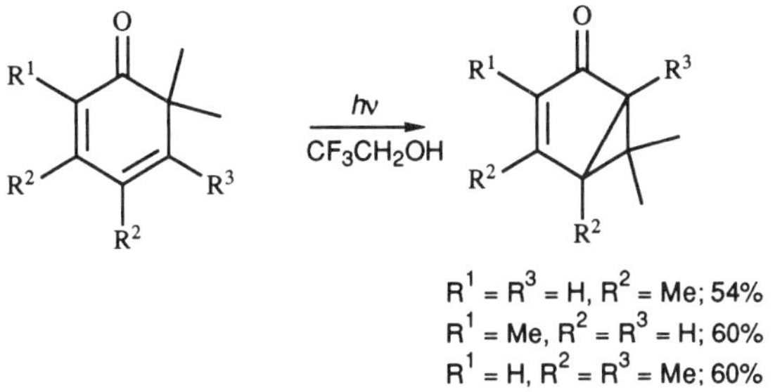

$R^1$ = $R^3$ = H, $R^2$ = Me; 54%
$R^1$ = Me, $R^2$ = $R^3$ = H; 60%
$R^1$ = H, $R^2$ = $R^3$ = Me; 60%

(211)

affords a bicyclic ketone [334]. Typically this conversion is carried out in trifluoroethanol and involves $\pi\pi^*$ excitation without intermediacy of the ketene. However, evidence has been put forward in support of $n\pi^*$ also being involved, in some cases (**212**), in the formation of the bicyclic ketone by

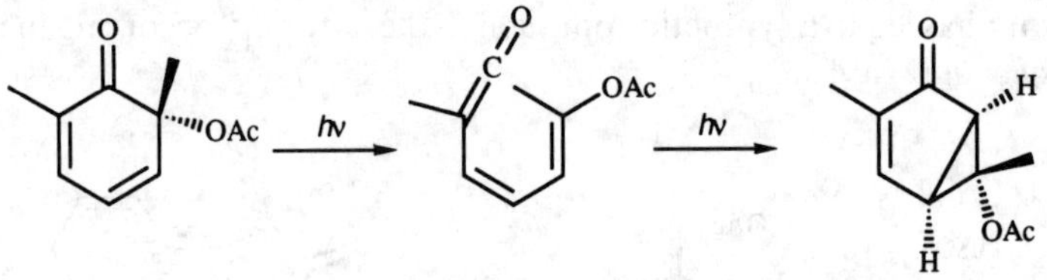

(212)

irradiative conversion of ketene transients [335]. Regardless, the route has some synthetic value in the synthesis of bicyclohexenones.

### 3.3.12 Quinones

Quinones, both *p*- and *o*-isomers, are a class of compounds which have been studied for many years [336]. They are usually highly coloured and are good materials for photochemical study.

#### *3.3.12.1* p-*Quinones*

Like enones and dienones these compounds undergo a variety of reactions such as hydrogen abstraction, oxetane formation with alkenes, and cyclobutane formation with ring double bonds. As a general rule the outcome of the reaction is dependent on the nature of the excited state and quinones such as *p*-benzoquinone, with an $n\pi^*$ triplet state, undergo hydrogen abstraction reactions and oxetane-forming addition reactions. Changes in substitution can influence the nature of the excited state and duroquinone has a triplet $\pi\pi^*$ state within which cyclobutane-forming reactions predominate.

##### *3.3.12.1.1 Hydrogen abstraction reactions*

The abstraction of hydrogen by *p*-chloranil (**213**) from *p*-xylene to afford the

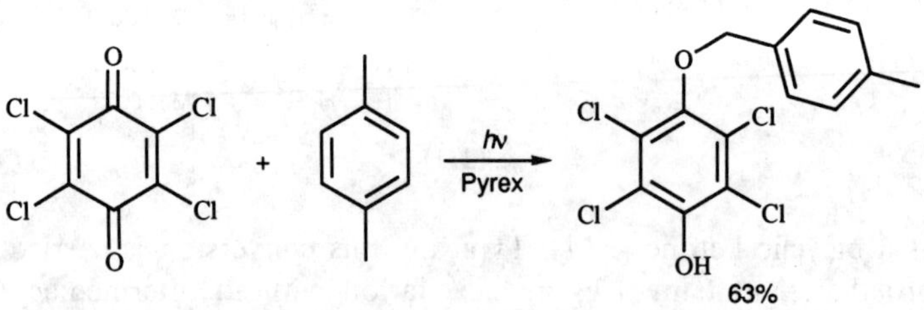

(213)

benzyloxyphenol [337] is typical of this reaction type. Hydrogen abstraction also occurs on reaction with aldehydes to afford products of *C*-acylation (**214**) [338]. Intramolecular hydrogen abstraction, a Norrish Type II process,

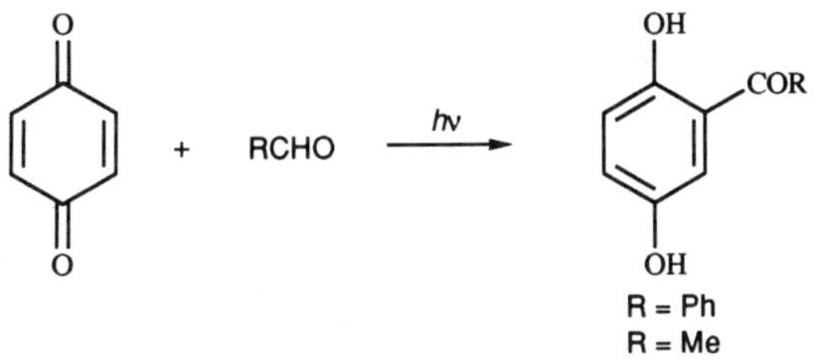

R = Ph
R = Me

(214)

can be used to activate sites on side chains. Thus irradiation of the benzoquinone (**215**) ultimately yields the cyclized product shown [339].

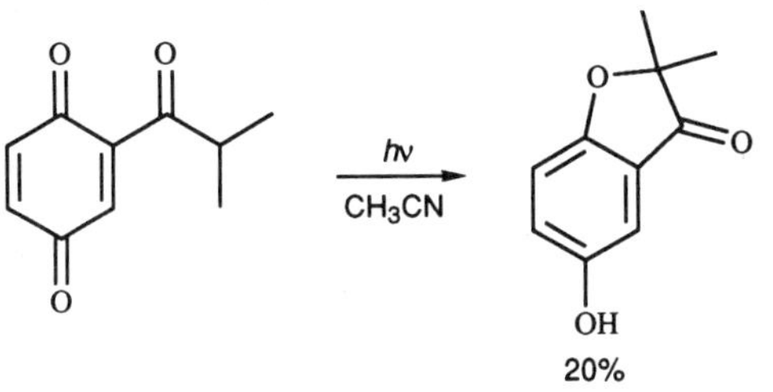

20%

(215)

Hydrogen abstraction from adjacent hydroxy functions also occurs, as in the conversion of the quinone (**216**) into the ketone [340]. A similar conversion

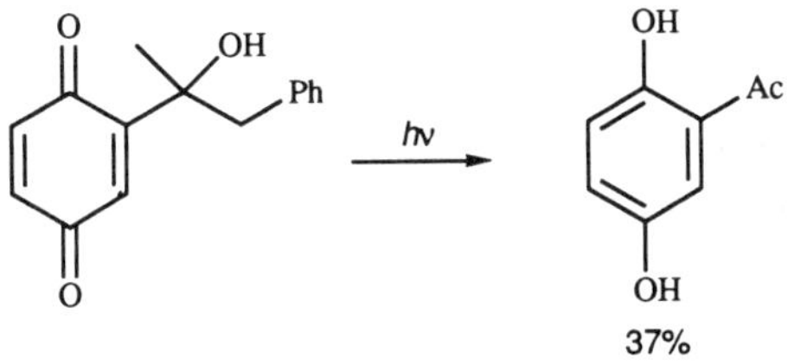

37%

(216)

is reported for the quinone (**217**) [340] when the acetophenone derivative and the acetone derivative are formed. The formation of the latter compound must arise by rearrangement of the side chain via the cyclopropyl intermediate formed by the abstraction of a benzylic hydrogen rather than the alcohol

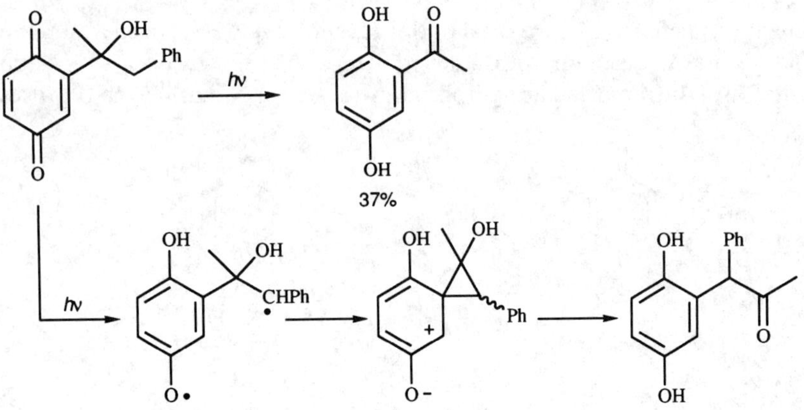

37%

(217)

hydrogen. Such side-chain rearrangements are fairly common in benzoquinone photochemistry [341].

*3.3.12.1.2 Cycloaddition reactions*

The dimerization of *p*-benzoquinone can be brought about by irradiation in molten maleic anhydride and affords a low yield of a cage compound (**218**) [342]. It is

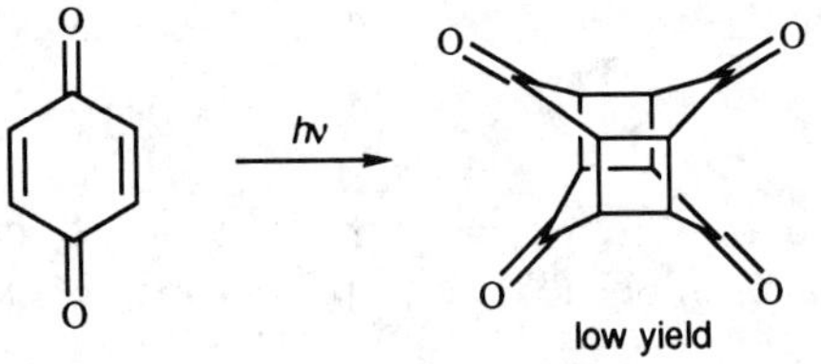

low yield

(218)

clear from this reaction that the cycloaddition affords a *cis-syn-cis*-adduct as the initial product, followed by a second (2 + 2)-addition. This process is also observed for the cycloaddition of the quinone **219**, which affords a dimer.

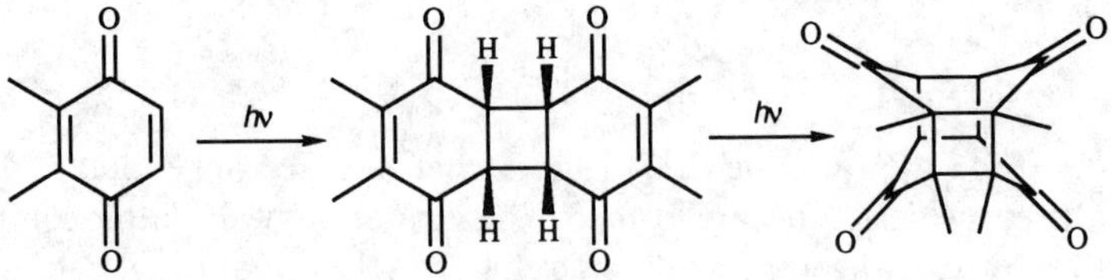

(219)

Further irradiation of this yields the cage compound [343]. More conventional intermolecular (2 + 2)-cycloaddition is also reported, such as the reaction between ethene and the chloronaphthoquinone, which yields the adduct shown [344a], or addition of 1,1-dimethylallene to duroquinone, affording a single adduct (**220**) [344b]. Other alkenes also undergo addition

hν

53% Ref [344a]

hν

low yield Ref [344b]

(220)

with the same regiospecificity. An interesting intramolecular cycloaddition has been reported for the quinones (**221**) which yields the quadricyclane

hν
λ>410 nm

$R^1$ = H, Me or MeO
$R^2$ = H or Me

*c.* 63% yield

(221)

derivatives in good chemical and quantum yields (0.21) [345]. The bridged quinones (**222**) are also photochemically reactive and sunlight irradiation of

$h\nu$, $H_2O$

$CH_2CO_2H$

$R^1 = R^2 = H$, 91%

$R^1 = H$, $R^2 = Me$, mixture of isomers 82%

(222)

the isomeric mixture in water affords a useful stereospecific synthesis of tricyclic compounds [346].

Cycloadditions involving the carbonyl oxygen are also common processes and can involve a variety of addends. Thus the photoreaction of cyclohexene with *p*-benzoquinone gives an almost quantitative yield of the adduct (**223**)

$h\nu$, $\lambda$>290 nm

quantitative

(223)

[347]. Additions to ketenimines are also efficient, yielding iminooxetanes (**224**) [348]. Allenes can also be useful addends in this photochemical process,

47%

(224)

as in the reaction of 1,1-diphenylallene with 2,3-dichloronaphthoquinone, affording spiroketones (**225**) in yields which can be as high as 85% [349].

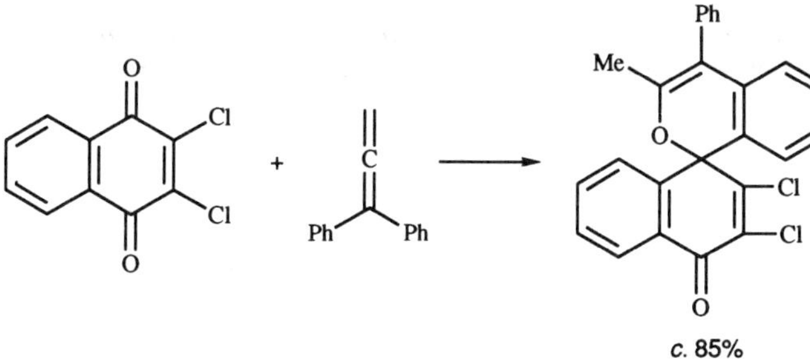

(225)

#### *3.3.12.2* **o-*Quinones***

As with the above *p*-isomers, *o*-quinones show a variety of reaction modes [350]. The compounds are again highly coloured and show absorptions in the 390 nm and 600 nm regions [351].

##### *3.3.12.2.1 Hydrogen abstraction reactions*

Photoexcited *o*-quinones can undergo hydrogen abstraction from many sources, such as aldehydes, alcohols, esters, ethers and alkenes, resulting in a 1,4-addition process across the two oxygens of the quinone. A typical reaction with an aldehyde is shown in **226**, where acetaldehyde reacts with

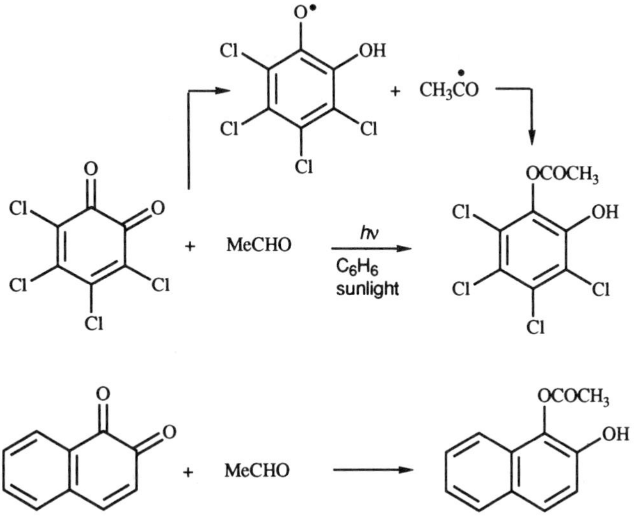

(226)

*o*-tetrachlorobenzo- and naphthoquinones [352]. The reaction of non-symmetrically substituted quinones (**227**) with acetaldehyde yields a mixture of

hν, MeCHO

20%

20%

$R^1$ = t-Bu, $R^2$ = H or vice versa

(227)

the two possible monoesters, as well as acetyl catechols. The latter products are presumed to arise by an out-of-cage process [353]. Glyoxals also undergo hydrogen abstraction when irradiated in the presence of phenanthro-9,10-quinone. In this example, the resultant acyl radical decarbonylates prior to formation of the monoesters (**228**) [354].

hν, RCOCHO

(228)

*3.3.12.2.2 Cycloaddition reactions*

One of the commonest types of reactions encountered in the photochemical reactions of *o*-quinones is cycloaddition reactions, usually in the presence of alkenes. The outcome of the reaction, which involves the quinone triplet state, is the formation of either a dioxene or a keto-oxetane. Occasionally hydrogen abstraction reactions are also observed. As an example of the complexity of these reactions, tetrachlorobenzo-1,2-quinone, a stable 1,2-quinone, adds photochemically to tetrachloroethene to yield a variety of potentially toxic heavily chlorinated compounds among which the dioxene was obtained in 17% yield (**229**) [355]. Naphtho-1,2-quinone also forms

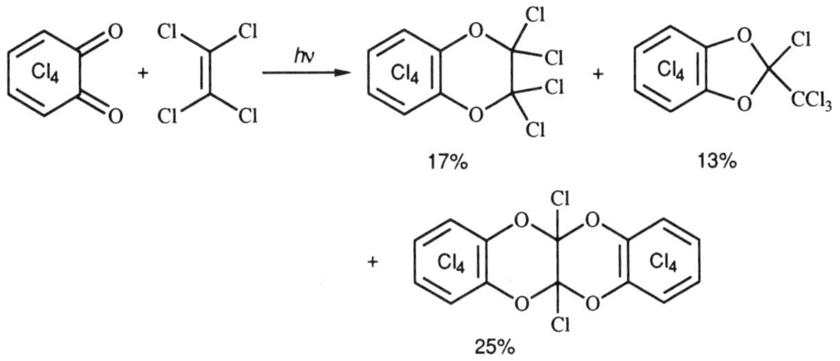

$h\nu$

17% 13%

25%

(229)

dioxenes on cycloaddition, as shown by the addition to *p*-dioxene (**230**) [356]

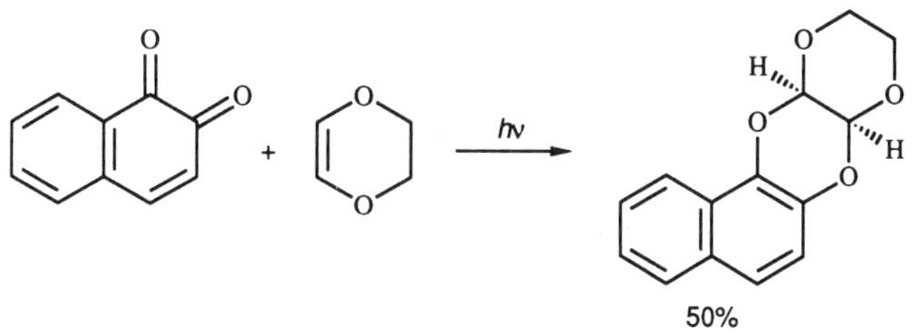

$h\nu$

50%

(230)

and to simple alkenes [357].

With electron-donating alkenes such as 1-ethoxyethene, the outcome of the addition is solvent-dependent. Thus in benzene the formation of dioxenes is dominant and the formation of an alternative adduct, a dihydrofuran, is minor. However, the situation is reversed in acetonitrile (**231**) [358].

$C_6H_6$, 87%
$CH_3CN$, 3%

$C_6H_6$, 6%
$CH_3CN$, 85%

(231)

The studies with 1,2-quinones have mainly used the stable phenanthro-9,10-quinone, and it is from this work that the majority of results have arisen. Thus reaction of phenanthro-9,10-quinone with either *cis*- or *trans*-but-2-ene is typical and affords the dioxene (**232**) [359], while addition to an oxazolone

via

either *cis* or *trans* isomer

40% *cis* isomer
26% *trans* isomer

(232)

affords the keto-oxetane (**233**) [360]. The reaction type is often dependent upon the alkene used and the substituents present. Cyclic alkenes, such as cyclopentene, cyclohexene, cycloheptene, *cis*-cyclo-octene, and cyclododecene

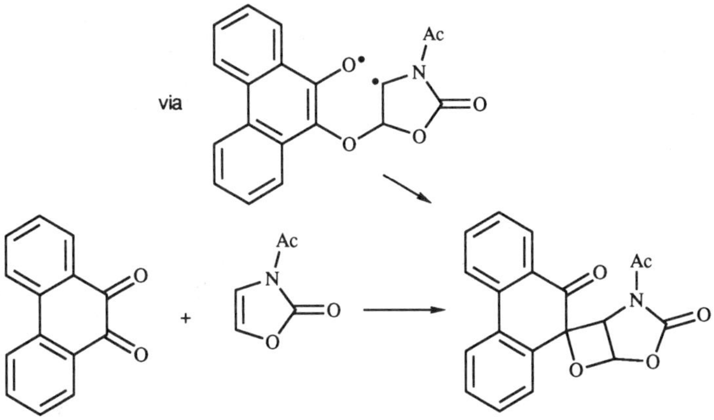

(233)

with phenanthro-9,10-quinone (**234**) [361] are common participants in the

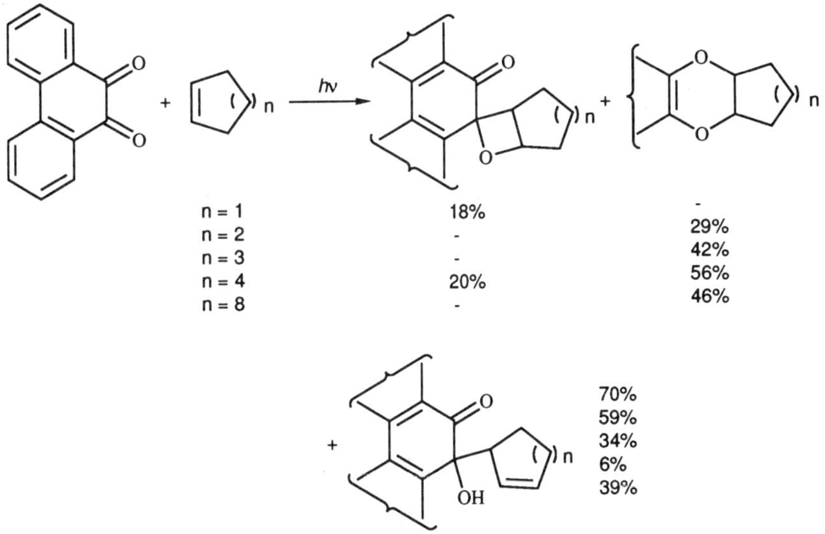

| n | | | |
|---|---|---|---|
| n = 1 | 18% | - | 70% |
| n = 2 | - | 29% | 59% |
| n = 3 | - | 42% | 34% |
| n = 4 | 20% | 56% | 6% |
| n = 8 | - | 46% | 39% |

(234)

cycloaddition reactions. Some ring-size effect is obvious from these results. The photo reaction with allenes has also been studied, and typically phenanthro-9,10-quinone yields a mixture of products, dioxenes, keto-oxetanes and hydrogen abstraction products (**235**) [362]. Acetylenes also

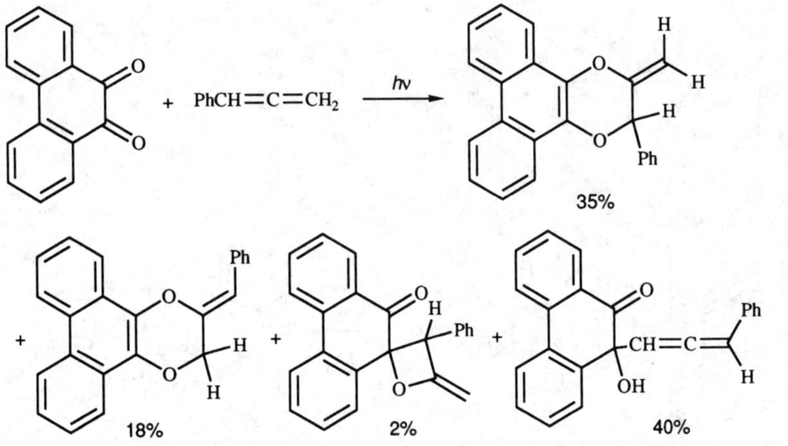

+ PhCH=C=CH$_2$ $\xrightarrow{h\nu}$ 35% + 18% + 2% + 40%

(235)

provide useful substrates and provide a route to the synthesis of dioxole derivatives on reaction of phenanthro-9,10-quinone with alkoxyalkynes (**236**) [363].

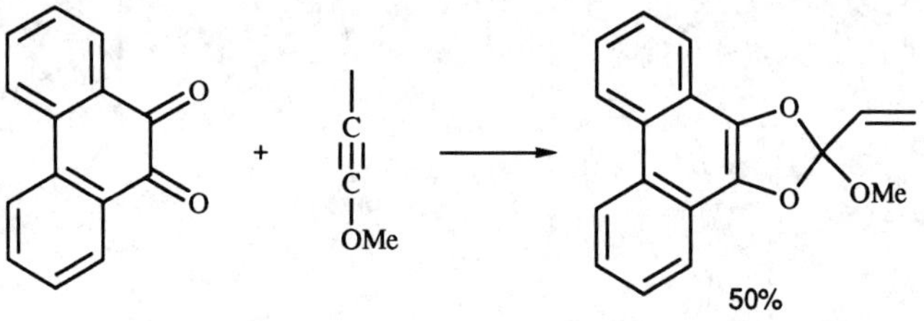

50%

(236)

## REFERENCES


[1] (a) C. von Sonntag and H.-P. Schuchmann, *Adv. Photochem.*, 1977, **10**, 59.

(b) C. von Sonntag and H.-P. Schuchmann, *The Chemistry of Ethers, Crown Ethers, Hydroxyl Groups, and their Sulphur Analogues*, Supplement E, ed. S. Patai, Wiley-Interscience, 1980, part 2, 903.

[2] J. G. Calvert and J. N. Pitts, jun., *Photochemistry*, Wiley, New York, 1966.

[3] H. Tsumbomura, K. Kimura, K. Kaya, J. Tanaka and S. Nagakura, *Bull. Chem. Soc. Jpn*, 1964, **37**, 417.

[4] M. B. Robin and N. A. Kuebler, *J. Electron Spectrosc.*, 1972, **1**, 13: M. B. Robin, *Higher Excited States of Polyatomic Molecules*, Academic Press, 1974, Vol 1, 254; C. Sandorfy, *Top. Curr. Chem.*, 1979, **86**, 92.
[5] J. C. Weeks, G. M. A. C. Meaburn and S. Gordon, *Radiat. Res.*, 1963, **19**, 559.
[6] J. Barrett, A. L. Mansell and M. F. Fox, *J. Chem. Soc. (B)*, 1971, 173
[7] D. Sanger, reported by C. von Sonntag and H.-P. Schuchmann, *Adv. Photochem.*, 1977, **10**, 59.
[8] A. J. Harrison, B. J. Cederholm and M. A. Terwilliger, *J. Chem. Phys.*, 1959, **30**, 355.
[9] H. Kaiser, reported by C. von Sonntag and H.-P. Schuchmann, *Adv. Photochem.*, 1977, **10**, 59.
[10] J. Doucet, P. Sauvageau and C. Sandorfy, *Chem. Phys. Lett.*, 1972, **17**, 316.
[11] H.-P. Schuchmann and C. von Sonntag, *Tetrahedron*, 1973, **29**, 1811.
[12] H. Kaiser reported in C. von Sonntag and H.-P. Schuchmann, *Adv. Photochem.*, 1977, **10**, 59.
[13] H.-P. Schuchmann and C. von Sonntag, *Tetrahedron*, 1973, **29**, 3351.
[14] G. Fleming, M. M. Anderson, A. J. Harrison and L. W. Pickett, *J. Chem. Phys.*, 1959, **30**, 351.
[15] L. W. Pickett, N. J. Hoeflich and T.-C. Liu, *J. Am. Chem. Soc.*, 1951, **73**, 4865; J. Doucet, P. Sauvageau and C. Sandorfy, *Chem. Phys. Letters*, 1972, **17**, 3351.
[16] Y. Ogata, K. Tomizawa and K. Furuta, in *The Chemistry of Peroxides*, ed S. Patai, Wiley, 1983, pp. 711–775.
[17] W. C. Schumb, C. N. Satterfield and R. L. Wentworth, *Hydrogen Peroxide*, Rainhold, New York, 1955.
[18] L. S. Silbert, in *Organic Peroxides*, Vol 2, ed. D. Swern, Wiley-Interscience, New York, 1971 pp 678 and 811.
[19] L. M. Toth and H. S. Johnston, *J. Am. Chem. Soc.*, 1969, **91**, 1276.
[20] A. Rieche, *Alkylperoxyde and Ozonide*, Steinkopff, Dresden, 1931.
[21] A. J. Everett and G. J. Minkoff, *Trans., Farad. Soc.*, 1953, **49**, 410.
[22] (a) O. J. Walker and G. L. E. Wild, *J. Chem. Soc.*, 1937, 1132.
(b) R. A. Sheldon and J. K. Kochi, *J. Am. Chem. Soc.*, 1970, **92**, 4395.
(c) B. C. Gilbert and A. J. Dobbs, in *Organic Peroxides*, ed. D. Swern, Wiley-Interscience, New York, 1971, Vol 3, p. 271.
[23] J. W. Breitenbach and J. Derkosch, *Monatsh. Chem.*, 1950, **81**, 530.
[24] L. H. Piette and W. C. Landgraf, *J. Chem. Phys.*, 1960, **32**, 1107.
[25] F. H. Dorer and S. N. Johnson, *J. Phys. Chem.*, 1971, **75**, 3651.

[26] N. C. Yang, D. P. C. Tang, D.-M. Thap and J. S. Sallo, *J. Am. Chem. Soc.*, 1966, **88**, 2851.
[27] C. von Sonntag, *Tetrahedron*, 1969, **25**, 5853: *Z. Phys. Chem.*, 1970, **69**, 292.
[28] C. von Sonntag, *Z. Naturforsch.*, 1972, **27B**, 41.
[29] D. Sanger and C. von Sonntag, *Tetrahedron*, 1970, **26**, 5489; *Z. Naturforsch*, 1970, **25B**, 1491.
[30] P. Wan and S. Muralidharan, *Can. J. Chem.*, 1986, **64**, 1949.
[31] P. Wan and S. Muralidharan, *J. Am. Chem. Soc.*, 1988, **110**, 4336.
[32] X. Ci and D. G. Whitten, *J. Am. Chem. Soc.*, 1987, **109**, 7215.
[33] X. Ci and D. G. Whitten, *J. Am. Chem. Soc.*, 1989, **111**, 3459.
[34] H.-P. Schuchmann, C. von Sonntag and G. Schomburg, *Tetrahedron*, 1972, **28**, 4333.
[35] H.-P. Schuchmann and C. von Sonntag, *Z. Naturforsch.*, 1975, **30B**, 399.
[36] R. Ford, H. P. Schuchmann and C. von Sonntag, *J. Chem. Soc., Perkin Trans. 2*, 1975, 1338.
[37] J. F. Meagher and R. B. Timmons, *J. Chem. Phys.*, 1972, **57**, 3175.
[38] C. A. F. Johnson and W. M. C. Lawson, *J. Chem. Soc., Perkin Trans. 2*, 1974, 353.
[39] J. D. Margerum, J. N. Pitts, jun., J. G. Rutgers and S. Searles, *J. Am. Chem. Soc.*, 1959, **81**, 1549.
[40] B. C. Roquitte, *J. Phys. Chem.*, 1966, **70**, 1334.
[41] B. C. Roquitte, *J. Am. Chem. Soc.*, 1969, **91**, 7664.
[42] H. P Schuchmann, P. Naderwitz and C. von Sonntag, *Z. Naturforsch.*, 1978, **33b**, 942.
[43] R. Gomer and W. A. Noyes, *J. Am. Chem. Soc.*, 1950, **72**, 101.
[44] B. C. Roquitte, *J. Phys. Chem.*, 1966, **70**, 2699.
[45] M. Bartok and K. L. Lang, in *The Chemistry of Ethers, Crown Ethers, Hydroxyl Groups and their Sulphur Analogues*, Supplement E, ed. S. Patai, Wiley-Interscience, 1980, Part 2, 609.
[46] S. G. Wilkinson, *Int. Rev. Sci., Org. Chem., Ser. 2*, 1975, **2**, 111.
[47] W. L. F. Armarego, in *Stereochemistry of Heterocyclic Compounds*, Part 2, Wiley, New York, 1977, pp. 12–36.
[48] G. W. Griffin, *Angew. Chem. Int. Ed. Engl.*, 1971, **10**, 537.
[49] G. W. Griffin and A. Padwa, in *Photochemistry of Heterocyclic Compounds*, ed. O. Buchardt, Wiley, New York, 1976.
[50] N. R. Bertoniere and G. W. Griffin. *Org. Photochem.*, 1973, **3**, 115.
[51] K. Tokumaru, *Bull. Chem. Soc. Jpn*, 1967, **42**, 242.
[52] (a) R. S. Becker, R. O. Bost, J. Kolc, N. R. Bertoniere, R. I. Smith

and G. W. Griffin, *J. Am. Chem. Soc.*, 1970, **92**, 1302.
(b) D. R. Paulson, A. S. Murray, D. Benett, E. Mills, jun., V. O. Terry and S. D. Lopez, *J. Org. Chem.*, 1977, **42**, 1252.

[53] J. K. Crandall and D. R. Paulson, *Tetrahedron Lett.*, 1969, 2751.

[54] H. Prinzbach and M. Klaus, *Angew. Chem. Int. Ed. Engl.*, 1969, **8**, 276.

[55] G. W. Griffin, K. Ishikawa and I. J. Lev, *J. Am. Chem. Soc.*, 1976, **98**, 5697.

[56] T. Do-Minh, A. M. Trozzolo and G. W. Griffin, *J. Am. Chem. Soc.*, 1970, **92**, 1402.

[57] D. R. Arnold and L. A. Karnischky, *J. Am. Chem. Soc.*, 1970, **92**, 1404.

[58] (a) R. Potter and W. Dilthey, *J. Prakt. Chem.*, 1937, **149**, 183; **150**, 40; J. M. Dunston and P. Yates, *Tetrahedron Lett.*, 1964, 505.
(b) E. F. Ullman, *J. Am. Chem. Soc.*, 1963, **85**, 3529; E. F. Ullman and J. E. Milks, *J. Am. Chem. Soc.*, 1964, **86**, 3814.
(c) V. M. Zolin, N. D. Dmitrieva, Yu. E. Gerasimenko and A. V. Zubkov, *Khim. Geterosikl., Soedin.*, 1984, 167.

[59] A. M. Trozzolo, W. A. Yager, G. W. Griffin, H. Kristinsson and I. Saktar, *J. Am. Chem. Soc.*, 1967, **89**, 3357.

[60] (a) R. S. Becker, J. Kolc, R. O. Bost, H. Dietrich, P. Petrellis and G. W. Griffin, *J. Am. Chem. Soc.*, 1968, **90**, 3292; R. S. Becker, R. O. Bost, J. Kolc, N. R. Bertoniere, R. L. Smith and G. W. Griffin, *J. Am. Chem. Soc.*, 1970, **92**, 1302.
(b) P. Petrellis, H. Dietrich, E. Meyer and G. W. Griffin, *J. Am. Chem. Soc.*, 1967, **89**, 1967.

[61] H. Kristinsson and G. W. Griffin, *Angew. Chem. Int. Ed. Engl.*, 1965, **4**, 868; *J. Am. Chem. Soc.*, 1966, **88**, 1579.

[62] (a) G. A. Lee, *J. Org. Chem.*, 1976, **41**, 2656.
(b) V. Markowski and R. Huisgen *Tetrahedron Lett.*, 1976, 4643.
(c) I. J. Lev, K. Ishikawa and G. W. Griffin, *J. Org. Chem.*, 1976, **41**, 2654.
(d) K. Maruyama, A. Osuka and K. Nakagawa, *Bull. Chem. Soc. Jpn*, 1987, **60**, 1021.

[63] (a) K. Ishii, M. Abe and M. Sakamoto, *J. Chem. Soc., Perkin Trans. 1*, 1987, 1937.
(b) U. Goldener, M. E. Scheller, P. Mathies, B. Frei and O. Jeger, *Helv. Chim. Acta*, 1985, **68**, 635; U. Goldener, B. Frei and O. Jeger, *Helv. Chim. Acta*, 1985, **68**, 919.

[64] A. Albini and D. R. Arnold, *Can. J. Chem.*, 1978, **56**, 2985.

[65] C. K. Johnson, B. Dominy and W. Reusch, *J. Am. Chem. Soc.*, 1963, **85**, 3894.

[66] (a) P. Morand and S. A. Samad, *Bangladesh J. Sci. Ind. Res.*, 1979, **14**, 265.
(b) P. Hallet, J. Muzart and J. P. Pete, *J. Org. Chem.*, 1981, **46**, 4275.
(c) M. Caus, H. Cerfontain and J. F. Piniella, *Recl.: J. R. Neth. Chem. Soc.*, 1983, **102**, 515.
(d) H. Wehrli, C. Lehmann, K. Schaffner and O. Jeger, *Helv. Chim. Acta*, 1964, **47**, 1336.

[67] W. J. Maguire and R. C. Pink, *Trans. Faraday Soc.*, 1967, **63**, 1097.

[68] S. A. Sojka, C. F. Poranski, jun. and W. B. Moniz, *J. Am. Chem. Soc.*, 1975, **97**, 5953; W. B. Moniz, S. A. Sojka, C. F. Poranski, jun. and D. L. Birkle, *J. Am. Chem. Soc.*, 1978, **100**, 7940.

[69] L. M. Dorfman and Z. W. Salsburg, *J. Am. Chem. Soc.*, 1951, **73**, 255.

[70] H. Paul, R. D. Small, jun. and J. C. Scaiano, *J. Am. Chem. Soc.*, 1978, **100**, 4520.

[71] P. J. Krusic and J. K. Kochi, *J. Am. Chem. Soc.*, 1968, **90**, 7155.

[72] (a) N. J. Turro and P. Lechten. *Tetrahedron Lett.*, 1973, 565.
(b) W. Adam, in *The Chemistry of Peroxides*, ed. S. Patai, Wiley, 1983, pp. 829–920.

[73] W. Adam and N. Duran. *J. Am. Chem. Soc.*, 1977, **99**, 2729.

[74] W. Adam and J. Sanabia, *J. Am. Chem. Soc.*, 1977, **99**, 2735.

[75] Reference [16] p. 736.

[76] V. I. Barchuk, A. A. Dubinskii, O. Y. Grinsberg and Y. S. Lebedev, *Chem. Phys. Lett.*, 1975, **34**, 476.

[77] M. Jones, jun. and M. R. DeCamp. *J. Org. Chem.*, 1971, **36**, 1536.

[78] O. L. Chapman, C. L. McIntosh and J. Pacansky, *J. Am. Chem. Soc.*, 1973, **95**, 4061.

[79] H. Ochiai, *Agr. Biol. Chem.*, 1971, **35**, 622 (*Chem. Abstr.*, 1967, **75**, 34132).

[80] D. R. Kearns, *J. Am. Chem. Soc.*, 1969, **91**, 6554.

[81] K. K. Maheshwari, P. de Mayo and D. Wiegand, *Can. J. Chem.*, 1970, **48**, 3265.

[82] R. Srinivasan, K. H. Brown, J. A. Oars, L. S. White and W. Adam, *J. Am. Chem. Soc.*, 1979, **101**, 7424.

[83] I. Saito and S. S. Nittala, in *The Chemistry of Peroxides*, ed. S. Patai, Wiley, 1983, pp. 311–374; M. Balci, *Chem. Rev.*, 1981, **81**, 91.

[84] J. Rigaudy, C. Deletang and J. J. Basselier, *Compt. Rend.*, 1969, **268**, 344.

[85] E. Paterno and G. Chieffi, *Gazz. Chim. Ital.*, 1909, **39**, 341; and also A. Schonberg, *Preparative Organic Photochemistry*, Springer-Verlag, New York, 1968, for an extensive coverage of the early literature.

[86] J. G. Calvert and J. N. Pitts, jun., *Photochemistry*, Wiley, New York, 1966, p. 368.
[87] J. B. Lambert, H. F. Shurvell, L. Verbit, R. G. Cooks and G. H. Stout, *Organic Structural Analysis*, Macmillan, New York, 1976, p. 350.
[88] R. F. Borkman and D. R. Kearns, *J. Chem. Phys.*, 1966, **44**, 945.
[89] C. A. Parker and T. A. Joyce, *J. Chem. Soc., Chem. Commun.*, 1968, 749.
[90] R. S. Becker, K. Inuzka and J. King, *J. Chem. Phys.*, 1970, **52**, 5164.
[91] J. G. Calvert and J. N. Pitts, jun., *Photochemistry*, Wiley, New York, 1966, p. 371.
[92] M. Kasha, *Radiation Research, Supplement 2*, 1960, 243.
[93] Reference [1b] p. 939.
[94] D. S. Herr and W. A. Noyes, jun., *J. Am. Chem. Soc.*, 1940, **62**, 2052.
[95] J. Caldwell and D. E. Hoare, *J. Am. Chem. Soc.*, 1962, **84**, 3987.
[96] U. Schmidt, *Angew. Chem. Int. Ed. Engl.*, 1965, **4**, 146, 239.
[97] J. N. Pitts, jun. and F. E. Blacet, *J. Am. Chem. Soc.*, 1950, **72**, 2810.
[98] J. W. Krauss and J. G. Calvert, *J. Am. Chem. Soc.*, 1957, **79**, 5921.
[99] N. C. Yang, E. D. Feit, M. H. Hui, N. J. Turro and J. C. Dalton, *J. Am. Chem. Soc.*, 1970, **92**, 6974.
[100] (a) P. S. Engel, *J. Am. Chem. Soc.*, 1970, **92**, 6074; W. K. Robbins and R. H. Eastman, *J. Am. Chem. Soc.*, 6076.
(b) M. Laeufer and H. Dreeskamp, *J. Mag. Reson.*, 1984, **28**, 405.
[101] (a) L. J. Johnston and J. C. Scaiano, *J. Am. Chem. Soc.*, 1987, **109**, 5487.
(b) B. N. Rao, M. S. Syamala, N. J. Turro and V. Ramamurthy, *J. Org. Chem.*, 1987, **52**, 5517.
(c) W. Adam, A. Berkessel, E.-M. Peters, K. Peters and H. G. von Schnering, *J. Org. Chem.*, 1985, **50**, 2811.
[102] S. S. Pappas and A. Chattopadhyay, *J. Am. Chem. Soc.*, 1973, **95**, 6484.
[103] F. D. Lewis and J. G. Magyar, *J. Am. Chem. Soc.*, 1973, **95**, 5973.
[104] H. G. Heine, *Tetrahedron Lett.*, 1971, 1473; F. D. Lewis, R. T. Lauterbach, H. G. Heine, W. Hartmann and H. Rudolph, *J. Am. Chem. Soc.*, 1975, **97**, 1519; J. C. Sheehan and R. M. Wilson, *J. Am. Chem. Soc.*, 1964, **86**, 5277.
[105] R. S. Givens and W. F. Oettle, *J. Org. Chem.*, 1972, **37**, 4325.
[106] N. J. Turro, B. Krautlerl and D. R. Anderson, *Tetrahedron Lett.*, 1980, **21**, 3; N. J. Turro, B. Krautler and D. R. Anderson, *J. Am. Chem. Soc.*, 1979, **101**, 7435; G. F. Lehr and N. J. Turro, *Tetrahedron*, 1981, **37**, 3411.

[107] K. Schaffner, *Chimia*, 1965, **19**, 575.
[108] B. Blank and H. Fischer, *Helv. Chim. Acta.*, 1973, **56**, 506.
[109] M. Cocivera and A. M. Trozzolo, *J. Am. Chem. Soc.*, 1970, **92**, 1772; G. L. Closs and D. R. Paulson, *J. Am. Chem. Soc.*, 1970, **92**, 7229.
[110] C. W. Funke and H. Cerfontain, *Tetrahedron Lett.*, 1975, 4061.
[111] H. E. Chen, A. Groen and M. Cocivera, *Can. J. Chem.*, 1973, **51**, 3032.
[112] H. E. Chen, S. P. Vaish and M. Cocivera, *J. Am. Chem. Soc.*, 1973, **95**, 7586; B. Blank, A. Henne and H. Fischer, *Helv. Chim. Acta*, 1974, **57**, 920.
[113] R. L. Whistler and K. H. Ong, *J. Org. Chem.*, 1971, **36**, 2575.
[114] (a) H. Kuentzel, H. Wolf and K. Schaffner, *Helv. Chim. Acta*, 1971, **54**, 868; K. Schaffner, H. Wolf, S. M. Rosenfeld, R. G. Lawler and H. R. Ward, *J. Am. Chem. Soc.*, 1972, **94**, 6553.
(b) H. Wolf, H.-U. Gonzenbach, K. Muller and K. Schaffner, *Helv. Chim. Acta*, 1972, **55**, 2919.
[115] (a) M. Oda, R. Breslow and J. Pecoraro, *Tetrahedron Lett.*, 1972, 4419.
(b) J. C. Sheehan and M. M. Nafissi, *J. Am. Chem. Soc.*, 1969, **91**, 1176.
[116] G. Quinkert, K. Optiz, W. Wiersdorff and J. Weinlich, *Tetrahedron Lett.*, 1963, 1863; D. C. Zecher and R. West, *J. Am. Chem. Soc.*, 1967, **89**, 153.
[117] E. K. C. Lee in *Excited State Chemistry*, ed. J. N. Pitts, jun., Gordon and Breach, New York, 1970, pp. 59–91.
[118] R. F. Klemm, *Can. J. Chem.*, 1970, **48**, 3320.
[119] D. C. England, *J. Am. Chem. Soc.*, 1961, **83**, 2205.
[120] (a) H. A. J. Carless, J. Metcalfe and E. K. C. Lee, *J. Am. Chem. Soc.*, 1972, **94**, 7221.
(b) H. G. Davies, S. S. Rahman, S. M. Roberts, B. J. Wakefield and J. A. Winders, *J. Chem. Soc., Perkin Trans. 1*, 1987, 85.
[121] N. J. Turro and R. M. Southam, *Tetrahedron Lett.*, 1967, 545.
[122] D. R. Morton and N. J. Turro, *Adv. Photochem.*, 1974, **9**, 197; W. D. Stohrer, P. Jacobs, K. H. Kaiser, G. Weich and G. Quinkert, *Topics Current Chem.*, 1974, **46**, 181.
[123] P. Dowd, A. Gold and K. Sachdev, *J. Am. Chem. Soc.*, 1970, **92**, 5724.
[124] (a) G. Quinkert, G. Cimbollek and G. Buhr, *Tetrahedron Lett.*, 1966, 4573.
(b) N. J. Turro and D. M. McDaniel, *J. Am. Chem. Soc.*, 1970, **92**, 5727.
[125] M. C. Pirrung and V. C. DeAmicis, *Heterocycles*, 1987, **25**, 189.

[126] (a) M. C. Pirrung and V. C. DeAmicis, *Tetrahedron Lett.*, 1988, **29**, 159.
(b) C. C. Howard, R. F. Newton, D. P. Reynolds, A. H. Wadsworth and D. R. Kelly, *J. Chem. Soc., Perkin Trans. 1*, 1980, 852.
[127] P. Yates, A. K. Verma and J. C. L. Tam, *J. Chem. Soc., Chem. Commun.*, 1976, 933.
[128] P. M. Collins and F. Farnia, *J. Chem. Soc., Perkin Trans. 1*, 1985, 575.
[129] A. G. Brook, J. B. Pierce and J. M. Duff, *Can. J. Chem.*, 1975, **53**, 2874.
[130] P. Yates and J. C. L. Tam, *J. Chem. Soc., Chem. Commun.*, 1975, 737.
[131] A. A. Scala and D. G. Ballan, *Can. J. Chem.*, 1972, **50**, 3938.
[132] J. E. Starr and R. H. Eastman, *J. Org. Chem.*, 1966, **31**, 1393.
[133] R. S. Cooke and G. D. Lyon, *J. Am. Chem. Soc.*, 1971, **93**, 3840.
[134] (a) D. S. Weiss, *J. Am. Chem. Soc.*, 1975, **97**, 2550.
(b) E. E. Nunn and R. N. Warrener, *J. Chem. Soc., Chem. Commun.*, 1972, 818.
(c) A. Peyman, H. D. Beckhaus and C. Ruechardt, *Chem. Ber.*, 1988, **121**, 1027.
(d) B. Guerin, L. J. Johnston and T. Quach, *J. Org. Chem.*, 1988, **53**, 2826.
(e) D. S. Weiss, M. Haslanger and R. G. Lawton, *J. Am. Chem. Soc.*, 1976, **98**, 1050.
[135] J. C. Dalton, K. Dawes, N. J. Turro, D. S. Weiss, J. A. Barltrop and J. D. Coyle, *J. Am. Chem. Soc.*, 1971, **93**, 7213.
[136] C. C. Badcock, M. J. Perona, G. O. Prichardt and B. Rickborn, *J. Am. Chem. Soc.*, 1969, **91**, 543.
[137] J. D. Coyle, *J. Chem. Soc. B*, 1971, 1736.
[138] J. C. Dalton and N. J. Turro, *Ann. Rev. Phys. Chem.*, 1970, **21**, 499.
[139] (a) P. J. Wagner and R. W. Spoerke, *J. Am. Chem. Soc.*, 1969, **91**, 4437.
(b) W. C. Agosta and W. L. Schreiber, *J. Am. Chem. Soc.*, 1971, **93**, 3947.
[140] (a) J. A. Barltrop and J. D. Coyle, *J. Chem. Soc., Chem. Commun.*, 1969, 1081.
(b) N. C. Yang and R. H. K. Chen. *J. Am. Chem. Soc.*, 1971, **93**, 530.
[141] (a) J. C. Dalton, D. M. Pond, D. S. Weiss, F. D. Lewis and N. J. Turro, *J. Am. Chem. Soc.*, 1970, **92**, 2564.
(b) R. G. Carlson and E. L. Biersmith, *J. Chem. Soc., Chem. Commun.*, 1969, 1049.
(c) R. K. Murray and C. A. Andruskiewicz, *J. Org. Chem.*, 1977, **42**, 3994.

(d) P. Yates and S. Stiver, *Can. J. Chem.*, 1988, **66**, 476.
(e) P. Camps, R. Lozano and M. A. Miranda, *J. Chem. Research (S)*, 1986, 250.
[142] R. C. Cookson and N. S. Wariyar, *J. Chem. Soc.*, 1956, 2302.
[143] A. Moscowitz, K. Mislow, M. A. W. Glass and C. Djerassi, *J. Am. Chem. Soc.*, 1962, **84**, 1945.
[144] K. N. Houk, *Chem. Rev.*, 1976, **76**, 1.
[145] L. A. Paquette and R. F. Eizember, *J. Am. Chem. Soc.*, 1967, **89**, 6205; J. K. Crandall, J. P. Arrlington and J. Hen. *J. Am. Chem. Soc.*, 1967, **89**, 6208.
[146] G. Buchi and E. M. Burgess, *J. Am. Chem. Soc.*, 1960, **82**, 4333.
[147] K. Schaffner, *Pure Appl. Chem.*, 1973, **33**, 329.
[148] (a) R. L. Coffin, R. S. Givens, and R. G. Carlson, *J. Am. Chem. Soc.*, 1974, **96**, 7554.
(b) J. R. Williams and G. M. Sarkisian, *J. Org. Chem.*, 1980, **45**, 5088.
(c) Y. Uyehara, T. Furuta, Y. Kabasawa, J. Yamada, T. Kato and Y. Yamamoto, *J. Org. Chem.*, 1988, **53**, 3669.
[149] D. Bellus and P. Hrdlovic, *Chem. Rev.*, 1967, **67**, 599.
[150] V. I. Stenberg, *Org. Photochem.*, 1967, **1**, 127.
[151] D. Bellus, *Adv. Photochem.*, 1971, **8**, 109.
[152] K. Nozaki, Y. Yamaguti, T. Okada, R. Noyori and M. Kawanisi, *Tetrahedron*, 1967, **23**, 3993.
[153] See reference [151], p. 147.
[154] T. S. Cantrell, W. S. Haller and J. C. Williams, *J. Org. Chem.*, 1969, **34**, 509.
[155] M. Gorodetsky and Y. Mazur. *J. Am. Chem. Soc.*, 1964, **86**, 5213.
[156] A. Yogev and Y. Mazur, *J. Am. Chem. Soc.*, 1965, **87**, 3520.
[157] J. C. Anderson and C. B. Reese, *Proc. Chem. Soc., London*, 1960, 217.
[158] H. Kobsa, *J. Org. Chem.*, 1962, **27**, 2293.
[159] R. A. Finnegan and J. J. Mattice, *Tetrahedron*, 1965, **21**, 1015.
[160] J. Saltiel, *Surv. Prog. Chem.*, 1964, **2**, 312.
[161] G. M. Coppinger and E. R. Bell, *J. Phys. Chem.*, 1966, **70**, 3479.
[162] M. R. Sander, E. Hedaya and D. J. Trecker, *J. Am. Chem. Soc.*, 1968, **90**, 7249.
[163] D. A. Plank, *Tetrahedron Lett.*, 1968, 5423.
[164] J. W. Meyer and G. S. Hammond, *J. Am. Chem. Soc.*, 1972, **94**, 2219; F. A. Carroll and G. S. Hammond, *J. Am. Chem. Soc.*, 1972, **94**, 7151.
[165] W. Adam, J. A. de Sanabia and H. Fischer, *J. Org. Chem.*, 1973, **38**, 2571; W. Adam, *J. Chem. Soc., Chem. Commun.*, 1974, 289.
[166] (a) W. M. Horspool and P. L. Pauson, *J. Chem. Soc.*, 1965, 5162.

(b) A. Erndt. J. Sepiol, G. Pyrc and A. Krajewska, *Pol. J. Chem.*, 1979, **53**, 533.
(c) R. Martinez-Utrilla and M. A. Miranda, *Tetrahedron Lett.*, 1980, **21**, 2281.
[167] D. Elad, D. V. Rao and V. I. Stenberg, *J. Org. Chem.*, 1965, **30**, 3252.
[168] D. Bellus and K. Schaffner, *Helv. Chim. Acta*, 1968, **51**, 221.
[169] J. Stumpe, K. Schwetlick and M. G. Kuzmin, *J. Prakt. Chem.*, 1982, **324**, 400.
[170] See Chapter 4 for examples.
[171] H. Sakurai, H. Yoshida and M. Kira, *J. Chem. Soc., Chem. Commun.*, 1985, 1780.
[172] K. Praefcke and U. Schulze, *J. Organomet. Chem.*, 1980, **184**, 189.
[173] L. P. Tenney, D. W. Boykin, jun. and R. E. Lutz, *J. Am. Chem. Soc.*, 1966, **88**, 1835.
[174] S. Domb and K. Schaffner, *Helv. Chim. Acta*, 1970, **53**, 679; S. Domb, G. Bozzato, J. A. Saboz and K. Schaffner, *Helv. Chim. Acta*, 1969, **52**, 2436.
[175] (a) M. Demuth, P. R. Raghavan, C. Carter, K. Nakano and K. Schaffner, *Helv. Chim. Acta*, 1980, **63**, 2434.
(b) M. Demuth, S. Chandrasekhar, K. Nakano, P. R. Raghavan and K. Schaffner, *Helv. Chim. Acta*, 1980, **63**, 2440.
[176] Reported in K. Schaffner and M. Demuth, *Modern Synthetic Methods*, ed. R. Scheffold, Springer-Verlag, Berlin, 1986, Vol 4, p. 63.
[177] K. Schaffner, *Chimia*, 1965, **9**, 575.
[178] D. E. Poel, H. Wehrli, K. Schaffner and O. Jeger, *Chimia*, 1966, **20**, 110.
[179] G. Bozzato, K. Schaffner and O. Jeger, *Chimia*, 1966, **20**, 114; E. Baggiolini, H. Berscheid, G. Bozzato, E. Cavalieri, K. Schaffner and O. Jeger, *Helv. Chim. Acta*, 1971, **54**, 429.
[180] J. Hill, J. Iriarte, K. Schaffner and O. Jeger, *Helv. Chim. Acta*, 1966, **49**, 292.
[181] J. A. Saboz, T. Iizuka, H. Wehrli, K. Schaffner and O. Jeger, *Helv, Chim, Acta*, 1968, **51**, 1362.
[182] E. Baggiolini, H. P. Hamlow, K. Schaffner and O. Jeger, *Chimia*, 1969, **23**, 181.
[183] E. Pfenninger, D. E. Poel, C. Berse, H. Wehrli, K. Schaffner and O. Jeger, *Helv. Chim. Acta*, 1968, **51**, 772.
[184] E. Baggiolini, H. P. Hamlow and K. Schaffner, *J. Am. Chem. Soc.*, 1970, **92**, 4906.
[185] D. Armesto, F. Langa, J.-A. Fernandez Martin, R. Perez-Ossorio and W. M. Horspool, *J. Chem. Soc., Perkin Trans. 1*, 1987, 743; D.

Armesto, W. M. Horspool and F. Langa, *J. Chem. Soc., Perkin Trans. 2*, 1989, 903.
[186] D. Armesto, W. M. Horspool, J.-A. Fernandez Martin and R. Perez-Ossorio, *J. Chem. Research (S)*, 1986, 46.
[187] D. Armesto, W. M. Horspool and F. Langa, *J. Chem. Soc., Chem. Commun.*, 1987, 167.
[188] J. N. Pitts and I. Norman, *J. Am. Chem. Soc.*, 1954, **76**, 4815.
[189] W. G. Dauben and G. W. Shaffer, *Tetrahedron Lett.*, 1967, 4415.
[190] R. S. Carson, W. Cocker, S. M. Evans and P. V. R. Shannon, *J. Chem. Soc., Chem. Commun.*, 1969, 726.
[191] R. Beugelmans, *Bull. Soc. Chim. Fr.*, 1967, 244.
[192] H. E. Zimmerman and C. M. Moore, *J. Am. Chem. Soc.*, 1970, **92**, 2023; H. E. Zimmerman, S. S. Hixson and E. F. McBride, *J. Am. Chem. Soc.*, 1970, **92**, 2000.
[193] D. H. R. Barton, J. E. D. Levisalles and J. T. Pinhey, *J. Chem. Soc.*, 1962, 3472.
[194] R. O. Loutfy and P. Yates, *J. Am. Chem. Soc.*, 1979, **101**, 4694.
[195] See references [58, 62d, 65] for other examples.
[196] G. Ciamician and P. Silber, *Chem. Ber.*, 1900, **33**, 2911; W. D. Cohen, *Rec. Trav. Chim. Pays-Bas*, 1920, **39**, 243.
[197] P. J. Wagner, *J. Am. Chem. Soc.*, 1966, **88**, 5672.
[198] P. J. Wagner, *J. Am. Chem. Soc.*, 1967, **89**, 2503.
[199] N. J. Turro and D. M. McDaniel, *Mol. Photochem.*, 1970, **2**, 39.
[200] S. G. Cohen and J. I. Cohen, *Israel J. Chem.*, 1968, **6**, 757; S. G. Cohen and J. I. Cohen, *Tetrahedron Lett.*, 1968, 4823.
[201] P. S. Mariano and J. L. Stavinoha, in *Synthetic Organic Photochemistry*, ed. W. M. Horspool, Plenum Press, New York and London, 1984, 145.
[202] A. Beckett and G. Porter, *Trans. Faraday Soc.*, 1963, **59**, 2051; G. Porter and P. Suppan, *Proc. Chem. Soc.*, 1964, 191.
[203] G. A. Davis, P. A. Carapellucci, K. Szoc and J. D. Gresser, *J. Am. Chem. Soc.*, 1969, **91**, 2264.
[204] M. B. Rubin, *Topics in Current Chemistry. Photochemistry*, Springer-Verlag, Berlin, 1968, 13, 251; *Top. Curr. Chem.*, 1985, **129**, 1.
[205] R. G. W. Norrish, *Trans. Faraday Soc.*, 1939, **33**, 1521.
[206] H. C. Hill, *Introduction to Mass Spectrometry*, Heyden and Son Ltd., 1966, p. 66.
[207] P. J. Wagner, *Acc. Chem. Res.*, 1983, **16**, 461.
[208] N. J. Turro and D. S. Weiss, *J. Am. Chem. Soc.*, 1968, **90**, 2185.
[209] (a) N. C. Yang, S. P. Elliot and B. Kim, *J. Am. Chem. Soc.*, 1969,

**91**, 7551.
(b) R. M. Wilson, *Org. Photochem.*, 1985, **7**, 339.
(c) P. J. Wagner, *Acc. Chem. Res.*, 1971, **4**, 168.

[210] P. J. Wagner, A. E. Kemppainnen and H. N. Schott, *J. Am. Chem. Soc.*, 1970, **92**, 5280.

[211] R. D. Small, jun. and J. C. Scaiano, *Chem. Phys. Lett.*, 1978, **59**, 246.

[212] P. J. Wagner, *J. Am. Chem. Soc.*, 1967, **89**, 5898.

[213] J. C. Scaiano, *J. Am. Chem. Soc.*, 1977, **99**, 1494.

[214] P. J. Wagner, *Tetrahedron Lett.*, 1967, 1753.

[215] D. C. Neckers, R. M. Kellogg, W. L. Prins and B. Schoustra, *J. Org. Chem.*, 1971, **36**, 1838.

[216] (a) L. Cottier, G. Remy and G. Descotes, *Synthesis*, 1979, 711.
(b) G. Bernasconi, L. Cottier, G. Descotes and G. Remy, *Bull. Soc. Chim. Fr.*, 1979, 332.
(c) R. W. Binkley and H. F. Jarrcll, *J. Carbohydr. Nucleosides, Nucleotides*, 1980, **7**, 347.

[217] (a) D. H. R. Barton, M. Bolton, P. D. Magnus, K. G. Marathe, G. A. Poulton and P. J. West, *J. Chem. Soc., Perkin Trans. 1*, 1973, 1567.
(b) M. A. F. Leite, M. H. Sarragiotto, P. M. Imamura and A. J. Marsaioli, *J. Org. Chem.*, 1986, **51**, 5409.

[218] P. Hrdlovic and I. Lukac, *Dev. Polym. Degrad.*, 1982, **4**, 101.

[219] R. B. Gagosian, J. C. Dalton and N. J. Turro, *J. Am. Chem. Soc.*, 1975, **87**, 5189.

[220] (a) A. Osuka, H. Shimizu, H. Suzuki and K. Maruyama, *Chem. Lett.*, 1987, 1061; K. Maruyama, A. Osuka, K. Nakagawa, K. Tabuchi, H. Shimizu and H. Suzuki, *Chem. Lett.*, 1986, 1849.
(b) H. R. Sonawane, B. S. Nanjundiah, S. I. Pajput and M. U. Kumar, *Tetrahedron Lett.*, 1986, **27**, 6125.
(c) Y. L. Chow and B. Marciniak, *J. Org. Chem.*, 1983, **48**, 2910.
(d) J. Hill, M. M. Zakaria and D. Mumford, *J. Chem. Soc., Perkin Trans. 1*, 1983, 2455.

[221] A. Abdul-Baki, F. Rotter, T. Schrauth and H. J. Roth, *Arch. Pharm.*, 1978, **311**, 341.

[222] H. G. Henning, H. Haber and H. Buchholz, *Pharmazie*, 1981, **36**, 160.

[223] H. G. Henning, R. Berlinghoff, A. Mahlow, H. Koeppl and K. D. Schleintz, *J. Prakt. Chem.*, 1981, **323**, 914.

[224] G. Bernasconi, L. Cottier, G. Descotes, J. P. Praly, G. Remy, M. F. Gernier-Loustalot and F. Metras, *Carbohydr. Res.*, 1983, **115**, 105.

[225] G. Descotes, *Bull. Soc. Chim. Belg.*, 1982, **91**, 973.

[226] M. A. Meador and P. J. Wagner, *J. Am. Chem. Soc.*, 1983, **105**, 4484.

[227] P. J. Wagner, M. A. Meador, B. P. Giri and J. C. Scaiano, *J. Am. Chem. Soc.*, 1985, **107**, 1087.
[228] J. D. Coyle, in *Synthetic Organic Photochemistry*, ed. W. M. Horspool, Plenum Press, New York and London, 1984, 259.
[229] (a) M. A. Meador and P. J. Wagner, *J. Org. Chem.*, 1985, **50**, 419.
(b) H. A. J. Carless and G. K. Fekarurhobo, *Tetrahedron Lett.*, 1984, **25**, 5943.
[230] R. Breslow, *Acc. Chem. Res.*, 1980, **13**, 170.
[231] R. Breslow and M. Winnik, *J. Am. Chem. Soc.*, 1969, **91**, 3083.
[232] M. A. Winnik, C. K. Lee, S. Basu and D. S. Saunders, *J. Am. Chem. Soc.*, 1974, **96**, 6182.
[233] R. Breslow and S. W. Baldwin, *J. Am. Chem. Soc.*, 1970, **92**, 732.
[234] J. E. Baldwin, A. K. Bhatnagar and R. W. Harper, *J. Chem. Soc., Chem. Commun.*, 1970, 659.
[235] R. M. Wilson, *Org. Photochem.*, 1985, **7**, 373.
[236] N. C. Yang and C. Rivas, *J. Am. Chem. Soc.*, 1961, **83**, 2213.
[237] F. Nerdel and W. Brodowski, *Chem. Ber.*, 1968, **101**, 1398.
[238] G. Quinkert and H. Stark, *Angew. Chem. Int. Ed. Engl.*, 1983, **22**, 637.
[239] H. A. J. Carless, *Photochemistry in Organic Synthesis*, ed., J. D. Coyle, Royal Society of Chemistry, 1986, 95; H. A. J. Carless, *Synthetic Organic Photochemistry*, ed. W. M. Horspool, Plenum Press, 1984, 425; D. R. Arnold, *Adv. Photochem.*, 1968, **6**, 301; G. Jones, II, *Org. Photochem.*, 1981, **5**, 122.
[240] E. S. Albone, *J. Am. Chem. Soc.*, 1968, **90**, 4663; C. DeBoer, *Tetrahedron Lett.*, 1971, 4977.
[241] S. C. Freilich and K. S. Peters, *J. Am. Chem. Soc.*, 1985, **107**, 3819.
[242] R. A. Caldwell, G. W. Sovocol and R. P. Gajewski, *J. Am. Chem. Soc.*, 1973, **95**, 2549; A. Gupta and G. S. Hammond, *J. Am. Chem. Soc.*, 1976, **98**, 1218; N. C. Yang, M. H. Hui, D. M. Shold, N. J. Turro, R. R. Hautala, K. Dawes and J. C. Dalton, *J. Am. Chem. Soc.*, 1977, **99**, 3023.
[243] E. Paterno and G. Chieffi, *Gazz. Chim. Ital.*, 1909, **39**, 314.
[244] G. Buchi, C. G. Inman and E. S. Lipinsky, *J. Am. Chem. Soc.*, 1954, **76**, 4327.
[245] H. A. J. Carless, *Tetrahedron Lett.*, 1973, 834.
[246] D. R. Arnold, R. L. Hinman and A. H. Glick, *Tetrahedron Lett.*, 1964, 1425.
[247] (a) H. Itokawa, H. Matsumoto, T. Oshima and S. Mihashi, *Yakugaku Zasshi.*, 1987, **107**, 767.
(b) A. Nehrings, H.-D. Scharf and J. Runsink, *Angew. Chem. Int. Ed.*

*Engl.*, 1985, **24**, 877.
(c) N. C. Yang and W. Eisenhardt. *J. Am. Chem. Soc.*, 1971, **93**, 1277.
[248] D. R. Morton and R. A. Morge, *J. Org. Chem.*, 1978, **43**, 2093.
[249] (a) M. Weuthen, H.-D. Scharf, J. Runsink and R. Vassen, *Chem. Ber.*, 1988, **121**, 971.
(b) H. Koch, H.-D. Scharf, J. Runsink and H. Leissmann, *Chem. Ber.*, 1985, **118**, 1485.
[250] N. C. Yang, M. Nussim and D. R. Coulson, *Tetrahedron Lett.*, 1965, 1525.
[251] H. A. J. Carless and G. K. Fekarurhobo, *J. Chem. Soc., Chem. Commun.*, 1984, 667.
[252] H. A. J. Carless, J. Beanland and S. Mwesigye-Kilbende, *Tetrahedron Lett.*, 1987, **28**, 5933.
[253] R. R. Sauers, K. W. Kelly and B. R. Sickles, *J. Org. Chem.*, 1972, **37**, 537.
[254] R. R. Sauers, A. A. Hagedorn, tert., S. D. Van Arnum, R. P. Gomez and R. V. Moquin, *J. Org. Chem.*, 1987, **52**, 5501.
[255] (a) N. C. Yang, M. H. Hui, D. M. Shold, N. J. Turro, R. R. Hautala, K. Dawes and J. C. Dalton, *J. Am. Chem. Soc.*, 1977, **99**, 3023.
(b) J. Kossanyi, G. Daccord, S. Sabbah, B. Furth, P. Chaquin, J. C. Andre and M. Bouchy, *Nouv. J. Chem.*, 1980, **4**, 337.
(c) J. A. Barltrop and J. D. Coyle, *J. Am. Chem. Soc.*, 1972, **94**, 8761.
[256] P. Dowd, A. Gold and K. Sachdev, *J. Am. Chem. Soc.*, 1970, **92**, 5725.
[257] H. A. J. Carless and A. K. Maitra, *Tetrahedron Lett.*, 1977, 1411.
[258] R. O. Duthaler, R. S. Stingelin-Schmidt and C. Ganter, *Helv. Chim. Acta*, 1976, **59**, 307.
[259] K. Shima, T. Kubota and H. Sakurai, *Bull. Chem. Soc. Jpn*, 1976, **49**, 2567.
[260] J. A. Barltrop and J. D. Coyle, *J. Am. Chem. Soc.*, 1971, **93**, 4794.
[261] N. C. Yang and W. Chiang, *J. Am. Chem. Soc.*, 1977, **99**, 3163.
[262] (a) T. Kubota, K. Shima and H. Sakurai, *Chem. Lett.*, 1972, 343.
(b) A. A. Gorman and R. L. Leyland, *Tetrahedron Lett.*, 1972, 5345.
(c) A. A. Gorman, R. L. Leyland, M. A. J. Rodgers and P. G. Smith, *Tetrahedron Lett.*, 1973, 5085.
[263] A. J. G. Barwise, A. A. Gorman, R. L. Leyland, C. T. Parekh and P. G. Smith, *Tetrahedron*, 1980, **36**, 397.
[264] G. O. Schenck, W. Hartmann and R. Steinmetz, *Chem. Ber.*, 1963, **96**, 498.
[265] M. Ogata, H. Watanabe and H. Kano, *Tetrahedron Lett.*, 1967, 533; S. Toki and H. Sakurai, *Tetrahedron Lett.*, 1967, 4119.

[266] K. Shima and H. Sakurai, *Bull. Chem. Soc. Jpn*, 1966, **39**, 1806.
[267] (a) S. L. Schreiber, D. Desmaele and J. A. Porco, jun., *Tetrahedron Lett.*, 1988, **28**, 6689.
(b) S. L. Schreiber and K. Satake, *J. Am. Chem. Soc.*, 1984, **106**, 4186.
(c) W. H. Liu and H. J. Wu, *J. Chin. Chem. Soc. (Taipei)*, 1988, **35**, 241.
(d) S. L. Schreiber and A. H. Hoveyda, *J. Am. Chem. Soc.*, 1984, **106**, 7200.
[268] H. Gotthardt, R. Steinmetz and G. S. Hammond, *J. Org. Chem.*, 1968, **33**, 2774; D. R. Arnold and A. H. Glick, *J. Chem. Soc. Chem. Commun.*, 1966, 813.
[269] H. Hogeveen and P. J. Smit, *Rec. Trav. Chim. Pays-Bas*, 1966, **85**, 1188.
[270] L. A. Singer, R. E. Brown and G. A. Davis, *J. Am. Chem. Soc.*, 1973, **95**, 6838.
[271] L. A. Singer, G. A. Davis and V. P. Muralidharan, *J. Am. Chem. Soc.*, 1972, **94**, 1188.
[272] S. H. Schroeter and C. M. Orlando, *J. Org. Chem.*, 1969, **34**, 1181; K. Shima and H. Sakurai, *Bull. Chem. Soc. Jpn*, 1969, **42**, 849.
[273] N. J. Turro and P. A. Wriede, *J. Am. Chem. Soc.*, 1968, **90**, 6863; 1970, **92**, 320.
[274] Y. Araki, J. Nagasawa and Y. Ishido, *Carbohydrate Res.*, 1981, **91**, 77.
[275] H. A. J. Carless and D. J. Haywood, *J. Chem. Soc., Chem. Commun.*, 1980, 1067.
[276] Y. Araki, K. Senna, K. Matsuura and Y. Ishido, *Carbohydrate Res.*, 1978, **60**, 389.
[277] N. R. Lazear and J. H. Schauble, *J. Org. Chem.*, 1974, **39**, 2069.
[278] N. J. Turro, P. A. Wriede and J. C. Dalton, *J. Am. Chem. Soc.*, 1968, **90**, 3274; N. J. Turro, *Pure Appl. Chem.*, 1971, **27**, 679.
[279] J. A. Barltrop and H. A. J. Carless, *Tetrahedron Lett.*, 1968, 3901; W. C. Herndon, *Topics in Current Chem.*, 1974, **46**, 141.
[280] J. J. Beereboom and M. S. von Wittenau, *J. Org. Chem.*, 1965, **30**, 1231.
[281] N. J. Turro and G. L. Farrington, *J. Am. Chem. Soc.*, 1980, **102**, 6056.
[282] H. A. J. Carless, *Photochemistry in Organic Synthesis*, ed. J. D. Coyle, Royal Society of Chemistry, 1986, 118.
[283] D. I. Schuster, *Rearrangements in Ground and Excited States*, ed. P. de Mayo, Academic Press, 1980, **3**, 167.
[284] (a) A. C. Weedon, *Synthetic Organic Photochemistry*, ed. W. M. Horspool, Plenum Press, 1984, 61.
(b) O. L. Chapman and D. S. Weiss, *Org. Photochem.*, 1973, **3**, 197.

[285] O. L. Chapman, T. A. Rattig, A. A. Griswold, A. I. Dutton and P. Fitton, *Tetrahedron Lett.*, 1963, 2049.
[286] W. G. Dauben, G. W. Shaffer and N. D. Vietmeyer, *J. Org. Chem.*, 1968, **33**, 4060.
[287] O. L. Chapman, J. B. Sieja and W. J. Welstead, *J. Am. Chem. Soc.*, 1966, **88**, 161; D. I. Schuster, R. H. Brown and B. M. Resnick, *J. Am. Chem. Soc.*, 1978, **100**, 4504.
[288] D. I. Schuster and D. F. Brizzolara, *J. Am. Chem. Soc.*, 1970, **92**, 4357.
[289] P. Lupon, J. C. Ferrer, J. F. Piniella and J.-J. Bonet, *J. Chem. Soc., Chem. Commun.*, 1983, 718.
[290] H. E. Zimmerman and J. W. Wilson, *J. Am. Chem. Soc.*, 1964, **86**, 4036; H. E. Zimmerman and K. G. Hancock, *J. Am. Chem. Soc.*, 1968, **90**, 3749; H. A. Zimmerman, K. G. Hancock and G. C. Licke, *J. Am. Chem. Soc.*, 1968, **90**, 4892; H. E. Zimmerman and W. R. Elser, *J. Am. Chem. Soc.*, 1969, **91**, 887.
[291] H. E. Zimmerman, R. C. Hahn, H. Morrison and M. C. Wani, *J. Am. Chem. Soc.*, 1965, **87**, 1138; H. E. Zimmerman, R. D. Rieke and J. R. Scheffer, *J. Am. Chem. Soc.*, 1967, **89**, 2033; H. E. Zimmerman and N. Lewin, *J. Am. Chem. Soc.*, 1969, **91**, 879.
[292] H. E. Zimmerman and D. J. Sam, *J. Am. Chem. Soc.*, 1966, **88**, 4905.
[293] (a) R. A. Bunce and E. M. Holt, *J. Org. Chem.*, 1987, **52**, 1549.
(b) H. E. Zimmerman and R. D. Solomon, *J. Am. Chem. Soc.*, 1986, **108**, 6276.
[294] H. Hart, B. Chen and M. Jeffares, *J. Org. Chem.*, 1979, **44**, 2722.
[295] F. Nobs, U. Burger and K. Schaffner, *Helv. Chim. Acta*, 1977, **60**, 1607.
[296] Y. Tobe, T. Iseki, K. Kakiuchi and Y. Odaira, *Tetrahedron Lett.*, 1984, **25**, 3895.
[297] C. S. K. Wan and A. C. Weedon, *J. Chem. Soc., Chem. Commun.*, 1981, 1235.
[298] R. Ricard, P. Sauvage, C. S. K. Wan, A. C. Weedon and D. F. Wong, *J. Org. Chem.*, 1986, **50**, 62.
[299] O. Piva, F. Henin, J. Muzart and J.-P. Pete, *Tetrahedron Lett.*, 1987, **28**, 4825.
[300] (a) E. J. Corey and S. Nozoe, *J. Am. Chem. Soc.*, 1964, **86**, 1652; E. J. Corey, R. B. Mitra and H. Uda, *J. Am. Chem. Soc.*, 1964, **86**, 5570.
(b) D. I. Schuster, P. B. Brown, L. J. Capponi, C. A. Rhodes, J. C. Scaiano, P. C. Tucker and D. Weir, *J. Am. Chem. Soc.*, 1987, **109**, 2533; D. I. Schuster, G. E. Hiebel, P. B. Brown, N. J. Turro and C. V. Kumar, *J. Am. Chem. Soc.*, 1988, **110**, 8261.
(c) D. J. Hastings and A. C. Weedon, *J. Am. Chem. Soc.*, 1991, **113**, 8525.

(d) D. I. Schuster, G. E. Heibel and J. Woning, *Angew. Chem. Int. Ed. Engl.*, 1991, **30**, 1345.

[301] P. de Mayo, *Acc. Chem. Res.*, 1970, **4**, 41; R. O. Loutfy and P. de Mayo, *J. Am. Chem. Soc.*, 1977, **99**, 3559.

[302] P. de Mayo, A. A. Nicholson and M. F. Tchir, *Can. J. Chem.*, 1969, **47**, 711.

[303] A. J. Wexler, J. A. Hyatt, P. W. Reynolds, C. Cottrell and J. S. Swenton, *J. Am. Chem. Soc.*, 1978, **100**, 512.

[304] B. D. Challand and P. de Mayo, *J. Chem. Soc., Chem. Commun.*, 1968, 982.

[305] (a) J. S. Swenton and E. L. Fritzen, *Tetrahedron Lett.*, 1979, 1951; J. S. Swenton and A. Wexler, *J. Am. Chem. Soc.*, 1976, **98**, 1602; K. J. Crowley, K. L. Erickson, A. Eckell and J. Meinwald, *J. Chem. Soc., Perkin Trans. 1*, 1973, 2671.

(b) P. A. Wender and J. C. Lechleiter, *J. Am. Chem. Soc.*, 1978, **100**, 4321.

(c) P. Singh, *J. Org. Chem.*, 1971, **36**, 3334.

(d) J. R. Williams and J. R. Callahan, *Synth. Commun.*, 1981, **11**, 551; G. L. Lange and F. C. McCarthy, *Tetrahedron Lett.*, 1978, 4749; P. A. Wender and J. C. Lechleiter, *J. Am. Chem. Soc.*, 1977, **99**, 267.

[306] J. D. White and D. N. Gupta, *J. Am. Chem. Soc.*, 1966, **88**, 5364.

[307] R. C. Gueldner, A. C. Thomson and P. A. Hedin, *J. Org. Chem.*, 1972, **37**, 1854.

[308] E. J. Corey and S. Nozoe, *J. Am. Chem. Soc.*, 1965, **87**, 5733.

[309] (a) L. A. Paquette, H.-S. Lin and M. J. Coghlan, *Tetrahedron Lett.*, 1987, **28**, 5017.

(b) F. Audenaert, D. De Keukeleire and M. Vandewalle, *Tetrahedron*, 1987, **43**, 5593.

(c) T. A. Angela and A. R. Pinder, *Tetrahedron*, 1987, **43**, 5537.

[310] (a) R. B. Kelly, J. Zamecnik and B. A. Beckett, *Can. J. Chem.*, 1972, **50**, 3455.

(b) L. K. Sydnes and W. Stensen, *Acta Chem. Scand., Ser. B*, 1986, **B40**, 657.

(c) E. P. Serebryakov, S. D. Kulomzina-Pletneva and A. K. Margaryan, *Tetrahedron*, 1979, **35**, 77.

(d) M. Cavazza and M. Zandomeneghi, *Gazz. Chim. Ital.*, 1987, **117**, 17.

[311] P. de Mayo, H. Takeshita and A. B. M. A. Sattar, *Proc. Chem. Soc.*, 1962, 119.

[312] S. W. Baldwin and R. E. Gawley, *Tetrahedron Lett.*, 1975, 3969.

[313] B. W. Disanayaka and A. C. Weedon, *J. Org. Chem.*, 1987, **52**, 2905.

[314] G. Ciamician and P. Silber, *Chem. Ber.*, 1908, **41**, 1928; G. Buchi and I. M. Goldman, *J. Am. Chem. Soc.*, 1957, **79**, 4741.

[315] (a) W. F. Erman and T. W. Gibson, *Tetrahedron*, 1969, **25**, 2493.
(b) S. Wolff, S. Ayral-Kaloustian and W. C. Agosta, *J. Org. Chem.*, 1976, **41**, 2947.
(c) M. Fetizon, S. Lazare, C. Pascard and T. Prange, *J. Chem. Soc., Perkin Trans. 1*, 1979, 1407.
(d) M. T. Crimmins and L. D. Gould, *J. Am. Chem. Soc.*, 1987, **109**, 6199.
(e) G. Mehta and K. S. Rao, *J. Chem. Soc., Chem. Commun.*, 1987, 1578.

[316] (a) Kahler, *Arch Pharm.*, 1830, **34**, 318.
(b) D. H. R. Barton, P. de Mayo and M. Shafiq, *J. Chem. Soc.*, 1958, 140.

[317] K. Schaffner and M. Demuth, *Rearrangements in Ground and Excited States*, ed. P. de Mayo, Academic Press, New York, 1980, **3**, 281.

[318] H. E. Zimmerman, *Adv. Photochem.*, 1963, **1**, 183; H. E. Zimmerman and D. I. Schuster, *J. Am. Chem. Soc.*, 1962, **84**, 4527; H. E. Zimmerman and J. S. Swenton, *J. Am. Chem. Soc.*, 1967, **89**, 906.

[319] J. S. Swenton, E. Saurborn, R. Srinivasan and F. I. Sontag, *J. Am. Chem. Soc.*, 1968, **90**, 2990.

[320] D. I. Schuster and D. J. Patel, *J. Am. Chem. Soc.*, 1965, **87**, 2515; 1966, **88**, 1825; 1967, **89**, 184; 1968, **90**, 5137, 5145; D. I. Schuster and V. Y. Abraitys, *Chem. Commun.*, 1969, 419; D. I. Schuster, K. V. Prabhu, S. Adcock, J. van der Veen and H. Fujiwara, *J. Am. Chem. Soc.*, 1971, **93**, 1557; D. I. Schuster and K. Liu, *J. Am. Chem. Soc.*, 1971, **93**, 6711.

[321] D. I. Schuster and C. J. Polowczyk, *J. Am. Chem. Soc.*, 1964, **86**, 4502; 1966, **88**, 1722; D. I. Schuster and I. S. Krull, *J. Am. Chem. Soc.*, 1966, **88**, 3456.

[322] W. H. Pirkle, S. G. Smith and G. F. Koser, *J. Am. Chem. Soc.*, 1969, **91**, 1580; W. H. Pirkle and G. F. Koser, *Tetrahedron Lett.*, 1969, 129.

[323] A. C. Brisimitzakis, D. I. Schuster and J. M. van der Veen, *Can. J. Chem.*, 1985, **63**, 685.

[324] A. G. Taveras, jun., *Tetrahedron Lett.*, 1988, **29**, 1103.

[325] (a) D. Caine and J. T. Gupton, tert., *J. Org. Chem.*, 1975, **40**, 809.
(b) D. Caine and F. N. Tuller, *J. Org. Chem.*, 1973, **38**, 3663.
(c) D. Caine, A. A. Boucugnani, S. T. Chao, J. B. Dawson and P. F. Ingwalson, *J. Org. Chem.*, 1976, **41**, 1539.

[326] (a) A. G. Schultz and M. Plummer, *J. Org. Chem.*, 1989, **54**, 2112.
(b) A. G. Schultz, M. Macielag and M. Plummer, *J. Org. Chem.*, 1988, **53**, 391.
[327] D. H. R. Barton, P. de Mayo and M. Shafiq, *J. Chem. Soc.*, 1958, 140.
[328] G. Quinkert, *Pure Appl. Chem.*, 1973, **33**, 285.
[329] D. H. R. Barton and G. Quinkert, *J. Chem. Soc.*, 1960, 1.
[330] (a) G. Quinkert, U.-M. Bilhardt, E. F. Paulus, J. W. Bats and H. Fuess, *Angew. Chem. Int. Ed. Engl.*, 1984, **23**, 442.
(b) A. G. Schultz and T. S. Kulkarni, *J. Org. Chem.*, 1984, **49**, 5202.
(c) R. J. Bastani and H. Hart, *J. Org. Chem.*, 1972, **37**, 2830.
[331] G. Quinkert, K. R. Schmieder, G. Durner, K. Hache, A. Stegk and D. H. R. Barton, *Chem. Ber.*, 1977, **110**, 3582.
[332] G. Quinkert, G. Fischer, U.-M. Bilhardt, J. Glenneberg, U. Hertz, G. Durner, E. F. Paulus and J. W. Bats, *Angew. Chem. Int. Ed. Engl.*, 1984, **23**, 440.
[333] G. Quinkert, H. Heim, J. Glenneberg, U.-M. Bilhardt, V. Autze, J. W. Bats and G. Durner, *Angew. Chem. Int. Ed. Engl.*, 1987, **26**, 362.
[334] H. Hart and A. J. Waring, *Tetrahedron Lett.*, 1965, 325; H. Hart, P. M. Collins and A. J. Waring, *J. Am. Chem. Soc.*, 1968, **88**, 1005; J. Griffiths and H. Hart, *J. Am. Chem. Soc.*, 1968, **90**, 5296.
[335] G. Quinkert, F. Cech, E. Kleiner and D. Rehm, *Angew. Chem. Int. Ed. Engl.*, 1979, **18**, 557.
[336] H. Klinger and O. Standke, *Ber. Dtsch. Chem. Ges.*, 1891, **24**, 1340; H. Klinger, *Liebigs Ann. Chem.*, 1888, **249**, 137; H. Klinger, *Liebigs Ann. Chem.*, 1911, **382**, 211.
[337] R. F. Moore and W. A. Waters, *J. Chem. Soc.*, 1953, 3405.
[338] H. Klinger and W. Kolvenbach, *Ber. Dtsch. Chem. Ges.*, 1898, **31**, 1214.
[339] Y. Miyagi, K. Maruyama, N. Tanaka, M. Sato, T. Tozimu, Y. Isogawa and H. Kashiwano, *Bull. Chem. Soc. Jpn*, 1984, **57**, 791.
[340] J. M. Bruce, D. Creed and K. Dawes, *J. Chem. Soc., Chem. Commun.*, 1969, 594.
[341] C. M. Orlando, H. Mark, A. K. Bose and M. S. Manhas, *J. Org. Chem.*, 1968, **33**, 2512.
[342] D. Bryce-Smith and A. Gilbert, *J. Chem. Soc.*, 1964, 2428.
[343] J. M. Bruce, *J. Chem. Soc.*, 1962, 2782; R. C. Cookson, D. A. Cox and J. Hudec, *J. Chem. Soc.*, 1961, 4499.
[344] (a) T. Naito, Y. Makita and C. Kaneko, *Chem. Lett.*, 1984, 921.
(b) K. Ogino, T. Matsumoto, T. Kawai and S. Kozuka, *J. Org. Chem.*, 1979, **44**, 3352.

[345] T. Susuki, Y. Yamashita, T. Mukai and T. Miyashi, *Tetrahedron Lett.*, 1988, **29**, 1405.
[346] T. Tsuji, Y. Hienuki, M. Miyake and S. Nishida, *J. Chem. Soc., Chem. Commun.*, 1985, 471; M. Miyake, T. Tsuji, A. Furusaki and S. Nishida, *Chem. Lett.*, 1988, 47.
[347] J. M. Bruce, *Quart. Rev.*, 1967, **21**, 405.
[348] K. Ogino, Y. Takaharu, T. Matsumoto and S. Kozuka, *J. Chem. Soc., Perkin Trans. 1*, 1979, 1552.
[349] K. Maruyama and H. Imakori, *Chem. Lett.*, 1988, 725.
[350] M. B. Rubin, *Fortschr. Chem. For.*, 1969, **13**, 251.
[351] L. Horner and H. Lang, *Chem. Ber.*, 1956, **89**, 2768; N. A. Shcheglova, D. N. Shigorin and M. V. Gorelik, *Russ. J. Phys. Chem.*, 1965, **39**, 471.
[352] (a) A. Schonberg, N. Latif, R. Moubasher and A. Sina, *J. Chem. Soc.*, 1951, 1364.
(b) W. I. Awad and M. S. Hafez, *J. Am. Chem. Soc.*, 1958, **80**, 6057.
[353] A. Takuwa, H. Iwamoto, O. Soga and K. Maruyama, *Bull. Chem. Soc. Jpn*, 1982, **55**, 3657.
[354] K. Maruyama and G. Takahashi, *Chem. Lett.*, 1973, 295.
[355] S. Kumamoto, K. Somekawa and S. Ide, *Kagoshima Daigaku Kogakubu Kenkyu Hokoku*, 1976, **18**, 101.
[356] W. M. Horspool and G. D. Khandelwal. *J. Chem. Soc., Chem. Commun.*, 1967, 1203.
[357] A. Takuwa, *Chem. Lett.*, 1989, 5.
[358] A. Takuwa and M. Sumikawa, *Chem. Lett.*, 1989, 9.
[359] Y. L. Chow, T. C. Joseph, H. H. Quon and J. N. S. Tan, *Can. J. Chem.*, 1970, **48**, 3045.
[360] S. Sekretar, J. Kopecky and A. Martvon, *Collect. Czech. Chem. Commun.*, 1983, **48**, 2812.
[361] K. Maruyama, T. Iwai and Y. Naruta, *Chem. Lett.*, 1975, 1219; K. Maruyama, M. Muaoka and Y. Naruta, *J. Org. Chem.*, 1981. **46**, 983.
[362] R. J. C. Koster and H. J. T. Bos, *Rec. Trav. Chim. Pays-Bas*, 1975, **94**, 72.
[363] H. J. T. Bos, H. Polman and P. F. E. Montfort, *J. Chem. Soc., Chem. Commun.*, 1973. 188.

# 4

# Sulphur-containing compounds

This chapter deals with the reaction of the group of organic molecules containing at least one sulphur atom. As with the previous chapters, the saturated and unsaturated examples are discussed and the material includes some gas phase processes to allow for a broad discussion of the area in general.

## 4.1 ABSORPTION SPECTRA OF THIOLS, SULPHIDES, AND DISULPHIDES

### 4.1.1 Thiols

Thiols, like alcohols, have a first absorption band which is assigned to the $n\sigma^*$ transition [1] with the maxima commencing at longer wavelengths than the oxygen analogues. Other maxima shown at shorter wavelengths by these compounds are attributed to Rydberg transitions. Typically methanethiol is reported to show three maxima in the gas phase at $\lambda = 254$ ($\varepsilon = 60$ $dm^3$ $mol^{-1}$ $cm^{-1}$), 230 and 204 nm [2, 3]. Ethanethiol has absorptions at similar wavelengths in the gas phase ($\lambda = 254$, $\varepsilon = 80$, 230 and 202 nm) while in solution only the maximum of the third band undergoes a shift to 196 nm [3, 4].

### 4.1.2 Sulphides

The ultraviolet spectra of sulphides are also dominated by the $n\sigma^*$ transition. Like the thiols the simple sulphides show three absorption bands, e.g. dimethyl sulphide 240 ($\varepsilon = 10\ dm^3\ mol^{-1}\ cm^{-1}$), 220, and 205 nm and diethyl sulphide 240 ($\varepsilon = 30\ dm^3\ mol^{-1}\ cm^{-1}$), 220 and 205 nm [2]. The cyclic derivatives, thiirane (**A1**) (260, 254, $\varepsilon = 17\ dm^3\ mol^{-1}\ cm^{-1}$, and 209 nm) and thietane (**A2**) (260, 254, $\varepsilon = 12\ dm^3\ mol^{-1}\ cm^{-1}$, and 205 nm) also show three absorption bands in the vapour phase [2].

S S

A1 A2

(A)

### 4.1.3 Disulphides

Like the above the spectra are relatively simple and the disulphides show the same $n\sigma^*$ transition, with the principal absorptions occurring at 254 nm. Typically dimethyl disulphide has absorptions at 254 nm with $\varepsilon = 300\ dm^3\ mol^{-1}\ cm^{-1}$ in the vapour and 250 nm with $\varepsilon = 316\ dm^3\ mol^{-1}\ cm^{-1}$ in solution [5] whereas diethyl sulphide absorbs at 254 nm ($\varepsilon = 310\ dm^3\ mol^{-1}\ cm^{-1}$) and 255 nm in the gas phase.

## 4.2 PHOTOCHEMISTRY OF THIOLS, SULPHIDES, AND DISULPHIDES

### 4.2.1 Thiols

The photochemistry of these compounds has been the subject of several reviews [6]. Irradiation of thiols brings about both S—H and R—S fission to afford radicals as shown (**1**). These reactions are extremely efficient and

$$RSH \xrightarrow{h\nu} R\bullet + HS\bullet$$

$$RSH \xrightarrow{h\nu} RS\bullet + H\bullet$$

(1)

in the gas phase the sum of the quantum yields for these processes is unity for the photolysis of methanethiol and ethanethiol [7, 8]. The products generated by the photolysis are then the result of thermal reactions of the radicals produced in the primary processes; these are summarized in (**2**). In

$$H\bullet + RSH \longrightarrow H_2 + RS\bullet$$

$$RS\bullet + RS\bullet \longrightarrow RSSR$$

$$R\bullet + RSH \longrightarrow RH + RS\bullet$$

$$HS\bullet + RSH \longrightarrow H_2S + RS\bullet$$

(2)

aqueous solution thiols are present in equilibrium with the corresponding anions and readily undergo ionization. Under such conditions the photolysis involves ejection of an electron from the thiolate (**3**) [9].

$$RS^- \xrightarrow[H_2O]{h\nu} RS\bullet + e_{aq}$$

$$RSH + e_{aq} \longrightarrow HS^- + R\bullet$$

$$RS^- + RS\bullet \longrightarrow [RSSR]^{\dot{-}}$$

(3)

## 4.2.2 Sulphides

Photolysis of sulphides falls into two distinct categories: the acyclic and the cyclic compounds.

### *4.2.2.1 Acyclic sulphides*

With the acyclic compounds the principal, if not the only, reaction path is photochemical fission of an S—C bond. There is little evidence for the fission of C—C bonds [10–13] or of C—H bonds. In unsymmetrically substituted compounds such as methyl ethyl sulphide (**4**) the smaller methyl group is

$$CH_3CH_2—S—CH_3 \xrightarrow{h\nu} CH_4 + CH_3CHS$$

(4)

expelled in preference [11, 12]. The overall efficiency of the reactions of the acyclic sulphides is less than that of the thiols, and the sum of the quantum yields for the primary processes is only 0.5 [13]. Molecular processes have also been reported but these are minor compared with the C—S fission path. The photochemical behaviour of the dithiaacetals is reminiscent of sulphide photolysis where S—C bond fission is dominant. Typical of this is the behaviour of 1,1-bis(methylthio)cyclohexane (**5**) which affords cyclohexyl

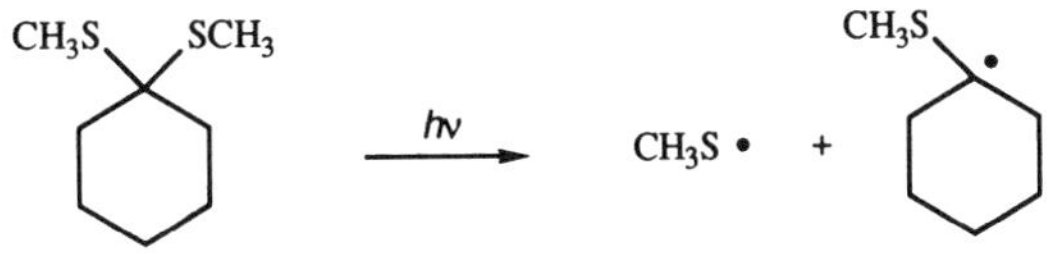

(5)

methyl sulphide and dimethyl disulphide as the principal products. The initial photolysis affords the radicals shown in **5** and dimerization of the methylthio radical affords the dimethyl disulphide [14]. The route to the other product is less straightforward and it could be formed by disproportionation or by a molecular path.

#### *4.2.2.2 Cyclic sulphides*

The photochemistry of the cyclic sulphides has been extensively reviewed [6]. The reactions encountered are markedly different from the open-chain systems, although again the principal reaction is the fission of an S—C bond, as shown in the photolysis of thiirane (**6**). Thiirane undergoes population of

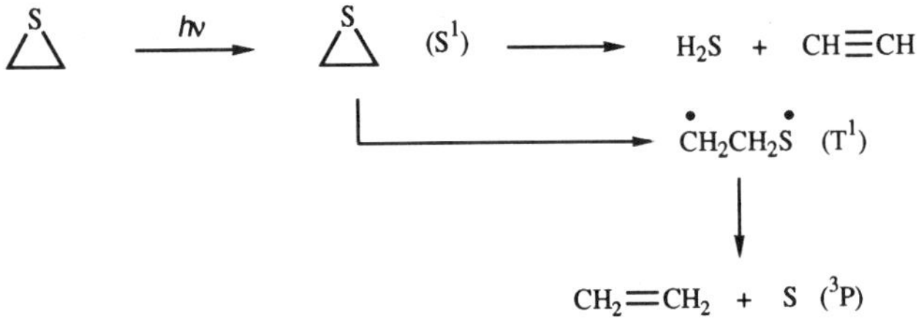

(6)

the singlet state on direct irradiation from which acetylene and hydrogen sulphide are formed. Intersystem crossing to the triplet state produces the triplet 1,3-biradical by S—C bond fission [15]. Thietane (**7**) also undergoes

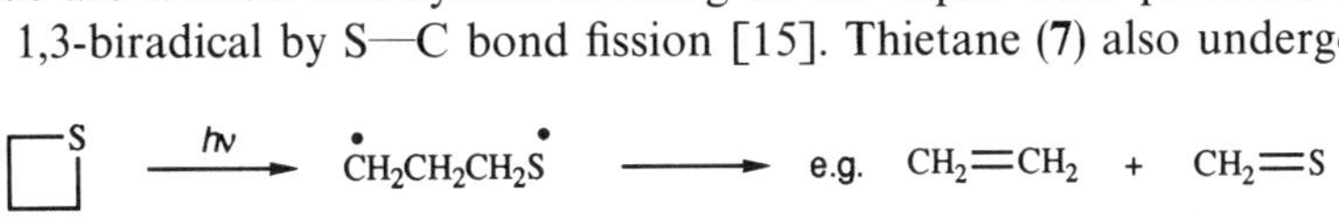

(7)

C—S bond fission on irradiation at 254 or 310 nm [16]. The 1,4-biradical shown in **7** accounts for the formation of the majority of products from the reaction. Ascending the series to thiolane (**8**) also finds the reactions

$$\text{(tetrahydrothiophene)} \xrightarrow{h\nu} \dot{C}H_2CH_2CH_2CH_2\dot{S} \xrightarrow{\text{e.g.}} CH_2{=}CH_2 + \dot{C}H_2CH_2\dot{S} \rightarrow \text{etc.}$$

$$\dot{C}H_2CH_2CH_2CH_2\dot{S} \xrightarrow{\text{e.g.}} \dot{C}H_2CH_2\dot{C}H_2 + H_2CS; \quad \dot{C}H_2CH_2\dot{C}H_2 \rightarrow \text{etc.}$$

(8)

dominatedby S—C bond fission. The final outcome of the reaction is dependent on the wavelength used, although irradiation at 254 or 214 nm indicates that the biradical plays an important role [17].

### 4.2.3 Disulphides

Several reviews have been devoted to this area of study [6]. In these compounds excitation produces an excited state which undergoes fission of either the S—S bond or the C—S bonds. Some reports suggest that by using $\lambda > 230$ nm only the S—S fission results [5, 18]. However, the results are inconclusive since there are reports of C—S fission resulting under similar conditions [19]. In the gas phase irradiation of dimethyl disulphide at 195 nm shows both processes occurring in a ratio of 1.35 : 1 [20]. In solution the C—S bond fission no longer occurs and the reactions are dominated by the S—S fission process. However, structural features can change the outcome of the reaction as with di-*t*-butyl disulphide (**9**) where C—S fission is important

$$Bu^t{-}S{-}S{-}Bu^t \xrightarrow[\substack{254\ \text{nm} \\ CCl_4}]{h\nu} S_2 + 2\ (CH_3)_3C\bullet \longrightarrow Bu^tCl \quad \text{90\% yield}$$

(9)

[21].

## 4.3 PHOTOCHEMISTRY OF COMPOUNDS WITH SULPHUR—HALOGEN BONDS

### 4.3.1 Alkyl sulphenyl halides

The ultraviolet/visible absorptions of the RSCl group of the sulphenyl halides is typified by methane sulphenyl chloride (**10**) which shows maxima at 355

$$CH_3SCl$$

(10)

($\varepsilon = 24.5$ dm$^3$ mol$^{-1}$ cm$^{-1}$) and 205 ($\varepsilon = 235$ md$^3$ mol$^{-1}$ cm$^{-1}$) nm [22]. Changes in substitution on the carbon do not appear to influence either the position or the intensity of the bands, as shown by **11**, which has low-intensity

$$CF_3SCl$$

(11)

maxima at 333 ($\varepsilon = 25$ dm$^3$ mol$^{-1}$ cm$^{-1}$) and 214 ($\varepsilon = 235$ dm$^3$ mol$^{-1}$ cm$^{-1}$) nm in the gas phase and maxima at 322 ($\varepsilon = 10$ dm$^3$ mol$^{-1}$ cm$^{-1}$) and 324 ($\varepsilon = 12$ dm$^3$ mol$^{-1}$ cm$^{-1}$) nm in solution [23]. The commonest photoreaction observed for the sulphenyl halides is fission of the S—halogen bond and the production of the corresponding sulphur-centred radical and halogen atoms. The chemistry thereafter is of free radical type. Tri-fluoromethanesulphenyl chloride (**11**) is typical and decomposes on irradiation to trifluoromethyl chloride and sulphur chlorides [23]. Other studies have used the free radical nature of the reaction, with the addition of the chlorine atom as the key step, to effect addition to alkenes (**12**) [24]. Haas and Klug [25]

$$CF_3SCl + CF_2{=}CHF \xrightarrow{h\nu} \underset{50\%}{CF_3SCF_2CHFCl} + \underset{11\%}{CF_3SCFHCF_2Cl}$$

$$CF_3SCl + ClCF{=}CF_2 \xrightarrow{h\nu} \underset{42\%}{CF_3SCFClCF_2Cl} + \underset{12\%}{CF_3SCF_2CFCl_2}$$

$$CF_3SCl + CH_2{=}CHCl \xrightarrow{h\nu} \underset{12\%}{CF_3SCH_2CHCl_2} + \underset{73\%}{CF_3SCHClCH_2Cl}$$

(12)

have also described the photochemical behaviour of $CF_3SCl$ and its use in the synthesis of disulphides. The photochemical reactions of trichloromethanesulphenyl chloride follow a similar reaction path and can be added to alkenes such as styrene, cyclohexene, and benzofuran, as illustrated in **13** [26–28]. The photochemical reactions of the sulphenyl chloride

$$\text{Ph-CH=CH}_2 + CCl_3SCl \xrightarrow{h\nu} \text{PhCH(SCCl}_3\text{)CH}_2\text{Cl} \quad 54\%$$

$$\text{cyclohexene} + CCl_3SCl \xrightarrow{h\nu} \text{2-chlorocyclohexyl-SCCl}_3 \quad 26.7\%$$

$$\text{benzofuran} + CCl_3SCl \xrightarrow{h\nu} \text{3-chloro-2-(SCCl}_3\text{)-2,3-dihydrobenzofuran} \quad 21.4\%$$

(13)

(**14**) also provide a route to sulphur-centred radicals by fission of the

ClSCN

(14)

S—Cl bond [29, 30]. The resultant thiyl radical adds efficiently to alkenes [29, 30]. Trichloromethanesulphenyl chloride also undergoes hydrogen abstraction reactions and in cyclohexane, for example, gives cyclohexyl chloride in 80% yield [31]. Subsequent work has shown that trichloromethane sulphenyl chloride is a highly selective chlorinating agent [32]. There are only a few reports on the photochemical behaviour of sulphenyl iodides since few have been obtained as isolable compounds [33].

### 4.3.2 Aryl sulphenyl halides

Pentachlorobenzenesulphenyl chloride (**15**) reacts in sunlight with cyclohexane

$$C_6Cl_5SCl + \text{cyclohexane} \xrightarrow{h\nu} \text{cyclohexyl-Cl} + \text{cyclohexyl-SAr} + ArSSAr + HCl \qquad Ar = C_6Cl_5$$

(15)

to yield the products shown. It is interesting to note that substitution by both chlorine atoms and arylsulphenyl radicals takes place but that the latter process is dominant. Different results are obtained when irradiation is carried out in toluene. Here the process is dominated by the formation of a sulphide, a preference explained by the involvement of a charge-transfer interaction between the sulphenyl chloride and and toluene [34]. Photoreaction of this sulphenyl chloride has also been studied in the presence of thiophene [35, 36].

The photochemical oxygen transfer reaction from an *o*-nitro group to the sulphur has some synthetic value and the general conversion process is shown in **16** [37, 38]. This reaction is not restricted to the dinitro derivatives and

SCl, $NO_2$, $NO_2$ —$h\nu$→ $SO_3H$, $NH_2$, $NO_2$

(16)

Pillai [39] has reported the conversion of 2-nitrobenzenesulphenyl chloride (**17**) into 2-aminobenzenesulphonic acid in 80% yield. This reaction is

$NO_2$, SCl → $NO_2$, SOH —$h\nu$→ N:, $SO_3H$ → $NH_2$, $SO_3H$ 80%

(17)

presumed to involve conversion of the sulphenyl halide into the corresponding sulphenic acid. Oxygen transfer from the nitrogen atom then follows ultimately to afford the observed compound.

### 4.3.3 Sulphonyl halides

The photochemical reactivity of sulphonyl chlorides is again dominated by the weak S—Cl bond. Irradiation of these results in fission into chlorine atoms and sulphonyl radicals. Typical results from such irradiations have been reported by Horowitz [40] in the irradiation of methane sulphonyl chloride in cyclohexane at 150°C. The principal gaseous products were identified as cyclohexyl chloride, methane, and sulphur dioxide with a trace of methyl chloride. Irradiation of cyclohexanesulphonyl chloride in the presence of oxygen brings about oxidation of the cyclohexylsulphonyl radical to yield cyclohexanesulphonic acid [41].

As with the sulphenyl halides, addition reactions to suitable substrates can also be brought about on irradiation of sulphonyl chlorides. Thus irradiation of benzenesulphonyl chloride (**18**) using a tungsten lamp in the presence of

$$PhSO_2Cl + CH_3CH{=}CHCH_3 \xrightarrow{h\nu}$$

PhSO2 Cl

(18)

either *cis*- or *trans*-but-2-ene affords 1:1 adducts [42, 43]. Photo-addition of *p*-toluenesulphonyl chloride (**19**) to norbornene yields the *trans*-adduct in

$$p\text{-}MeC_6H_4SO_2Cl + \xrightarrow{h\nu}$$

SO2Tolyl

64% Cl

(19)

64% yield. This process is selective and does not bring about structural rearrangement of the norbornane skeleton nor is the *cis*-adduct formed [44]. However, skeletal rearrangement does occur on the photoaddition of benzenesulphonyl chloride (**20**) to norbornadiene [45] yielding the products

$$PhSO_2Cl + \longrightarrow$$

SO2Ph Cl SO2Ph

Cl

trace 90%

(20)

shown. The reactivity of benzenesulphonyl chloride is poor, and better yields of adducts are obtained using the corresponding bromo- and iodo-sulphonyl compounds. Reviews of the synthetic utility of these reactions have been published [46, 47]. Other applications have also been reported, such as the irradiation of trichloromethanesulphonyl chloride as a means of halogenating alkanes via the trichloromethanesulphonyl radical. The system was found to be more specific than sulphuryl chloride [48–50].

As mentioned above, the sulphonyl iodides are often more reactive than the corresponding chlorides. Again the radicals, an iodine atom and a sulphonyl radical, can be trapped in a reaction which has some synthetic utility. Typical of this is the irradiation of *p*-toluenesulphonyl iodide (**21**) in

$p$-$MeC_6H_4SO_2I$ $\xrightarrow{h\nu}$ (with butadiene) $p$-$MeC_6H_4SO_2CH_2CH{=}CHCH_2I$ 89%; (with acrylonitrile) $p$-$MeC_6H_4SO_2CH_2CH(CN)I$ 87%

(21)

the presence of acrylonitrile and butadiene, affording the adducts shown [51]. Truce *et al.* [52] have also utilized this reaction mode in a study of the light-induced addition of sulphonyl iodides to allenes.

## 4.4 PHOTOCHEMISTRY OF COMPOUNDS WITH SULPHUR—NITROGEN BONDS

### 4.4.1 Sulphenamides

Sulphenamides (**22**) are also photochemically reactive and undergo homolytic

$$(CH_3)_3C{-}N(H)SAr \xrightarrow{h\nu} (CH_3)_3C{-}\dot{N}H + ArS\bullet$$

(22)

fission of the S—N bond, resulting in the production of sulphur- and nitrogen-centred radicals [53, 54]. This conversion into radicals can be used to obtain rearrangement products, as in the irradiation of the sulphenamides (**23**) where photo-Fries-type products (see section 4.4.3.3) are formed by the

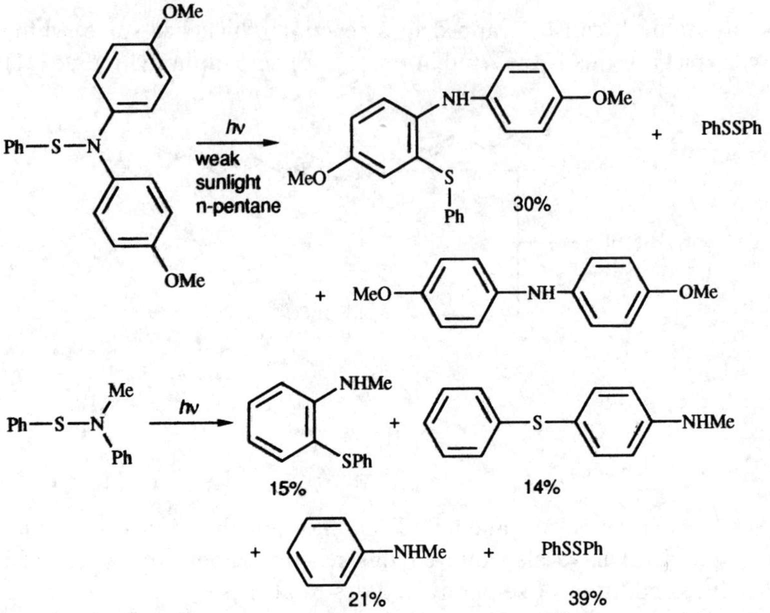

(23)

lateral migration of the sulphur radical on the aryl group [55]. Changes of substitution on the aryl rings can affect the reaction and when *o*-nitro groups are present as in **24** then an oxygen-transfer reaction from the neighbouring

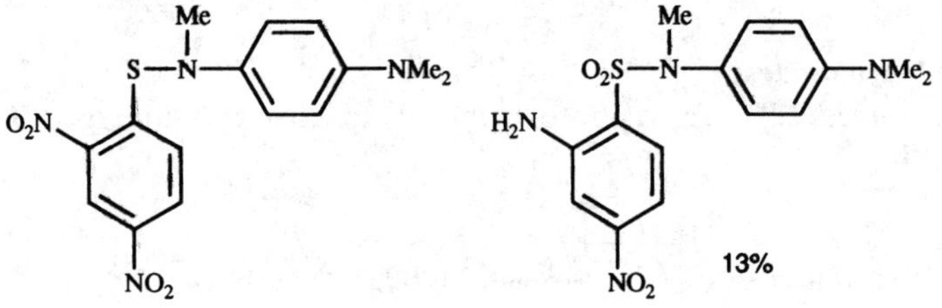

(24)

nitro group is observed, affording sulphonamides [56]. This reaction is reminiscent of the process described previously for the sulphenyl halides.

### 4.4.2 Isothiazolonones

The isothiazole group of compounds is related to the sulphenamides. The outcome of irradiation of such systems is again dominated by the weakest bond and usually fission of the S—N bond results in the formation of radicals from which the products are obtained. A typical reaction is shown for the irradiation of the isothiazol-3(2*H*)-ones (**25**), which affords the isomeric

| R | % yield |
|---|---|
| Ph | 88 |
| $PhCH_2$ | 80 |
| $ClCH_2CH_2$ | 71 |
| $Bu^t$ | 30 |
| $C_6H_{11}$ | 70 |

(25)

compounds. The initially formed 1,5-biradical ring closes to the aziridinone from which the final products are obtained [57]. Analogous behaviour is reported for the 2-aryl-1,2-benzisothiazol-3(2*H*)-ones (**26**) which, on ir-

a: R = OMe
b: R = Me
c: R = H
d: R = Cl
e: R = CN

yield (%)
a: 47
b: 57
c: 31
d: 13
e: 9

(26)

radiation through Corex or Pyrex, rearrange as shown [58]. This free radical path, cyclization, and 1,7-hydrogen migration is common to a variety of derivatives of **26** such as the pyridyl [59], pyrazinyl [59], and naphthyl [60] analogues. All of these rearrange efficiently, yielding the corresponding thiazepinones.

## 4.4.3 Sulphonamides

### *4.4.3.1 Photodeprotection*

One of the commoner photochemical reactions of sulphonamides and their derivatives is conversion to the free amine, a process referred to as photochemical deprotection of functional groups. The synthetic usefulness of this reaction sequence has been reviewed by Pillai [61]. A typical example of this type of behaviour is that of the SET (single electron transfer) reactivity of the sulphonamides (**27**). These compounds can be readily detosylated on

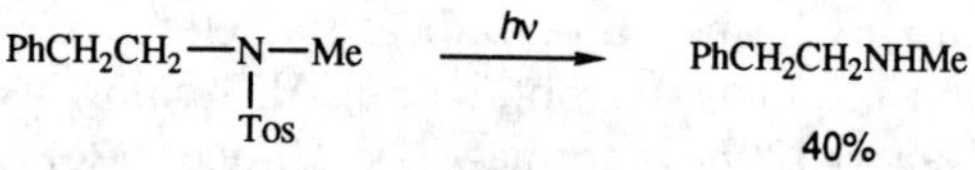

$$PhCH_2CH_2{-}N(Tos){-}Me \xrightarrow{h\nu} PhCH_2CH_2NHMe \quad 40\%$$

(27)

irradiation in the presence of electron–donating sensitizers (1,2-dimethoxybenzene or 1,4-dimethoxybenzene) and reductants such as ammonia, borane, hydrazine or sodium tetrahydroborate, giving high yields of the corresponding amines [62, 63]. SET photochemistry of the toluenesulphonamide (**28**) using 1,5-dimethoxynaphthalene as the electron donor brings

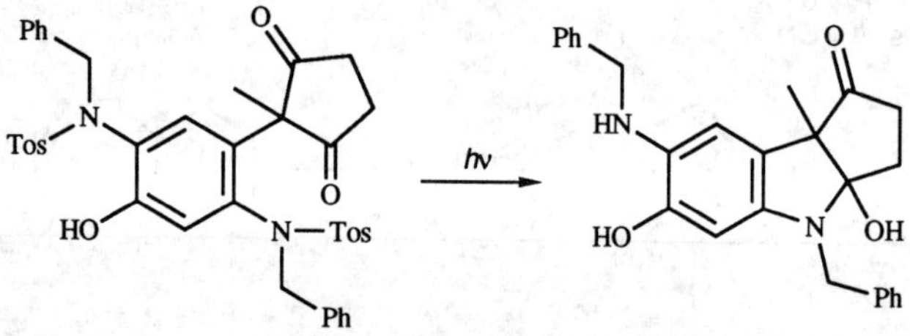

(28)

about double detosylation and cyclization to yield the product illustrated [64]. Efficient photodetosylation of the cyclic sulphonamides (**29**) can be

a: R = H
b: R = Me
c: R = Ph
d: R = $PhCH_2$

a: 83%
b: 77%
c: 71%
d: 99%

(29)

effected by irradiation in a mixture of ethanol/sodium carbonate and sodium borohydride. The free amines are obtained in good to excellent yield [65–67]. The presence of a reductant is important and its absence leads to products of elimination and oxidation. Deprotection is also effective on direct irradiation, as reported by Pincock and Jurgens [68]. A variety of derivatives was studied using the conditions shown in **30** and **31**. The yields of amine

$$PhCH_2SO_2{-}NR^1R^2 \xrightarrow[Pr^iOH]{h\nu} Ph\dot{C}H_2 + \bullet SO_2NR^1R^2 \longrightarrow R^1R^2N{-}H$$

a: $R^1 = R^2 = C_4H_9$
b: $R^1 = H$, $R^2 = (CH_2)_8CH_3$
c: $R^1 = H$, $R^2 = C_6H_{11}$
d: $R^1 = H$, $R^2 = Ph$

yield (%)
a: 78
b: 81
c: 98 (in MeOH)
d: 10

(30)

$$PhSO_2{-}NR^1R^2 \xrightarrow[Pr^iOH]{h\nu} Ph\dot{S}O_2 + \bullet NR^1R^2 \longrightarrow R^1R^2N{-}H$$

a: $R^1 = R^2 = C_4H_9$
b: $R^1 = H$, $R^2 = C_6H_{11}$

yield (%)
a: 81
b: 96

(31)

obtained vary from 10–98 % yield. Two mechanistic paths are proposed. One involves C—S bond fission for sulphonamides (**30**) with the formation of a radical pair which then undergoes loss of $SO_2$. The other path is followed by the sulphonamides (**31**) where S—N bond fission is operative. Others have also demonstrated the ease of deprotecting amines by the photochemical cleavage of a series of *p*-toluenesulphonamides in ether [69, 70].

### *4.4.3.2 Loss of sulphur dioxide*

As mentioned above, irradiation of some sulphonamides brings about the extrusion of sulphur dioxide. One example of such a process is shown in **32**

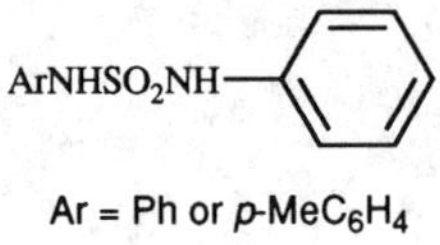

$ArNHSO_2NH-C_6H_5$

Ar = Ph or p-$MeC_6H_4$

(32)

where irradiation brings about the formation of azobenzene and aniline. The former product is thought to arise by an intramolecular path since the photolysis of the tolyl derivative yields only 3-methylazobenzene and no mixed derivatives [71]. Pete and his coworkers have also shown that an intramolecular reaction is involved in the photoconversion of the enones (**33**)

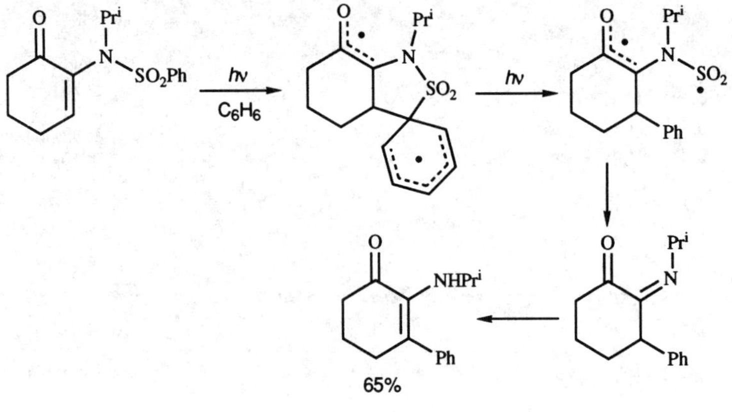

hν, $C_6H_6$ → hν → → 65%

(33)

into the 3-substituted enones [72, 73]. Further studies have shown that the outcome of the reaction is dependent upon the type of substituent on the sulphonamide side chain. If hydrogen abstraction is possible, a Norrish Type II process, competing reactions occur affording the azetidine derivatives (**34**)

a: R = Me, Ar = Tolyl
b: R = Ph, Ar = Tolyl
c: R = Tolyl, Ar = Tolyl

a: 30% a: - a: 25%
b: 50% b: - b: 10%
c: 40% c: 35% c: 15%

(34)

[74, 75]. In small ring systems photochemical loss of sulphur dioxide can also occur efficiently. Thus the photochemical decomposition of the thiazete (**35**) in benzene at 30°C affords carbazole by loss of $SO_2$ and cyclization

(35)

within the resultant biradical [76]. Loss of sulphur dioxide also occurs on irradiation of the sultam (**36**) yielding the quinomethane imine which ring

300 nm

62%

(36)

closes to an azetidine [77]. Another example of some synthetic value is shown in **37** [78, 79].

R = H or Me

Major product

(37)

Rupture of the N—S bond is the dominant photoreaction of the sultam (**38**), affording the intermediate shown. In many respects this ring-opening

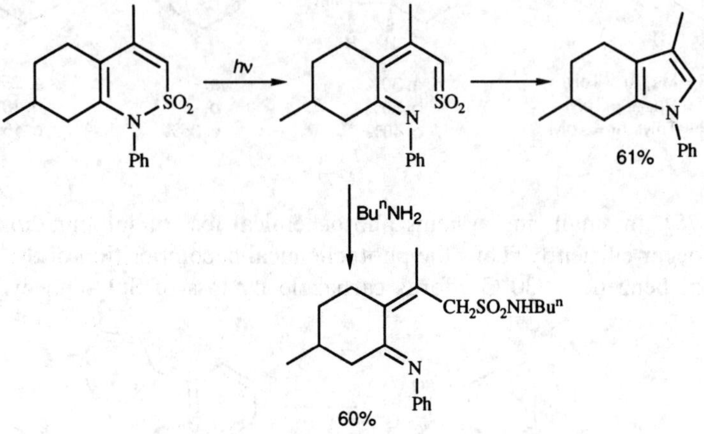

$SO_2$ N Ph $h\nu$ $SO_2$ N Ph N Ph 61%

$Bu^nNH_2$

$CH_2SO_2NHBu^n$ N Ph 60%

(38)

process is analogous to those described for dienes (Chapter 2) and linearly conjugated dienones (Chapter 3). In the absence of an external nucleophile, ring closure affords a pyrrole, but when *n*-butylamine is added to the reaction mixture intermolecular trapping affords an open-chain sulphonamide [80]. The sulphonamides (**39**) also undergo ring opening on irradiation. However, in benzene or methanol, ring expansion by a photo-Fries-type 1,3-migration occurs to produce the observed products, following a 1,3-hydrogen migration, in yields of 50–70% [81].

hν

50-70%

$R^1$ = H, Me, OMe or *p*-$C_6H_4CO_2Me$
$R^2$ = H or Me

(39)

#### *4.4.3.3 Photo-Fries reactions*

As described previously, in Chapters 2 and 3, the photo-Fries reaction of a substituted aromatic compound results in the lateral migration of a group following fission of a side chain [82]. The examples recorded here involve the migration of sulphur groups and a typical example of such a process is the rearrangement of the *N*-phenylsulphonamides (**40**), which on irradiation

R–N–$SO_2$Tolyl   hν   NHR, $SO_2$Tolyl + NHR, $SO_2$Tolyl

6-35%   13-35%

(40)

are converted into the *o*- and *p*-amino-substituted sulphones in the yields shown [83]. Other examples involving a variety of aromatic compounds are illustrated in **41** [84–87].

$ArSO_2NH$–C₆H₂($R^1$,$R^2$,$R^3$) $\xrightarrow{h\nu}$ $ArSO_2$–C₆H₂($R^1$,$R^2$)–$NH_2$ + $NH_2$–C₆H₂($R^1$,$R^2$,$R^3$)

| | | |
|---|---|---|
| Ar = *p*-Tolyl, $R^1 = R^2 = R^3 = H$ | 25% | 68% |
| Ar = Ph, $R^1 = R^2 = R^3 = H$ | 12% | 30% |
| Ar = Ph, $R^1$ = Me, $R^2 = R^3 = H$ | 14% | 25% |
| Ar = Ph, $R^2$ = Me, $R^1 = R^3 = H$ | 6% | 43% |
| Ar = Ph, $R^3$ = Me, $R^1 = R^2 = H$ | - | 41% |

Ref [84]

$R^1O$, $R^1O$, N—Tos, $R^2$ $\xrightarrow{h\nu}$ $R^1O$, $R^1O$, N—H, Tos

$R^1$ = Me, $R^2$ = H, D or Cl 40% Ref [85]

N-$SO_2$Tolyl indole $\xrightarrow[EtOH]{h\nu}$ 3-$SO_2$Tolyl indole (N–H) + $R^1$, $R^2$ indole (N–H)

24%

$R^1$ = H, $R^2$ = $SO_2$Tolyl, 18%
$R^1$ = $SO_2$Tolyl, $R^2$ = H, 2% Ref [86]

N-$SO_2R$ carbazole → $R^1$, $R^2$ carbazole (N–H)

R = Ph, *p*-Tolyl or Me

$R^1 = SO_2R$, $R^2$ = H: 35%, R = Ph; 38%, R = *p*-Tolyl; 48%, R = Me
$R^1$ = H, $R^2 = SO_2R$: 45%, R = Ph; 45%, R = *p*-Tolyl; 30%, R = Me

Ref [87]

(41)

## 4.5 PHOTOCHEMISTRY OF COMPOUNDS WITH SULPHUR—OXYGEN BONDS

### 4.5.1 Sulphenates

Sulphenate esters (**42a**) exhibit weak absorptions at 265.5 ($\varepsilon = 70.5\ dm^3\ mol^{-1}\ cm^{-1}$) nm [88]. While the irradiation of these has not been reported, the trichloromethyl derivative (**42b**) is reported to undergo fission by rupture of the S—O bond [89, 90]. Aryl derivatives of sulphenates have been studied in more detail and irradiation of the compound (**42c**) has shown that these,

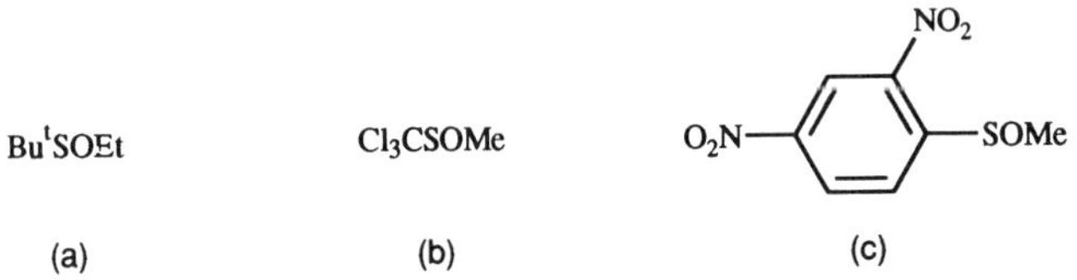

$Bu^tSOEt$ (a)  $Cl_3CSOMe$ (b)  2,4-$(O_2N)_2C_6H_3SOMe$ (c)

(42)

which are considered to be more like ethers than esters [88], are inert to irradiation [38]. However, change of substitution on sulphur, as in **43**,

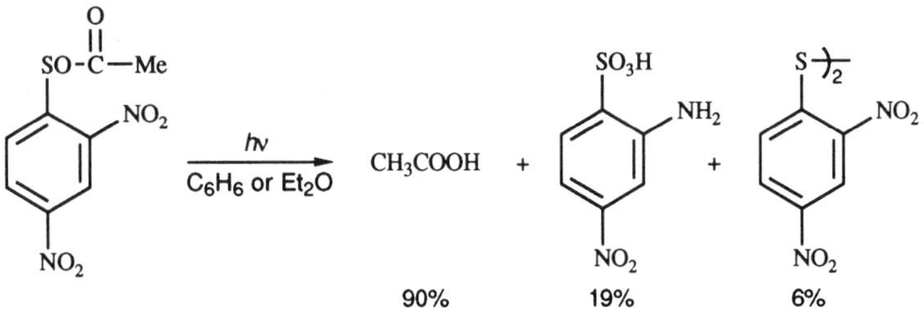

2,4-$(NO_2)_2C_6H_3$SO–C(=O)–Me $\xrightarrow[C_6H_6 \text{ or } Et_2O]{h\nu}$ $CH_3COOH$ (90%) + 2-$NH_2$-4-$NO_2C_6H_3SO_3H$ (19%) + $(2,4\text{-}(NO_2)_2C_6H_3S)_2$ (6%)

(43)

reintroduces photochemical reactivity. The important product in each case was the corresponding aminosulphonic acid where oxygen had been transferred intramolecularly from the *o*-nitro function to the sulphur [55]. However, others have reported different reactivity (see scheme **45**) This type of reactivity is analogous to that described by Kaluza and Perold [37] for the reaction of 2,4-dinitrobenzenesulphenyl chloride described earlier. Other derivatives (**44**) are also photolabile on irradiation in benzene or ether, and again

$SO{-}C(=O){-}CH_2Ph$, $NO_2$, $NO_2$ $\xrightarrow{h\nu}$ $PhCH_2CO_2H$ 98%

(44)

again the products resulting from the cleavage of the SO bond provide excellent yields of the corresponding carboxylic acid [38, 91]. The fission process observed in these compounds is a heterolysis generating ions rather than radicals. This proposal is supported by the irradiation of 2,4-dinitrobenzenesulphenyl acetate in a mixture of benzene and anisole, where no 2,4-dinitrophenyl sulphide was obtained and the product was the methoxy derivative resulting from trapping the sulphur cation by anisole in an electrophilic reaction. The scope of this electrophilic process has been studied and some results are shown in **45** [55]. The *o*-nitrobenzenesulphenyl acetate

NO$_2$ S Ph hν $C_6H_6$ O$_2$N 73%

NO$_2$ S hν PhCl SOCOMe O$_2$N Cl O$_2$N 14%

NO$_2$ NO$_2$ S hν O$_2$N 40%

NO$_2$ S S hν S O$_2$N 13%

(45)

(**46**) is also photochemically reactive, and irradiation in benzene affords a

SOCOMe NO$_2$ hν $C_6H_6$ S)$_2$ NO$_2$

NO$_2$ S hν OMe OMe

(46)

disulphide, whereas in anisole irradiation follows the previously described path [55].

### 4.5.2 Rearrangement of sulphoxides

Racemization is one of the commoner reactions of this category of compounds and the process is brought about by homolytic fission and rebonding of an S—C bond [92]. Alternative paths are possible, such as rebonding to yield a sulphenate by a process formally analogous to the formation of oxacarbenes from ketones (Chapter 3) [93]. Secondary photolysis of the sulphenate would result in S—O bond fission and, via the resultant biradicals, provide a route to a variety of products. The involvement of sulphenates was initially observed on irradiation of the sulphoxide (**47**) [94]. Schultz and Schlessinger [95] also

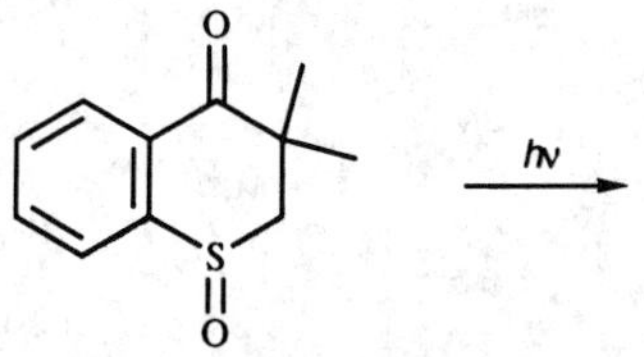

(47)

observed that direct irradiation of the *cis*-sulphoxide (**48**) yielded the

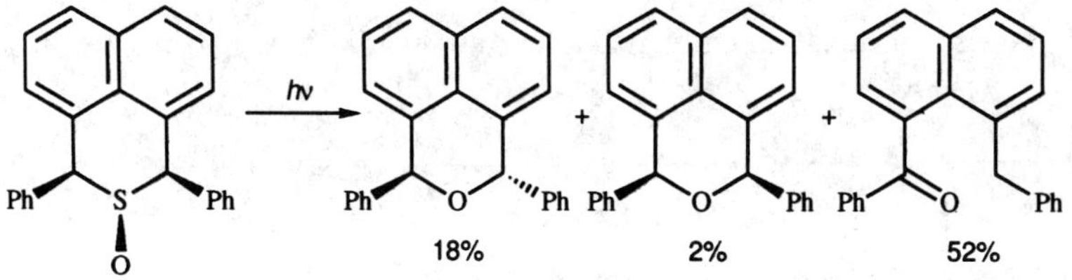

(48)

*trans*-pyran, the *cis*-pyran, and the ketone in 18, 2, and 52% yields respectively. The *trans*-sulphoxide yields the same three products but in different yields (2, 8, 13% respectively). The pyrans arise from the transformation of the sulphoxide into the sulphenate followed by secondary irradiation in which loss of sulphur occurs. Several examples (**49**) of the rearrangement

hν

Ref [96]

R = H
$C_6H_6$

+ $PhCO_2H$
15% 25%

R = D

CHO
+
25%
Ref [97]

CDO + D + COOH
D D D

hν

Ar Ar CHO + COAr H

Ar = $p$-$ClC_6H_4$ 38% 56%
Ar = Ph 32% 46%
Ar = $p$-$MeC_6H_4$ 79% 13%
Ref [98-100]

Ph H hν
$CO_2H$
Ref [101]

(49)

of sulphoxides via the sulphenate esters are recorded [96–101]. The episulphoxide (**50**) is also photoreactive and on irradiation is converted into benzil and thiobenzil via conversion into the sulphenate prior to fragmentation [102, 103].

(50)

### 4.5.3 Photochemistry of sulphonates

#### *4.5.3.1 Open–chain systems*

The photoreactions of sulphonates have been shown to involve the singlet excited state. The reactions for the conversion of the esters into the free sulphonic acids are reasonably efficient, with quantum yields in the 0.02 to 0.07 range. The processes involve homolytic fission, a postulate substantiated by the work of Izawa and Kuromiya [104] who demonstrated that irradiation of the sulphonate ester (**51**) in methanol affords benzene, biphenyl and a trace

$$PhSO_3Me \xrightarrow[MeOH]{h\nu} \left[\begin{matrix} Ph\dot{S}O_2 + Me\dot{O} \\ Ph^\bullet + \dot{S}O_3Me \end{matrix}\right] \longrightarrow Ph^\bullet + SO_2 + MeO^\bullet \longrightarrow \underset{24\%}{C_6H_6} + \underset{12\%}{PhPh} + \underset{trace}{PhOMe}$$

(51)

of anisole. The formation of products is thought to involve two free radical paths, i.e. either S—O or S—C bond fission, as shown. The S—C bond fission path has been exploited as a method for photodeprotection of alcohols. The earliest examples were reported by Pete and his group [105–107] who demonstrated that the free alcohol could be obtained in reasonable yields on irradiation of tosyloxy derivatives by the route shown, where fission of the C—S bonds yields a radical pair and loss of $SO_2$ affords an alkoxy radical from which the free alcohol is produced (**52**). Many examples of this

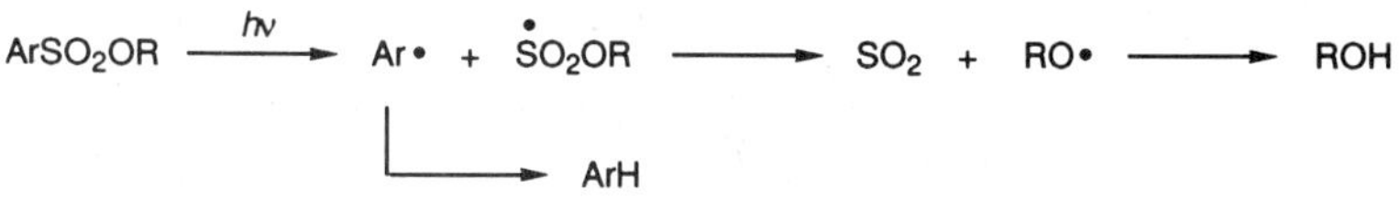

$$ArSO_2OR \xrightarrow{h\nu} Ar\bullet + \dot{S}O_2OR \longrightarrow SO_2 + RO\bullet \longrightarrow ROH$$
$$Ar\bullet \longrightarrow ArH$$

(52)

deprotection path have been reported over the years [108–110] and it has been used extensively on carbohydrate derivatives (**53**) which are readily

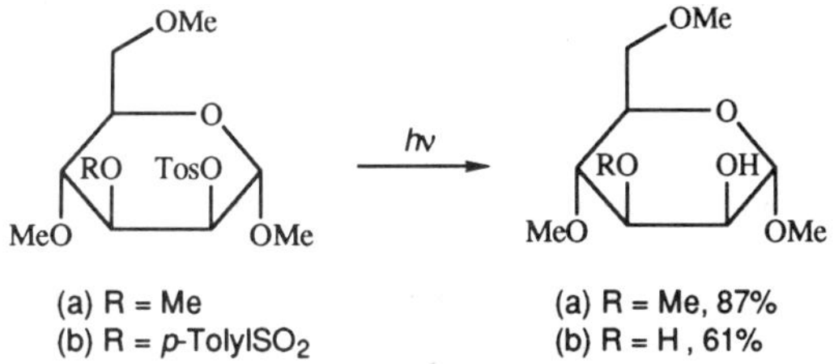

(a) R = Me
(b) R = $p$-TolylSO$_2$

(a) R = Me, 87%
(b) R = H, 61%

(53)

converted into the free alcohols on irradiation in yields up to 87% [111, 112]. Even higher yields can be obtained as with the irradiation of **54** which

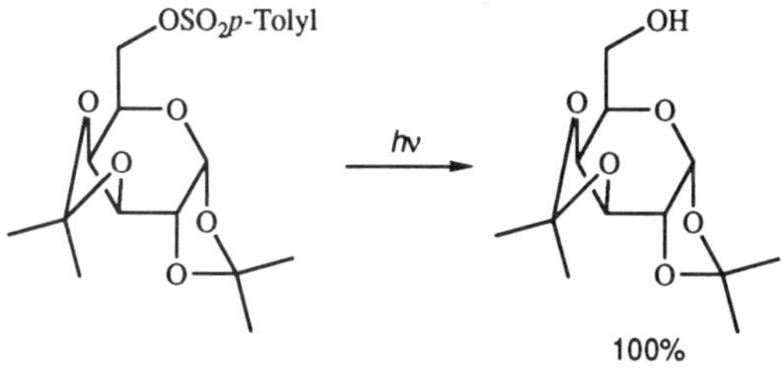

100%

(54)

affords 100% detosylation [113]. Many other examples of synthetic use have been reported [114]. A change of solvent plays an important role in the success of the reaction. Of particular value is the use of a hexamethylphosphoric triamide/water system. Under these conditions and using light of 254 nm the efficient conversions shown in **55** were effected [115, 116]. No

R = $MeSO_2$
R = *p*-Tolyl$SO_2$

*hν*

OMe
$NHCO_2Me$

trace
-

86%
91%

(55)

reaction was detected on irradiation at 300 nm. Irradiation of steroidal tosylates in the presence of sodium borohydride can also bring about deprotection, with the formation of the corresponding alcohols [117]. Binkley [118] has demonstrated that under such conditions an ionic and not a free radical mechanism is involved. Thus the fate of the alkoxy radical is dependent upon the conditions under which the reaction is carried out, and in the presence of base such as $Et_3N$ or $MeO^-$ a SET is involved and the alkoxy radical is converted to an alkoxide. The alkoxide path is substantiated by the detosylation of carbohydrate derivatives in base without epimerization at carbon (**56**). Other studies have shown the efficient detosylation of the

*hν*
25 h/quartz
MeOH/NaOH

85%

*hν*

70%

(56)

carbohydrate derivatives (**57**) in the presence of benzyl groups [119].

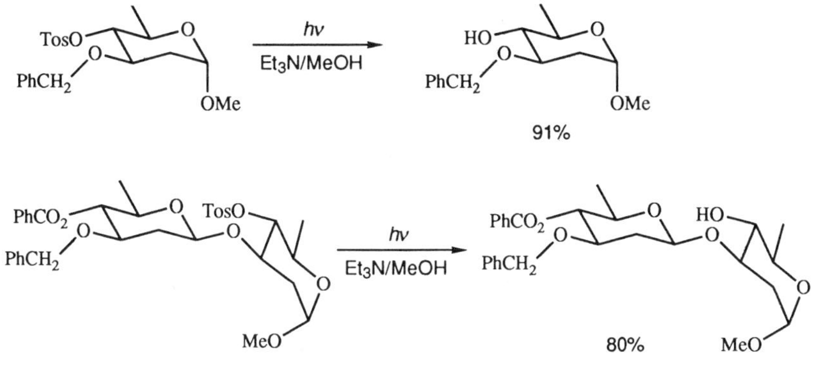

(57)

Alternative modes of fission of the S—C bond are found, such as heterolysis in the photochemical reactivity of the ketotosylate (**58**) [120]. Irradiation in

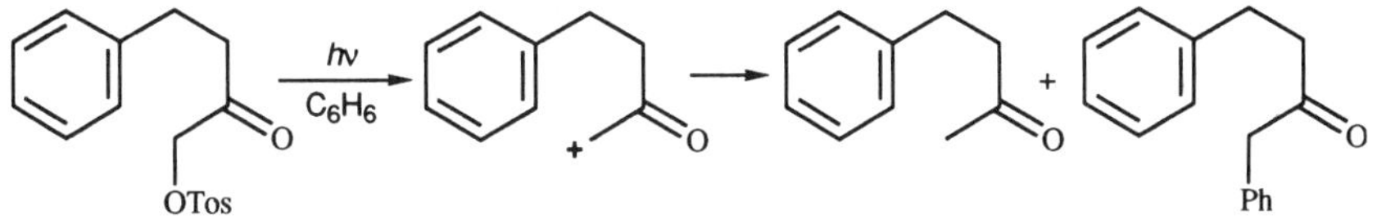

(58)

benzene gives *p*-toluenesulphonic acid in 74% yield. The other products formed from this reaction are the ketones shown. The reaction is presumed to proceed via an intermediate carbocation formed by S—C bond heterolysis. Similar reactivity is shown by the tosylate (**59**) which is more able to undergo

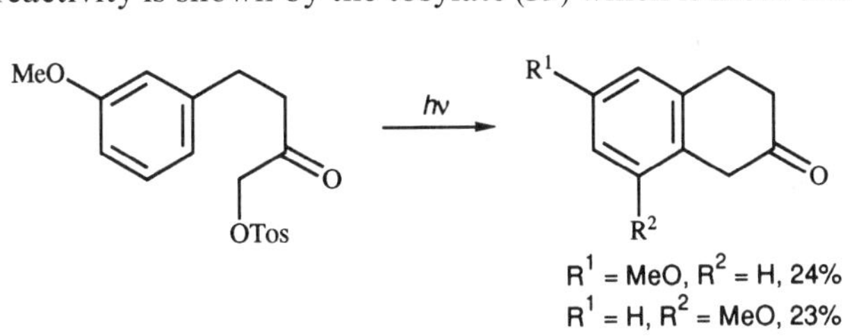

$R^1$ = MeO, $R^2$ = H, 24%
$R^1$ = H, $R^2$ = MeO, 23%

(59)

intramolecular addition owing to the electron-donating methoxy group and gives ketonic products in 24 and 23% yields. An analogous mechanism is involved in the conversion of the sulphonate (**60**) on irradiation in benzene

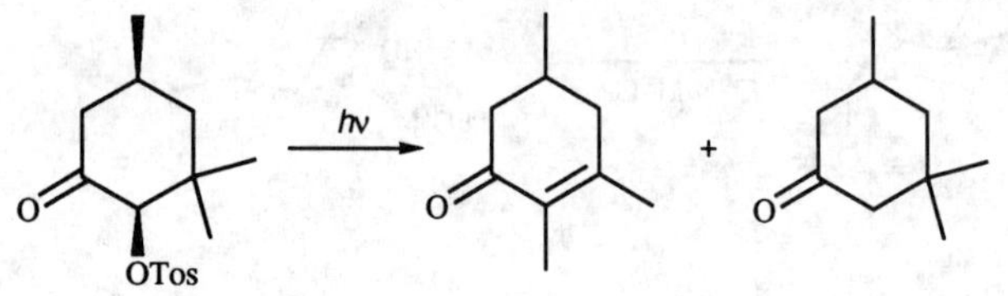

OTos

hv

+

(60)

into the two products in a ratio of 4:1 [121]. The formation of the major product presumably involves the heterolytic fission of an O—C bond to afford a cation which undergoes a methyl migration. Cationic paths are reported to be involved in the phototransformation of the steroidal sulphonates (**61**) [122–124].

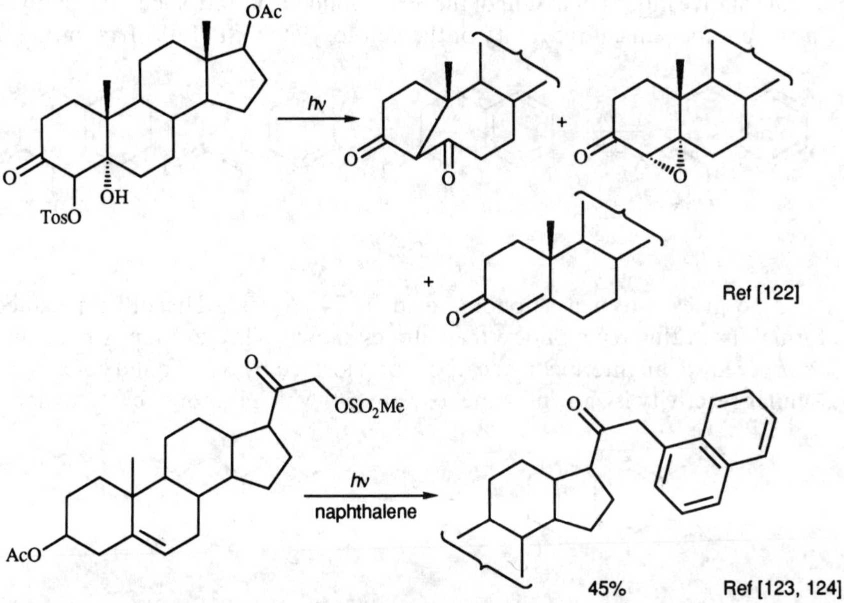

OAc

TosO

OH

hv

+

+

Ref [122]

$OSO_2Me$

AcO

hv

naphthalene

45%

Ref [123, 124]

(61)

Earlier in this chapter the elimination of sulphur dioxide along with the migration of an aryl group in the irradiation of sulphonamides was described. This reaction is also observed with sulphonates (**62**) [125–127]. Others have

$R = H$, 40%
$R = Me$, 50%
$R = MeO$, 50%
Ref [125-127]

(62)

demonstrated that the same type of reaction occurs with the cyclopentenone (**63**), affording the product shown [128].

Ref [128]

(63)

*4.5.3.2 Cyclic sulphonates*

Extrusion of sulphur dioxide also occurs on irradiation of some cyclic sulphonates, such as the sulphone (**64**), which yields an *o*-quinomethide [129].

(64)

Fission of the S—O is also the dominant reaction when the sultones (**65**) are

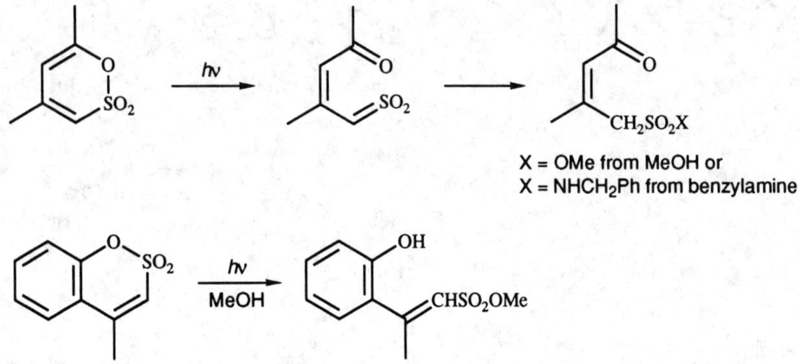
O
SO2
hν
O
SO2
O
CH2SO2X
X = OMe from MeOH or
X = NHCH2Ph from benzylamine
O
SO2
hν
MeOH
OH
CHSO2OMe

(65)

irradiated [130–133]. These compounds ring-open in a manner analogous to linearly conjugated cyclohexadienones (Chapter 3), affording a sulphene intermediate which can be readily trapped by alcohols or amines, affording the products shown.

### *4.5.3.3 Photo-Fries reactions*

In an earlier section of this chapter the photo-Fries reaction of sulphonamides was discussed. Sulphonates show similar reactivity on irradiation and typical of this is the rearrangement of phenyl *p*-toluenesulphonate into 2-hydroxyphenyl and 4-hydroxyphenyl *p*-tolylsulphones and phenol (**66**) [134].

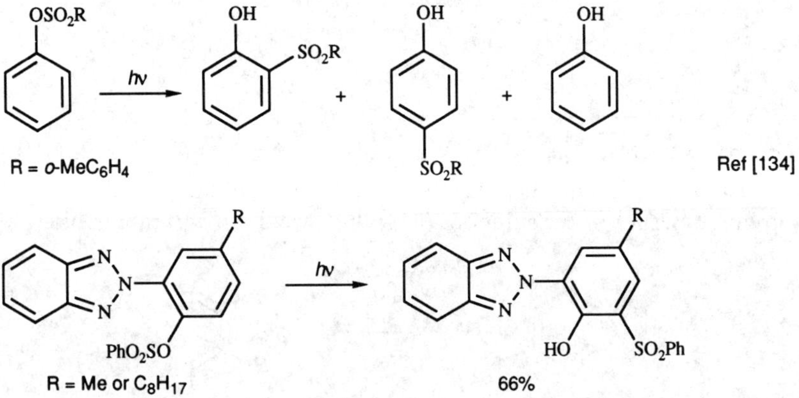
OSO2R
hν
OH
SO2R
+
OH
SO2R
+
OH
R = o-MeC6H4
Ref [134]
R
N
N
N
PhO2SO
R = Me or C8H17
hν
R
N
N
N
HO
SO2Ph
66%

(66)

The conversion of phenyl benzenesulphonate into 2-hydroxyphenyl and 4-hydroxyphenyl phenylsulphones as well as phenol, small amounts of diphenyl ether and polymer is also typical of this class of rearrangement [135]. Irradiation (330 nm) of the sulphonates (**66**) in ethyl acetate affords the hydroxysulphones in 66% yield [136].

## 4.6 SULPHONES AND SULTONES

### 4.6.1 Spectroscopic data of sulphones and sultones

The u.v. absorptions of sulphones and sultones are dependent upon the type of substituent attached to the sulphonyl group. Thus dimethyl sulphone shows an absorption around 180 nm [137]. Diphenyl sulphone, however, shows absorptions at 201 ($\varepsilon = 31\,700$ dm$^3$ mol$^{-1}$ cm$^{-1}$), 235 ($\varepsilon = 15\,500$ dm$^3$ mol$^{-1}$ cm$^{-1}$), 260 ($\varepsilon = 1740$ dm$^3$ mol$^{-1}$ cm$^{-1}$), 266 ($\varepsilon = 2140$ dm$^3$ mol$^{-1}$ cm$^{-1}$) and 274 ($\varepsilon = 1390$ dm$^3$ mol$^{-1}$ cm$^{-1}$) [138, 139]. Sultones also exhibit changes in band position due to the type of substitution as shown by the sultone (**67a**) with a high-intensity absorption at 291 nm ($\varepsilon = 33\,400$ dm$^3$ mol$^{-1}$ cm$^{-1}$) [140] and the naphthalene derivative (**67b**) has absorptions at 225 nm ($\varepsilon = 38\,000$ dm$^3$ mol$^{-1}$ cm$^{-1}$), 274 ($\varepsilon = 4300$ dm$^3$ mol$^{-1}$ cm$^{-1}$), 286 ($\varepsilon = 5000$ dm$^3$ mol$^{-1}$ cm$^{-1}$) and 316 ($\varepsilon = 410$ dm$^3$ mol$^{-1}$ cm$^{-1}$) [141]. These values show that a variety of conditions can be used to effect excitation of such compounds.

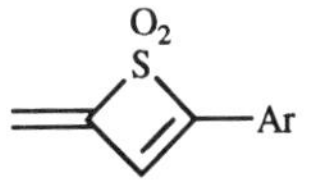

$O_2$
S
Ar

(a) Ar = Ph
(b) Ar = naphthyl

(67)

### 4.6.2 Photochemistry of sulphones and sultones

There are many examples of the photochemical behaviour of these classes of compounds [142–144]. The photochemistry is dominated by the fission of a C—S bond affording a carbon-centred radical and a sulphonyl radical.

#### *4.6.2.1 Sulphones*

Arene sulphones are a good source of phenyl radicals and for example the irradiation of diphenyl sulphone at 254 nm leads to phenyl radicals and $SO_2$ [145]. These phenyl radicals can be used to arylate a variety of substrates, and one example of this is the irradiation in pyridine which affords the phenylated pyridines shown (**68**) [146]. Irradiation of other symmetrically

PhSO$_2$Ph —hν, 30 h, pyridine→ 2-phenylpyridine (28.1%) + 3-phenylpyridine (66.1%) + 4-phenylpyridine (5.8%) + PhSO$_2$H

(68)

substituted aromatic sulphones in aromatic solvents provides an efficient route to biphenyls [147, 148].

The photochemistry of the sulphones (**69**) is also dominated by the

Ph—C≡C—SO$_2$Ar

Ar = Ph or p-MeC$_6$H$_4$

hν

SO$_2$Ph

R

SO$_2$Ph

R

SO$_2$Ph

R

R = C≡C—Ph

yields 50-70%

(69)

homolysis of the C—S bond and the formation of the corresponding radicals. The radicals formed by this reaction path add efficiently to a variety of alkenes to afford 1,2-addition products [149]. In all cases addition takes place without skeletal rearrangement of the alkene, and the adduct has the *trans*-arrangement reminiscent of the additions reported in a previous section. The sulphonyl cyanide (**70**) also undergoes 1,2-addition to alkenes (hex-1-ene

$(CH_2)_3CH_3$

NC $(CH_2)_3CH_3$

Tosyl

77%

$SO_2CN$ $h\nu$

CN

Tosyl

72%

(70)

or cyclohexene) to afford 1 : 1 adducts in high yield [150].

Benzyl sulphones (**71**) also undergo homolysis, which results in the

$$PhCH_2SO_2CH_2Ph \xrightarrow{h\nu} PhCH_2CH_2Ph$$

(71)

photochemical extrusion of sulphur dioxide, affording bibenzyl [151]. Studies have shown that benzyl and $\beta$-naphthyl systems react mainly from the singlet state while $\alpha$-naphthyl derivatives react from the triplet state [152]. The free radicals produced on decomposition from the triplet state escape from the solvent cage while the singlet radicals react within the cage [153, 154]. The reaction has been studied from a variety of standpoints and in a variety of media such as micelles [155]. The reaction path of fission and recombination is quite common and two examples are shown in **72** [156, 157]. Homolytic

$SO_2$ — $h\nu$ — quantitative

$PhCH_2SO_2CH_2-C_6H_4-Cl \xrightarrow[C_6H_6]{h\nu} PhCH_2CH_2Ph + PhCH_2CH_2-C_6H_4-Cl$

$+ Cl-C_6H_4-CH_2-CH_2-C_6H_4-Cl$

Ratio 1.0 :1.8 : 0.9

(72)

fission is also involved in the conversion of the sulphones (**73**) which provides

$$PhCH_2SO_2R \xrightarrow{h\nu} Ph\dot{C}H_2 + R\dot{S}O_2 \longrightarrow RSO_2H + PhCH_2CH_2Ph$$

| | |
|---|---|
| R = Me | 63% |
| R = Et | 59% |
| R = $Pr^n$ | 60% |
| R = $CH_2CH_2Cl$ | 40% |
| R = $PhCH_2CH_2$ | 54% |

(73)

an efficient route to sulphinic acids [158]. Photochemical S—C fission is also found to be efficient in carbohydrate derivatives, reviewed by Binkley [159], such as those shown in **74** [160, 161].

OAc, $R^2$, OAc, $R^1$, AcO, OAc — $h\nu$, $C_6H_6$ — OAc, H, OAc, H, AcO, OAc

$R^1$ = H, $R^2$ = $PhSO_2$
$R^1$ = $PhSO_2$, $R^2$ = H

(74)

In some instances the fission of the S—C bond can occur by heterolysis. Thus, under irradiation, the sulphones (**75**) undergo C–S heterolysis with the

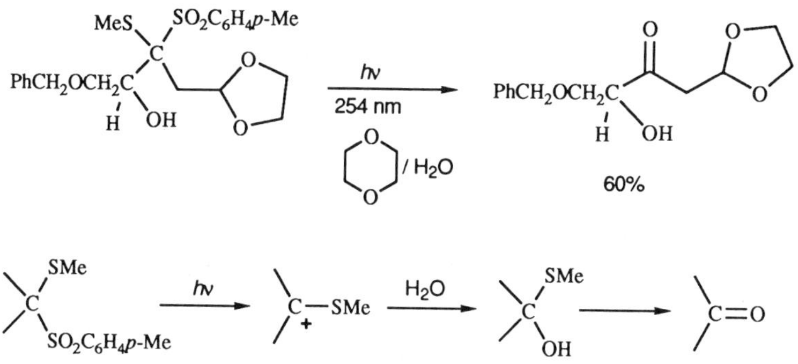

(75)

expulsion of sulphinate and the formation of a carbocation. In water, this is trapped and undergoes elimination to yield ketones and aldehydes [162].

*4.6.2.1.1 Cyclic sulphones*

Much of the photochemistry carried out on the cyclic sulphones has had a leaning towards the synthesis of novel compounds [163, 164]. Typical examples of the efficiency of the reaction is the sulphur dioxide extrusion from phenylthiirane 1,1-dioxide (**76**) and 2,3-diphenylthiirene-1,1-dioxide

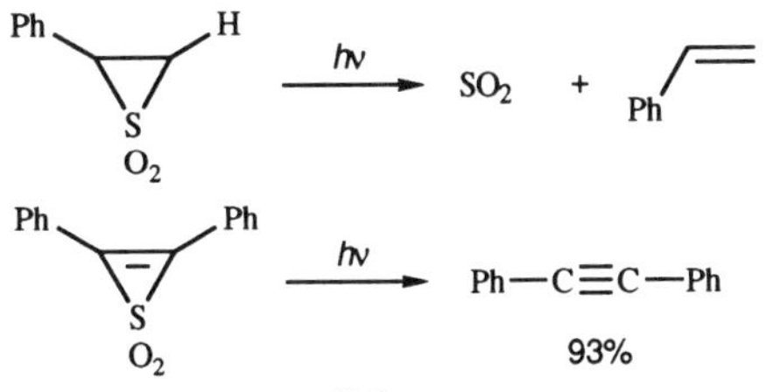

(76)

[165, 166].

Irradiation of the cyclobutylsulphone (**77a**) at 147.0, 123.6 and 106.7–104.8 nm

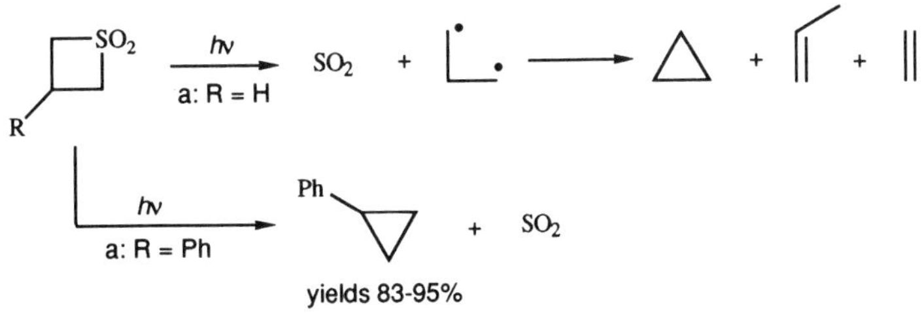

(77)

results in decomposition where again the principal step is the extrusion of sulphur dioxide and the formation of a biradical from which cyclopropane, propene and ethylene are formed [167]. Substitution does not appear to have an adverse effect on the extrusion as shown by Durst *et al.* [168, 169] in the formation of phenylcyclopropane from the sultone (**77b**) and indeed this reaction process provides a useful route to cyclopropanes [170].

Loss of $SO_2$ from tetramethylenesultone on irradiation at 147 nm [171] follows the pattern already established. The formation of dienes from the photochemical extrusion of sulphur dioxide from sulpholenes (**78**) has been

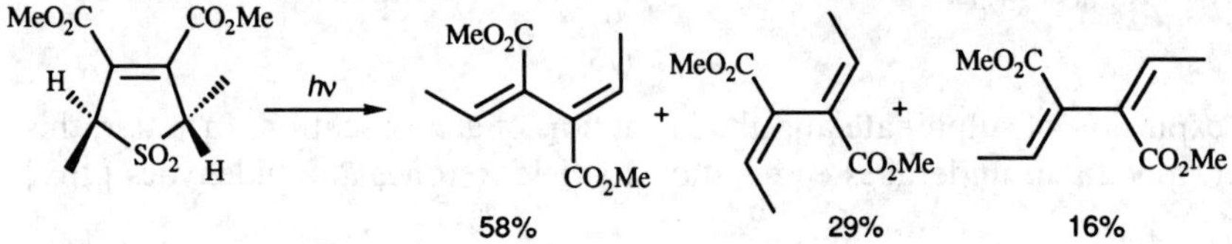

(78)

studied in detail [172, 173]. While the foregoing examples provide a route to dienes, the irradiation of other derivatives (**79**) can be used as a synthetic

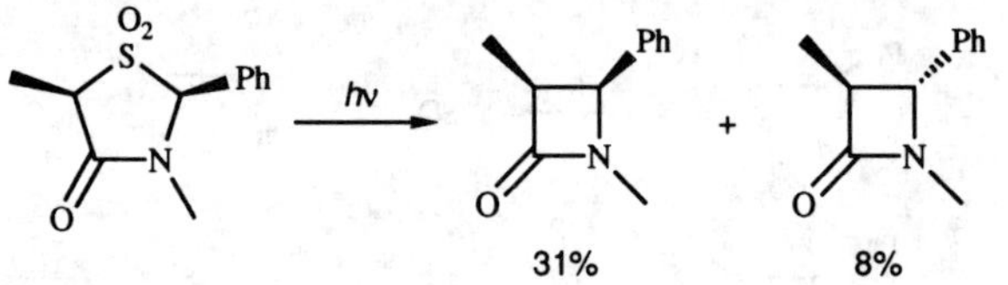

(79)

route to $\beta$-lactams [174]. The fact that both the *cis*- and the *trans*-isomers are formed from the *cis* starting material is indicative of a free radical process [175].

The bridged sulphones (**80**) all undergo photochemical loss of sulphur

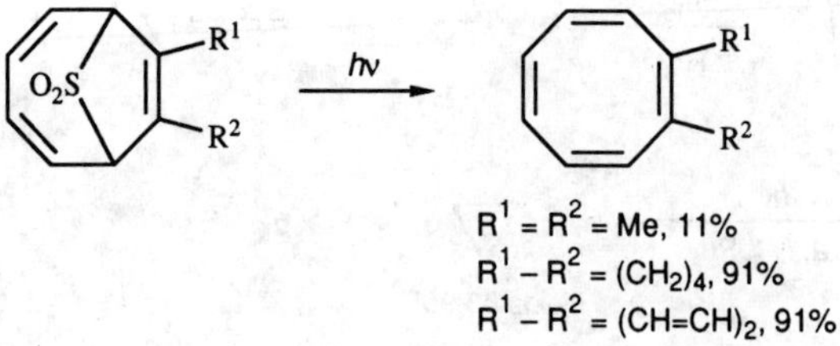

$R^1 = R^2 = Me$, 11%
$R^1 - R^2 = (CH_2)_4$, 91%
$R^1 - R^2 = (CH{=}CH)_2$, 91%

(80)

dioxide on irradiation in acetone to yield cyclo-octatetraene derivatives [176]. The benzylic sulphones (**81**) are also photochemically reactive and using

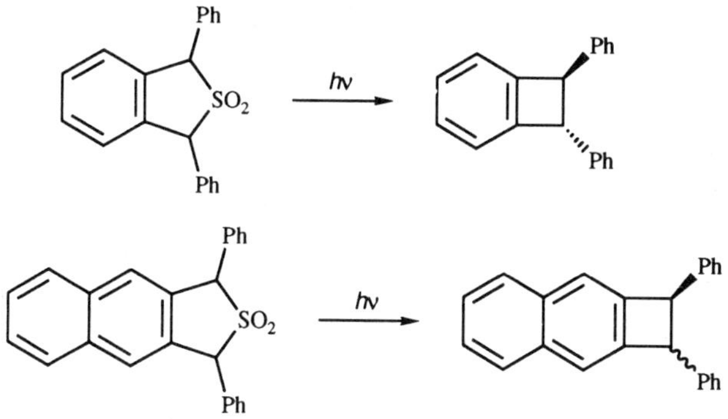

(81)

280–320 nm irradiation affords dimers where extrusion of sulphur dioxide has occurred [177–179].

One of the most interesting synthetic uses for the sulphur dioxide extrusion process has been the application to the synthesis of cyclophanes, a subject reviewed by Boekelheide [180] and Givens [181]. The earliest example of cyclophane formation by this route was the irradiation of the disulphone to afford a good yield of the cyclophane (**82**) [182]. Other examples of this

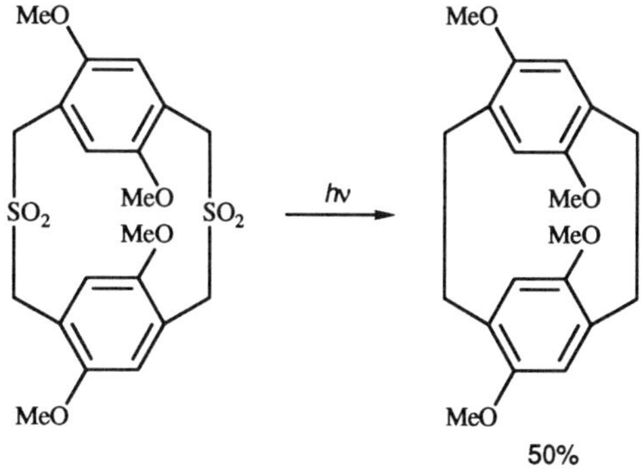

(82)

well-tried path are shown in **83** [183–185].

$SO_2$ $SO_2$ $h\nu$ 30% Ref [183]

$SO_2$ $SO_2$ $h\nu$ 61% Ref [184, 185]

$SO_2$ $SO_2$ $h\nu$ 32% Ref [184, 185]

(83)

#### *4.6.2.2 (2 + 2)-Cycloaddition reactions of sulphones*

(2 + 2)-Cycloaddition reactions can occur on the irradiation of some $\alpha,\beta$-unsaturated sulphones. Usually the alkene component is constrained within a ring since open-chain $\alpha,\beta$-unsaturated sulphones undergo readily *cis*–

*trans*-isomerism (**84**) as an energy-wasting step [186]. However, the more

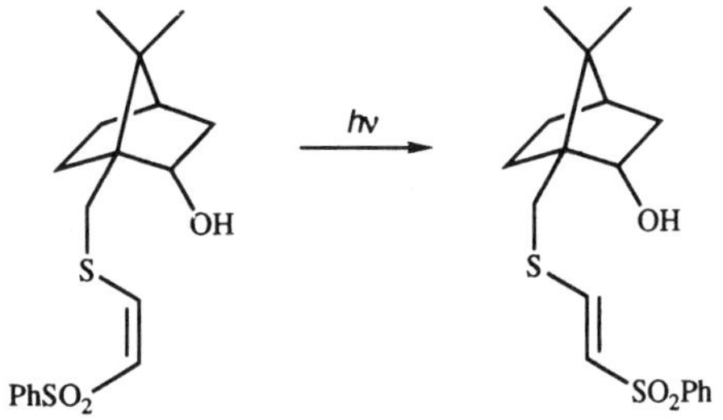

(84)

rigid sulphones (**85**) do undergo cycloaddition to afford adducts on irradiation

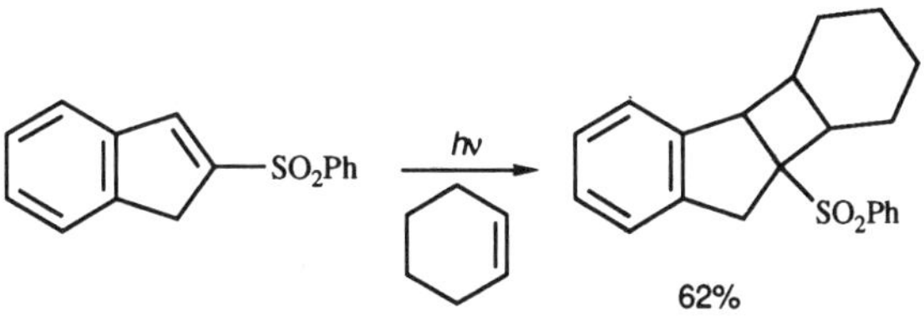

(85)

in cyclohexene [187]. This cycloaddition path can be applied to the synthesis of thiepinones as shown by the addition of thiophene dioxides to cyclohexene and ring opening of the resultant cyclobutane derivative (**86**) [188]. Other

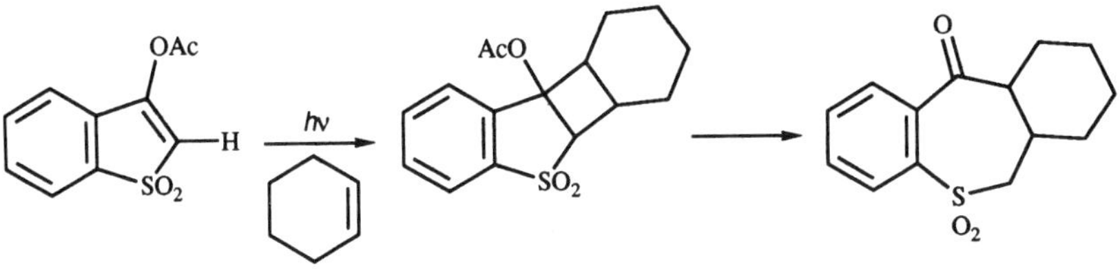

(86)

sulphones have been reported to undergo dimerization such as the sunlight irradiation of the derivatives (**87**) affording either the *head-to-head* or

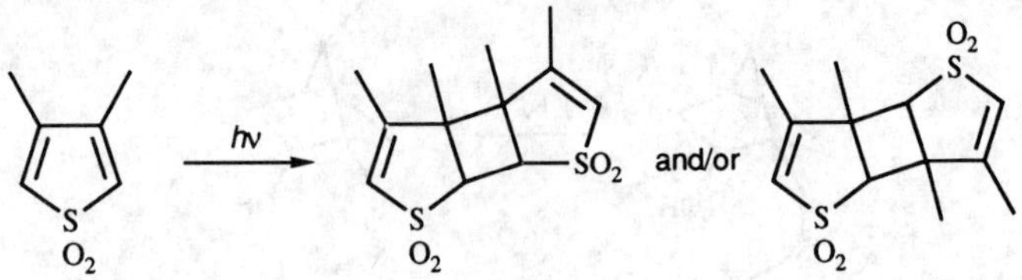

(87)

*head-to-tail* dimers [189, 190]. Cage compounds are also available by this route as shown in (**88**) [191] and some examples of norbornadiene/quadricyclane transformations (**89**) have been studied [192–195].

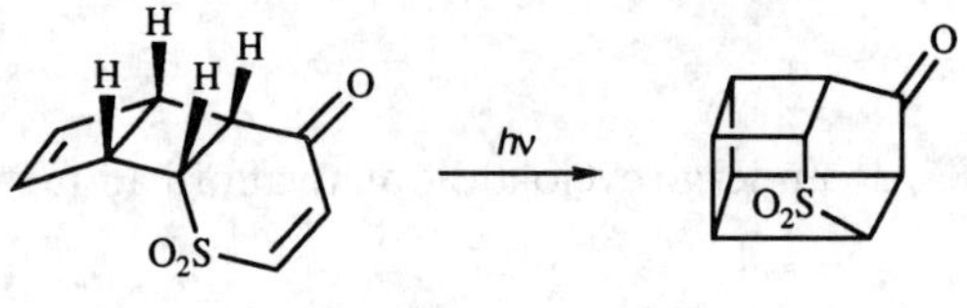

(88)

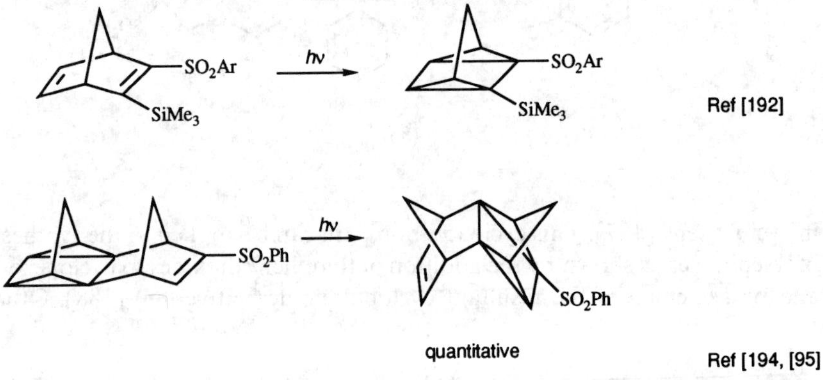

Ref [192]

quantitative

Ref [194, [95]

(89)

## 4.7 THIOCARBONYL COMPOUNDS

Thiocarbonyl compounds [196] formally analogous to carbonyl compounds (Chapter 3), are considerably more reactive and less stable than their oxygen analogues.

### 4.7.1 Spectra of thiocarbonyl compounds

Typically these compounds show absorptions in the 200–700 nm region and are coloured [197]. The longest wavelength absorption has been assigned as the $n\pi^*$ triplet involving a $S_0 \rightarrow T_1$ direct excitation of a non-bonding electron on sulphur. In some thiones the $S_0 \rightarrow S_1$ transitions are stronger and it can be difficult to identify the $S_0 \rightarrow T_1$ in the spectra. However, these absorption bands are generally weak [198] and details of a few examples are shown in Table 4.1 [197]. The assignment of this absorption to a directly populated triplet state has been confirmed by Zeeman-effect measurements [199]. The shorter wavelength absorptions around 350 nm in thiobenzophenone for example are $\pi\pi^*$ transitions and excitation into this band (the K band) results in a $S_0 \rightarrow S_2$ electronic promotion and the population of the second singlet state. As a result of the inclusion of a sulphur atom giving a lower electronegativity and a greater polarizability of the C=S double bond, there are marked differences in reactivity. Within the thione group the order of stability is aryl thiones > alicyclic thiones > aliphatic thiones.

**Table 4.1** Thioketones: spectroscopic data

| Thioketone | $\lambda_{max}$ | $S_2$ | $\lambda_{max}$ | $S_1$ | $\lambda_{max}$ | $S_0 \Rightarrow T_1$ |
|---|---|---|---|---|---|---|
| Thiobenzophenone | 316.5 | (4.20) | 599 | (2.26) | 685 | (1.95)[b] |
| 4,4′-Dimethoxy-thiobenzophenone | 352.5 | (4.30) | 577 | (2.46) | — | — |
| Thiocamphor | 244 | (4.06) | 493 | (1.09) | 545 | (0.30) |
| Adamantanethione | 340 | (4.11) | 486 | (1.16) | 545 | (0.70) |

[a] Recorded in cyclohexane solution.
[b] All intensities quoted in brackets are for log $\varepsilon$.

### 4.7.2 Cleavage reactions

Unlike ketones (Chapter 3) cyclic and alicyclic thioketones are resistant to $\alpha$-cleavage reactions. The failure to observe such reactivity could be due to electronic factors, although thermodynamic considerations may also be important. Despite extensive research, only cyclobutane thiones (**90**) [200,

hν

$T_1$

$+ S_8$

| | | |
|---|---|---|
| cyclohexane | 20% | 40% |
| methanol | 22% | 37% |

hν [$T_1$] cyclohexane

MeOH

via

OMe

$(CH_2)_n$

hν [$T_1$] $C_6H_6$

MeOH

n = 5, 20% 5%
n = 4, 20% 15%

OMe

n = 5, 10% 5% 10%
n = 4, 15% 10% 5%

(90)

201] and cyclopropene thiones (**91**) [202, 203] have been shown to undergo such a reaction. The cleavage reaction of the 2,2,4,4-tetramethylthione (**90**) arises from the lowest triplet state, which either reacts with solvent to afford the reduced dimer or undergoes fission to a biradical [200]. If this biradical rearranges to a carbene, in this instance, the species does not react with methanol in the conventional manner since no thioacetals are formed. However, the biradical is trapped by interaction with ground-state thione. The behaviour of the 2,2,4,4-tetramethyl-1,3-dithione (**90**) is somewhat different and in this case the thiocarbene is involved, as demonstrated by the isolation of a thioacetal on reaction in methanol. There is a close analogy to the reactions of carbonyl groups, and cleavage to the biradical is followed by migration of the α-carbon atom to the thiyl radical. The ring-expansion process is probably favoured by the involvement of the sulphur atom and its ability to stabilize the resultant carbene. Similar ring-expansion products are obtained from irradiation of the spiro-1,3-dithiones also illustrated in **90** [200].

Ph Ph S Ph Ph S Ph Ph S Ph

$C_6H_6$, 30%
MeOH, 15%

Ph Ph S $h\nu$ $[T_1]$ Ph Ph S Ph Ph S MeOH Ph Ph S H OMe 10%

Ph Ph C S MeOH Ph Ph C S H OMe Ph MeO S OMe Ph 50%

(91)

The cyclopropenethione (**91**) also undergoes α-fission involving the lowest triplet state [202]. Again fission to the biradical is the key photochemical step. In benzene this biradical is trapped by ground-state thione to yield the thieno[3,2-*b*]thiophene, while in methanol trapping of the biradical occurs in competition with ring expansion to the carbene or formation of an open-chain carbene. The presence of both of these accounts for the formation of the observed products.

### 4.7.3 Hydrogen abstraction reactions

#### *4.7.3.1 Intramolecular*

Like carbonyl compounds, which undergo the Norrish Type II reaction by abstraction of a γ-hydrogen, the diaryl thiones undergo intramolecular hydrogen abstraction on irradiation into the $n\pi^*$ band. Typical of this reactivity is the formation of a photoenol, confirmed by spectroscopic and trapping studies, by irradiation of the thiobenzophenone derivative (**92**) [203]. Interestingly, naphthyl and other polycyclic aromatic

Ph S Ph $h\nu$ $[T_1]$ Ph H SH Ph EC≡CE Ph H E E SH $CH_2N_2$ Ph H SMe Ph + Ph H Ph SMe Ph H S E E

E = $CO_2Me$

(92)

compounds (**93**) which have vacant *peri* positions also undergo an inefficient

hν
$S_1$ reactivity
51%
hν
59%

(93)

cyclization reaction involving attack by sulphur and a hydrogen-transfer process. In the case of the naphthyl derivative the reaction arises from the $S_1$ state and intersystem crossing is inefficient [204]. The hydrogen-migration step is thought to involve a dipolar intermediate rather than a biradical.

Aryl alkyl thiones behave differently from their carbonyl compound counterparts in that, although they undergo hydrogen abstraction, the preferred site for abstraction is the $\delta$-carbon rather than the $\gamma$-carbon. Typical of this is the irradiation of the thione **94**, which is converted into a

hν
$S_2$ excitation
high yield

$R^1 = R^2 = R^3 = H$
$R^1 = H, R^2 = R^3 = Me$
$R^1 = R^2 = H, R^3 = Ph$
$R^1 = R^2 = Me, R^3 = H$

(94)

tetrahydrothiophene via a singlet biradical with a lifetime of $2 \times 10^{-9}$ s [205]. The reaction is inefficient and occurs from the $S_2$ state, a belief confirmed by the failure to quench or to sensitize the reaction. Furthermore excitation to the $S_1$ state fails to yield the same products [206]. In the absence of $\delta$-hydrogens the hydrogen abstraction fails [207]. The incorporation of an oxygen atom into the side chain complicates the photochemistry, and hydrogen abstraction can occur from both the $\gamma$- and the $\varepsilon$-site, as shown for **95** [207]. In this instance both $S_1$ and $S_2$ are reactive. On excitation of

S Ph O hν Ph SH O + HS O Ph + O

40% 45% 5%

(95)

the $S_1$ state, hydrogen abstraction occurs only from the $\gamma$-carbon, affording the cyclobutane thiol and the elimination product, whereas excitation of the $S_2$ state yields both of these products and a tetrahydropyran formed from hydrogen abstraction from the $\varepsilon$-carbon. In some instances hydrogen abstraction has been found to take place from the $\beta$-carbon, affording cyclopropane thiols (**96**) [208]. Cyclopropane thiols are also formed on

S R HS $R^1$ Ph $R^2$

R = Ph, 4-$MeOC_6H_4$,
4-$CNC_6H_4$, 1-naphthyl
or SMe

$R^1$ = H, $R^2$ = R
or vice versa
ca. 80%

(96)

irradiation of the bicyclic thiones (**97**) [209].

R R

S

R = Me or H

$h\nu$

$S_2$

R R

SH

(97)

#### *4.7.3.2 Intermolecular*

Thiobenzophene (**98**) undergoes photochemical reduction in alcohols. The

S

Ph Ph

$h\nu$

$\lambda > 500$ nm

$S_2$ state

in ROH

$Ph_2CH-SS-CHPh_2$ + $Ph_2CH-SH$ + $Ph_2CH_2$ + $Ph_2CH-S-CHPh_2$

+ $Ph_2CH-SS-CPh_2-SS-CHPh_2$

(98)

process is wavelength-dependent, since the $n\pi^*$ state of the C=S group has insufficient energy to abstract hydrogen. However, both the $S_1$ and $S_2$ states are reactive, producing the same general mixture of products, by the intermediacy of the radicals, although the former excited state is significantly slower [210, 211]. Photochemical reduction of thiobenzophenone takes place with a variety of substrates such as benzhydrylthiol [211], dihydrobenzene and other dihydro-aromatics [212], hydrocarbons and ethers [213].

The dialkyl thiones such as adamantanethione (**99**) [214] and di-*t*-

quantitative

$R^1 = C_6H_{11}, R^2 = H$
$R^1 = H, R^2 = C_6H_{11}$

(99)

butylthioketone (**100**) [215] both undergo reduction on irradiation. Ada-

$h\nu$
540 nm

, $C_6H_6$ 72%
or $PhCH_2OH, C_6H_6$ 82%
or $Et_3N, C_6H_6$ 80%

(100)

mantanethione is photoreduced from the $T_1$ level in the presence of adamantanethiol [216]. The reaction is efficient and involves a chain process. Photoreduction also occurs from the $S_2$ state [215] and in cyclohexane a dimer is formed as well as products derived by hydrogen abstraction from solvent. Apparently the abstraction of hydrogen from hydrocarbons by adamantanethione is indiscriminate. Di-*t*-butylthioketone is also photoreactive in the $S_2$ state [215]. Furthermore, this thione affords a thiol and a sulphide on direct or triplet-sensitized irradiation in the presence of an aliphatic amine. The mechanism in this instance is presumed to involve the formation of a radical–ion pair by transfer of an electron from the amine to the thione [215].

### 4.7.4 Cycloaddition reactions

Excitation of diaryl thiones by irradiation of the $S_1$ state ($n\pi^*$ triplet state) in the presence of electron-rich and electron-poor alkenes affords good yields of thietanes [217, 218]. The formation of thietanes in such addition reactions had been implicated earlier following the irradiation of thiobenzophenone in the presence of but-2-ene and hex-1-ene when 1,1-diphenyl-substituted alkenes were obtained by the fragmentation of the initially formed thietanes [217]. Detailed study has indicated that the type of reactivity is dependent upon the nature of the alkene component. Thus thiobenzophenone (**101**)



(101)

reacts with electron-rich alkenes on excitation to the $S_1$ or $S_2$ state to yield 1,4-dithianes or thietanes. The reaction with electron-rich alkenes is apparently concentration-dependent and, for example, styrene yields the 1,4-dithiane at high olefin concentration, while at lower concentration the thietane is formed. A complication in any analysis of reaction type is the observation that some 1,4-dithianes decompose to thietanes and thioketone. It is likely that the mechanism for the formation of the products involves the formation of a 1,4-biradical, similar to oxetane formation (Chapter 3), which is in competition with either ring closing or trapping by excess thioketone.

Excitation of either excited state of thiobenzophenone (**102**) with electron-

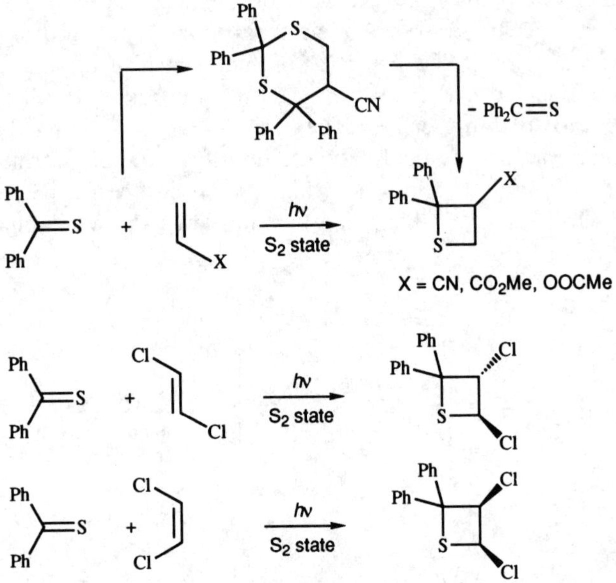

(102)

poor alkenes (methyl acrylate, acrylonitrile, or acetoxyethene) yields only thietanes. The reaction of thiobenzophenone with electron-poor alkenes is regiospecific, affording thietanes in which the geometry of the starting alkene is retained. The reaction arises from the $S_2$ excited state. Interestingly the thietane is not the primary product but is formed by collapse of an intermediate 1,3-dithiane formed from two molecules of thioketone and one of alkene [219].

Dialkyl thioketones also undergo (2 + 2)-cycloaddition from the $S_1$ state, with rapid intersystem crossing to the triplet state, with both electron-poor and electron-rich alkenes [220]. Examples of the addition to adamantanethione are shown in **103** [221]. A 1,4-biradical is involved and some evidence in support of this is obtained from the formation of the 'ene' product from the reaction with 1-methylstyrene. The $S_2$ of adamantanethione (**103**) is also reactive, again yielding thietanes but with retention of the alkene geometry [222, 223].

A variety of products can be obtained from the addition of diaryl thiones and dialkyl thiones to alkynes via the triplet excited state populated by irradiation into the $S_1$ band [224]. The involvement of a 1,4-biradical appears

R-R = adamantane

via $T_1$

Ratio 15:1 > 69%

$n\pi^*$

Ratio 1:1

$h\nu$

$S_2$ state

R-R = adamantane

quantitative

44%

31%

(103)

to be implicated (**104**). Addition reactions also occur with dienes, affording

$h\nu$

$n\pi^*$

$R^1$ = COOH, Ph or $CH_2OH$

(104)

the products shown in **105** by a 1,4-addition and 1,2-addition processes [225–230].

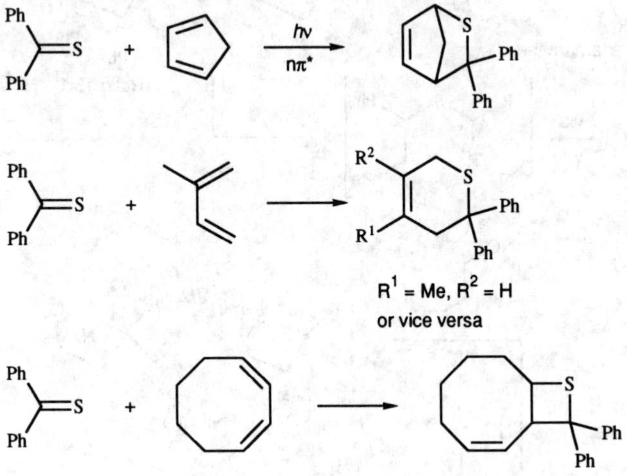

(105)

## 4.8 PHOTOCHEMISTRY OF THIOPHENES AND RELATED AROMATIC COMPOUNDS

### 4.8.1 Thiophenes

Substituted thiophenes have been demonstrated to undergo apparent group transpositions on irradiation [231]. Thiophenes with alkyl or aryl groups undergo this reaction, and typical examples of this are shown in **106** [232]

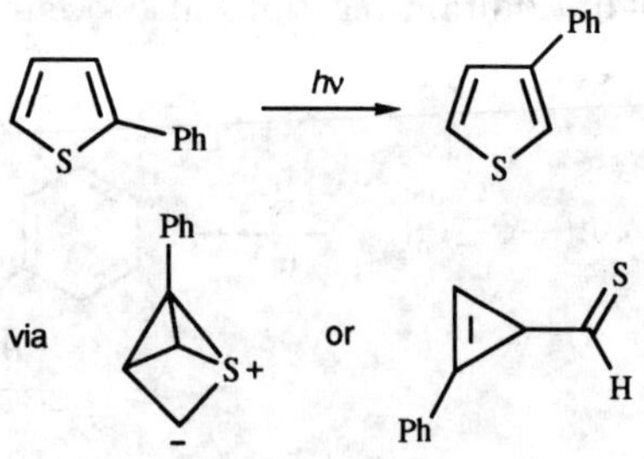

(106)

and **107** [233] respectively. Several proposals to account for these transforma-

hν
$RNH_2$
N
R
+
N
R
Ratio 1:8

via
S
or
S
H

(107)

tions were postulated such as cyclopropenes [233], akin to intermediates found in furan chemistry, in the rearrangements shown in **106** or tricyclic intermediates [233, 234] also in **106**, but in neither was an intermediate detected. The most significant advance in an understanding of the rearrangement process was obtained on irradiation of the substituted thiophene (**108**) from which a stable, isolable intermediate identified as a Dewar form

$F_3C$ $CF_3$
$F_3C$ S $CF_3$
hν
$F_3C$ $CF_3$
$F_3C$ S $CF_3$

(108)

was obtained [235]. The proof of structure for this intermediate relied on a variety of spectroscopic techniques as well as on chemical reaction, and it was shown that the Dewar isomer could react thermally with furan or 2,3-dimethylbuta-1,3-diene to yield the products shown in **109** [236, 237].

$F_3C$ $CF_3$
$F_3C$ S $CF_3$
hν
$F_3C$ $CF_3$
$F_3C$ S $CF_3$
O
O
$CF_3$
$CF_3$
S
$CF_3$
$CF_3$
$CF_3$
$CF_3$
S
$CF_3$
$CF_3$

(109)

Dewar isomers have also been implicated in the photorearrangement of the isomeric cyanothiophenes (**110**). These intermediates can be trapped by dienes such as furan [238].

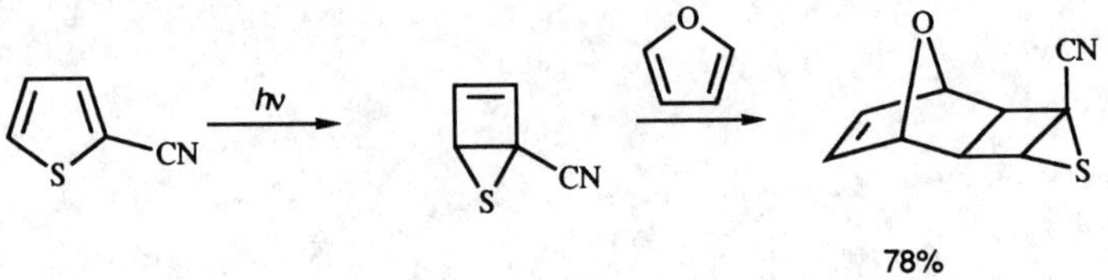

78%

(110)

### 4.8.2 Isothiazoles

Substituted isothiazoles also undergo group transpositions similar to those encountered with the thiophene derivatives. Thus irradiation at 254 nm of the isothiazoles affords the isomeric thiazoles shown in (**111**) [239, 240].

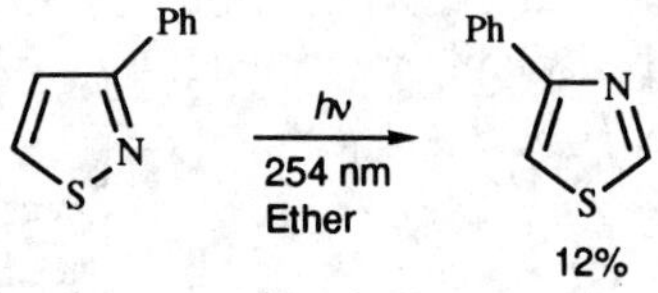

hν 254 nm Ether 12%

(111)

Other detailed studies on the photochemistry of isothiazoles and thiazoles have shown that there is a decreasing order of reactivity as follows: 2-phenylthiazole or 5-phenylisothiazole > 5-phenylthiazole or 3-phenylisothiazole > 4-phenylthiazole or 4-phenylisothiazole [241]. The methylisothiazole derivatives are also photoreactive and afford a mixture of products dependent on the position of the substituent. Thus irradiation of 3-methylisothiazole yields 2-methylthiazole (**112**) [242]; 4-methylisothiazole

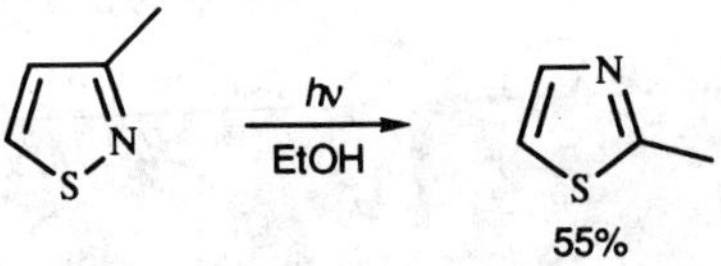

hν EtOH 55%

(112)

yields 4-methylthiazole and 5-methylisothiazole; and 5-methylisothiazole yields 5-methylthiazole, 4-methylisothiazole and 3-methylisothiazole. In general disubstitution gives better yields of product [239], as shown in the study of 4-methyl-5-phenylisothiazole and 4-methyl-3-phenylisothiazole derivatives (**113**) [243]. The details of the mechanism for the rearrangement

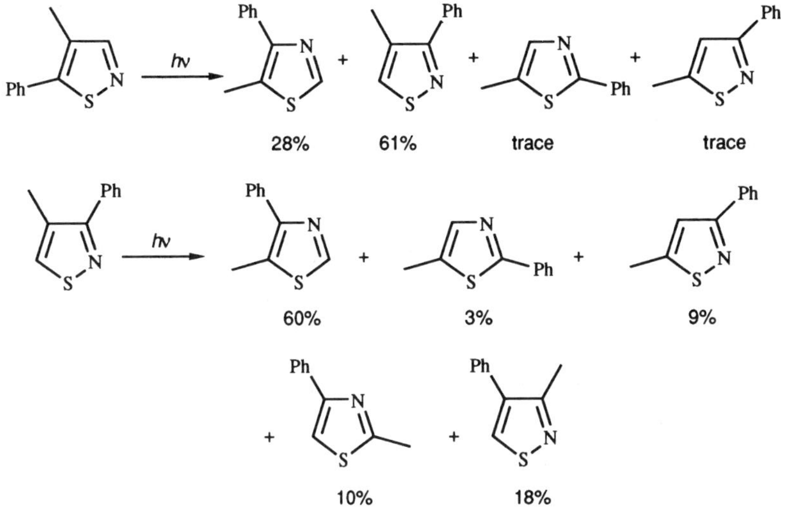

28% 61% trace trace

60% 3% 9%

10% 18%

(113)

are not clear, although Dewar-type intermediates with a sulphur atom walk could account for the products, e.g. **114**. The intervention of dipolar species

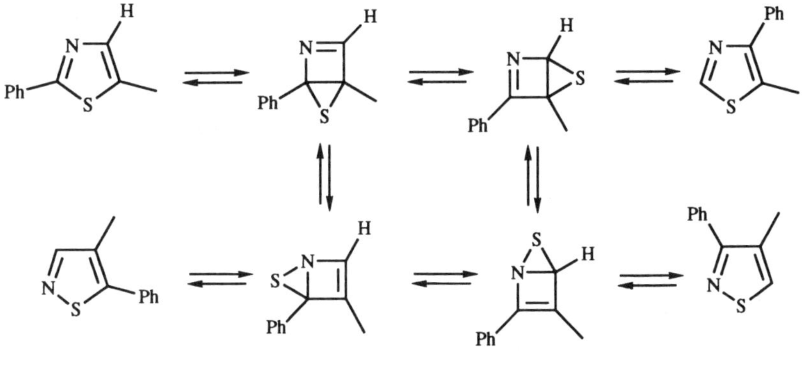

(114)

such as the tricyclic sulphonium species shown in **115** has also to be taken

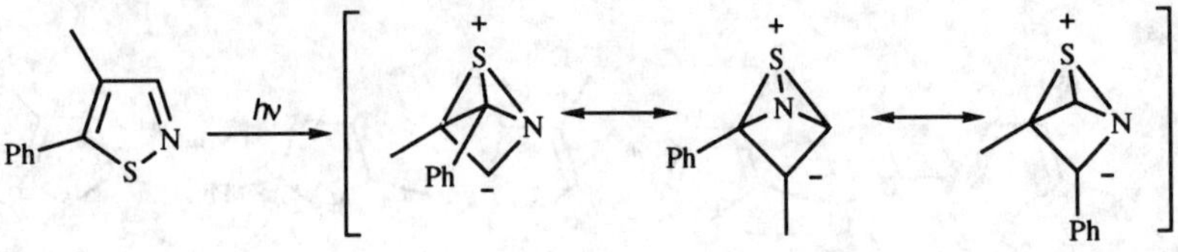

(115)

into consideration following deuterium labelling studies [239, 244, 245]

### 4.8.3 Benzoisothiazoles

Fusion of the isothiazole to a benzene ring re-introduces the S—N fission path, and irradiation in ether of the benzoisothiazole affords the disulphide via a biradical intermediate (**116**) [246]. Evidence for the involvement of a

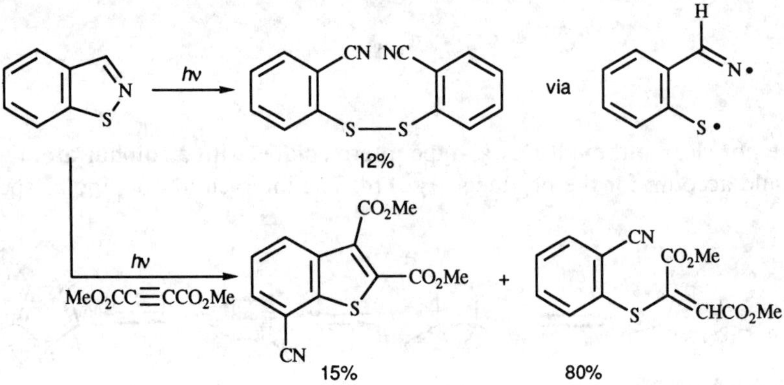

(116)

biradical is also found in trapping experiments such as the formation of the adducts from the irradiation of benzoisothiazole in the presence of dimethyl acetylenedicarboxylate (**116**) [247]. Alkenes can also take part in such addition, resulting in the formation of the benzothiepines (**117**). In this case

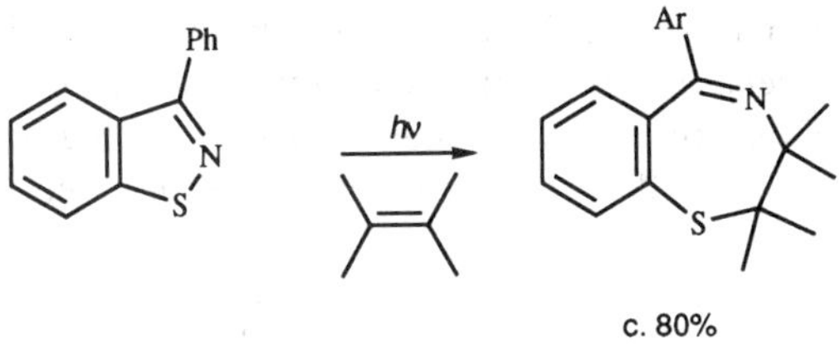

(117)

the mechanism of the addition is more complex and it is proposed that the reaction involves a solvent-sensitive exciplex which dissociates to a radical cation/radical anion pair which subsequently collapses to afford the final products in a regio- and stereo-specific fashion [248, 249], although, alternatively, it has been proposed that a charge-separated resonance form could contribute to the process, and addition to this would lead to product in which the stereochemistry of the alkene was preseved [250]. In at least one instance the S—N bond of the benzoisothiazole initially remains intact and irradiation of 3-phenyl-1,2-benzoisothiazole in the presence of electron-donating alkynes (ethoxyethyne and diethylaminoethyne) affords the bicyclic componds by (2 + 2)-addition and rearrangement (**118**) [251].

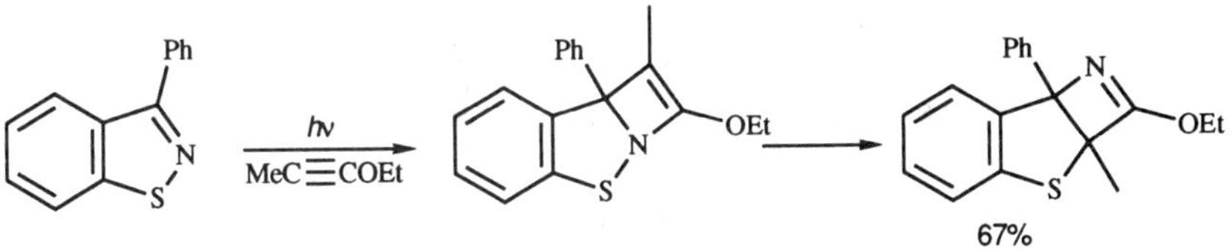

(118)

#### 4.8.4 1,2,3-Thiadiazoles

The 1,2,3-thiadiazole system is prone to undergo photochemical loss of nitrogen. Krantz and Laureni and coworkers [252] have studied the photochemical reactivity of the parent 1,2,3-thiadiazole (**119**) in an argon

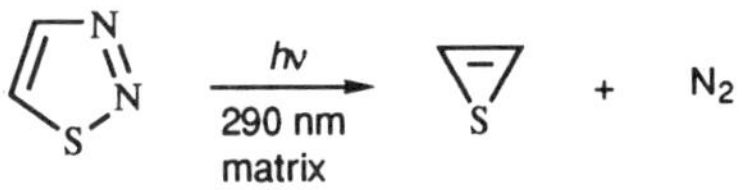

(119)

matrix where irradiation at 290 nm brought about loss of nitrogen and the formation of thiirene [252]. Some synthetic use of the photochemical decomposition of substituted 1,2,3-thiadiazoles has being made [253] as a route to dithiafulvenes (**120**). Other studies have illustrated the influence of

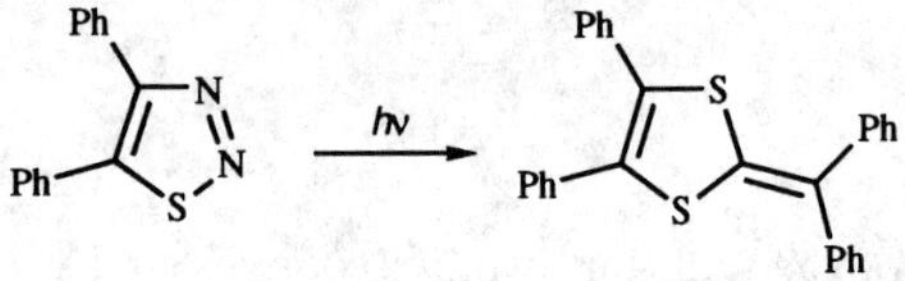

(120)

substituents on the outcome of the reaction (**121**) [254]. The influence of

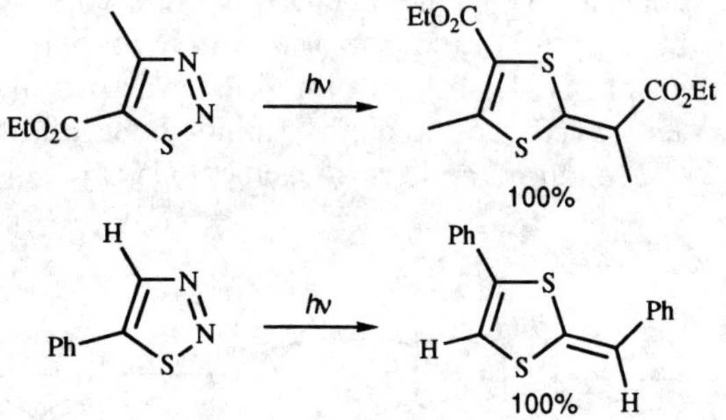

(121)

ring size on the process has also been studied and the products obtained are shown in **122** [255].

9% 66%

7%

12% 31%

27%

(122)

## REFERENCES


[1] M. B. Robin, in *Higher Excited States of Polyatomic Molecules*, Vol 1 Academic Press, 1974, p. 276.

[2] J. G. Calvert and J. N. Pitts, jun., *Photochemistry*, Wiley, New York, 1966, p. 489.

[3] L. B. Clark and W. T. Simpson, *J. Chem. Phys.*, 1965, **43**, 3666.

[4] *UV Atlas of Organic Compounds,* Spectrum I/6, Butterworths, 1971.

[5] D. D. Carlson and A. R. Knight, *Can. J. Chem.*, 1973, **51**, 1410.

[6] C. von Sonntag and H.-P. Schuchmann, in *The Chemistry of Functional Groups: Supplement E, The Chemistry of Ethers, Crown Ethers, Hydroxyl Groups and their Sulphur Analogues,* Pt 2, ed. S. Patai, Wiley, 1980, p. 923; A. R. Knight in *The Chemistry of the Thiol Group* ed. S. Patai, Wiley, 1974, p. 455; W. A. Pryor, *Mechanisms of Sulfur Reactions*, McGraw-Hill, New York, 1962; O. P. Strausz, H. E. Gunning, and J. W. Lown, in *Comprehensive Chemical Kinetics*,

Vol. 5 (eds C. H. Bamford and C. F. H. Tipper), Elsevier, Amsterdam, 1972, p. 697; E. Block, *Quart. Rep. Sulfur Chem.*, 1969, **4**, 239, 283; S. Braslavsky and J. Heicklen, *Chem. Rev.*, 1977, **77**, 473; H. Durr in *Houben–Weyl, Methoden der Organischen Chemie*, Vol 4/5b (ed. E. Muller), Thieme, Stuttgart, 1975, p. 1008; A. Padwa, *Int. J. Sulfur Chem. (B)*, 1972, **7**, 331.

[7] L. Bridges and J. M. White, *J. Phys. Chem.*, 1973, **77**, 295.

[8] L. Bridges, G. L. Hemphill and J. M. White, *J. Phys. Chem.*, 1972, **76**, 2668.

[9] T.-L. Tung and J. A. Stone, *J. Phys. Chem.*, 1974, **78**, 1130; T.-L. Tung and J. A. Stone, *Can. J. Chem.*, 1975, **53**, 3153.

[10] C. S. Smith and A. R. Knight, *Can, J. Chem.*, 1973, **51**, 780.

[11] W. E. Haines, G. L. Cook and J. S. Ball, *J. Am. Chem. Soc.*, 1956, **78**, 5213.

[12] D. R. Tycholiz and A. R. Knight, *Can. J. Chem.*, 1972, **50**, 1734.

[13] C. S. Smith and A. R. Knight, *Can. J. Chem.*, 1976, **54**, 1290.

[14] R. E. Kohrman and G. A. Berchtold, *J. Org. Chem.*, 1971, **36**, 3971.

[15] P. Fowles, M. DeSorgo, A. J. Yarwood, O. P. Strausz and H. E. Gunning, *J. Am. Chem. Soc.*, 1967, **89**, 1352.

[16] D. R. Dice and R. P. Steer, *Can. J. Chem.*, 1975, **53**, 1744.

[17] S. Bralavsky and J. Heicklen, *Can. J. Chem.*, 1971, **49**, 1316.

[18] P. M. Rao and A. R. Knight, *Can. J. Chem.*, 1967, **45**, 1369; K. Sayamol and A. R. Knight, *Can. J. Chem.*, 1968, **46**, 999; P. M. Rao and A. R. Knight, *Can. J. Chem.*, 1968, **46**, 2462.

[19] F. C. Adam and A. J. Elliot, *Can. J. Chem.*, 1977, **55**, 1546.

[20] A. B. Callear and D. R. Dickson, *Trans. Faraday Soc.*, 1970, **66**, 1987.

[21] G. W. Byers, H. Gruen, H. G. Giles, H. N. Schott and J. A. Kampmeier, *J. Am. Chem. Soc.*, 1972, **94**, 1016.

[22] J. M. White, *Spectroscopic Lett.*, 1969, **2**, 301.

[23] R. N. Haszeldine and J. M. Kidd, *J. Chem. Soc.*, 1953, 3219.

[24] J. F. Harris, *J. Am. Chem. Soc.*, 1962, **84**, 3148.

[25] A. Haas and W. Klug, *Chem. Ber.*, 1968, **101**, 2617; A. Haas, H. Reinke and J. Sommerhoff, *Angew. Chem. Int. Edn.*, 1970, **9**, 466.

[26] V. Prey and E. Gutschik, *Monatsh.*, 1960, **90**, 551.

[27] H. Kloosterziel, *Quarterly Reports on Sulfur Chemistry*, 1967, **2**, 353.

[28] V. Prey, E. Gutschik and H. Berbalk, *Monatsh.*, 1960, **91**, 794.

[29] R. G. R. Bacon, R. G. Guy, R. S. Irwin and T. A. Robinson, *Proc. Chem. Soc.*, 1959, 304.

[30] R. G. R. Bacon and R. G. Guy, *J. Chem. Soc.*, 1961, 2428.

[31] V. Prey, E. Gutschik and H. Berbalk, *Monatsh.*, 1960, **91**, 556.

[32] H. Kloosterziel, *Rec. Trav. Chim. Pays-Bas*, 1963, **82**, 497.
[33] L. Field and J. E. White, *Internat. J. Sulfur Chem.*, 1976, 539; L. Field and J. E. White, *Proc. Nat. Acad. Sci. U.S.*, 1973, **70**, 328.
[34] N. S. Kharasch and Z. S. Ariyan, *Chem. Ind.*, 1964, 929.
[35] N. S. Kharasch and Z. S. Ariyan, *Chem. Ind.,* 1964, 302.
[36] N. Kharasch and A. J. Khodair, in *The Chemistry of Sulfides*, ed. A. V. Tobolsky, Interscience, New York, 1968, p. 105.
[37] F. Kaluza and G. W. Perold, *J. S. African Chem. Inst.*, 1960, **13**, 89 (*Chem. Abstr.*, 1961, **55**, 11 346).
[38] D. H. R. Barton, Y. L. Chow, A. Cox and G. W. Kirby, *Tetrahedron Lett.*, 1962, 1055; *idem, J. Chem. Soc.*, 1965, 3571.
[39] V. N. R. Pillai, *Chem. and Ind.*, 1976, 456.
[40] A. Horowitz, *Int. J. Chem. Kinet.*, 1975, **7**, 927 (*Chem. Abstr.*, 1975, **83**, 192 199).
[41] M. A. Saitova and R. F. Khalimov, *Izv. Akad. Nauk SSSR, Ser. Khim.*, 1979, 1642 (*Chem. Abstr.*, 1979, **91**, 174 469).
[42] P. S. Skell and J. H. McNamara, *J. Am. Chem. Soc.*, 1957, **79**, 85.
[43] P. S. Skell, R. C. Woodworth and J. H. McNamara, *J. Am. Chem. Soc.*, 1957, **79**, 1253.
[44] S. J. Cristol and J. A. Reeder, *J. Org. Chem.*, 1961, **26**, 2181.
[45] S. J. Cristol and D. I. Davies, *J. Org. Chem.*, 1964, **29**, 1282.
[46] G. Sosnovsky, in *Free Radicals in Preparative Organic Chemistry*, Macmillan, New York, 1964, p. 110.
[47] G. Sosnovsky, *Intra-Science Chemistry Reports*, 1967, **1**, 1.
[48] E. S. Huyser, *J. Am. Chem. Soc.*, 1960, **82**, 5246.
[49] E. S. Huyser, H. Schimke and R. L. Burham, *J. Org. Chem.*, 1963, **28**, 2141.
[50] E. S. Huyser and R. Giddings, *J. Org. Chem.*, 1962, **27**, 3391.
[51] C. M. M. Da Silva Correa and W. A. Waters, *J. Chem. Soc (C)*, 1968, 1874.
[52] W. E. Truce, D. L. Heuring and G. C. Wolf, *J. Org. Chem.*, 1974, **39**, 238.
[53] Y. Miura, H. Asada and M. Kinoshita, *Bull. Chem. Soc. Jpn*, 1977, **50**, 1855.
[54] R. F. Bayfield and E. R. Cole, *Phosphorus and Sulphur*, 1976, **1**, 19.
[55] T. Anda, M. Nojima and N. Tokura, *J. Chem. Soc., Perkin, Trans, 1*, 1977, 2227.
[56] D. H. R. Barton, T. Nakano and P. G. Sammes, *J. Chem. Soc. (C).*, 1968, 322.
[57] J. Rokach and P. Hamel, *J. Chem. Soc., Chem. Commun.*, 1979, 786.

[58] N. Kamigata, S. Hashimoto, S. Fujie and M. Kobayashi, *J. Chem. Soc., Chem. Commun.*, 1983, 765; N. Kamigata, S. Hashimoto and M. Kobayashi, *Sulfur Lett.*, 1984, **2**, 17; N. Kamigata, S. Hashimoto, M. Kobayashi and H. Nakanishi, *Bull. Chem. Soc. Jpn*, 1985, **58**, 3131.
[59] N. Kamigata, H. Iizuka and M. Kobayashi, *Bull. Chem. Soc. Jpn*, 1986, **59**, 1601.
[60] N. Kamigata, H. Iizuka and M. Kobayashi, *Heterocycles*, 1986, **24**, 919.
[61] V. N. R. Pillai, *Org. Photochem.*, 1987, **9**, 225.
[62] T. Hamada, A. Nishida and O. Yonemitsu, *J. Am. Chem. Soc.*, 1986, **108**, 140.
[63] T. Hamada, A. Nishida, Y. Matsumoto and O. Yonemitsu, *J. Am. Chem. Soc.*, 1980, **102**, 3978.
[64] K. Oda, T. Ohnuma and Y. Ban, *J. Org. Chem.*, 1984, **49**, 953.
[65] B. Umezawa, O. Hoshino and S. Sawaki, *Chem. Pharm. Bull.*, 1969, **17**, 1120 (*Chem. Abstr.*, 1969, **71**, 081 115).
[66] B. Umezawa, O. Hoshino and S. Sawaki, *Chem. Pharm. Bull.*, 1969, **17**, 1115 (*Chem. Abstr.*, 1969, **71**, 081 114).
[67] T. Hamada, A. Nishida and O. Yonemitsu, *Heterocycles*, 1979, **12**, 647.
[68] J. A. Pincock and A. Jurgens, *Tetrahedron Lett.*, 1979, 1029.
[69] A, Abad, D. Mellier, J. P. Pete and C. Portella, *Tetrahedron Lett.*, 1971, 4555.
[70] J. P. Pete and C. Portella, *J. Chem. Res. (S)*, 1979, 20.
[71] D. L. Forster, T. L. Gilchrist and C. W. Rees, *J. Chem. Soc. C.*, 1971, 993.
[72] J. Cossy and J. P. Pete, *Tetrahedron*, 1981, **37**, 2287.
[73] J. C. Arnould, J. Cossy and J. P. Pete, *Tetrahedron Lett.*, 1976, 3919.
[74] J. C. Arnould, J. Cossy and J. P. Pete, *Tetrahedron*, 1980, **36**, 1585.
[75] J. C. Arnould and J. P. Pete, *Tetrahedron Lett.*, 1975, 2463.
[76] M. S. Ao and E. M. Burgess, *J. Am. Chem. Soc.*, 1971, **93**, 5298.
[77] M. Lancaster and D. J. H. Smith, *J. Chem. Soc., Chem. Commun.*, 1980, 471.
[78] N. Kamigata, T. Saegusa, S. Fujie and M. Kobayashi, *Chem. Lett.*, 1979, 9.
[79] R. W. Hoffmann, W. Sieber and G. Guhn, *Chem. Ber.*, 1965, **98**, 3470.
[80] T. Durst and J. F. King, *Can. J. Chem.*, 1966, **44**, 1869.
[81] C. V. Kumar, K. R. Gopidas, K. Bhattacharyya, P. K. Das and M. V. George, *J. Org. Chem.*, 1986, **51**, 1967.
[82] D. Bellus, *Adv. Photochem.*, 1971, **8**, 109.
[83] B. Weiss, H. Durr and H. J. Hass, *Angew. Chem. Int. Ed. Engl.*, 1980,

**19**, 648.

[84] H. Nozaki, T. Okada, R. Noyori and M. Kawanisi, *Tetrahedron*, 1966, **22**, 2177.

[85] A. Graftieaux and J. Gardent, *Tetrahedron Lett.*, 1972, 3321.

[86] M. Somei and M. Natsume, *Tetrahedron Lett.*, 1973, 2451.

[87] A. Chakrabarti, G. K. Biswas and D. P. Chakrabotry, *Tetrahedron*, 1989, **46**, 5059.

[88] J. A. Barltrop, P. M. Hayes and M. Calvin, *J. Am. Chem. Soc.*, 1954, **76**, 4348; H. Reinholdt and E. Motzkus, *Ber.*, 1939, **74**, 657.

[89] T. Kawamura, P. J. Krusic and J. K. Kochi, *Tetrahedron Lett.*, 1972, 4075.

[90] W. B. Gara, B. P. Roberts, B. C. Gilbert, C. M. Kirk and R. O. C. Norman, *J. Chem. Research (S)*, 1977, 152.

[91] R. W. Binkley and T. W. Flechtner, in *Synthetic Organic Photochemistry*, ed. W. M Horspool, Plenum, N. Y. 1984, pp. 375–423.

[92] A. G. Schultz and R. H. Schlessinger, *J. Chem. Soc., Chem. Commun.*, 1969, 1483; A. G. Schultz, C. D. DeBoer and R. H. Schlessinger, *J. Am. Chem. Soc.*, 1968, **90**, 5314; A. G. Schultz and R. H. Schlessinger, *J. Chem. Soc., Chem. Commun.*, 1970, 1051.

[93] W. M. Horspool, *Aspects of Organic Photochemistry*, Academic Press, London, 1976.

[94] I. J. W. Still and M. T. Thomas, *Tetrahedron Lett.*, 1970, 4225.

[95] A. G. Schultz and R. H. Schlessinger, *Tetrahedron Lett.*, 1973, 3605.

[96] I. J. W. Still, M. S. Chauhan and M. T. Thomas, *Tetrahedron Lett.*, 1973, 1311; I. W. J. Still, P. C. Arora, M. S. Chauhan, M. H. Kwan and M. T. Thomas, *Can. J. Chem.*, 1976, **54**, 455.

[97] I. J. W. Still, P. C. Arora, M. S. Chauhan, M. H. Kwan, and M. T. Thomas, *Can. J. Chem.*, 1976, **54**, 455.

[98] K. Kobayashi and K. Mutai, *Tetrahedron Lett.*, 1981, **22**, 5201.

[99] K. Kobayashi and K. Mutai, *Phosphorus Sulfur*, 1985, **25**, 43.

[100] C. L. Gajurel, *Indian J. Chem. Sect. B.*, 1986, **25B**, 319.

[101] B. S. Larsen, J. Kolc and S.-O. Lawesson, *Tetrahedron*, 1971, **27**, 5163.

[102] D. C. Dittmer, G. C. Levy and G. E. Kuhlman, *J. Am. Chem. Soc.*, 1967, **89**, 2793.

[103] D. C. Dittmer, G. E. Kuhlman and G. C. Levy, *J. Org. Chem.*, 1970, **35**, 3676.

[104] Y. Izawa and N. Kuromiya, *Bull. Chem. Soc. Jpn*, 1975, **48**, 3197.

[105] A. Abad, D. Mellier, J. P. Pete, and C. Portella, *Tetrahedron Lett.*, 1971, 4555.

[106] D. Mellier, J. P. Pete and C. Portella, *Tetrahedron Lett.*, 1971, 4559.

[107] J. P. Pete and C. Portella, *Bull. Soc. Chim. Fr.*, 1980, 275.
[108] R. W. Binkley and T. W. Flechtner, in *Synthetic Organic Photochemistry*, Ed. W. M. Horspool, Plenum Press, 1984, p. 375.
[109] R. W. Binkley, *Modern Carbohydrate Chemistry*, Marcel Dekker, Inc., New York, 1988.
[110] R. W. Binkley, in *Advances in Carbohydr. Chem. and Biochem.*, eds R. S. Tipson and D. Horton, Academic Press, 1981, Vol 38, p. 105.
[111] F. R. Seymour, *Carbohydr. Res.*, 1974, **34**, 65.
[112] F. R. Seymour, M. E. Slodki, R. D. Plattner and L. W. Tjarks, *Carbohydr. Res.*, 1976, **46**, 189.
[113] S. Zen, S. Tashima and S. Koto, *Bull. Chem. Soc. Jpn*, 1968, **41**, 3025.
[114] W. A. Szarek, R. G. S. Ritchie and D. M. Vyas, *Carbohydr. Res.*, 1978, **62**, 89; A. D. Barford, A. B. Foster and J. H. Westwood, *Carbohydr. Res.*, 1970, **13**, 189; A. D. Bradford, A. B. Foster, J. H. Westwood, L. D. Hall and R. N. Johnson, *Carbohydr. Res.*, 1971, **19**, 49; L. Vegh and E. Hardegger, *Helv. Chim. Acta,* 1973, **56**, 2020; R. A. Boigegrain and B. Gross, *Carbohydr. Res.*, 1975, **41**, 135; C. D. Chang and T. L. Hullar, *Carbohydr. Res.*, 1977, **54**, 217.
[115] T. Kishi, T. Tsuchiya and S. Umezawa, *Bull. Chem. Soc. Jpn*, 1979, **52**, 3015.
[116] T. Tsuchiya, F. Nakamura and S. Umezawa, *Tetrahedron Lett.*, 1979, 2805.
[117] Y. Kondo, K. Hosoyama and T. Takemoto, *Chem. Pharm. Bull.*, 1975, **23**, 2167 (*Chem. Abstr.*, 1979, **84**, 017600).
[118] R. W. Binkley, *J. Org. Chem.*, 1985, **50**, 5649; R. W. Binkley and D. J. Koholic, *J. Org. Chem.*, 1989, **54**, 3577.
[119] J. Masnovi, D. J. Koholic, R. J. Berki and R. W. Binkley, *J. Am. Chem. Soc.*, 1987, **109**, 2851.
[120] J. L. Charlton, H. K. Lai and G. N. Lypka, *Can. J. Chem.*, 1980, **58**, 458.
[121] S. Iwasaki and K. Schaffner, *Helv. Chim. Acta*, 1968, **51**, 557.
[122] G. Hueppi, G. Eggart, S. Iwasaki, H. Wehrli, K. Schaffner and O. Jeger, *Helv. Chim. Acta*, 1966, **49**, 1986.
[123] A. Tuinman, S. Iwasaki, K. Schaffner and O. Jeger, *Helv. Chim. Acta.*, 1968, **51**, 1778.
[124] K. Schaffner, *Pure Appl. Chem.*, 1968, **16**, 75.
[125] A. Feigenbaum, J. P. Pete and D. Scholler, *Tetrahedron Lett.*, 1979, 537.
[126] A. Feigenbaum, J. P. Pete and D. Scholler, *J. Org. Chem.*, 1984, **49**, 2355.

[127] A. L. Poquet, A. Feigenbaum and J. P. Pete, *Tetrahedron Lett.*, 1986, **27**, 2975.
[128] K. Tomari, K. Machiya, I. Ichimoto and H. Ueda, *Agric. Biol. Chem.*, 1980, **44**, 2135.
[129] O. L. Chapman and C. L. McIntosh, *J. Chem. Soc. D*, 1971, 383.
[130] L. A. Levy, *Tetrahedron Lett.*, 1972, 3289.
[131] E. Henmo, P. de Mayo, A. B. M. A. Sattar and A. Stoessl, *Proc. Chem. Soc.*, 1961, 238.
[132] J. F. King, P. de Mayo, E. Morkved, A. B. M. A. Sattar and A. Stoessl, *Can. J. Chem.*, 1963, **41**, 100.
[133] J. L. Charlton and P. De Mayo, *Can. J. Chem.*, 1968, **46**, 55.
[134] J. L. Stratenus and E. Havinga, *Rec. Trav. Chim. Pays-Bas*, 1966, **85**, 434.
[135] Y. Ogata, K. Takagi and S. Yamada, *J. Chem. Soc., Perkin Trans. 2*, 1977, 1629.
[136] D. R. Olson, *J. Appl. Polym. Sci.*, 1983, **28**, 1159.
[137] M. Prochazka and M. Paleck, *Coll. Czech. Chem. Commun.*, 1967, **32**, 3049.
[138] C. C. Price and S. Oae, in *Sulfur Bonding*, Ronald Press, New York, 1962.
[139] C. W. N. Cumper, J. F. Read and A. I. Vogel, *J. Chem. Soc., (A)*, 1966, 239.
[140] L. A. Paquette and M. Rosen, *J. Org. Chem.*, 1968, **33**, 3027.
[141] R. W. Hoffmann and W. Sieber, *Justus Liebigs Ann. Chem.*, 1967, **703**, 96.
[142] A. Mustafa, *Adv. Photochem.*, 1964, **2**, 63.
[143] J. D. Coyle, *Chem. Soc. Rev.*, 1975, **4**, 523.
[144] S. T. Reid, *Adv. in Heterocyclic Chem.*, 1970, **11**, 1.
[145] T. Nakabayashi, Y. Nagata and J. Tsurugi, *Int. J. Sulf. Chem., Part A*, 1971, **1**, 54 (*Chem. Abstr.*, 1971, **75**, 097 971).
[146] T. Nakabayashi, Y. Abe and T. Horii, *Phosphorus Sulfur*, 1976, **1**, 285 (*Chem. Abstr.*, 1977, **86**, 139787).
[147] A. I. Khodair, T. Nakabayashi and N. Kharasch, *Int. J. Sulf. Chem.*, 1973, **8**, 37.
[148] N. Kharasch and A. I. A. Khodair, *Chem. Commun.*, 1967, 98.
[149] O. De Lucchi, G. Licini, L. Pasquato and M. Senta, *J. Chem. Soc., Chem. Commun.*, 1985, 1597.
[150] R. G. Pews and T. E. Evans, *J. Chem. Soc. D*, 1971, 1397.
[151] T. Sato, Y. Goto, T. Tohyama, S. Hayashi and K. Hata, *Bull. Chem. Soc. Jpn*, 1967, **40**, 2975.

[152] R. S. Givens, B. Hrinczenko, J. H. S. Liu, B. Matuszewski and J. Tholen-Collinson, *J. Am. Chem. Soc.*, 1984, **106**, 1779.
[153] R. S. Givens and B. Matuszewski, *Tetrahedron Lett.*, 1978, 861.
[154] R. S. Givens, W. K. Chae, J. H. Liu, C. V. Neywich, W. D. Gillaspey, R. J. Olsen and L. P. Wylie, *Prepr. Div. Pet. Chem., Am. Chem. Soc.*, 1979, 24 (*Chem. Abstr.*, 1981, **94**, 055 807).
[155] I. R. Gould, C. H. Tung, N. J. Turro, R. S. Givens and B. Matuszewski, *J. Am. Chem. Soc.*, 1984, **106**, 1789.
[156] A. S. Amin and J. M. Mellor, *J. Photochem.*, 1978, **9**, 571.
[157] G. E. Robinson and J. M. Vernon, *J. Chem. Soc., Perkin Trans. 1*, 1977, 1682.
[158] R. F. Langler, Z. A. Marini and J. A. Pincock, *Can. J. Chem.*, 1978, **56**, 903.
[159] R. W. Binkley, in *Adv. in Carbohydrate Chem. and Biochem.*, eds R. S. Tipson and D. Horton, Academic Press, New York, 1981, Vol 38, p. 105.
[160] P. M. Collins and B. R. Whitton, *J. Chem. Soc., Perkin Trans. 1*, 1974, 1069.
[161] P. M. Collins and B. R. Whitton, *Carbohydr. Res.*, 1974, **36**, 293 (*Chem. Abstr.*, 1974, **81**, 152531).
[162] K. Ogura, K. Ohtsuki, M. Nakamura, N. Yahata, K. Takahashi and H. Iida, *Tetrahedron Lett.*, 1985, **26**, 2455; K. Ogura, T. Tsuruda, K. Takahashi and H. Iida, *Tetrahedron Lett.*, 1986, **27**, 3665; K. Ogura, T. Iihama, S. Kiuchi, T. Kajiki, O. Koshikawa, K. Takahashi and H. Iida, *J. Org. Chem.*, 1986, **51**, 700.
[163] I. W. J. Still, in *The Chemistry of Sulphones and Sulphoxides*, ed. S. Patai, Wiley, New York, 1988, pp. 873–887.
[164] W. M. Horspool, in *The Chemistry of Sulphonic Acids, Esters and their Derivatives*, ed. S. Patai, Wiley, New York, 1991, pp. 501–551
[165] F. G. Bordwell, J. M. Williams, jun., E. B. Hoyt, jun. and B. B. Jarvis, *J. Am. Chem. Soc.*, 1968, **90**, 429.
[166] F. G. Bordwell, J. M. Williams, jun. and B. B. Jarvis, *J. Org. Chem.*, 1968, **33**, 2026.
[167] A. A. Scala and I. Colon, *J. Phys. Chem.*, 1979, **83**, 2025.
[168] T. Durst, J. C. Huang, N. K. Sharma and D. J. H. Smith, *Can. J. Chem.*, 1978, **56**, 512.
[169] N. K. Sharma, F. Jung and T. Durst, *Tetrahedron Lett.*, 1973, 2863.
[170] J. D. Finlay, D. J. H. Smith and T. Durst, *Synthesis*, 1978, 579.
[171] A. A. Scala, I. Colon and W. Rourke, *J. Phys. Chem.*, 1981, **85**, 3603.
[172] W. L. Prins and R. M. Kellogg, *Tetrahedron Lett.*, 1973, 2833.

[173] R. M. Kellogg and W. L. Prins, *J. Org. Chem.*, 1974, **39**, 2366.
[174] M. R. Johnson, M. J. Fazio, L. D. Ward and L. R. Sousa, *J. Org. Chem.*, 1983, **48**, 494.
[175] R. M. Kellogg, in *Photochemistry of Heterocyclic Compounds*, ed. O. Buchardt, Wiley, New York, 1976, p. 367.
[176] L. A. Paquette, R. H. Meisinger and R. E. Wingar, jun., *J. Am. Chem. Soc.*, 1973, **95**, 2230.
[177] M. P. Cava, R. H. Schlessinger and J. P. Van Meter, *J. Am. Chem. Soc.*, 1964, **86**, 3173.
[178] M. P. Cava and R. H. Schlessinger, *Tetrahedron*, 1965, **21**, 3065, 3073.
[179] M. P. Cava, *NASA Accession No. N 6620840* (*Chem. Abstr.,* 1967, **66**, 75511).
[180] V. Boekelheide, *Acc. Chem. Res.*, 1980, **13**, 65.
[181] R. S. Givens, *Org. Photochem.*, 1981, **5**, 227.
[182] W. Rebafka and H. A. Staab, *Angew. Chem. Int. Ed. Engl.*, 1973, **12**, 776.
[183] R. Gray, L. G. Harruff, J. Krymowski, J. Peterson and V. Boekelheide, *J. Am. Chem. Soc.*, 1978, **100**, 2892.
[184] R. S. Givens and P. L. Wylie, *Tetrahedron Lett.*, 1978, 865.
[185] R. S. Givens, R. J. Olsen and P. L. Wylie, *J. Org. Chem.*, 1979, **44**, 1608.
[186] O. De Lucchi, V. Lucchini, C. Marchiori and G. Modena, *Tetrahedron Lett.*, 1985, **26**, 4539.
[187] M. A. A. M. El Tabei, N. V. Kirby and S. T. Reid, *Tetrahedron Lett.*, 1980, **21**, 565.
[188] N. V. Kirby and S. T. Reid, *J. Chem. Soc., Chem. Commun.*, 1980, 150.
[189] W. Davies and F. C. James, *J. Chem. Soc.*, 1955, 315.
[190] A. Mustafa, *Nature*, 1955, **175**, 992.
[191] H.-D. Scharf and F. Korte, *Angew. Chem. Int. Ed. Engl.,* 1965, **4**, 429.
[192] A. Padwa and M. W. Wannamaker, *J. Chem. Soc., Chem. Commun.*, 1987, 1742.
[193] *Jpn Kokai Tokyo Koho,* **JP 59,231,063** (*Chem. Abstr.*, 1985, **102**, 220500).
[194] L. A. Paquette and H. Kuenzer, *J. Am. Chem. Soc.*, 1986, **108**, 7431.
[195] L. A. Paquette, H. Kuenzer and M. A. Kesselmeyer, *J. Am. Chem. Soc.*, 1988, **110**, 6521.
[196] V. Ramamurthy, *Org. Photochem.*, 1985, **7**, 231.
[197] R. P. Steer, *Rev. Chem. Intermediates*, 1981, **4**, 1.
[198] A. Safarzadehi-Amiri, R. E. Verrall and R. P. Steer, *Can. J. Chem.*, 1983, **61**, 894.
[199] A. H. Maki, P. Suejda and J. R. Huber, *Chem. Phys.*, 1980, **32**, 369;

M. R. Taherian and A. H. Maki, *Chem. Phys.*, 1982, **68**, 179.
[200] K. Muthuramu and V. Ramamurthy, *J. Org. Chem.*, 1980, **45**, 4532; K. Muthuramu, B. Sundari and V. Ramamurthy, *Indian J. Chem.*, 1981, **20B**, 797; K. Muthuramu, B. Sundari and V. Ramamurthy, *J. Org. Chem.*, 1983, **48**, 4482.
[201] K. Muthuramu and V. Ramamurthy, *Chem. Lett.*, 1981, 1261; K. Kimura, Y. Fukuda, T. Negoro and Y. Odaira, *Bull. Chem. Soc. Jpn*, 1981, **54**, 1901.
[202] S. Sharat, M. M. Bhadbhade, V. Venkatesan and V. Ramamurthy, *J. Org. Chem.*, 1982, **47**, 3550; A. Schonberg and M. Mamluk, *Tetrahedron Lett.*, 1971, 4993; B. M. Trost and R. Atkins, *Tetrahedron Lett.*, 1968, 1225.
[203] N. Kito and A. Ohno, *J. Chem. Soc., Chem. Commun.*, 1971, 1338; N. Kito and A. Ohno, *Int. J. Sulphur Chem.*, 1973, **8**, 427.
[204] A. Cox, D. R. Kemp, R. Lapouyade, P. de Mayo, J. Joussot-Dubien and R. Bonneau, *Can. J. Chem.*, 1975, **53**, 2386.
[205] P. de Mayo and R. Suau, *J. Am. Chem. Soc.*, 1974, **96**, 6807.
[206] A. Couture, K. Ho, M. Hoshino, P. de Mayo, R. Suau and W. R. Ware, *J. Am. Chem. Soc.*, 1976, **98**, 6218.
[207] S. Basu, A. Couture, K. W. Ho, M. Hoshino, P. de Mayo and R. Suau, *Can. J. Chem.*, 1981, **59**, 246.
[208] A. Couture, M. Hoshino and P. de Mayo, *J. Chem. Soc., Chem. Commun.*, 1976, 131; A. Couture, J. Gomez and P. de Mayo, *J. Org. Chem.*, 1981, **46**, 2010.
[209] D. S. L. Blackwell and P. de Mayo, *J. Chem. Soc., Chem. Commun.*, 1973, 130; D. S. L. Blackwell, K. H. Lee, P. de Mayo, G. L. R. Petrasiunas and G. Reverdy, *Nouv. J. Chim.*, 1979, **3**, 123.
[210] G. Oster, L. Citarel and M. Goodman, *J. Am. Chem. Soc.*, 1962, **84**, 7036.
[211] A. Ohno and N. Kito, *Int. J. Sulphur Chem. A.*, 1971, **1**, 26.
[212] Y. Ohnishi and A. Ohno, *Bull. Chem. Soc. Jpn*, 1973, **46**, 3868.
[213] N. Kito and A. Ohno, *Bull. Chem. Soc. Jpn*, 1973, **46**, 2487; S. J. Formosinho, *J. Chem. Soc., Faraday Trans. II*, 1976, 1332.
[214] J. R. Bolton, K. S. Chen, A. H. Lawrence and P. de Mayo, *J. Am. Chem. Soc.*, 1975, **97**, 1832.
[215] A. Ohno, M. Uohama, K. Nakamura and S. Oka, *Bull. Chem. Soc. Jpn*, 1979, **52**, 1521; V. J. Rao and V. Ramamurthy, *Indian J. Chem.*, 1979, **18B**, 265.
[216] K. Y. Law and P. de Mayo, *J. Am. Chem. Soc.*, 1977, **99**, 5813; 1979, **101**, 3251.

[217] E. T. Kaisers and T. F. Wulfers, *J. Am. Chem. Soc.*, 1964, **86**, 1897.
[218] A. Ohno, *Int. J. Sulphur Chem. B.*, 1971, **6**, 183; G. Tsuchihashi, M. Yamauichi and M. Fukuyama, *Tetrahedron Lett.*, 1967, 1971; A. Ohno, Y. Ohnishi, M. Fukuyama and G. Tsuchihashi, *J. Am. Chem. Soc.*, 1968, **90**, 7038; A. Ohno, Y. Ohnishi and G. Tsuchihashi, *Tetrahedron Lett.*, 1969, 283; A. Ohno, Y. Ohnishi and G. Tsuchihashi, *J. Am. Chem. Soc.*, 1969, **91**, 5038.
[219] P. de Mayo and A. A. Nicholson, *Isr. J. Chem.*, 1972, **10**, 341.
[220] C. C. Liao and P. de Mayo, *J. Chem. Soc., Chem. Commun.*, 1971, 1525.
[221] A. H. Lawrence, C. C. Liao, P. de Mayo and V. Ramamurthy, *J. Am. Chem. Soc.*, 1976, **98**, 2219.
[222] A. H. Lawrence, C. C. Liao, P. de Mayo and V. Ramamurthy, *J. Am. Chem. Soc.*, 1976, **98**, 3572.
[223] R. Rajee and V. Ramamurthy, *Tetrahedron Lett.*, 1978, 3463.
[224] A. Ohno, T. Koizumi, Y. Ohnishi and G. Tsuchiahashi, *Tetrahedron Lett.*, 1970, 2025; A. Ohno, T. Koizumi and G. Tsuchiahashi, *Bull. Chem. Soc. Jpn*, 1971, **44**, 2511.
[225] Y. Omote, M. Yoshioka, K. Yamada and N. Sugiyama, *J. Org. Chem.*, 1968, **33**, 1240.
[226] O. Ohno, N. Kito and T. Koizumi, *Tetrahedron Lett.*, 1971, 2421.
[227] R. G. Visser, J. P. B. Baaij, A. C. Brouwer and H. J. T. Bos, *Tetrahedron Lett.*, 1977, 434.
[228] H. J. T. Bos, H. Schinkel and Th. C. M. Wijsman, *Tetrahedron Lett.*, 1971, 3905; H. Gotthardt, *Tetrahedron Lett.*, 1971, 2345; G. Hofstra, J. Kamphuis and H. J. T. Bos, *Tetrahedron Lett.*, 1984, **25**, 873; R. G. Visser and H. J. T. Bos, *Tetrahedron Lett.*, 1979, 4857.
[229] R. G. Visser, E. A. Oostereen and H. J. T. Bos, *Tetrahedron Lett.*, 1981, 1139.
[230] T. S. Cantrell, *J. Org. Chem.*, 1974, **39**, 853.
[231] Y. Kobayashi and I. Kumadaki, in *Adv. Heterocyclic Chem.*, 1982, **31**, 169.
[232] H. Wynberg, R. M. Kellogg, H. van Driel and G. E. Beekhis, *J. Am. Chem. Soc.*, 1967, **89**, 3501.
[233] A. Couture and A. Lablache-Combier, *Tetrahedron*, 1971, **27**, 1059.
[234] R. M. Kellogg, *Tetrahedron Lett.*, 1972, 1429.
[235] H. A. Wiebe, S. Braslavsky and J. Heicklen, *Can. J. Chem.*, 1972, **50**, 2721; Y. Kobayashi, I. Kumadaki, A. Ohsawa, Y. Sekine and M. Mochizuki, *Chem. Pharm. Bull.*, 1975, **23**, 2773.
[236] Y. Kobayashi, I. Kumadaki, A. Ohsawa and Y. Sekine, *Tetrahedron*

*Lett.*, 1974, 2841.

[237] Y. Kobayashi, A. Ando, K. Kawada and I. Kumadaki, *J. Chem. Soc., Chem. Commun.*, 1981, 1289; *ibid. J. Am. Chem. Soc.*, 1981, **103**, 3958.

[238] J. A. Barltrop., A. C. Day and E. Irving, *J. Chem. Soc., Chem. Commun.*, 1979, 881, 966.

[239] M. Ohashi, A. Iio and T. Yonezawa, *J. Chem. Soc., Chem. Commun.*, 1970, 1148.

[240] G. Vernin, J.-C. Poite, J. Metzger, J.-P. Aune and H. J. M. Dou, *Bull. Soc. Chim. Fr.*, 1971, 1103.

[241] G. Vernin, C. Riou, H. J. M. Dou, L. Bouscasse, J. Metzger and G. Loridan, *Bull. Soc. Chim. Fr.*, 1973, 1743.

[242] J. P. Catteau, A. Lablache-Combier and A. Pollet, *Tetrahedron*, 1972, **28**, 3141; A. Lablache-Combier, in *Photochemistry of Heterocyclic Compounds*, ed. O. Buchardt, Wiley, New York, 1976, p. 123.

[243] C. Riou, J.-C. Poite, G. Vernin and J. Metzger, *Tetrahedron*, 1074, **30**, 879.

[244] M. Maeda, A. Kawahara, M. Kai and M. Kojima, *Heterocycles*, 1975, **3**, 389.

[245] M. Maeda and K. Kojima, *Tetrahedron Lett.*, 1973, 3523.

[246] M. Ohashi, A. Ezaki and T. Yonezawa, *J. Chem. Soc., Chem. Commun.*, 1974, 617.

[247] M. Sindler-Kulyk and D. C. Neckers, *Tetrahedron Lett.*, 1981, **22**, 525.

[248] M. Sindler-Kulyk, D. C. Neckers and J. R. Blount, *Tetrahedron*, 1981, **37**, 3377.

[249] M. Sindler-Kulyk and D. C. Neckers, *Tetrahedron Lett.*, 1981, **22**, 529.

[250] J. J. McCullough, *Chem. Rev.*, 1987, **87**, 811.

[251] M. Sindler-Kulyk and D. C. Neckers, *J. Org. Chem.*, 1983, **48**, 1275.

[252] A. Krantz and J. Laureni, *J. Am. Chem. Soc.*, 1977, **99**, 4842; A. Krantz and J. Laureni, *J. Am. Chem. Soc.*, 1981, **103**, 486; J. Laureni and A. Krantz, *Ber. Bunsengesellschaft Phys. Chem.*, 1978, **82**, 13; J. Laureni, A. Krantz and R. A. Hajdu, *J. Am. Chem. Soc.*, 1976, **98**, 7872; A. Krantz, J. Laureni and R. A. Hajdu, *J. Am. Chem. Soc.*, 1974, **96**, 6768.

[253] W. Kirmse and L. Horner, *Liebigs Ann. Chem.*, 1958, **614**, 4.

[254] K.-P. Zeller, H. Meier and E. Mueller, *Tetrahedron Lett.*, 1971, 537; K.-P. Zeller, H. Meier and E. Mueller, *Liebigs Ann. Chem.*, 1972, **766**, 32.

[255] H. Buehl, U. Timm and H. Meier, *Chem. Ber.*, 1979, **112**, 3728.

# 5

# Nitrogen-containing compounds

The photochemistry discussed in this chapter describes the reactions of the various classes of compounds containing at least one nitrogen atom. Thus the photochemistry of a variety of groups such as nitro, nitrite and cyano as well as *N*-oxides and aromatic nitrogen-containing compounds is discussed. In the main, the reactions are carried out in solution phase, since it is these which are most usually of synthetic value.

## 5.1 ABSORPTION SPECTRA OF IMINES AND RELATED COMPOUNDS

The u.v. absorption spectra of saturated imines show two bands at 170–180 nm and 230–260 nm [1, 2]. The band at lower wavelength exhibits high $\varepsilon$ values (*c.* $10^4$ $dm^3$ $mol^{-1}$ $cm^{-1}$) and has been assigned to a $\pi\pi^*$ transition [1, 2] while the lower energy band at 230–260 nm arises from an $n\pi^*$ transition [1, 2].

The conjugation of the C=N double bond with olefinic or aryl groups affects the u.v. spectra of these compounds, as might be expected, shifting the absorptions to longer wavelengths. On the other hand, the spectra became more complicated, exhibiting more absorptions that are less easy to assign. This is thought to be due to partial delocalization of the nitrogen lone pair

[1, 2]. In addition to this, the spectra, very often, show absorptions that can be assigned to charge-transfer complexes [2].

In the case of iminium salts, which are currently of great interest, the situation is completely different, since the nitrogen atom does not have a lone pair of electrons, thus eliminating the possibility of $n\pi^*$ transitions. These compounds absorb in the 250–330 nm region depending on substitution, and the absorptions have been assigned to a $\pi\pi^*$ transition [3].

## 5.2 PHOTOCHEMICAL REACTIVITY OF IMINES AND RELATED COMPOUNDS

The photochemistry of the C=N bond has not been as extensively studied as that of the C=O bond. However, in the last 10 to 15 years there has been increasing interest in this class of compounds.

In a simple approach it could be expected that the C=N chromophore should behave in a similar way to the C=O group since both functions have a doubly bonded carbon atom attached to a more electronegative atom. Furthermore, both the oxygen and the nitrogen atoms have lone pairs of electrons. Finally, the two chromophores should have quite similar $n\pi^*$ and $\pi\pi^*$ excited states. However, although there are similarities between the photochemical reactivity of these two functional groups, C=N and C=O, the photochemistry of the imines and related systems have peculiarities that merit an independent study. It should also be borne in mind that a C=N functional group can be considered as structurally related to a C=C bond and, therefore, some of the reactions of the imines and related compounds will be similar to those observed for alkenes.

Probably one of the reasons for the growing interest in the photochemistry of imines and related compounds is that they show versatile reactivity. Thus isolated C=N functional groups usually react by concerted or biradical mechanisms. However, in the presence of good electron acceptors, intra- and intermolecular single-electron transfer processes from the nitrogen lone pair to the acceptor have been observed. On the other hand, when the nitrogen lone pair of electrons is coordinated with a proton or a Lewis acid, C=N functional groups undergo frequently photochemical intra- and intermolecular single-electron transfer in the presence of electron donors. This versatile reactivity has enabled the observation of reactions that do not have equivalents in the photochemistry of alkenes and carbonyl compounds.

## 5.2.1 Reactions analogous to alkenes

### *5.2.1.1* E,Z-*Isomerization*

There are many examples of *E,Z*-isomerization in imines and related compounds. Thus, the irradiation of the *E*-imines, both under direct or sensitized conditions, brings about the formation of the *Z*-isomers (**1**) [1, 2].

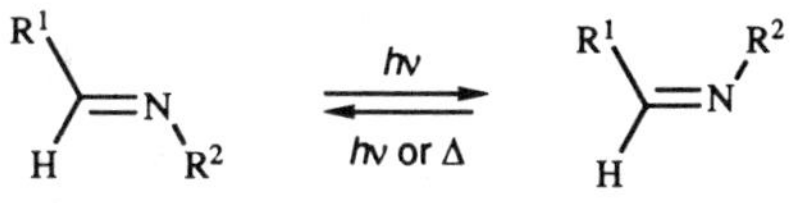

(1)

However, from the synthetic viewpoint this reaction is considerably less interesting than the analogous one for alkenes. The reason for this is that the energy barrier for the thermal conversion of the less stable diastereoisomer into the more stable one is very low (*c.* 58–75 kJ $mol^{-1}$) [4]. As a result, it is usually impossible to isolate the *Z*-isomer since it reverts to starting material during work-up. In many cases the excited state of a simple imine is deactivated via isomerization around the C=N double bond. This accounts for the lack of photochemical reactivity of some acyclic compounds in which the chromophore is a C=N functional group.

The situation is totally different in the case of the oximes and oxime derivatives. In these cases, where the energy barrier between the two isomers is greater, then the *Z*-isomer is usually thermally stable. Therefore, the reaction has some synthetic utility; for instance, the pharmacologically active *anti*- or *Z*-isonicotinaldehyde oxime (**2**) has been obtained by irradiation of

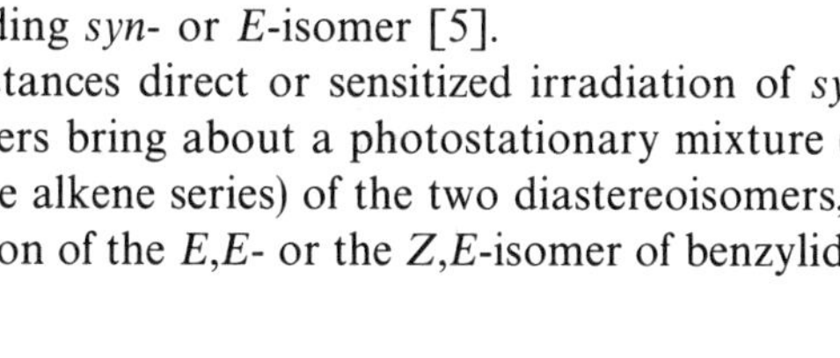

(2)

the corresponding *syn*- or *E*-isomer [5].

In many instances direct or sensitized irradiation of *syn*- or *anti*-oximes and oxime ethers bring about a photostationary mixture (see Chapter 2 for examples in the alkene series) of the two diastereoisomers, like, for instance, in the irradiation of the *E,E*- or the *Z,E*-isomer of benzylideneacetone oxime

*O*-methyl ether (**3**) [6]. This example allows a comparison between the

Ph H H N OMe $\xrightarrow{h\nu}$ fast Ph H H OMe N
hν slow hν slow
H H Ph N OMe $\xrightarrow{h\nu}$ fast H H Ph OMe N

(3)

photochemical reactivity of the C=C and the C=N towards geometric isomerization. The irradiation results in rapid carbon–nitrogen double bond isomerization accompanied by slower carbon–carbon double bond isomerization.

However, other photochemical reactions can also take place in the irradiation of oximes. The most important competing process is usually the Beckmann rearrangement, which will be discussed later (see section 5.12). Nevertheless, the *syn–anti*-isomerization is usually more efficient than other competing reaction paths.

Other C=N double bond derivatives also undergo *E–Z*-isomerization as, for example, hydrazone (**4**) [7], semicarbazone (**5**) [8], triphenylformazone

$NPh_2$ N O $\xrightarrow{h\nu}$ N $NPh_2$ O

(4)

$O_2N$ O N—$NHCONH_2$ $\xrightarrow[\text{visible}]{h\nu}$ $O_2N$ O N $NHCONH_2$

(5)

**(6)** [9], and azine **(7)** [10]. Again in this last case an alternative photochemical

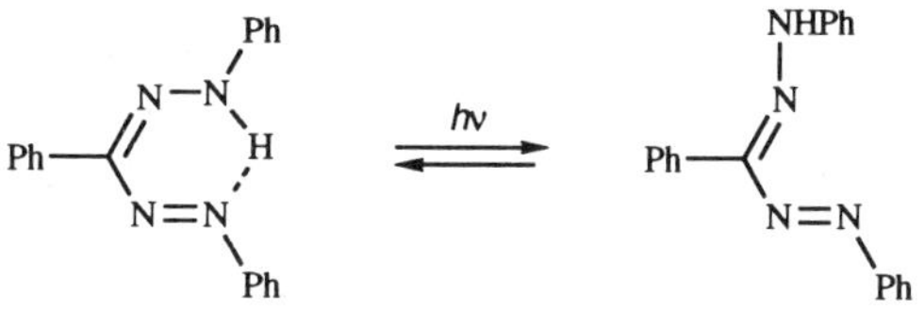

(6)

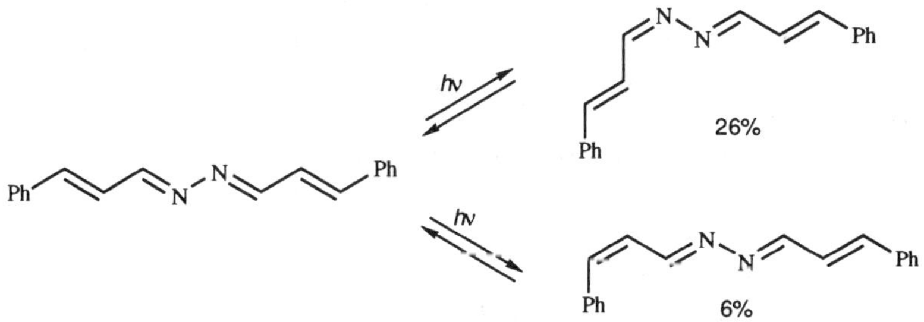

(7)

pathway is the $E$–$Z$-isomerization around the C=C double bond.

Apart from the synthetic potential of the $Z$-$E$-photoisomerization in unsaturated imines and related compounds, this photochemical reaction is of importance because of its implication in the vision process. Thus, one of the key steps in the transformation of visible light into electrical pulses transmitted to the brain by the optic nerve, is the photoisomerization of the C-11 *cis*-iminium salt to the all-*trans*-isomer **(8)** [11].

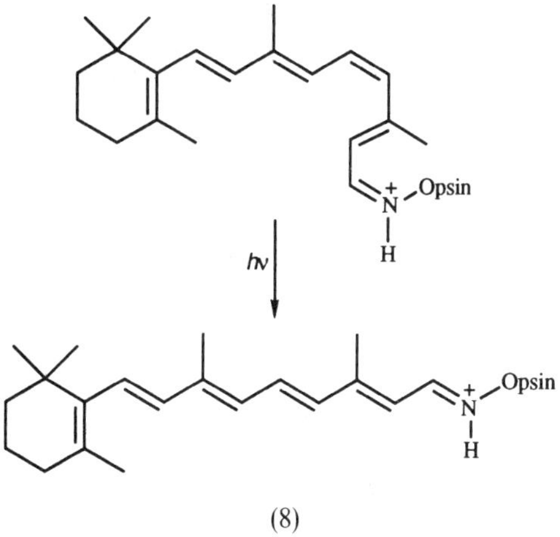

(8)

The geometrical photoisomerization of the C=N can take place by two different mechanisms. One of them is rotation around the C=N double bond, in a similar way to the isomerization of C=C. This mechanism has been postulated in cases in which the first excited state of the C=N bond has $\pi\pi^*$ character. However, the C=N bond could also isomerize by an inversion mechanism consisting of a change in the hybridization of the nitrogen lone pair of electrons from $sp^2$ to $sp$. This mechanism probably occurs via the $n\pi^*$ excited state [12].

In general terms, the *Z*,*E*-photoisomerization of carbon–nitrogen double bonds is usually very efficient. Therefore, it should be expected that one of the main routes of deactivation of the excited state of C=N double bond derivatives would be the *E*–*Z*-isomerization around the double bond, provided that this isomerization is structurally possible.

### *5.2.1.2 Photocyclizations*

Many substituted *N*-phenylbenzylimines undergo stilbene-type cyclizations (see Chapter 2, section 5.7) to yield heterocyclic compounds, usually in the presence of acids and oxidants, as shown in **9** for some Schiff bases [13–15].

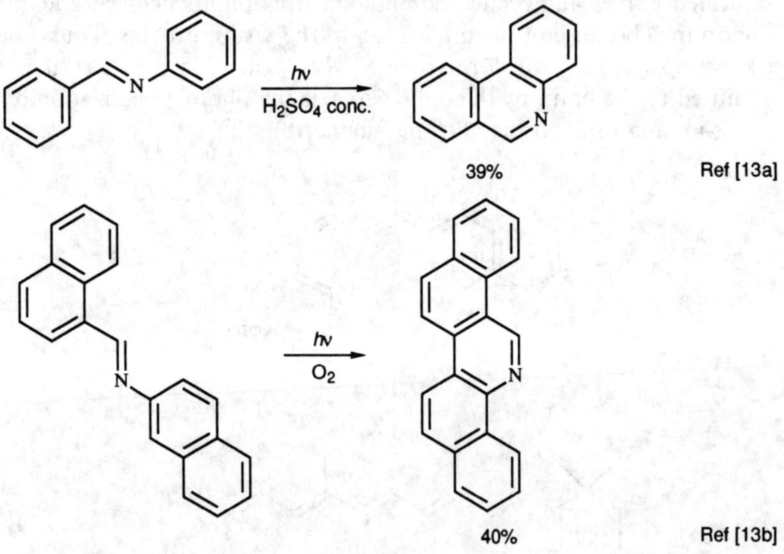

$h\nu$, $H_2SO_4$ conc.

39% Ref [13a]

$h\nu$, $O_2$

40% Ref [13b]

(9)

OMe
hν
$O_2$
OMe
56%
Ref [14]

Ph
Ph
Ph
hν
$I_2$
Ph
46%
Ref [15]

(9 continued)

The reaction usually takes place via the iminium salt and can be interpreted as a typical six-electron photochemical conrotatory ring closure followed by aromatization (**10**).

hν
hν
6e⁻
conrotatory
1) -H⁺
2) [O]

(10)

In the absence of acid, the quantum efficiency of the reaction is usually quite low, probably owing to deactivation of a high percentage of the molecules in the excited state by *syn–anti*-isomerization via the $n\pi^*$ singlet excited state. The co-ordination of the nitrogen lone pair in a strong acid medium suppresses this mode of excitation and the reaction takes place via the $\pi\pi^*$ singlet excited state in a way analogous to the photocyclization of stilbenes. Similar six-electron electrocyclic cyclizations have been observed in the photochemistry of aza-1,3-dienes, such as, for example, the 1-aza-1,3-dienes shown in **11** [16]. In this case direct irradiation using a Pyrex

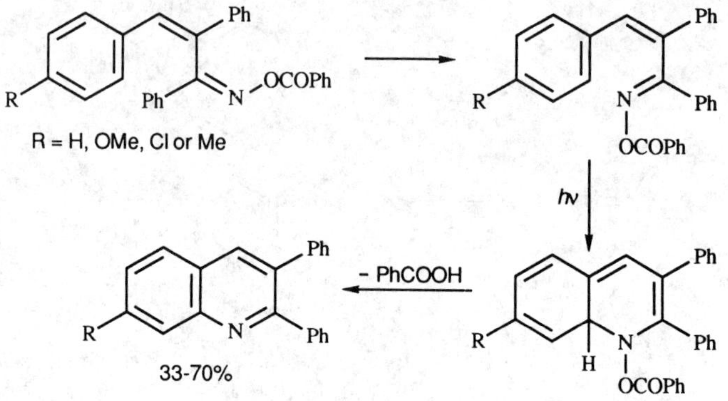

R = H, OMe, Cl or Me

33-70%

(11)

filter affords the corresponding C-7-substituted quinolines that are difficult to synthesize by conventional means. The reaction takes place in the absence of acids or oxidants. In this example the cyclization affords a dihydro-intermediate which will readily aromatize by elimination of benzoic acid to afford the final product. Another efficient reaction of this type has been observed in the irradiation of the 1-azadiene shown in **12** that yields a

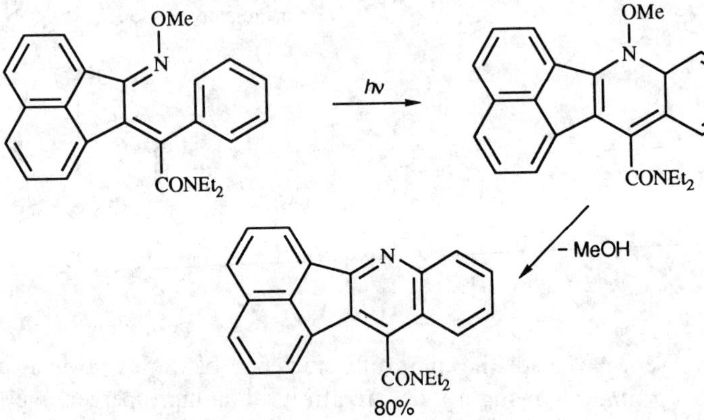

80%

(12)

pentacyclic compound [17].

Iminium salts derived from aza-1,3-dienes also undergo six-electron photochemical cyclizations yielding different heterocyclic systems. Thus, direct irradiation of 4-acyloxy-2-azabuta-1,3-dienes in the presence of acids affords substituted isoquinolin-4-(1*H*)-ones (**13**) [18]. Another interesting

$R^3$ $R^1$ Ph ArCO$_2$ N + H $R^2$ $ClO_4^-$ → hν → $R^3$ H $R^1$ Ph + ArCO$_2$ N H $R^2$ → $-H^+$ → $R^3$ $R^1$ Ph ArCO$_2$ N H $R^2$

$R^1$ = Ph or Me
$R^2$ = Ph, $p$-$MeC_6H_4$, $p$-$CNC_6H_4$, $p$-$MeOC_6H_4$
$R^3$ = H, $p$-CN, $p$-Me, $p$-MeO, $m$-MeO, $m$-CN

1) $H_2O$
2) [O]

$R^3$ $R^1$ Ph O N $R^2$

20-92%

(13)

example of a reaction of this type is the cyclization of 1-styrylpyridinium salts to benzo[*a*]quinoliziniùm salts (**14**) [19]. Some diaza-1,3-dienes also

$X^-$ + N ⇌ hν ⇌ $X^-$ + N → hν, $O_2$ → $X^-$ + N

60%

(14)

undergo similar cyclizations. For instance, direct or sensitized irradiation of imines from benzil mono-oxime esters affords the quinoxaline derivative in another example of a six-electron cyclization (**15**) [20]. Interestingly a different

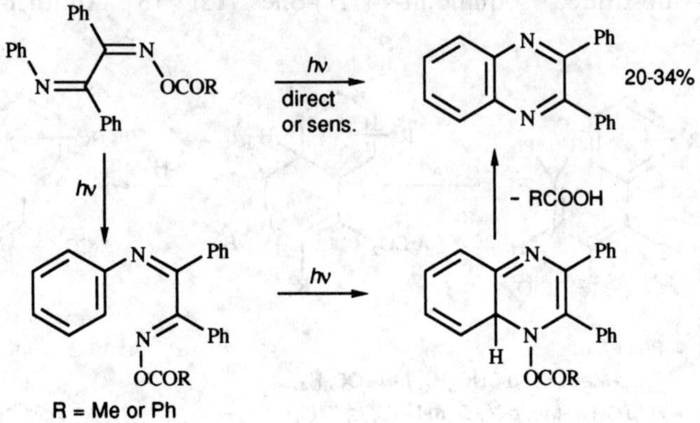

(15)

reaction path is followed when the irradiation of these diazadienes is carried out in the presence of $BF_3$-etherate. Under these conditions the only observed product is a phenanthridine derivative [20]. This change in reaction mode can be due to the co-ordination of the Lewis acid with the nitrogen lone pair that promotes the *cis*-stilbene type cyclization (**16**).

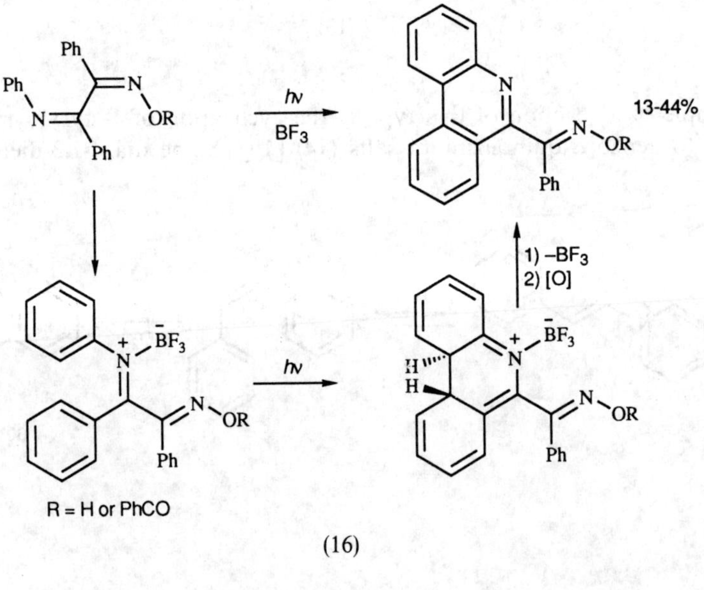

(16)

These 1,4-diaza-1,3-dienes undergo a completely different reaction when the substituent on the imine nitrogen has hydrogen atoms on the carbon attached directly to the nitrogen, as in the case shown in **17**. In this instance

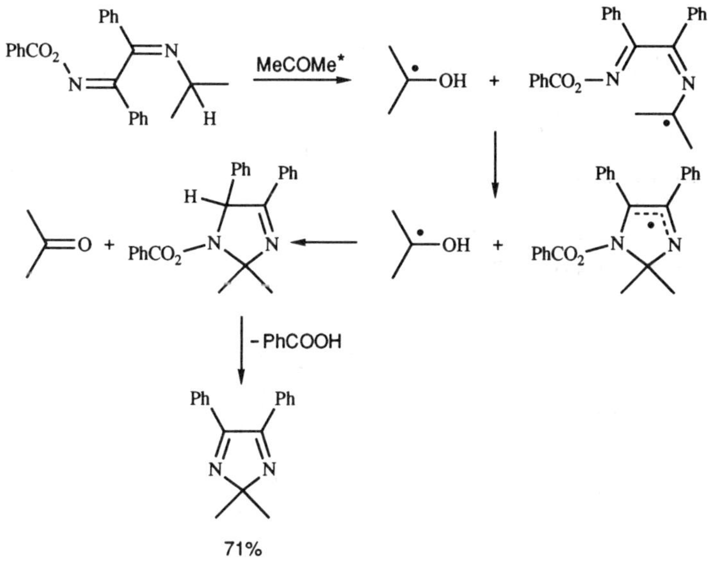

(17)

acetone-sensitized irradiation gives a high yield of 2,2-dimethyl-4,5-diphenyl-(2*H*)-imidazole. The formation of the imidazole can be interpreted in terms of hydrogen abstraction by excited-state acetone to yield a radical. Cyclization of this, followed by elimination of a benzoyloxy radical, affords the observed product [20].

The commonest ring closures in six- and seven-membered heterocycles containing C═N double bonds in conjugation with an alkene moiety arise by $4\pi$-electrocyclic pathways, although there are also examples of $6\pi$-ring closures. Some of these examples are shown in **18** [21–24].

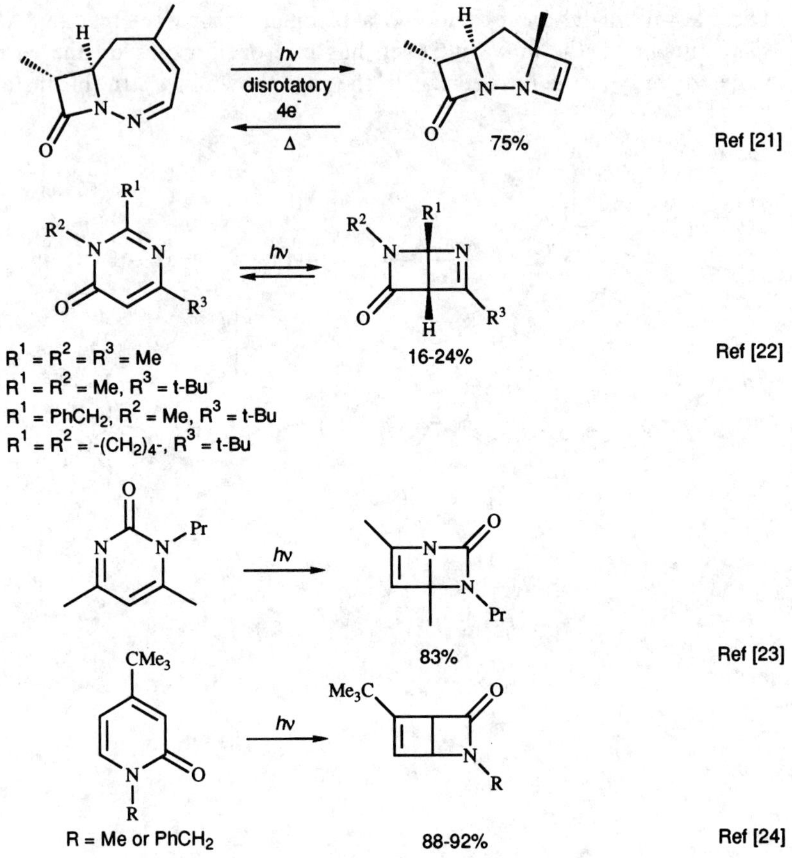

$h\nu$ disrotatory $4e^-$ / $\Delta$

75% Ref [21]

$h\nu$

$R^1 = R^2 = R^3 = Me$
$R^1 = R^2 = Me$, $R^3 = t\text{-}Bu$
$R^1 = PhCH_2$, $R^2 = Me$, $R^3 = t\text{-}Bu$
$R^1 = R^2 = -(CH_2)_4-$, $R^3 = t\text{-}Bu$

16-24% Ref [22]

$h\nu$

83% Ref [23]

$h\nu$

R = Me or $PhCH_2$

88-92% Ref [24]

(18)

Most of the photochemical electrocyclic reactions described in the literature are used as a synthetic route to cyclized products. This contrasts with the thermal counterpart, in which case the percentage of the open chain and the cyclic product, in the reaction mixture, is controlled by the relative stability of the two products. However, the composition of the photostationary mixture depends on the extinction coefficient of the two compounds at the wavelength used in the irradiation. The main component in the photomixture is always the one that absorbs less radiation at the exciting wavelength. Usually this is the cyclic compound, since the conjugation in the cyclic product is totally or partially lost. However, photochemical ring openings have also been described, particularly in cases in which the open-chain product is trapped, resulting in a shift of the equilibrium. Such is the case represented in **19** [25].



(19)

Alternatively the photoproduct reacts thermally to give a new compound that cannot be transformed photochemically into the starting material, as, for example, in the irradiation of the tricyclic compound that yields 1,4-oxazepines (**20**) [26].



(20)

Photochromism, shown by many systems, is due very often to electrocyclic processes such as those discussed above. A review of this topic has been published recently [27].

### *5.2.1.3 (2 + 2)-Cycloadditions*

There are very few examples of (2 + 2)-cycloadditions in the photochemistry of C═N compounds. This situation is in clear contrast with the photochemistry of alkenes and carbonyl compounds in which this reaction is very common, yielding cyclobutanes and oxetanes respectively (see sections 2.1.3.4, 2.1.4.5.1, 2.2.6 and 3.3.7). Attempts made to carry out a (2 + 2)-cycloaddition in simple imines such as the benzaldehyde *N*-cyclohexylimine were unsuccessful. The only observed product in the irradiation of this imine was the reduced dimer (**21**) [28]. However, the first example of a (2 + 2)-cycloaddition

(21)

of this type was described in the photoaddition of 2,5-diphenyloxadiazole to indene and furan (**22**) [29]. More examples have been described which usually

26% Ref [29b]

30% Ref [29a]

(22)

involve the cycloaddition of cyclic C=N double bonds to alkenes, as shown in **23** [30–34].

Some of the factors that control the (2 + 2)-cycloaddition to C=N bonds can be summarized as follows:

(i) In most cases the reaction takes place with cyclic imines in which the C=N group is conjugated with an electron-withdrawing group.
(ii) Normally electron-deficient alkenes do not undergo the cycloaddition.
(iii) The reaction proceeds with a high degree of regioselectivity. The regioselectivity observed is usually similar to that described for the cycloaddition of alkenes to ketones (see section 3.3.8.2).

hν

78% Ref [30]

hν EtOH

$R^1$ = Me or CN
$R^2$ = $p$-MeO$C_6H_4$
$R^3$ = Me, PhO, H

42-84% Ref [31]

hν

60% Ref [32]

hν

50% Ref [33]

hν

60-70% Ref [34]

(23)

Two different mechanisms have been formulated to account for the regioselectivity. Thus, the regiochemical outcome of these reactions could be controlled by the orientation of the two molecules in an exciplex between the alkene and the compound that contains the C=N functional group, both in the triplet [35] and singlet states [36]. Alternatively, as has been demonstrated in some cases, a stepwise mechanism via radical intermediates could be operative (**24**).

1) $h\nu$
2) ISC
3) $R_2C{=}CH_2$

(24)

Recent studies have shown that in some cases the (2 + 2)-cycloaddition could also take place using electron-deficient alkenes. Thus, pteridine-2,4,7-triones undergo regiospecific addition to electron-deficient and neutral alkenes to afford azetines (**25**) [37]. Further examples of this are shown below

$R^1$ = Me or Ph
$R^2$ = H, CN, Ph or $CO_2Me$
$R^3$ = H, Me or Ph
$R^4$ = CN, $CO_2Me$ or Ph

$h\nu$

30-95%

(25)

in the intramolecular cycloaddition of an oxime ether to yield an azapropellane (**26**) [38] and in the intermolecular reaction of 1,4-benzoxazin-2-ones

71%

(26)

with differently substituted alkenes (**27**) [39].

$R^1$ = Me, Bu or $CH_2CH_2Pr^i$
$R^2$ = Me, CN, H or $CO_2Me$
$R^3$ = Me, CN, H, $CO_2Me$ or $CO_2CH{=}CH_2$

11-75%

(27)

The cycloadditions between two C=N bonds are less frequently encountered although some examples of reactions of this type have been described, for instance in the dimerization of 2-(4-fluorophenyl)benzoxazole (**28**) [40],

Ar = $p$-$FC_6H_4$

41%

(28)

in the irradiation (in the solid state) of the cyclohexylidene (**29**) [41] and in

~100%

(29)

the acetone-sensitized irradiation of the imine (**30**) [42]. In all the cases

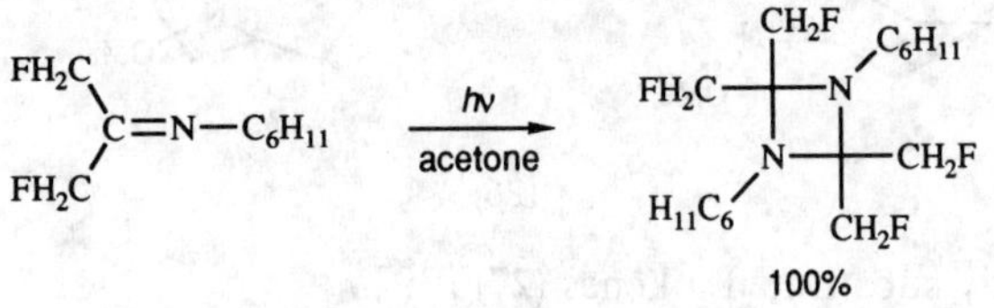

(30)

described the only regio-isomer formed is the corresponding 1,3-diazetine.

There are very few examples of (2 + 2)-cycloadditions of alkenes to iminium salts. These compounds usually react by more efficient alternative modes involving electron transfer (see section 5.2.3.3) [43].

To date, the (2 + 2)-cycloaddition of a C=N double bond to a C=O group has not been reported.

### *5.2.1.4 The aza-di-π-methane rearrangement*

The di-π-methane reaction of 1,4-alkenes (section 2.1.4.4.1) and β,γ-unsaturated ketones (section 3.3.5) has been known for many years [45]. However, the extension of the rearrangement to 1-aza-1,4-dienes is relatively recent. The first two examples of this reaction were described in the sensitized irradiation of a β,γ-unsaturated imine that yielded exclusively the corresponding cyclopropyl imine (**31**) [45], which hydrolysed to the corresponding

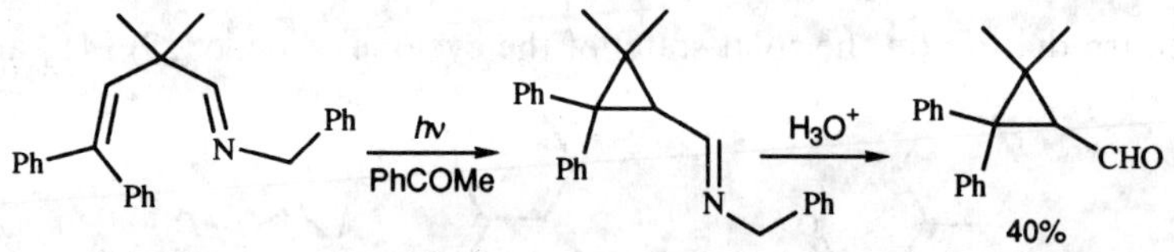

(31)

aldehyde on isolation, and in the cyclization of the cyclic oxime that gives the tricyclic derivative (**32**) [46]. While the aza-di-π-methane reaction of

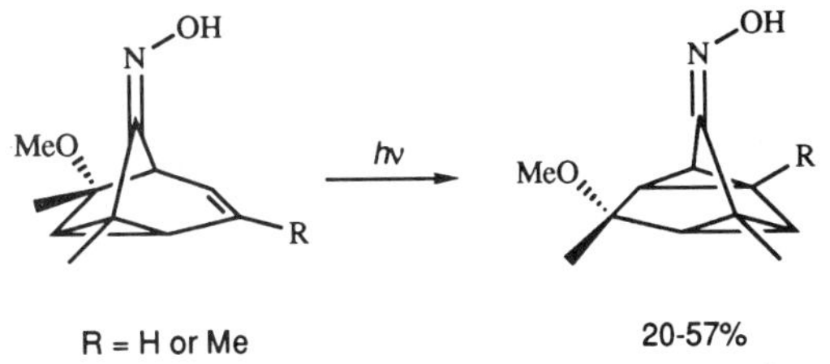

R = H or Me

20-57%

(32)

$\beta,\gamma$-unsaturated imines has proved to be very general, the corresponding rearrangement of the oximes has only been observed for the compounds shown in **32** and acyclic $\beta,\gamma$-unsaturated oximes do not undergo the cyclization.

The aza-di-$\pi$-methane rearrangement occurs with a variety of substituted $\beta,\gamma$-unsaturated imines (**33**) [47] and has been demonstrated to take place

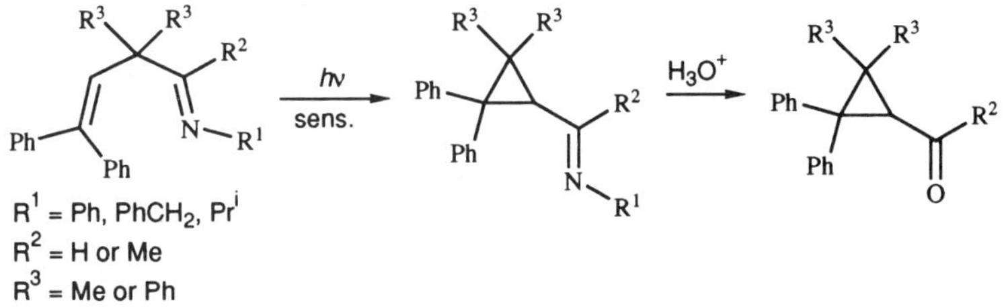

$R^1$ = Ph, $PhCH_2$, $Pr^i$
$R^2$ = H or Me
$R^3$ = Me or Ph

(33)

via the triplet excited state [48]. The regiospecificity observed in the reaction can be explained by means of a classical biradical mechanism in which the 1,3-cyclopropyl biradical, formed by bridging, ring-opens to the more stable biradical (bond b rupture) instead of giving the alternative intermediate, which is considerably less stable (**34**). However, studies on the influence of substitution on the efficiency of the reaction have proved that some of the molecules in the triplet excited state are deactivated by single-electron transfer from the nitrogen

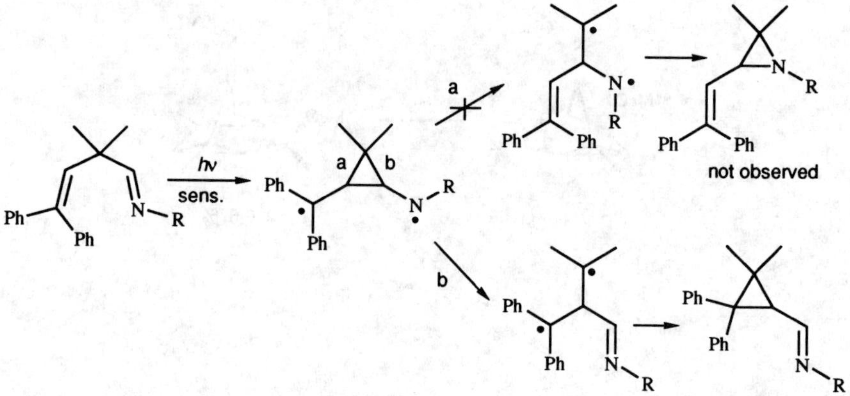

(34)

lone pair to the alkene moiety (**35**) [49].

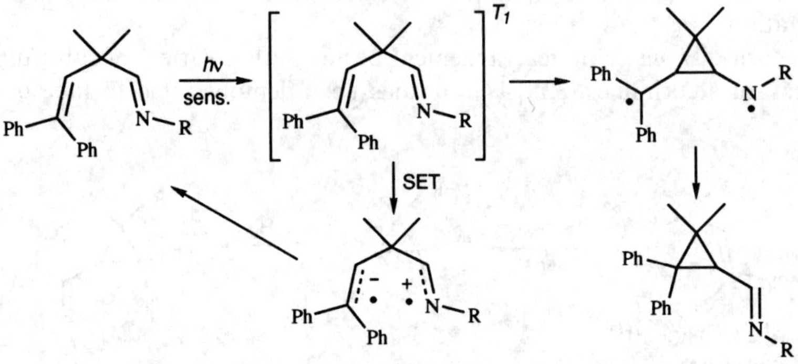

(35)

The aza-di-$\pi$-methane rearrangement is more general than the oxa-di-$\pi$-methane reaction (section 3.3.5). Thus, the normal photochemical reactivity of $\beta,\gamma$-unsaturated aldehydes is decarbonylation, and many $\beta,\gamma$-unsaturated ketones do not undergo the oxa-di-$\pi$-methane rearrangement [50]. The generality of the aza-di-$\pi$-methane reaction provides an indirect route by which aldehydes and ketones that do not undergo the oxa-di-$\pi$-methane rearrangement can be transformed into the corresponding cyclopropyl carbonyl compounds by a simple conversion into the corresponding imine, followed by sensitized irradiation and quantitative hydrolysis of the cyclopropyl imine (**36**).

The only inconvenience of this reaction is the instability of the imines that hydrolyse very readily in the presence of traces of moisture. Attempts made to extend the reaction to stable derivatives of aldehydes and ketones such

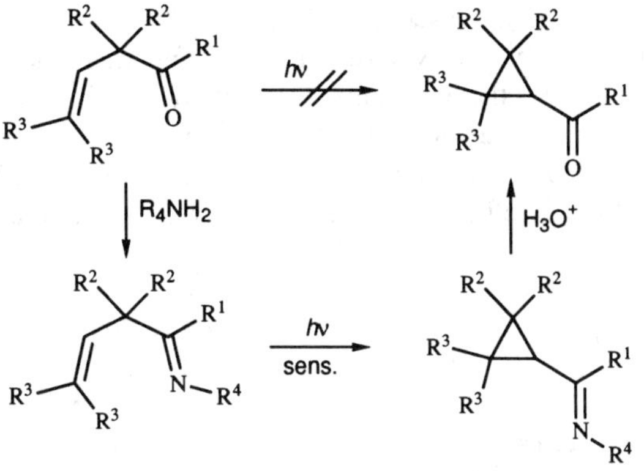

(36)

as the oxime [47] and oxime ethers [51] were unsuccessful (**37**). This problem

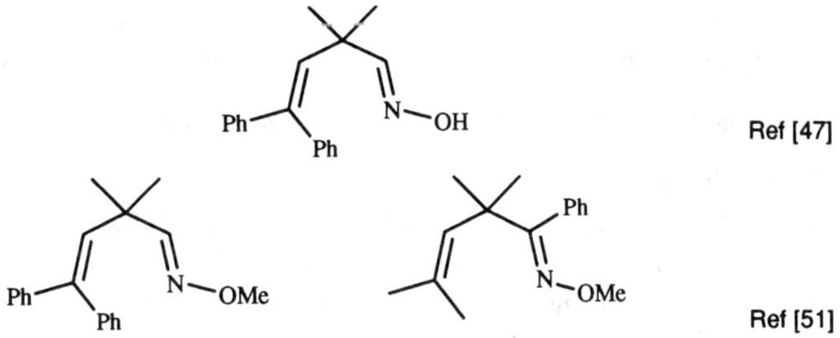

Ref [47]

Ref [51]

(37)

was overcome by using a derivative in which the oxygen atom of the oxime is attached to an electron-withdrawing group such as an acetate [52]. This suppresses the adverse intramolecular single-electron transfer from the nitrogen lone pair of electrons to the alkene moiety, which has been postulated as responsible for the lack of reactivity of the oximes and oxime ethers [52]. The aza-di-$\pi$ methane rearrangement of the oxime acetates (**38**) has proved

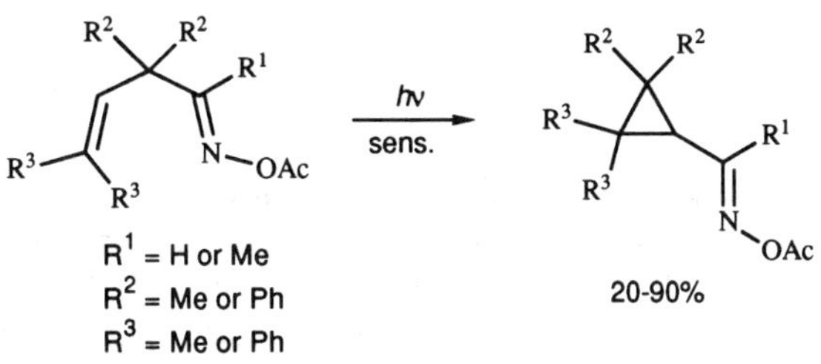

$R^1$ = H or Me
$R^2$ = Me or Ph
$R^3$ = Me or Ph

20-90%

(38)

to be as general as in the case of the imines (**33**) [53–55]. Further to the advantage of using stable derivatives, the rearrangement of oxime acetates is considerably more efficient than that of the imines (**33**) [53]. The aza-di-$\pi$-methane rearrangement has been extended to other C=N stable derivatives such as other oxime esters, semicarbazones, and benzoyl hydrazones (**39**) [56]. The efficiency of the aza-di-$\pi$-methane rearrangement of

hν, sens. → 20-90%

X = OCOPh
X = NHCOPh
X = $NHCONH_2$

(39)

some of these stable derivatives of $\beta,\gamma$-unsaturated carbonyl compounds is even higher than that observed for comparable 1,4-dienes [53].

## 5.2.2 Reactions analogous to carbonyl compounds

### *5.2.2.1 Cleavage*

The C=N chromophore can undergo fragmentation similar to the Norrish Type I reaction of carbonyl compounds (section 3.3.2). However, this reactivity is uncommon and has been observed only in cases in which at least one of the radicals formed is highly stabilized. For instance, irradiation of *N*-benzyl-1-methyl-2-phenylethylimine brings about the formation of acetonitrile, toluene, and bibenzyl. The formation of these compounds can be interpreted by $\alpha$-cleavage followed by either hydrogen abstraction or recombination (**40**) [57]. Some aldimines undergo a different type of

$$PhCH_2(CH_3)C{=}N\text{–}CH_2Ph \xrightarrow{h\nu} CH_3\text{–}\dot{C}{=}N\text{–}CH_2Ph + Ph\dot{C}H_2 \xrightarrow{Ph\dot{C}H_2} Ph\text{–}CH_2CH_2\text{–}Ph$$

$$CH_3\text{–}C{\equiv}N + Ph\dot{C}H_2 \xrightarrow{\text{H donor}} Ph\text{–}Me$$

(40)

fragmentation promoted by hydrogen abstraction by excited ketones as, for example, in the benzophenone-sensitized irradiation of *N*-*t*-butylbenzylimine, using benzene as solvent. In this case the reaction affords benzonitrile and *t*-butylbenzene. The formation of these two products is interpreted by the mechanism shown in **41** [58].

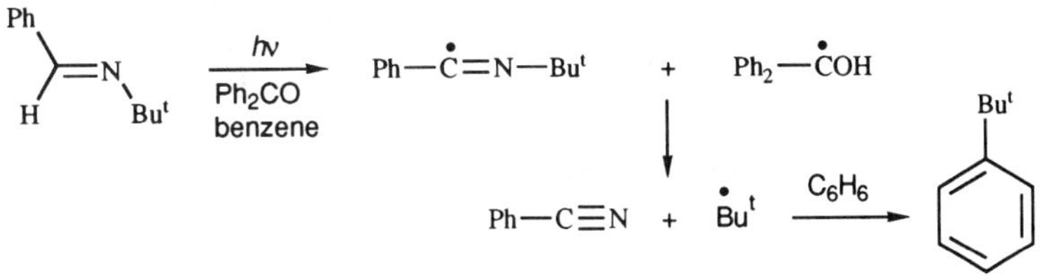

(41)

Fragmentations similar to the Norrish Type II reaction of carbonyl compounds (section 3.3.7.3) have also been observed in the irradiation of cyclic imines in which deactivation of the excited state by *syn–anti*-isomerization is impossible. Thus, for example, acetone-sensitized irradiation of 4-alkyl-substituted pyrimidines brings about the formation of 4-methylpyrimidine (**42**) by a reaction path involving $\gamma$-hydrogen abstraction by the

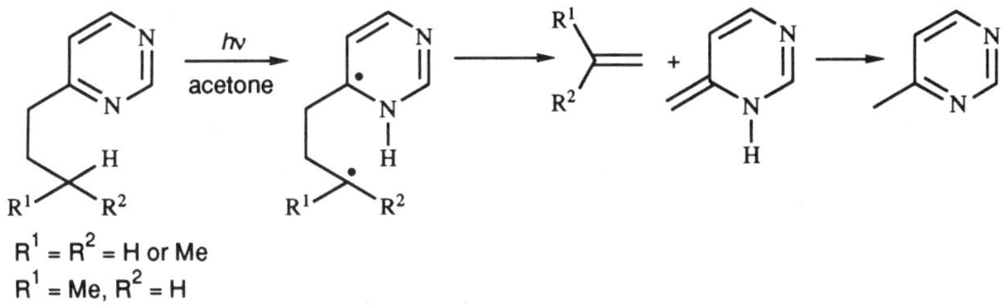

$R^1 = R^2 = H$ or Me
$R^1 = Me, R^2 = H$

(42)

C=N group and fragmentation [59]. Another example of this type of reactivity in imines is the photocyclization of the substituted dihydroisoquinoline to yield a spirobenzylisoquinoline (**43**) [60]. In this example the biradical

6%

(43)

formed cannot fragment and gives the cyclic product. It should be pointed out that the number of photofragmentations of imines and related compounds described in the literature is small by comparison with the large number of reactions of this type observed in carbonyl compounds. This is probably due to the fact that in most of the acyclic imines the excited state is deactivated by *syn–anti*-isomerization.

### 5.2.2.2 *Photoreductions*

In the presence of good hydrogen-donating solvents, many *N*-alkyl-arylimines undergo photoreduction. In general terms ketimines are reduced to the corresponding amines, whereas aldimines yield dihydro photodimers (**44**).

$$Ph{-}CH(R^1){-}NH{-}R^2 \xleftarrow[\text{H donor}]{h\nu} Ph{-}C(R^1){=}N{-}R^2 \xrightarrow[\text{H donor},\ R^1 = H]{h\nu} \begin{array}{l} Ph{-}CH{-}NH{-}R^2 \\ \quad | \\ Ph{-}CH{-}NH{-}R^2 \end{array}$$

(44)

However, Padwa [61] and Fischer [62] have demonstrated that the reaction does not take place by a mechanism similar to the one observed for the photoreduction of ketones. Based on some of the features shown by the reaction these authors have proposed that the reduction arises by a 'chemical sensitization' mechanism instead of a mechanism analogous to the photoreduction of ketones (**45**). According to this mechanism the photoreduction

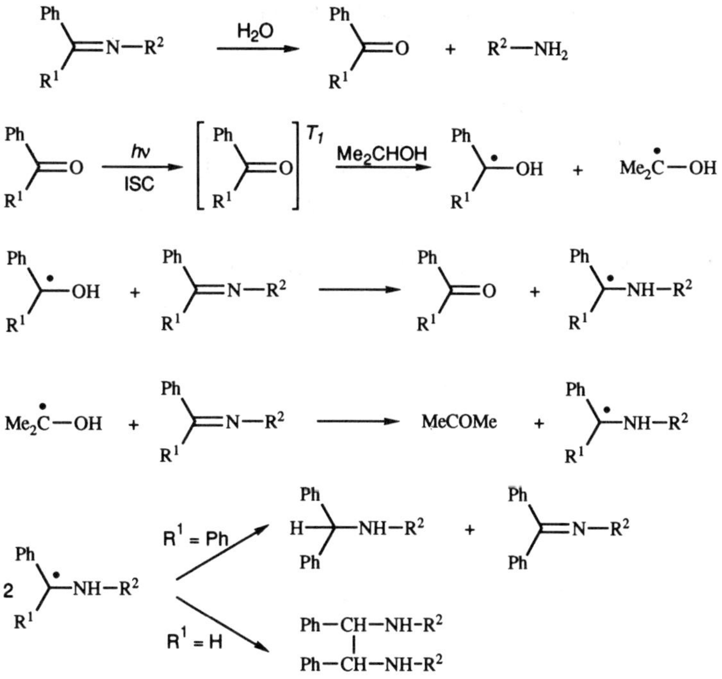

(45)

of imines does not involve the excited state of the imine but is brought about by the excited carbonyl group in the ketone-sensitized irradiations or generated by hydrolysis of the imine in the direct irradiations. By this mechanism it is possible to explain all the experimental observations, such as, for example, the fact that *N*-alkylbenzylimines are not photoreduced using 2-propanol as solvent, conditions which are very efficient in photoreducing benzophenone. However, the process takes place very readily when the reaction is carried out in the presence of water, which promotes the hydrolysis of the imine. There are also some examples of photoreductions of heteroaromatic iminium salts, although this class of compound tends to give alkylated products when irradiated in the presence of a good hydrogen donor, as will be discussed in the next section.

### 5.2.3 Miscellaneous reactions of the C=N double bond

#### *5.2.3.1 Photoalkylations*

There are many examples of reactions in which a C=N undergoes C-alkylation by irradiation in the presence of good hydrogen donors such as a primary alcohol or diethyl ether. In many cases the reaction takes place by hydrogen abstraction from the solvent by the excited C=N. This brings about the formation of a radical pair, and coupling yields a $\beta$-amino-alcohol (**46**). Many heterocyclic compounds undergo this reaction, and some of the

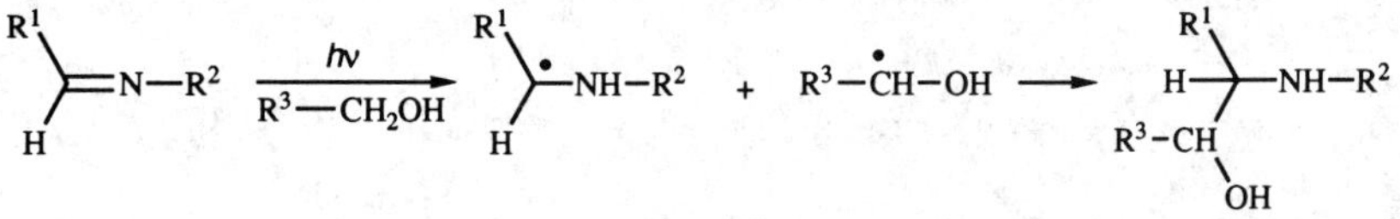

(46)

cases reported in the literature are shown in **47** [63–66].

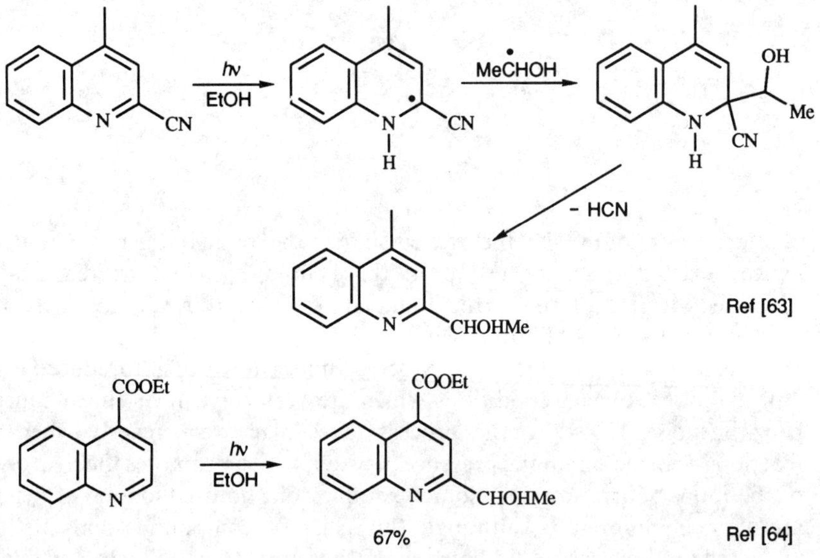

Ref [63]

67%

Ref [64]

(47)

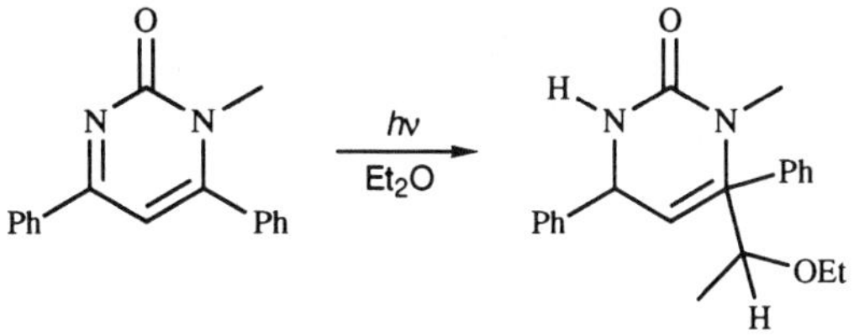

hν Et$_2$O

56%

Ref [65]

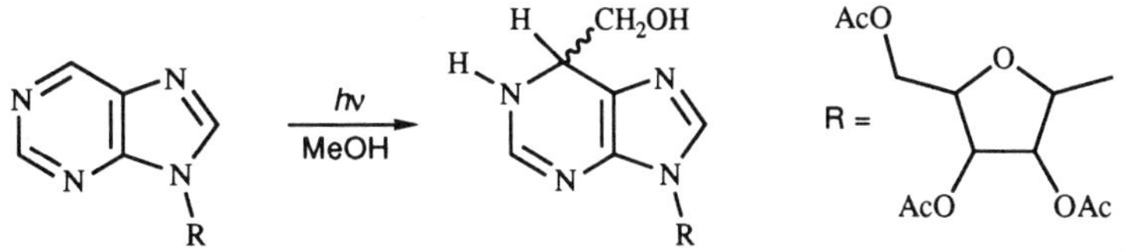

hν MeOH

R =

75%

Ref [66]

(47 continued)

In triplet-sensitized photoalkylations a 'chemical sensitization' mechanism involving the initial abstraction of a hydrogen from the alcohol by the sensitizer has been postulated in some cases as outlined in **48**. An example

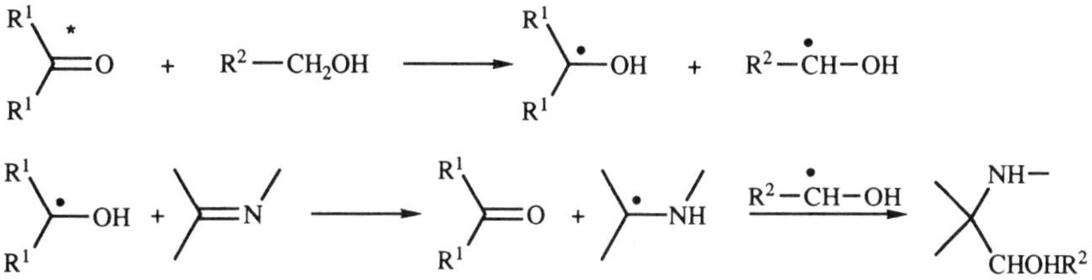

(48)

of a photoalkylation that takes place by this mechanism has been observed in the irradiation of *theophylline* in diethyl ether/acetone. This yields two alkylated products that re-aromatize on isolation to the observed products (**49**) [67].

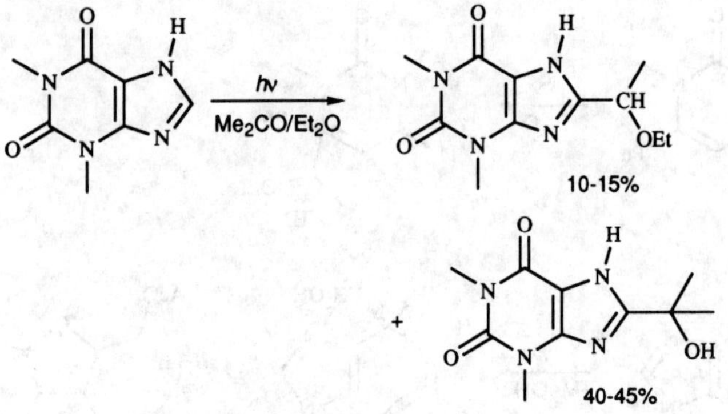

hv
$Me_2CO/Et_2O$
10-15%
+
40-45%

(49)

Many heteroaromatic compounds undergo alkylation in acidic medium. In these cases alkylation takes place via the corresponding iminium salt and involves a single electron transfer mechanism as outlined in **50**. A typical

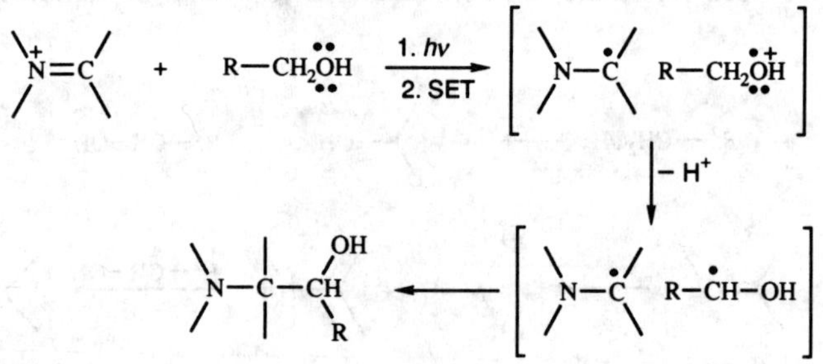

1. hv
2. SET
– $H^+$

(50)

example of this is the irradiation of an ethanol solution of phenanthridine in acidic medium. This treatment yields 10-ethylphenanthridine by a route involving the formation of the alkylated product followed by dehydration and isomerization (**51**) [68]. Other reasonably efficient examples of alkylations

(51)

of iminium salts are shown in **52** [69–71].



(52)

Another route the alkylation of iminium salts is by the use of carboxylic acids or carboxylate anions as alkylating agents. These reactions can be rationalized again by a mechanism involving a single-electron transfer from the carboxylate anion to the iminium salt. Decarboxylation within the resultant radical pair followed by radical combination affords the observed products (**53**). Additional examples of this reaction are collected in **54** [77–75].

$$\ce{>N^+=C< + R-COO^- ->[h\nu][SET] [>N-\dot{C}< \quad R-\dot{C}OO] ->[-CO_2] [>N-\dot{C}< \quad R\cdot] -> >N-C(R)<}$$

(53)

R = $CH_2OH$, CHMeOH, $CMe_2OH$ — Ref [72]

Quinoline $\xrightarrow[Me_2CHCH_2CO_2H]{h\nu}$ 2-($CH_2CHMe_2$)quinoline, 37% — Ref [73]

(54)

$CH_2(CH_2)_nCH_2COOH$ — hν, pyridine → $(CH_2)_n$ ... N—H

n = 4, 10%
n = 5, 7%

Ref [74]

Ph, $ClO_4^-$ + $CO_2H$ — hν, $H_2O$, pH 7 → Ph

68%

Ref [75]

(54 continued)

#### *5.2.3.2 Photochemistry of azirines*

The photochemistry of azirines has been studied for many years. It has been shown that the irradiation of 2*H*-aryl azirines yields nitrile ylides as reactive intermediates that can be trapped by a large variety of dipolarophiles to form five-membered ring heterocycles as shown in **55** [76]. Padwa and his

Ar, N, $R^1$, $R^2$ — hν → [Ar—C≡N⁺—C⁻($R^1$)($R^2$) ⟷ Ar—C⁻=N⁺=C($R^1$)($R^2$)]

X=Y

Ar, N, $R^1$, $R^2$, X—Y

(55)

group [76] have demonstrated that the reaction takes place via the $n\pi^*$ excited state. The reaction has found application in the synthesis of a large variety of heterocyclic systems; some of these are shown in **56** [77–80]. The nitrile

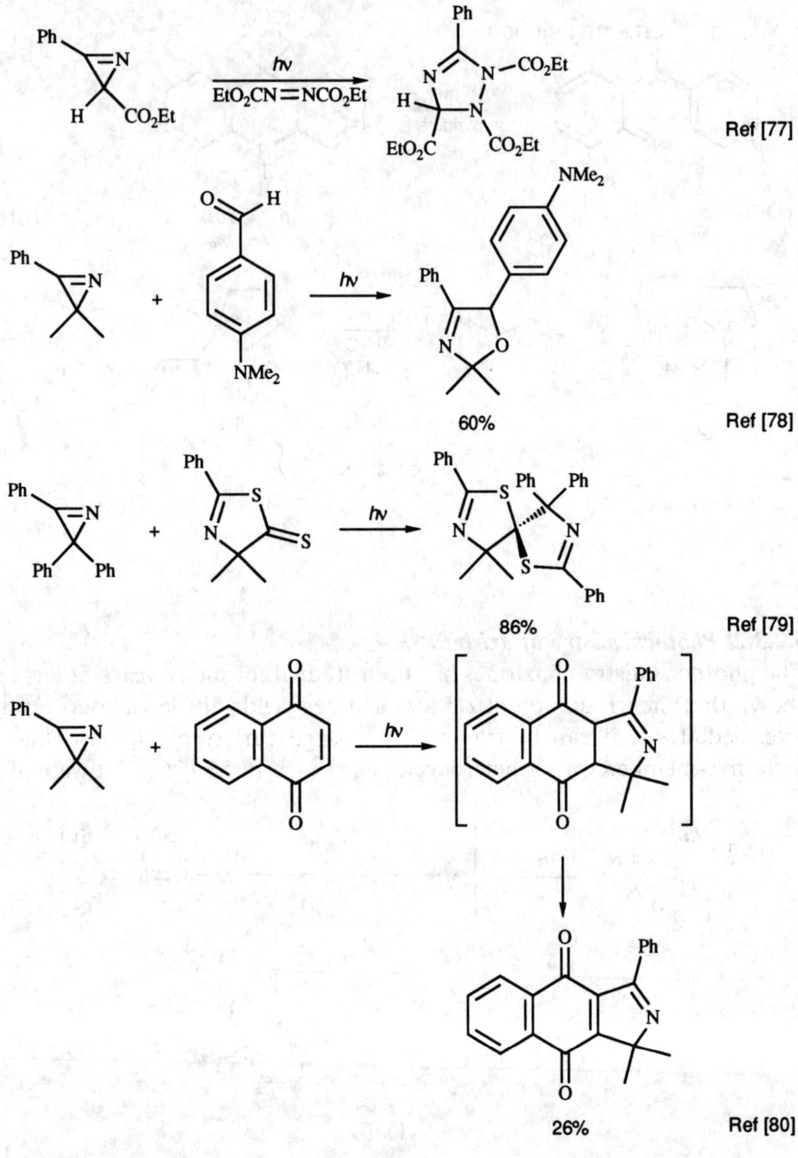

Ref [77]

60% Ref [78]

86% Ref [79]

26% Ref [80]

(56)

ylide generated by irradiation of azirines can also be trapped intramolecularly, as seen in **57** [81, 82].

hν 80% Ref [81]

hν 100% Ref [82]

(57)

#### *5.2.3.3 Electron-transfer photochemistry of non-aromatic iminium salts*

The photochemistry of iminium salts has received considerable attention in recent years, partially owing to current interest in single-electron transfer processes and also because of potential applications in organic synthesis. The iminium chromophore can be considered to be very similar to an alkene (section 5.1) since the coordination of the nitrogen lone pair suppresses the $n\pi^*$ transition that occurs in imines and other C=N derivatives. Therefore, the most likely excited state available for population is $\pi\pi^*$, as is the case in the alkenes. On the other hand, iminium salts are ideal substrates for electron-transfer processes in the excited state. Both the HOMO and the LUMO orbitals of iminium salts are lower in energy than the corresponding alkenes. It is this low-lying LUMO that makes the iminium salts perfect acceptors in single-electron transfer processes from a great variety of electron donors such as electron-rich alkenes, arenes, alcohols, and ethers [3]. This single-electron transfer process can take place either from the interaction between the excited iminium salt and the ground state donor, or by interaction between the excited electron-rich donor and the iminium salt ground state. The outcome of the reaction promoted by single-electron transfers depends on the chemical behaviour of the radical cations derived from the electron donor.

Four main reaction paths have been observed in the electron-transfer photochemistry of iminium salts yielding products arising from addition and cyclization, dependent on the nature of the electron donor, as shown schematically in **58**. In recent years Mariano and coworkers [3, 83] have

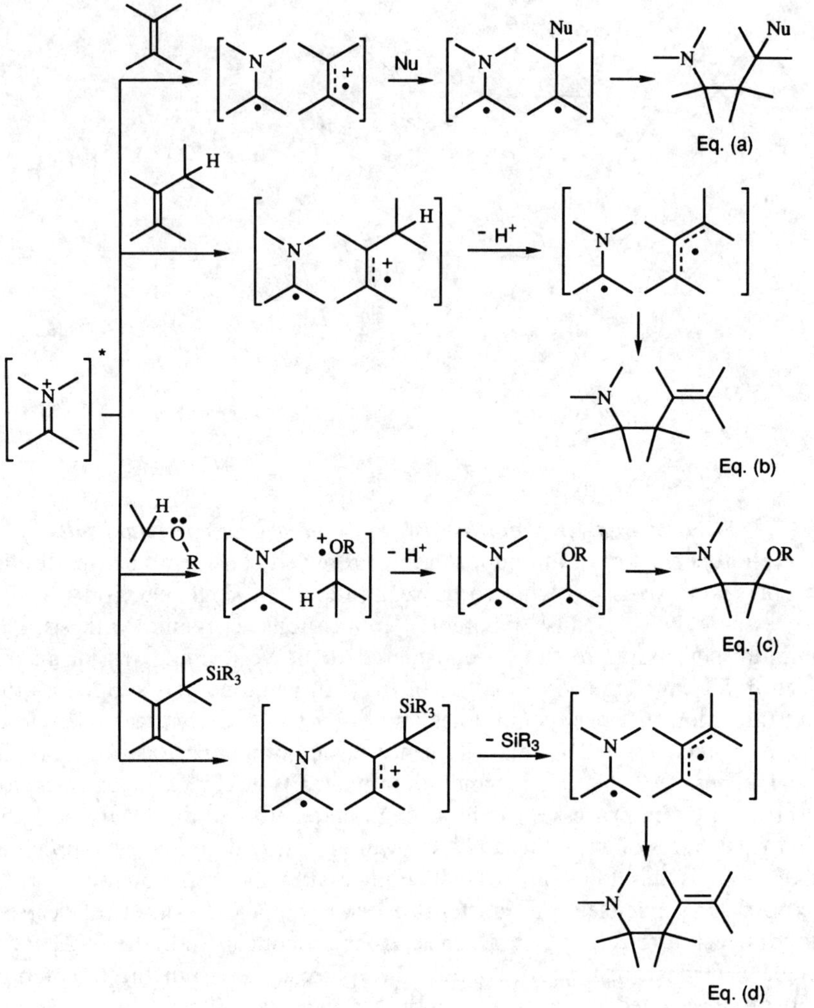

N; Nu; N; Nu; N; Nu

Eq. (a)

H; N; H; $-H^+$; N; N

Eq. (b)

H; Ö; R; N; OR; H; $-H^+$; N; OR; N; OR

Eq. (c)

$SiR_3$; N; $SiR_3$; $-SiR_3$; N; N

Eq. (d)

(58)

described a large number of these reactions and have demonstrated their application in organic synthesis.

*5.2.3.3.1 Photoaddition reactions*

A typical example of the reaction described by equation **58a** is the addition of methylpropene to 2-phenyl-1-pyrrolinium perchlorate (**59**) [84]. The

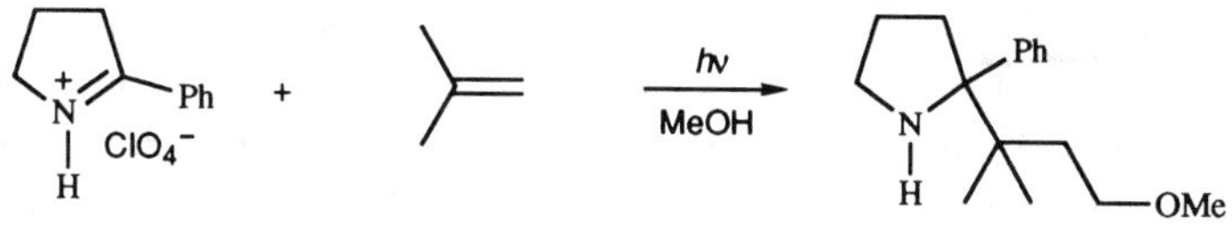

hν MeOH

(59)

synthetic utility of this reaction involving both intra- and inter-molecular processes is illustrated in the examples shown in **60** [85, 86].

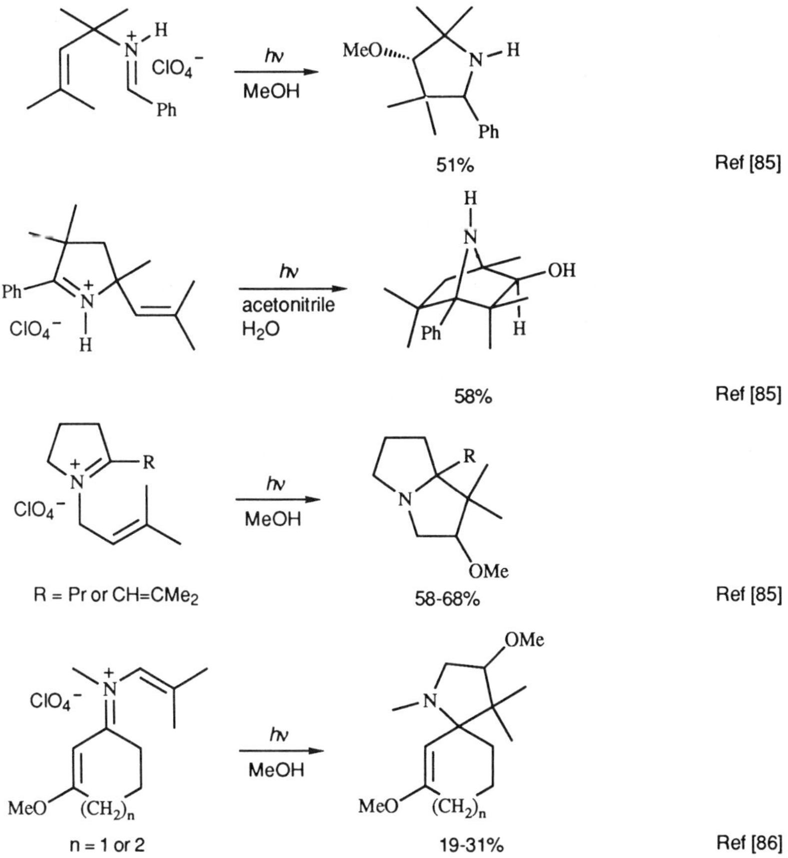

hν MeOH 51% Ref [85]

hν acetonitrile $H_2O$ 58% Ref [85]

R = Pr or CH=CMe$_2$ hν MeOH 58-68% Ref [85]

n = 1 or 2 hν MeOH 19-31% Ref [86]

(60)

*5.2.3.3.2 Sequential electron–proton transfer reactions*

Reactions of the type illustrated in **58b** and **58c** have been observed in alkenes with hydrogens in the allylic or benzylic position, and also with alcohols and ethers in which the radical cation formed by single-electron transfer loses a proton to give a stabilized radical. This latter type of reaction has already been discussed as part of the alkylation reactions (see section 5.2.3.1). An example of a reaction of the type represented in equation **58b** is shown in **61** [87].

$ClO_4^-$ hν PhCH$_2$CH$_2$Ph

28% 15% 4%

(61)

*5.2.3.3.3 Allylsilane and stannane photoadditions to iminium salts*

As represented in equation **58d**, allylsilane and stannane cation radicals can also undergo facile elimination of the silyl or stannyl group in the presence of weak nucleophiles to yield stabilized radicals. This reaction provides a useful path for the formation of new C—C bonds and has been used successfully in the high-yield synthesis of the protoberberine alkaloid *xylopinine* (**62**) [88]. Other examples of this type of C—C bond formation,

(62)

represented schematically by equation **58d**, are shown in **63** [89–91].

Ref [89]

Ref [90]

Ref [91]

(63)

## 5.3 PHOTOCHEMISTRY OF ENAMIDES

Enamides are also photochemically reactive, and two main reaction paths have been observed on irradiation. Simple enamides undergo a very efficient 1,3-acyl migration to yield enamines in a reaction reminiscent of the photo-Fries rearrangement (Chapters 2, 3, 4). Thus, irradiation of enamides affords the isomeric products in almost quantitative yield (**64**) [92]. In most

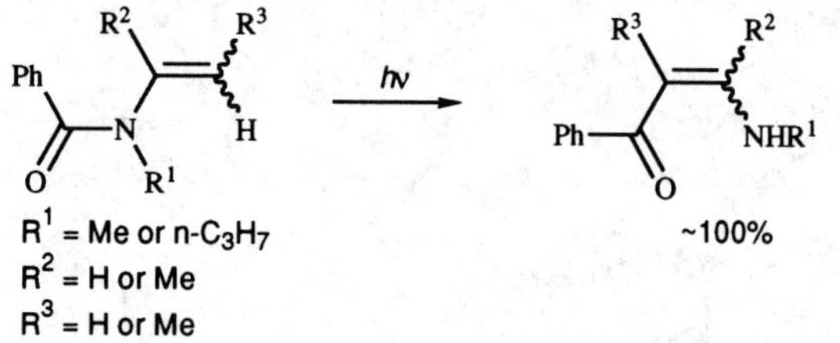

$R^1$ = Me or n-$C_3H_7$
$R^2$ = H or Me
$R^3$ = H or Me

~100%

(64)

of the cases of 1,3-acyl migration in enamides the nitrogen atom is tertiary. Usually the irradiation of secondary enamides yields unchanged starting material or products arising from degradation or polymerization. However, there are some exceptions as, for example, in the irradiation of the cyclic dienamide that yields the corresponding 1,3-migrated product (**65**) [93].

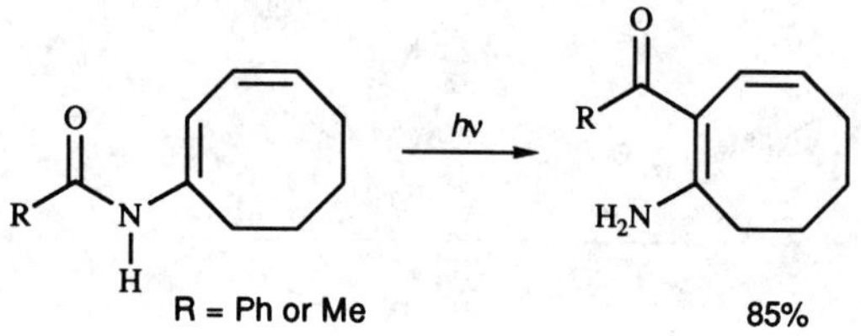

R = Ph or Me

85%

(65)

An alternative reaction mode has been observed in the photochemical reactivity of some dienamides and involves addition of the carbonyl oxygen atom to the distant double bond of the conjugated diene system. This affords a biradical which on intramolecular hydrogen abstraction affords a spiro product. Several examples of such reactivity have been reported (**66**) [93, 94].

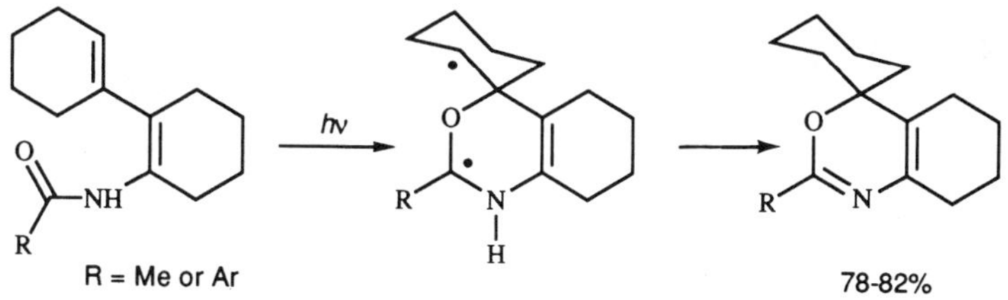

R = Me or Ar

78-82%

(66)

From a synthetic point of view, the most important photochemical reaction of enamides is cyclization of compounds of the type shown in **67**. These

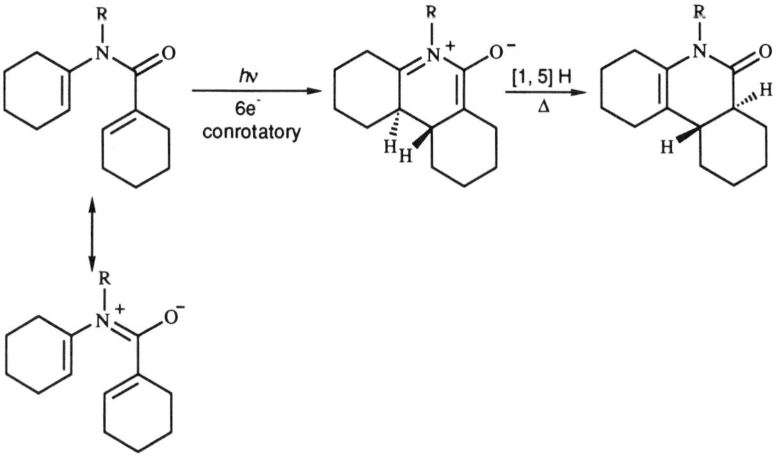

(67)

enamides undergo a conrotatory six-electron electrocyclic cyclization to generate a dihydro intermediate which yields the *trans*-fused cyclic product by a [1,5]-suprafacial hydrogen migration. This cyclization has been used in photochemical syntheses of some alkaloids, particularly of the benzylisoquinoline and indole type. This photochemical path to natural products competes favourably with thermal alternatives. Some examples of the synthetic utility are illustrated in **68** [95–97].

$R^1$ = Ar
$R^2$ = H or Cl

hv MeOH    $I_2$

81-89%    Ref [95]

hv

27%    Ref [96]

hv - MeOH

58%    Ref [97]

(68)

When the dihydro intermediate is fairly long-lived it is possible to carry out the reductive photocyclization of enamides using sodium borohydride as the reducing agent. This alternative reaction path is of particular value in the synthesis of some alkaloids. Thus, the cyclization of an enamide to a furanoquinolizine has been employed in the synthesis of ipecac and hetero-yohimbine alkaloids (**69**) [98].

## 5.4 NITRILES

The electronically excited C≡N triple bond does not participate in many photochemical reactions of interest to the organic photochemist because the nitrile chromophore absorbs at short wavelength (*ca.* 160 nm) [99]. On the other hand, there are not many cases in which nitriles are attacked by other species in the excited state. This lack of reactivity makes nitriles highly suitable polar solvents for photochemical reactions. Nevertheless, some photochemi-

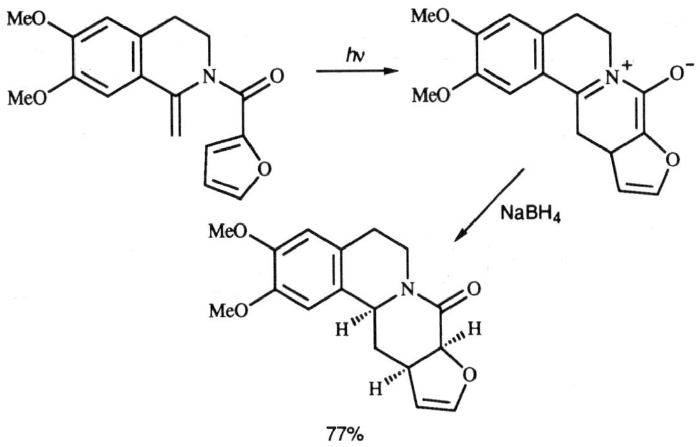

(69)

cal reactions of nitriles have been described. For instance, nitriles undergo (2 + 2)-cycloadditions to alkenes, although this reaction is even less frequently encountered than in the case of the C=N double bond. The cycloaddition of acetonitrile to tetracyanoethylene takes place via an electron-transfer mechanism. This yields an azetine which ring opens in the reaction medium to give a 2-azadiene that is hydrolysed to an enamine on isolation (**70**) [100].

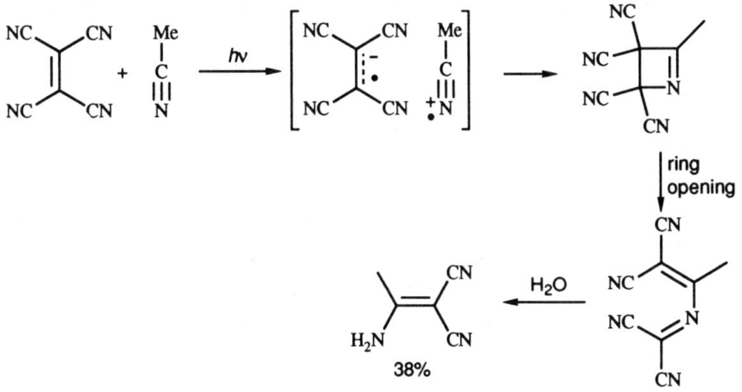

(70)

Irradiation of benzonitrile in the presence of electron-rich alkenes also gives cycloaddition products that are thermally unstable and are transformed into the final products as shown in **71** [101]. In some cases azetines can be isolated

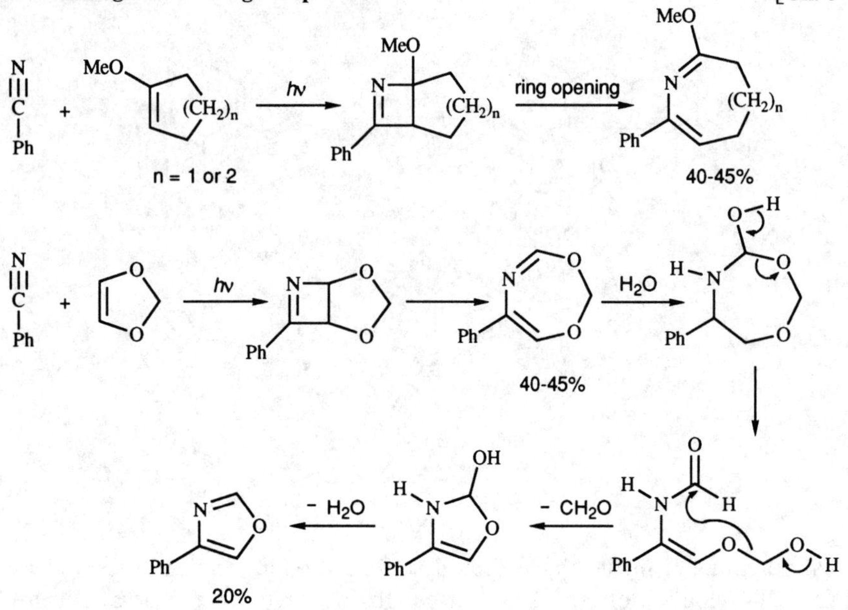

(71)

from the (2 + 2)-cycloaddition of nitriles to simple alkenes, as for example in the reaction of benzonitrile with tetramethylethene and 1,2-dimethylcyclohexene (**72**) [102, 103]. Some arenes also undergo (2 + 2)-cycloaddition to

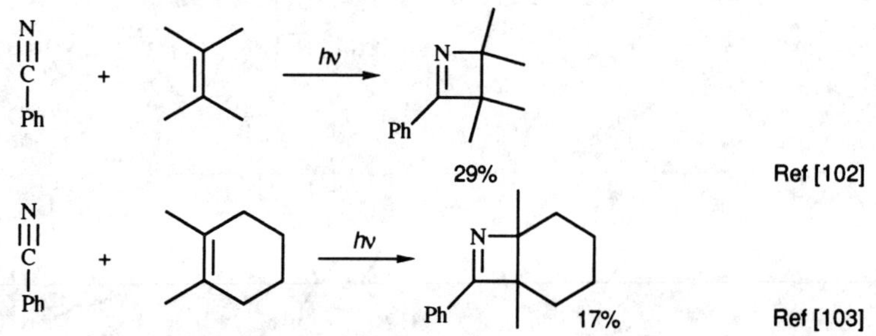

(72)

nitriles. Thus, benzonitrile adds photochemically to phenols affording azocin-2(1*H*)-ones (**73**) [104].

OH
$h\nu$
Ph—C≡N
R
OH
N
Ph
R
OH
N
Ph
R
O
N
H
R
Ph
R = H, Me or MeO
15-20%

(73)

Cyano groups can also be eliminated on irradiation of heterocyclic systems in the presence of secondary or tertiary amines (**74**) [105]. An electron-transfer

MeO
MeO
N
N
CN
CN
$h\nu$
$Et_3N$
MeO
MeO
N
N
CN
80%
$h\nu$
$Et_3N$
MeO
MeO
N
N
46%

(74)

mechanism has been postulated to take account of these results (**75**).

(75)

## 5.5 PHOTOCHEMISTRY OF THE N=N SYSTEM AND RELATED COMPOUNDS

### 5.5.1 Absorption spectra of azo compounds, diazo compounds, diazonium salts, and azides

The absorption spectra of aliphatic azo compounds show a maximum around 350–380 nm, assigned as an $n\pi^*$ transition, with an extinction coefficient of approximately 150 $dm^3\ mol^{-1}\ cm^{-1}$ for the *E*-isomer. The less stable isomer usually absorbs at longer wavelengths, but the position and intensity of the absorption band depends on the degree of steric hindrance. Aromatic azo compounds absorb at longer wavelengths as a result of conjugation of the N=N double bond with the aromatic rings. Usually two bands are observed, a moderately intense absorption in the visible region of the spectrum and a strong one in the near ultraviolet region. For instance, the spectrum of *E*-azobenzene shows a moderately intense $n\pi^*$ transition at 443 nm ($\varepsilon = 500\ dm^3\ mol^{-1}\ cm^{-1}$) and a $\pi\pi^*$ transition at 320 nm ($\varepsilon = 20\,000\ dm^3\ mol^{-1}\ cm^{-1}$). *Z*-Azobenzene also shows two absorptions at 433 nm ($\varepsilon = 1500\ dm^3\ mol^{-1}\ cm^{-1}$) and at 281 nm ($\varepsilon = 5200\ dm^3\ mol^{-1}\ cm^{-1}$) [106].

### 5.5.2 Photochemistry of azo compounds

The excited state of most acyclic azo compounds is deactivated by *syn–anti-*

isomerization in a similar way to the imines and other C=N derivatives (**76**)

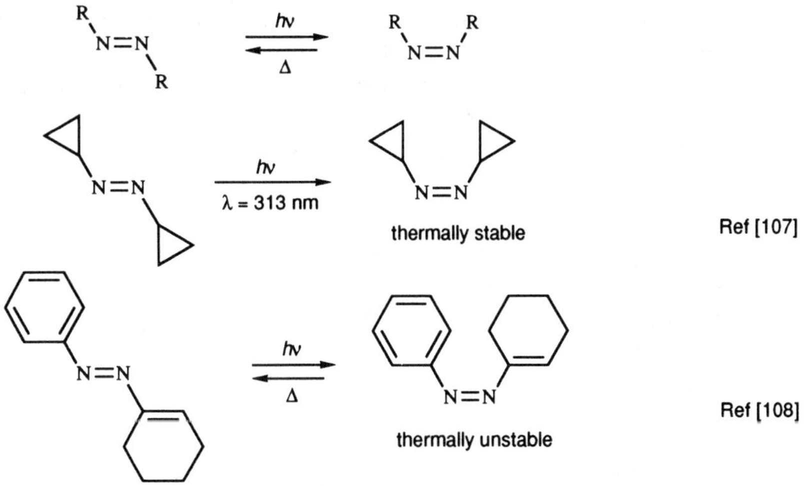

Ref [107]

thermally stable

thermally unstable

Ref [108]

(76)

[107, 108]. Recent studies have demonstrated that the *E*,*Z*-isomerization of azo compounds takes place by a nitrogen inversion mechanism in the lower excited states and via rotation around the N=N double bond in the higher excited states [109]. This reaction has found applications in the synthesis of photoresponsive crown ethers containing an intra-annular azobenzene substituent (**77**). Only the *Z*-isomer of the compound is able to interact with the

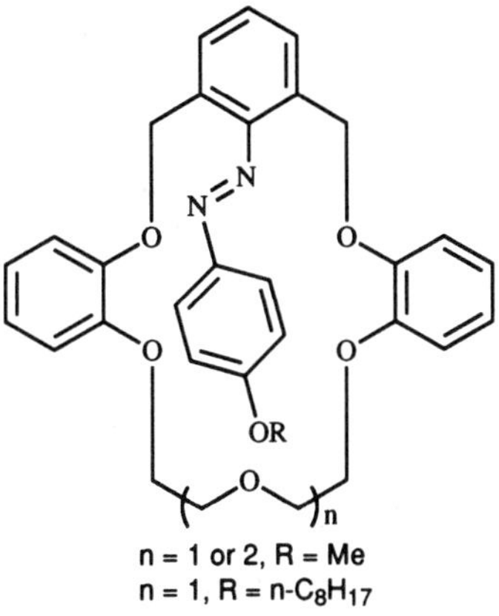

n = 1 or 2, R = Me
n = 1, R = n-$C_8H_{17}$

(77)

metal. This system has been used to transport sodium ions across a liquid membrane, a process that only occurs on irradiation [110].

The most important photochemical reaction of azo compounds is the elimination of nitrogen to yield radicals. This reaction is particularly important in cyclic azo derivatives that cannot dissipate the excitation energy by *E*,*Z*-isomerization. Many strained ring systems have been synthesized by this route, as illustrated in the examples collected in **78** [111–115].

R = H or Me — $h\nu$ → 50-66% Ref [111]

$h\nu$ → 100% Ref [112]

$h\nu$, low temp. → 40-90% Ref [113]

$R^1$ = Me or Ph
$R^2$ = H or Me
$R^3$ = Me
$R^4$ = Ph, $PhCH_2$ or Me

$h\nu$ → 100% Ref [114]

$h\nu$, PhCOPh → 96% Ref [115]

(78)

Diazirines are a special type of diazo compound which provide a simple and efficient way of generating carbenes by the elimination of nitrogen. This route has become increasingly popular and has advantages over other routes in that diazirines are quite stable thermally and the carbenes are generated under mild conditions. Some of the synthetic applications of carbenes generated by this route are the synthesis of substituted 2,4-didehydronoradamantane (**79**) [116] and the trapping of the carbene by pyridine to

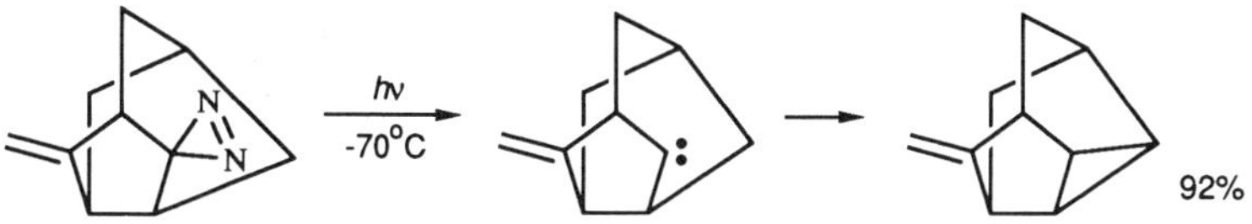

hν, -70°C; 92%

(79)

give a pyridinium ylide, which is subsequently trapped by acetylene dicarboxylate to yield a heterocycle (**80**) [117]. Other examples of the

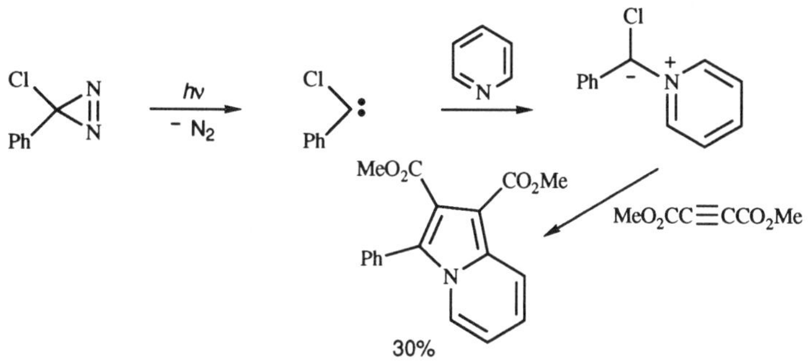

Cl, Ph, N, N; hν, $-N_2$; Cl, Ph; N; Cl, Ph, $N^+$; $MeO_2CC{\equiv}CCO_2Me$; $MeO_2C$, $CO_2Me$, Ph, N; 30%

(80)

photochemical generation of carbenes from diazirines are shown in **81** [118, 119].

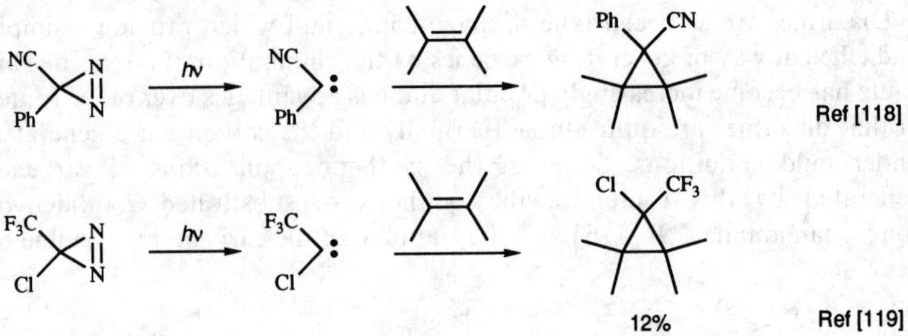

NC
N
N
Ph
hν
NC
Ph
Ph
CN
Ref [118]
F3C
N
N
Cl
hν
F3C
Cl
Cl
CF3
12%
Ref [119]


(81)

Some (2 + 2)-cycloadditions of N=N double bonds to alkenes have been observed but only in rigid molecules in which the azo-bond is in close proximity to a C=C double bond. In such systems even a [6 + 2]-cycloaddition of an N=N double bond to a benzene ring has been described. Some examples of these reactions are shown in **82** [120, 121]. Acyclic azo

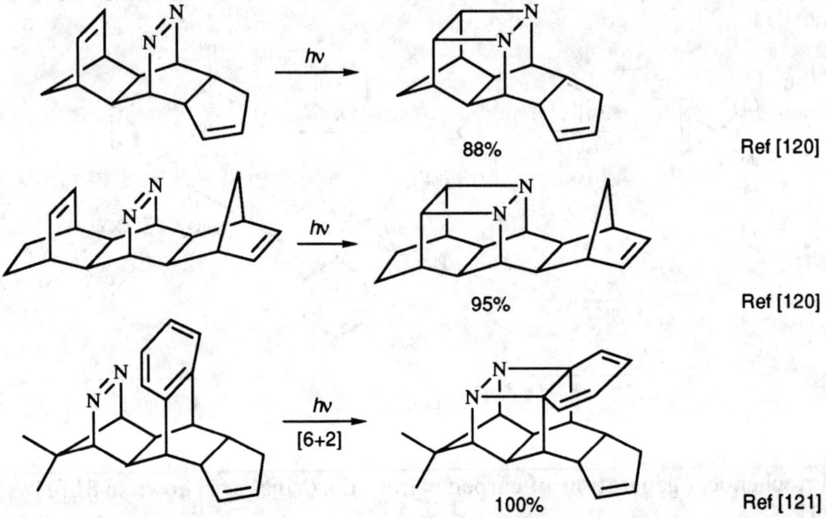

N
N
hν
N
N
88%
Ref [120]
N
N
hν
N
N
95%
Ref [120]
N
N
hν
[6+2]
N
N
100%
Ref [121]


(82)

compounds also undergo intermolecular cycloadditions with conjugated dienes. In these examples the additions are of the [4 + 2] type, as seen in **83** [122, 123].

OAc
OAc
+
$MeO_2C$–N=N–$CO_2Me$
hν
$CO_2Me$
$MeO_2C$
N
N
$CO_2Me$
$CO_2Me$
75% Ref [122]

+
N
N
N—Ph
O
O
hν
acetone
N
N
N
O
O
Ph
62% Ref [123]

(83)

### 5.5.3 Diazo compounds

The irradiation of diazoalkenes is another convenient way of generating carbenes. These compounds absorb with low intensity, in an $n\pi^*$ transition, in the 400–500 nm region [124]. The absorption is due to an $n\pi^*$ transition. Direct irradiation usually brings about the formation of singlet carbenes whereas triplet sensitization yields triplet carbenes that can be trapped intra- or inter-molecularly as in the examples collected in **84** [125, 126]. Irradiation

N
N
$N_2$
hν
$- N_2$
N
N
Ph
N
N
Ph
54% Ref [125]

$CHN_2$
N
hν
$- N_2$
CH
N
=C=
N
Ref [126]

(84)

of α-diazoketones also affords carbenes that can undergo different reactions, such as an Arndt–Eistert rearrangement and cyclization, as illustrated in (**85**) [127, 128].

$COCHN_2$ $\xrightarrow[MeOH]{h\nu}$ $CO\ddot{C}H$ $\longrightarrow$ $CH_2CO_2Me$

86% Ref [127]

$(CH_2)_n$–C(=O)–$CHN_2$, OMe $\xrightarrow{h\nu}$ $(CH_2)_n$–$CO\ddot{C}H$, OMe

n = 1 cyclization → 40%

n = 2, $H_2O$ → COOH, OMe 10%

n = 0 → 17% Ref [128]

(85)

Another way of generating carbenes photochemically is the irradiation of sodium and lithium salts of arylsulphonylhydrazones. The photodecomposition of these salts yields a diazo intermediate that, on further irradiation, gives a carbene. Thus, for example, direct irradiation of the sodium salt of the sulphonylhydrazone in methanol yields the diazo derivative that transforms into an ether via the singlet carbene. However, the triplet carbene generated in the sensitized irradiation yields the cyclopropane derivatives (**86**) [129].

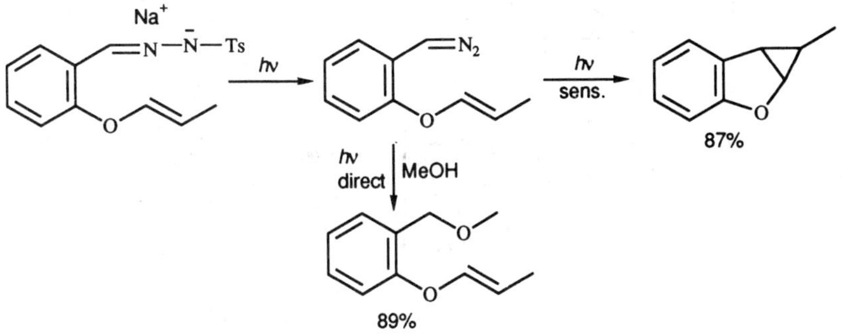

(86)

### 5.5.4 Diazonium salts

Most aryl diazonium salts are yellow and absorb in the region 360–450 nm. This absorption is assigned to a $\pi\pi^*$ transition [124]. Irradiation of aromatic diazonium salts results in an efficient elimination of nitrogen, yielding substitution products. The mechanism of the reaction is dependent on the solvent and it has been postulated that in aqueous medium the elimination of nitrogen brings about the formation of a cation which is trapped by water or anions. However, there is also evidence that in ethanol solutions the reaction yields preferentially a radical that usually abstracts a hydrogen or else can be trapped intermolecularly using halogen, as illustrated in **87**

OH
$N_2^+$ $X^-$
*hν*
$H_2SO_4/H_2O$
$H_2O$
$NO_2$ 84%
$NO_2$
*hν*
EtOH
H donor
77%
$I_2$
I
$NO_2$ 96%

(87)

[130]. In the specific example of the photochemistry of tetrafluoroborate diazonium salts, fluoride is the trapping agent, as shown in **88** [131].

F X
N N–H
$N_2^+$ $BF_4^-$
*hν*
F X
N N–H
F
X = $CO_2Et$, 53%
X = $CONH_2$, 28%

(88)

The photo-elimination of nitrogen from aromatic diazonium salts has been used in the synthesis of alkaloids by a photochemical version of the Pschorr reaction that competes successfully with thermal counterparts as illustrated (**89**) [132]. Some of the side reactions observed in the thermal reactivity of

(89)

these diazonium salts, such as diazo coupling and loss of protecting groups, do not occur in the photo-Pschorr reaction. Another example of the synthetic application of the photo-Pschorr cyclization is the synthesis of substituted fluorenones and azafluorenones (**90**) [133].



(90)

### 5.5.5 Azides

Irradiation of azides brings about the facile elimination of nitrogen, yielding a nitrene which can undergo rearrangement, insertion or addition reactions. Aromatic nitrenes usually undergo ring expansion to seven-membered heterocycles. This reactivity can be rationalized by addition of the singlet nitrene to the aromatic ring to give an azirine that is in equilibrium with an azacycloheptatetraene (**91**). Either of these two possible intermediates can be



(91)

trapped by nucleophiles present in the reaction medium to yield the observed products. An early study of this reaction involved the azirine exclusively in the formation of product. However, more recent work has demonstrated the existence of the azacyclotetraene as an intermediate [134]. Examples of ring expansion of this type are illustrated in the irradiation, in the presence of sodium methoxide, of 3-azidopyridines. This treatment yields 4-methoxy-5*H*-1,3-diazepines [135]. Under similar conditions the quinolines are converted into 3-methoxy-3*H*-1,4-benzodiazepines (**92**) [136]. An alternative ring

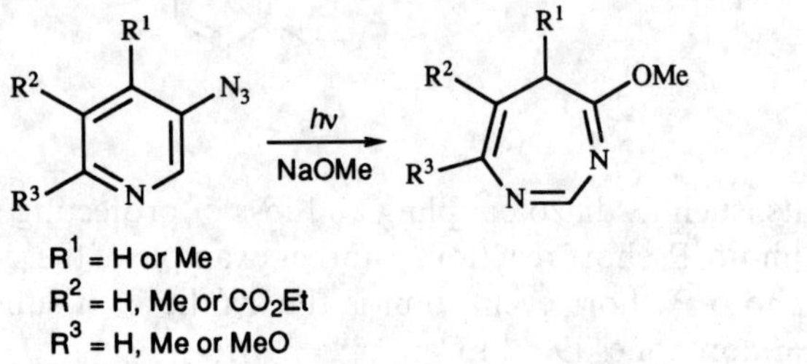

$R^1$ = H or Me
$R^2$ = H, Me or $CO_2Et$
$R^3$ = H, Me or MeO

Ref [135]

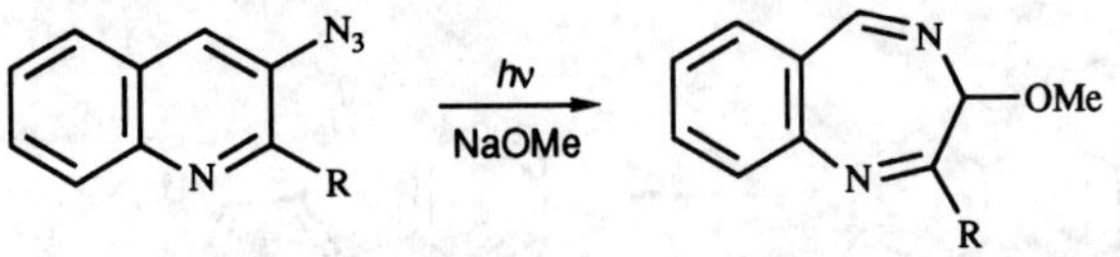

Ref [136]

(92)

opening of the intermediate azirine by nucleophiles is shown in **93** [137].

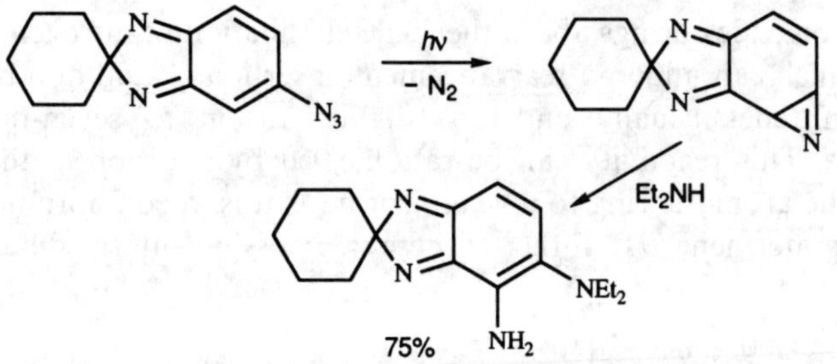

75%

(93)

Nitrenes generated from aliphatic azides can undergo different reactions from those described above. Among the processes encountered are rearrangement and inter- or intra-molecular addition to double bonds, which are shown in **94** [138–143].

$h\nu$
$-N_2$

$CO_2Me$ $N_3$ $CO_2Me$

24% $NHCO_2Me$ $CO_2Me$

Ref [138]

$R-CH=C(CO_2Et)N_3$ $\xrightarrow[-N_2]{h\nu}$ trimerization

R = $o$-$MeC_6H_4$, $o$-$PhCH_2C_6H_4$, Ph or fluoren-1-yl

40-66% Ref [139]

$h\nu$

78% Ref [140]

$Et-O-C(=O)-N_3$ + OSiMe₃ OEt $\xrightarrow{h\nu}$ $CO_2Et$ $OSiMe_3$ OEt $\longrightarrow$ $CO_2Et$ $NHCO_2Et$

75% Ref [141]

$N_3$ $\xrightarrow{h\nu}$ 65%

Ref [142]

$N_3$ $N_3$ $\xrightarrow{h\nu}$ $\xrightarrow{h\nu}$

(94)

48% Ref [143]

## 5.6 THE N=O GROUP AND RELATED COMPOUNDS

### 5.6.1 Nitrites

The absorption spectra of aliphatic nitrites show two bands at 220–230 nm ($\varepsilon$ 1000–1500 $dm^3$ $mol^{-1}$ $cm^{-1}$) and at 310–385 nm ($\varepsilon$ 20–80 $dm^3$ $mol^{-1}$ $cm^{-1}$) (four or five bands) that have been assigned to an $n\pi^*$ transition [144]. Irradiation into the long-wavelength absorption band results in highly efficient fission of the O—NO bond. When the alkoxy radical thus formed can react via intramolecular hydrogen abstraction, utilizing a six-membered transition state, the reaction yields nitrosoalcohols, as illustrated for the irradiation of nitrite (**95**) [145]. This photochemical method has been used

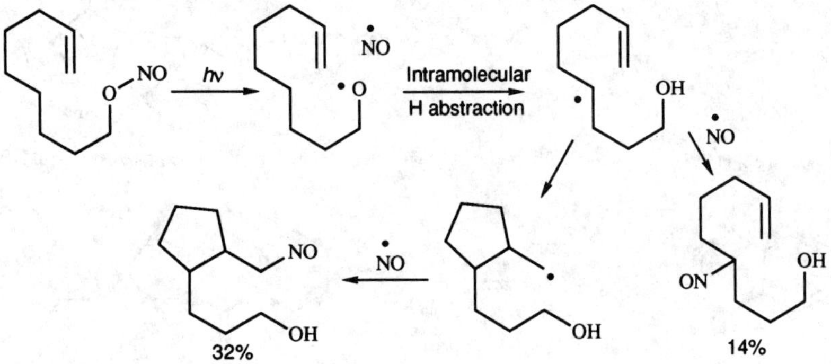

(95)

widely as a method for achieving remote functionalization of non-activated sites. This reaction, known as the Barton Reaction, has found synthetic application in the functionalization of positions 18 and 19 in steroidal systems. In this type of compound C-18 is attacked by alkoxy radicals at C-20, C-8, C-15, and C-11. The attack at C-19 takes place by alkoxy radicals at C-11, C-6, C-4 or C-2 (**96**) [144]. The first example of this reaction was reported

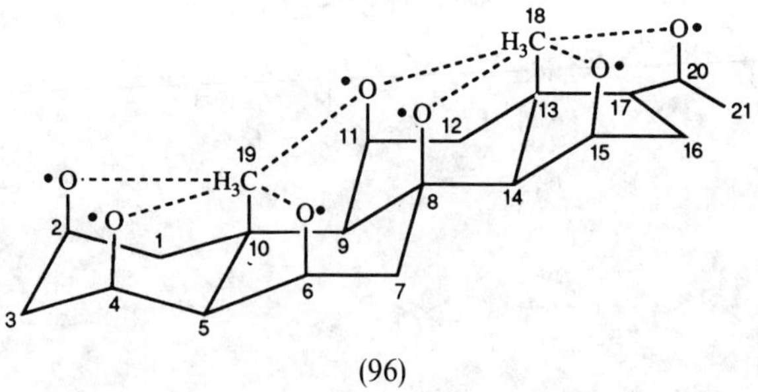

(96)

by Barton and coworkers [146a] in the photochemical conversion of 3b-acetoxy-5a-pregnan-20b-yl nitrite into the corresponding oxime (**97**).

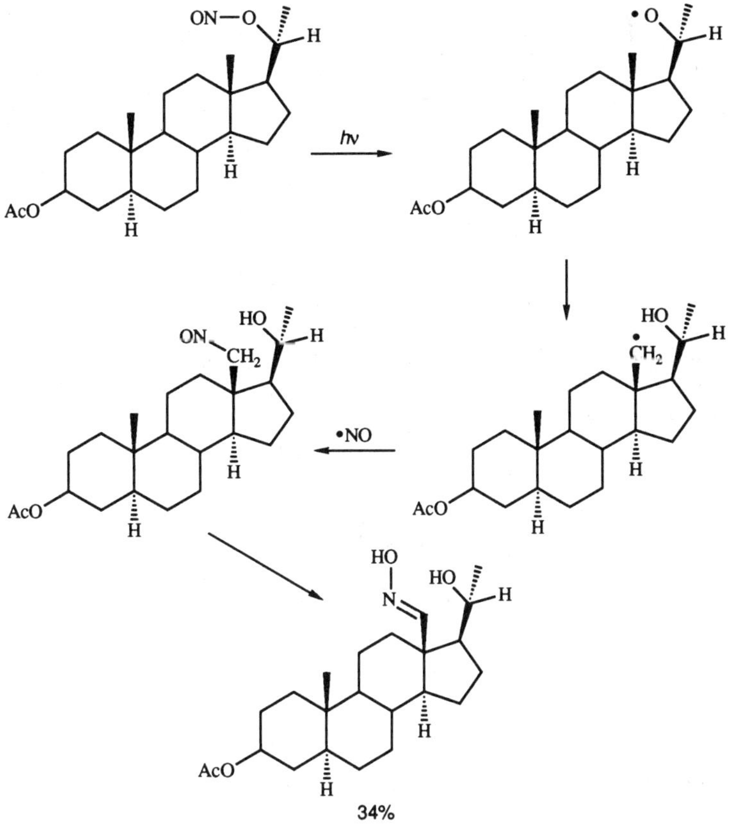

hν

•NO

34%

(97)

Examples of the Barton Reaction are shown in **98** [146b] and **99** [147, 148].

$C_8H_{17}$ AcO H ONO $h\nu$ ON AcO H OH 67% $Pr^iOH$ Δ OH N AcO H OH

(98)

$C_8H_{17}$ ONO H $h\nu$ •O •NO O H •NO O H NO O H HO N 38% Ref [147]

ONO $h\nu$ O Me O• •NO O O CHO O 100% Ref [148]

(99)

In many instances the starting material of this reaction is a nitro derivative that undergoes a photochemical nitro–nitrite rearrangement followed by O—NO bond fission, as will be discussed in the next section.

### 5.6.2 Nitro compounds

Aliphatic nitro compounds show two absorption bands in the u.v. spectra: An intense absorption around 210 nm ($\varepsilon$ *c.* 10000 dm$^3$ mol$^{-1}$ cm$^{-1}$) and low intensity band around 270 nm ($\varepsilon$ *c.* 10000 dm$^3$ mol$^{-1}$ cm$^{-1}$). The latter absorption has been assigned to an $n\pi^*$ transition. In aromatic nitro compounds the $n\pi^*$ band is usually difficult to locate because of the high intensity of the $\pi\pi^*$ absorption that appears at approximately the same region of the spectrum [149].

Unconjugated nitro groups usually rearrange photochemically to the corresponding nitrites via C—N bond fission to give alkyl and $NO_2$ radicals and recombination (**100**). Thc nitrite usually is the primary photochemical

$$R{-}NO_2 \xrightarrow{h\nu} R^\bullet + {}^\bullet NO_2 \longrightarrow R{-}O{-}N{=}O$$

(100)

product but undergoes secondary photolysis, giving products from O—NO bond fission as discussed previously. For example, the irradiation of either of the two isomers shown in **101** yields the two expected products arising

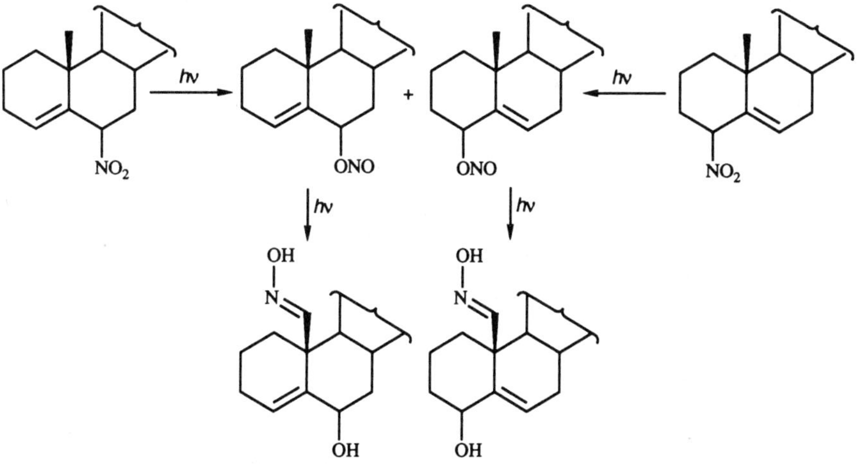

(101)

from the Barton Reaction of the nitrite [150].

In some cases the C—$NO_2$ bond fission can give rise to different reactions, such as, for example, in the irradiation of the dinitro compound (**102**) [151].

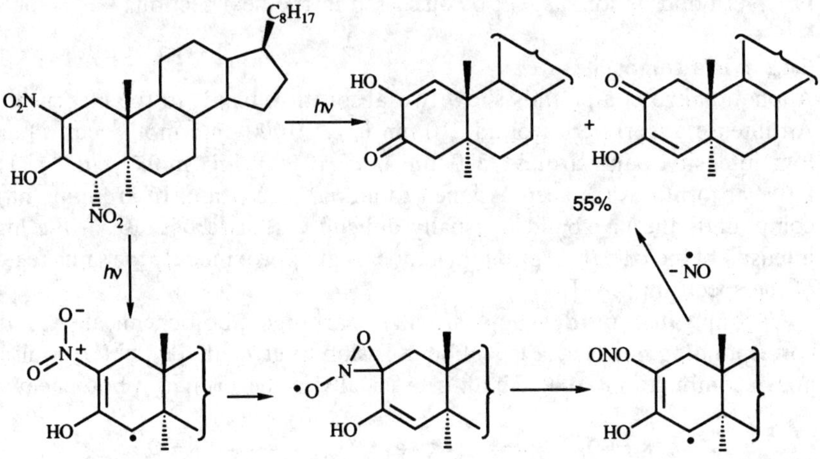

(102)

This process yields a mixture of two ketones in a total yield of 55% formed as outlined (**102**).

$\alpha,\beta$-Unsaturated nitro compounds also undergo nitro–nitrite rearrangement, giving $\alpha$-keto-oximes as outlined in **103** [152]. Another example of a

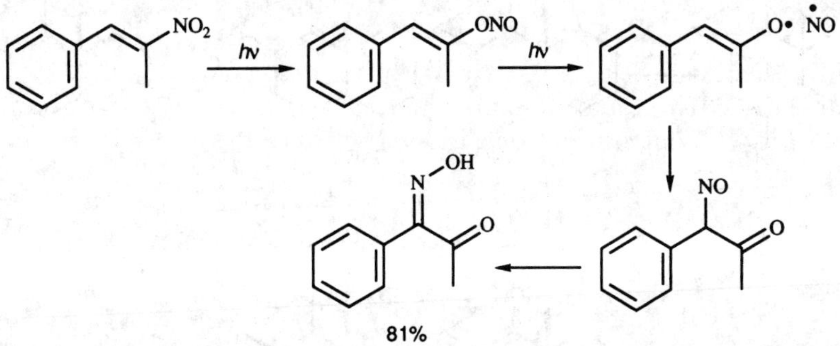

(103)

reaction of this type is the photoconversion of *N*-substituted 2-methyl-5-nitro-1*H*-imidazoles into oxadiazole-3-carboxamides by irradiation in water. This involves initial rearrangement to the corresponding oxime, hydration, ring-opening, and recyclization (**104**) [153]. Another reaction path

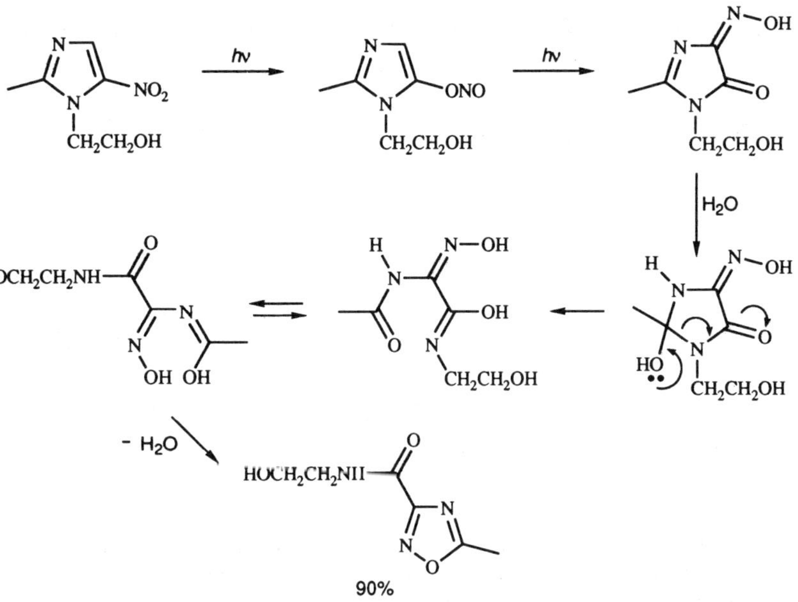

(104)

observed in the photochemistry of α,β-unsaturated nitro compounds is intramolecular (2 + 2)-cycloadditions of an NO bond to an alkene yielding oxazete *N*-oxides, a reaction that is firmly established in the photochemistry of this type of compound. The oxazete *N*-oxides are unstable and undergo a variety of thermal reactions. For instance, irradiation of 1-methyl-2-nitrocyclohexene yields an oxazete *N*-oxide that ring-opens to the 1,3-dipolar nitrile oxide. This can be trapped by methyl acrylate to yield an isoxazoline (**105**) [154]. A different outcome is observed when the reactions are carried

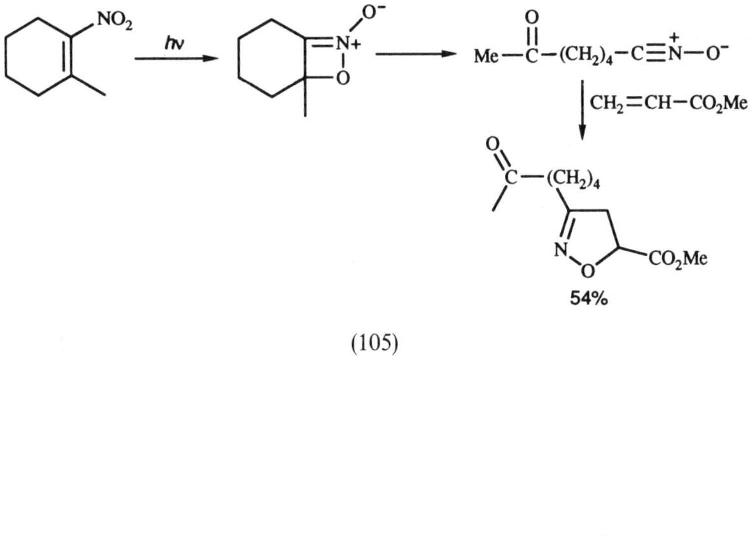

(105)

out in the presence of methanol, as in the case illustrated in **106** [155].

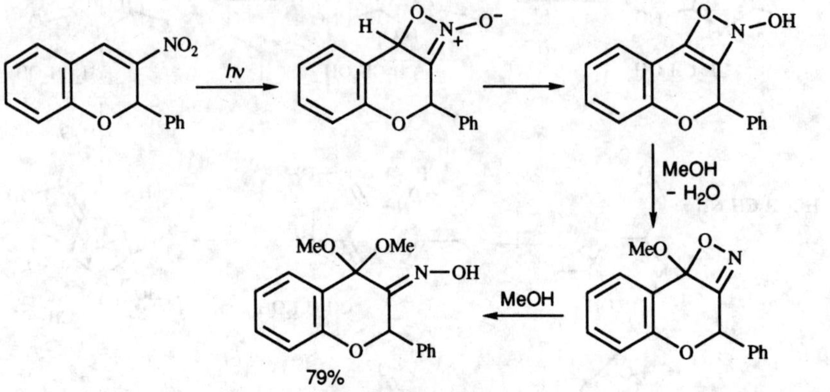

(106)

Aromatic nitro compounds with ring substituents at C-2 that can act as hydrogen donors undergo intramolecular hydrogen abstraction reminiscent of Norrish Type II reactivity. This reaction mode accounts for the well-documented photoisomerization of *o*-nitrobenzaldehyde to *o*-nitrosobenzoic acid (**107**). The photochromism observed in 2-nitrotoluenes is also due to an

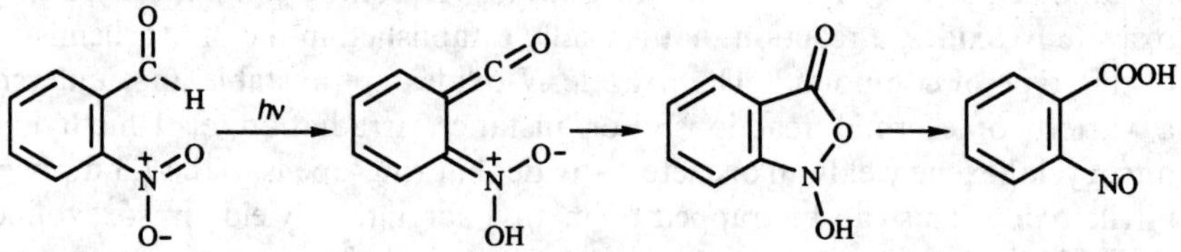

(107)

intramolecular hydrogen abstraction by the excited nitro group (**108**) from

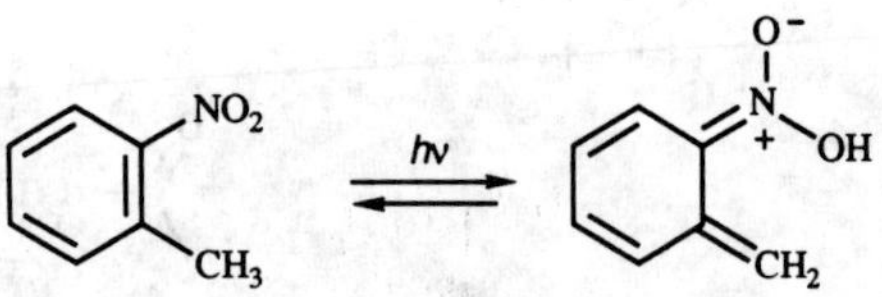

(108)

a neighbouring group [156]. Another example of a reaction of a nitro derivative that is promoted by hydrogen abstraction is shown in **109** [157].

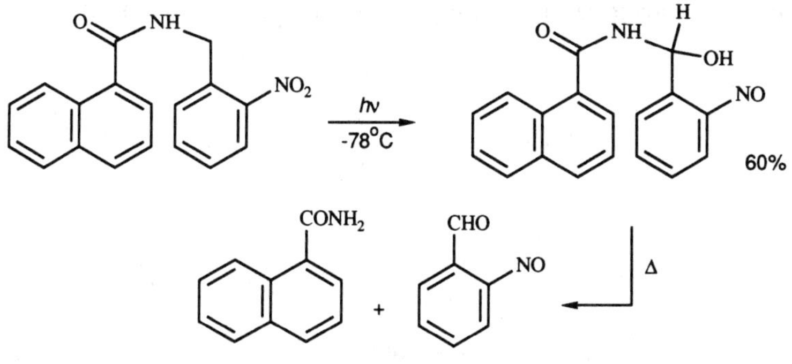

O NH $NO_2$ $h\nu$ -78°C O NH H OH NO 60%

$CONH_2$ + CHO NO Δ

(109)

This reaction has found application in the photochemical deprotection of some functional groups [158]. For instance 2-nitrophenylethyleneglycol has been used as a photoremovable protecting group for aldehydes and ketones. Thus, reaction of the carbonyl compound with 2-nitrophenylethyleneglycol gives the corresponding acetals. Irradiation of the acetals at 350 nm removes the protecting group efficiently by the path shown (**110**) [159]. This method

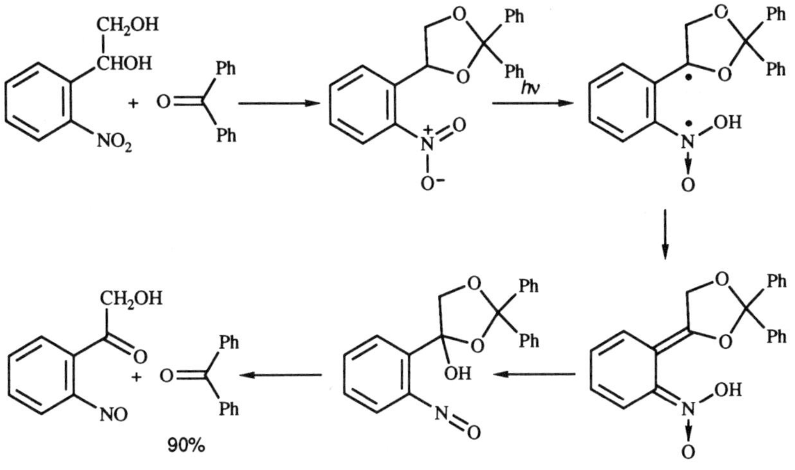

$CH_2OH$ CHOH $NO_2$ + O= Ph Ph → O Ph O Ph $\overset{+}{N}$=O $O^-$ $h\nu$ → O Ph O Ph • • N OH O

$CH_2OH$ O NO + O= Ph Ph 90% ← O Ph O Ph OH N=O ← O Ph O Ph N OH O

(110)

permits the selective removal of an acetal group under anhydrous and chemically inert conditions.

Nitroalkanes are also capable of undergoing single-electron transfer processes. This provides a convenient photoreductive path to oximes on irradiation in acetone in the presence of triethylamine (**111**) [160].

$R^1 =$ cyclohexenyl, cycloheptenyl, $PhCH_2$, indolyl

$R^2 = H, EtO_2CCH_2CH_2\text{-}$ or $NCCH_2CH_2\text{-}$

30-71%

(111)

The photochemical reactivity of anions of nitroalkanes, called nitronates, is also interesting and gives a route to hydroxamic acids, presumably via an intermediate oxazirine as illustrated in **112** [161]. The reaction can be highly

$Me(CH_2)_6CH_2NO_2$ → ($MeO^-$ or $MeNH_2$) $Me(CH_2)_6{-}CH{=}N^+(O^-)O^-$ → ($h\nu$) oxaziridine → $Me(CH_2)_6{-}CH(O^-){-}N{=}O$ → ($H^+$) $Me(CH_2)_6{-}CH(OH){-}N{=}O$ ⇌ $Me(CH_2)_6{-}C(OH){=}N{-}OH$ → $Me(CH_2)_6{-}C(=O){-}NHOH$ 85%

(112)

regio- and stereo- selective and has been extended to a large number of nitrocycloalkanes such as, for example, in the transformation of cyclohexyl or steroidal nitro derivatives (**113**) [162, 163].

$NO_2$ $CH_2CO_2Me$ H hν MeONa/MeOH OH O N $CH_2CO_2Me$ H

79% Ref [162]

$NO_2$ AcO H hν EtONa OH N O AcO H 55% Ref [163]

(113)

## 5.7 OXIMES, OXAZIRIDINES, NITRONES, AND HETEROCYCLIC *N*-OXIDES

Oximes, nitrones, and heterocyclic *N*-oxides have the common photochemical feature that they all undergo cyclization to oxaziridines on irradiation (**114**).

$R^1$ $R^2$ N OH hν $R^1$ $R^2$ N—H O

$R^1$ $R^2$ $R^3$ N O⁻ hν $R^1$ $R^2$ N—$R^3$ O

N O- hν N O

(114)

The oxaziridines are usually unstable and can rearrange thermally or photochemically to yield a variety of compounds. The normal photochemical reactivity of oxaziridines is ring opening to amides but other reaction modes, such as extrusion of oxygen, have also been observed.

The primary photochemical process in the majority of the oximes is *syn–anti* isomerization, as happens in all the C=N derivatives (see section 5.2.1.1).

However, on prolonged irradiation many oximes undergo a photo-Beckmann rearrangement. This reaction proceeds via the intermediacy of an oxaziridine that converts into the observed amide. The mechanism of the reaction is outlined in **115**. While, generally speaking, the thermal Beckmann

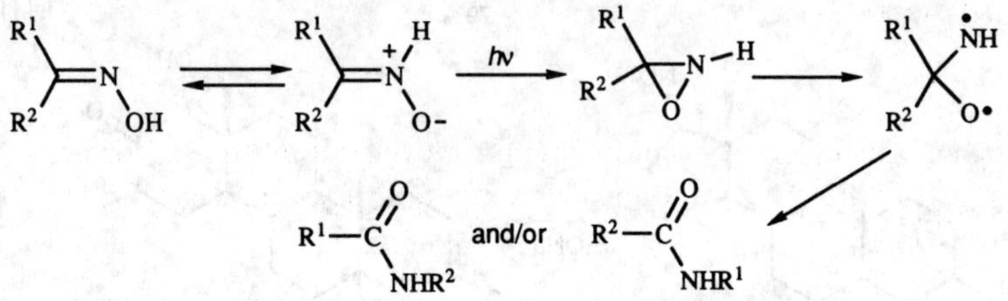

(115)

rearrangement is restricted to keto-oximes, the photochemical process is also operative with arylaldoximes that are converted into unsubstituted amides (**116**) [164]. The rearrangement of aryl aldoximes can lead to aryl carboxam-

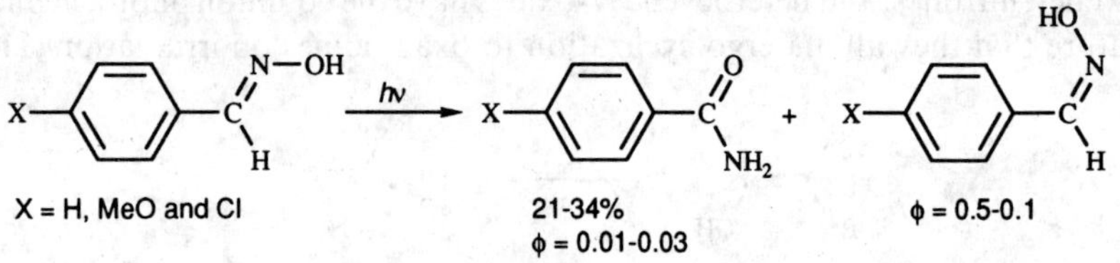

X = H, MeO and Cl

21-34%
$\phi$ = 0.01-0.03

$\phi$ = 0.5-0.1

(116)

ides or to *N*-arylformamides depending on whether the migration involves hydrogen or the aryl group. However, in all the cases studied the major or, very often, the only product observed is the carboxamide (**117**) [164]. The

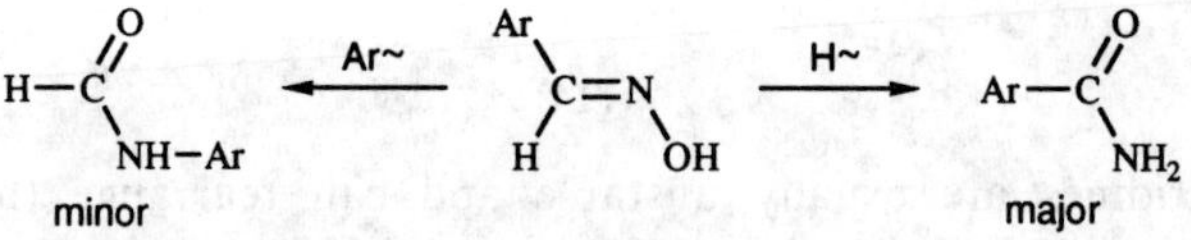

minor

major

(117)

photo-Beckmann rearrangement of (+)-camphor oxime further exemplifies this reaction (**118**) [165].

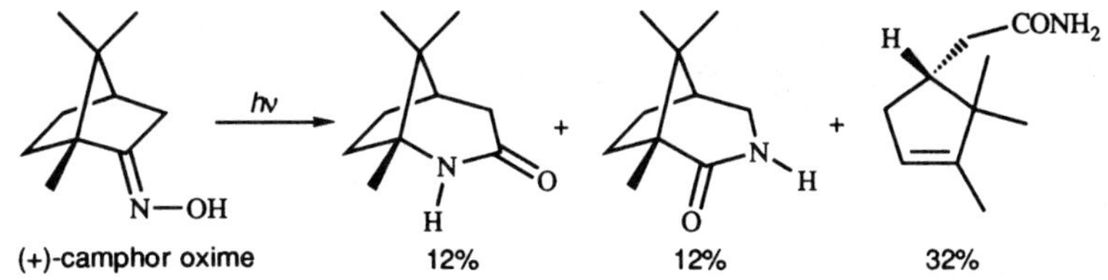

(+)-camphor oxime $\xrightarrow{h\nu}$ 12% + 12% + 32%

(118)

Nitrones also undergo photochemical cyclization to oxaziridines that rearrange to different compounds, as shown in the examples collected in **119** [166–168].

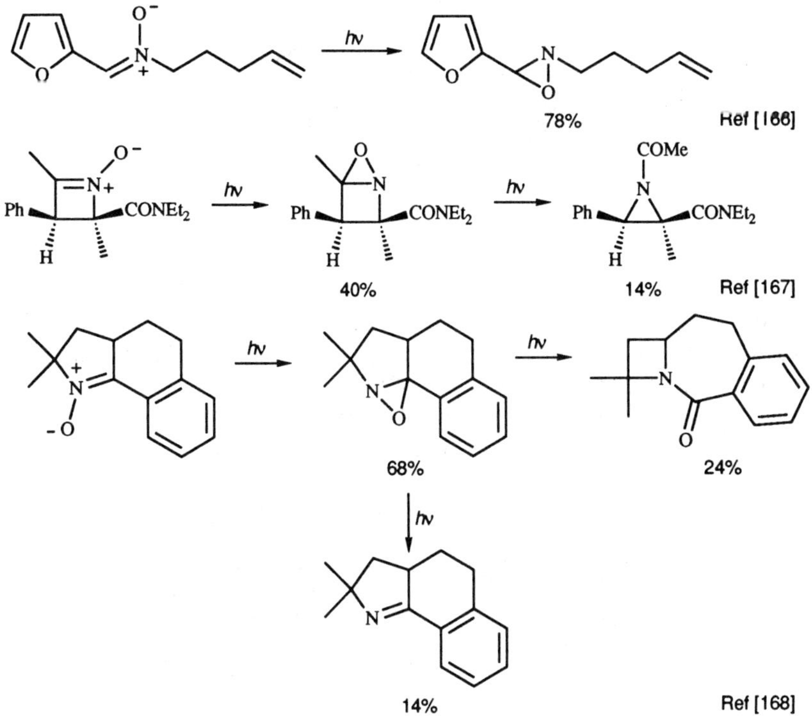

78% Ref [166]

40% 14% Ref [167]

68% 24%

14% Ref [168]

(119)

The irradiation of pyridine *N*-oxides in solution gives a mixture of heterocyclic compounds that can be rationalized in terms of the initial formation of an oxaziridine (**120**) [169]. The oxaziridine has not been isolated

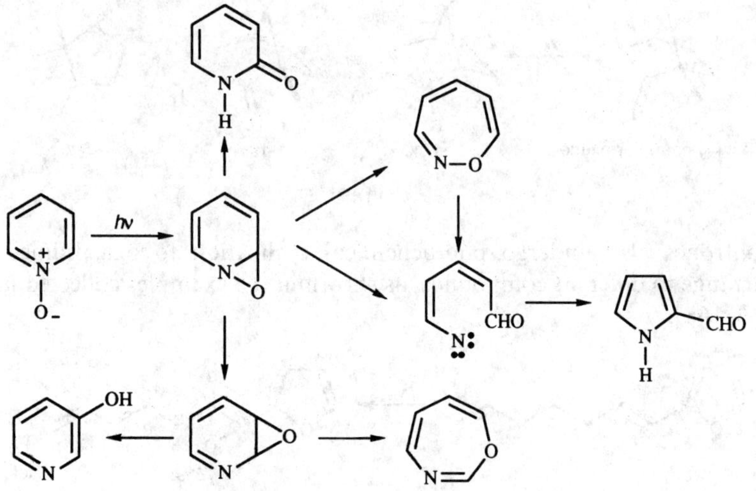

(120)

in any of the cases studied, but its formation is analogous to that of oxaziridines from nitrones. It has recently been proposed that biradicals play a part in the mechanism [170]. 1,2-Oxazepines, illustrated in **120**, are not generally isolated, although dibenzo derivatives can be obtained from acridine *N*-oxides in reasonable yields (**121**) [171]. Acylpyrroles, which may be

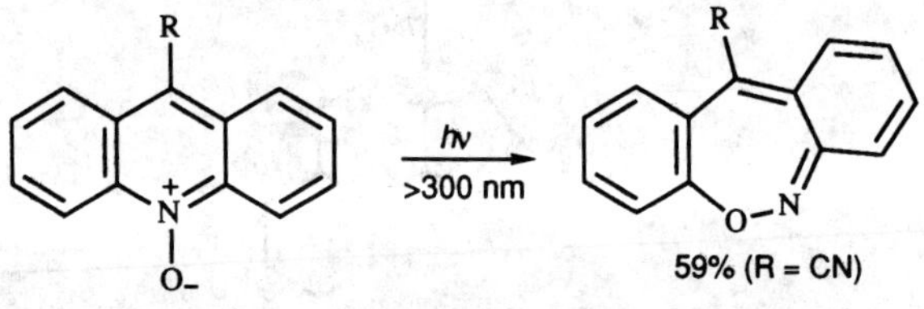

(121)

produced by way of 1,2-oxazepines, are often present in minor amounts and their formation can be enhanced by irradiation in the presence of copper(II) salts (**122**) [172]. The production of 2-pyridones (**120**) or related compounds

R
N+ O−
hν
$CuSO_4$
R
CHO
N H

40% (R = OMe, CN)

(122)

such as 2-quinolones (**123**) [173], can be a major reaction route in hydroxylic

R
N+ O−
hν
MeOH
R
N H O

60-70%

(123)

solvents whereas 1,3-oxazepines are formed in high yield in other solvents (**124**) [174]. Irradiation of quinoline *N*-oxide is the most convenient way of

Ph
Ph
Ph
N+ O−
hν
350 nm
Ph
Ph
O
N
Ph

80%

(124)

making 3,1-benzoxazepine (**125**) [175], and similar reactions occur for

N+ O−
hν
Pyrex
O
N

50%

(125)

phenanthridine *N*-oxides (**126**) [176] or for phenazine bis-*N*-oxide (**127**) [177].

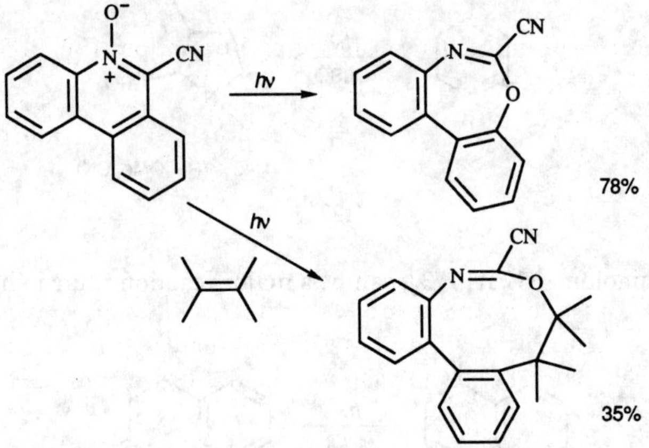

(126)

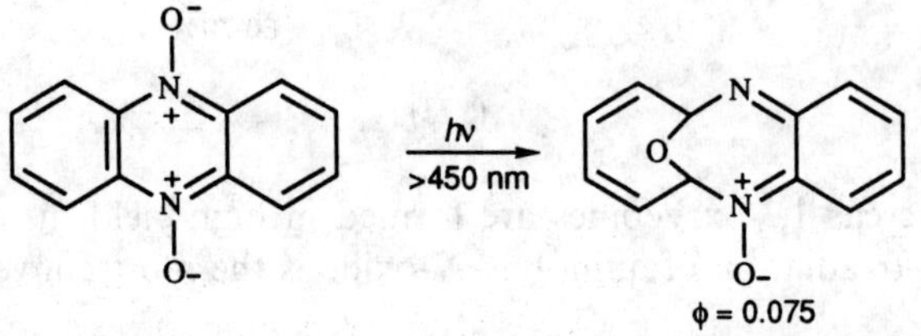

(127)

In the case of the phenanthridine derivatives a larger ring system can be formed if the irradiations are carried out in the presence of alkenes. Pyridine *N*-oxide itself gives a complex mixture of products on irradiation in moist solvents, but in the presence of base [178] or of secondary amines [179], ring-opening occurs to give a substituted dienenitrile (**128**), possibly by way

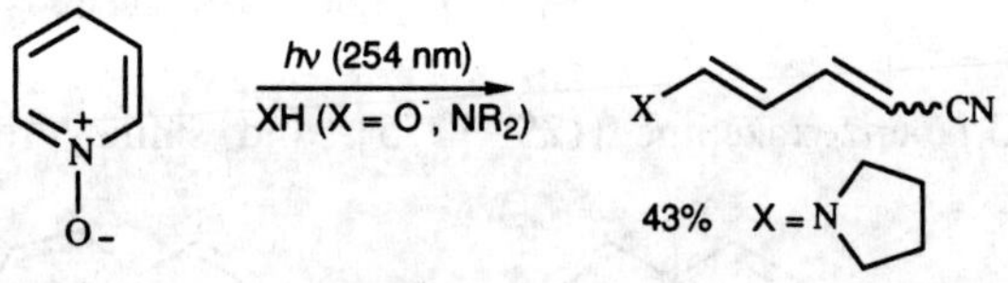

(128)

of the 1,2-oxazepine as an intermediate. The gas-phase irradiation of *N*-oxides gives substantial amounts of the deoxygenated heterocycle, and many aromatic *N*-oxides act as oxygen-transfer agents when irradiated in the presence of acceptors such as alkenes. The involvement of atomic oxygen continues to be debated. The process can be enhanced by added $BF_3$, as illustrated in the intramolecular aromatic hydroxylation (**129**) [180].

hν, Pyrex; $\bar{O}BF_3$; 43%; OH

(129)

The heteroaromatic *N*-imides are related to *N*-oxides and again they are photoreactive. Thus, irradiation of pyridine *N*-imides provides a route to 1,2-diazepines (**130**) [181], which unlike the 1,2-oxazepines, can often be

hν, Pyrex; - NR; N—N; R; 90% (R = $CO_2Pr^i$)

(130)

isolated readily. Quinoline *N*-imides generally give rise to 1,3-benzodiazepines (**131**) [182], and isoquinoline *N*-imides form the same compounds (**132**) [183].

MeO; - $\bar{N}CO_2Et$; hν, Pyrex; MeO; N—$CO_2Et$; 50%

(131)

R; $\bar{N}CO_2Et$; hν, Pyrex; R; N—$CO_2Et$; 20% (R = H); 40% (R = OMe)

(132)

## 5.8 AROMATIC HETEROCYCLIC COMPOUNDS

Five and six-membered aromatic heterocyclic systems undergo photochemical reactions such as ring scrambling, substitutions, and cycloadditions that are similar to those described for aromatic hydrocarbons (see section 2.2).

Five-membered aromatic nitrogen heterocycles undergo scrambling of the atoms of the ring to give isomeric compounds. For example, irradiation of *N*-methyl-2-(trimethylsilyl)pyrrole gives the 3-isomer (**133**) in 84% yield. The reaction is interpreted by formation of the Dewar intermediate followed by 1,3-migration and ring-opening (**133**) [184]. Other five-membered ring

$h\nu$ — 1,3 shift — $SiMe_3$ — 84%

(133)

heterocycles undergo similar reactions and 2-cyanopyrrole is converted to the 3-cyano isomer in reasonable yields (**134**) on irradiation [185]. However,

$h\nu$, 254 nm — 55%

(134)

a study [186] of methyl-substituted derivatives (e.g. **135**) shows that the

$h\nu$, 254 nm — 20-60% + minor

(135)

picture is more complex and involves 1,3- as well as 1,2-transpositions of ring atoms. A mechanism has been proposed which involves a 5-azabicyclopentene that undergoes one or two isomerizations involving migration of nitrogen before ring-opening occurs (**136**). Support for this type of intermediate is found in results from trapping experiments with methanol or with furan [187]. The respective adducts obtained from 2-cyano-1-methylpyrrole are shown in equation **137**.

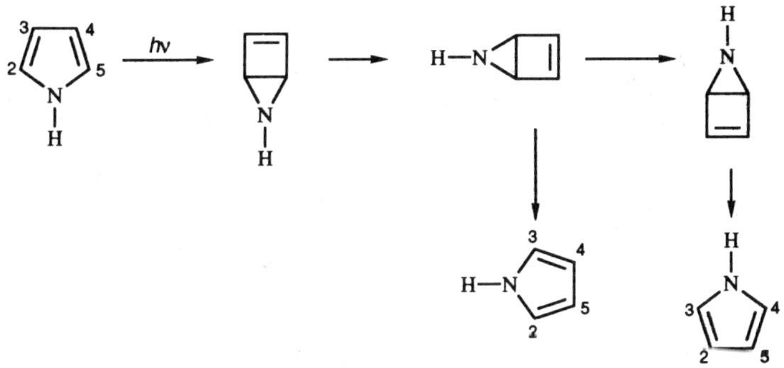

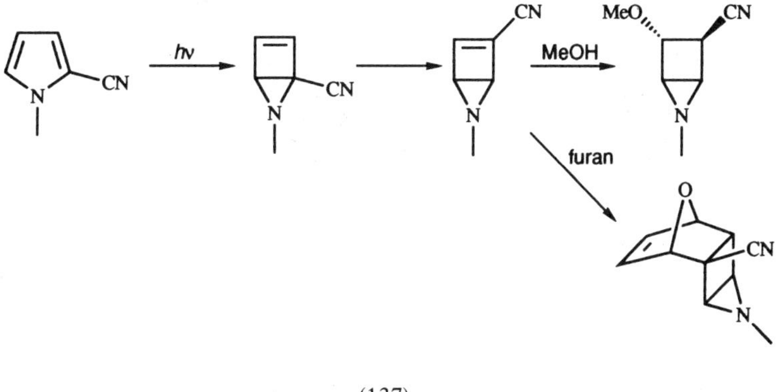

(136)

(137)

A great deal of work has been reported concerning the photoisomerization of five-membered heterocycles with two hetero-atoms and **138** shows some of the results reported [188–197]. In general the products can be accounted for by either the mechanism mentioned above or by a process involving an

$CF_3$ hν → $F_3C$ N N 41% Ref [188]

hν → N N 26% Ref [189]

N N— hν → N N 96% Ref [189]

N N hν → N N 40% Ref [190]

$CO_2Me$ N O hν → N O $CO_2Me$ 5.4% Ref [191]

O Ph Ph N O hν → O N Ph O Ph Ref [192]

$H_2N$ O N hν → N $NH_2$ O 72% Ref [193]

N O hν → N O 70% Ref [194]

(138)

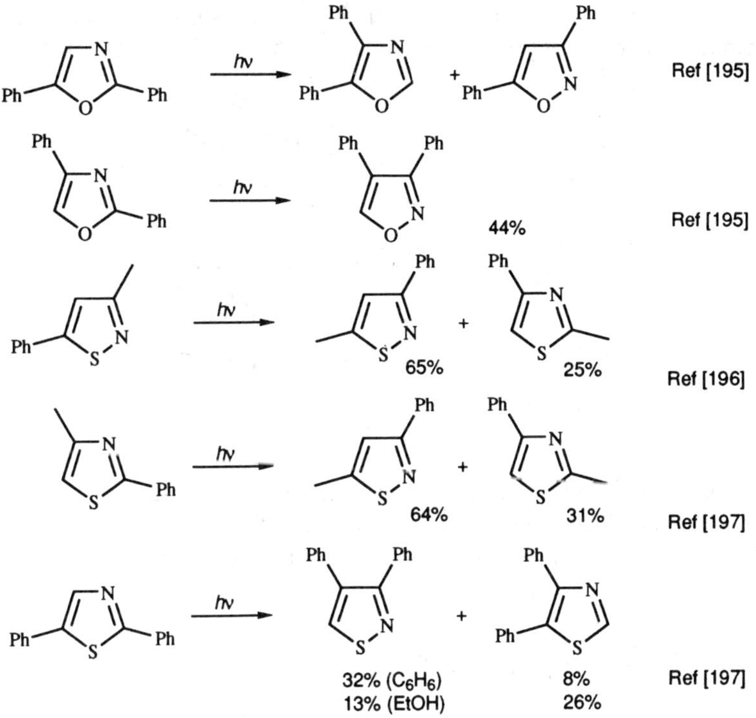

Ref [195]

44% Ref [195]

65% 25% Ref [196]

64% 31% Ref [197]

32% ($C_6H_6$) 8% Ref [197]
13% (EtOH) 26%

(138 continued)

analogue of the acylcyclopropene (from furans) or of the thiabicyclopentene (from thiophenes) (see sections 2.2.2 and 4.8.1). Often both mechanisms operate, as shown for a pyrazole rearrangement (**139**) [188]. Acylazirines

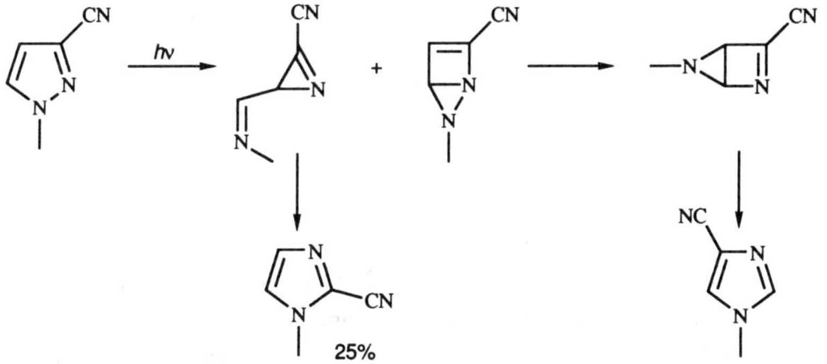

25%

(139)

have been isolated from both oxazole [198] and isoxazole systems, and the formation of different products according to the wavelength of light used (**140**)

Ph
hν
(<313 nm)
Ph
Ph
100% (φ = 0.29)
hν
(>334 nm)
Ph
Ph
(φ = 0.35)

(140)

points to the involvement of different excited states of the acylazirine in the second stage of the overall photoisomerization [199]. Additional intermediates such as zwitterions (**141**) in some isoxazole–oxazole interconversions

hν
N≡CH

(141)

[200] and enaminonitriles in some pyrazole–imidazole reactions (**142**) [201]

NC
hν
MeNHCH=C(CN)CN
32%
+
NC
38%

(142)

are undoubtedly involved in some of the photochemical transformations.

Photosubstitution reactions of heteroaromatic compounds have not been widely reported [202]. Illustration (**143**) [203–208] contains a selection of examples of different mechanistic types.

hν
CN⁻/$C_6H_4(CN)_2$
CN
R
Ref [203]

hν
$Et_2NH$
F
$NEt_2$
Ref [204]

Br
hν
$C_6H_6$
Ph
Ref [205]

hν
$O^-$
Br
O
95%
Ref [206a]

hν
$KCH_2CN$
Br
CN
75%
Ref [206b]

hν
PhOH
Cl
OPh
Ref [207]

COOMe
hν
$MeOH/O_2/H_2SO_4$
COOMe
OMe
Ref [208]

(143)

With six-membered nitrogen heterocycles, 1,3-transpositions of ring atoms are commonly encountered (e.g. **144**) [209], and these are best accounted for

hν
+
via
or

(144)

in terms of an azabicyclohexadiene mechanism. Pyridazines react in a similar way to give pyrazines (**145**) [210], and in some cases the intermediate

$(R^1, R^4 = Cl, F; R^2, R^3 = Cl)$ up to 80% $(\phi = 0.015)$

(145)

diazabicyclohexadienes have been isolated [211]. Pyrazines can react further to give pyrimidines by a mechanism involving an overall 1,2-shift, and the conversion of the parent compound (**146**) is the subject of a patented process [212].

$h\nu$ (262 nm), Hg vapour; $\phi = 0.02$

(146)

For the 1,2-transposition of pyrazines a diazabenzvalene mechanism can be invoked, but it should be remembered that in these as in other cases there is no direct evidence for the intermediates suggested. The fact that several pathways could lead to the same overall result, especially in systems where some ring atoms are indistinguishable because they carry the same substituents, led to an analysis based on the 12 possible permutation patterns for the transformation of a six-membered ring [213].

Pyridine also is reported to undergo *Dewar* pyridine formation but this is unstable and yields a ring-opened product (**147**) [214]. *N*-methylpyridinium

(147)

chloride [215], however, gives a 6-azabicyclo[3,1,0]hexene on irradiation in aqueous alkali (**148**).

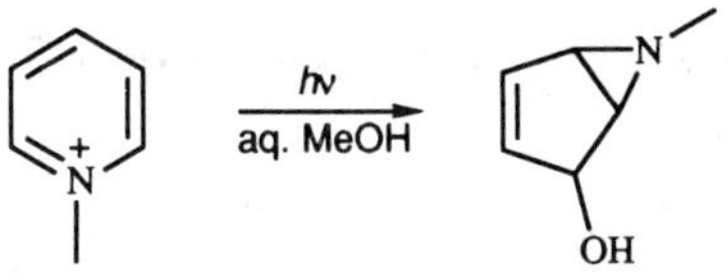

(148)

Irradiation of 2-pyridones yields bicyclic $\beta$-lactams as well as (2 + 2)-cyclodimers and this has been used as a model system [216] for the preparation of $\beta$-lactams with a *cis*-arrangement of substituents (**149**) as in

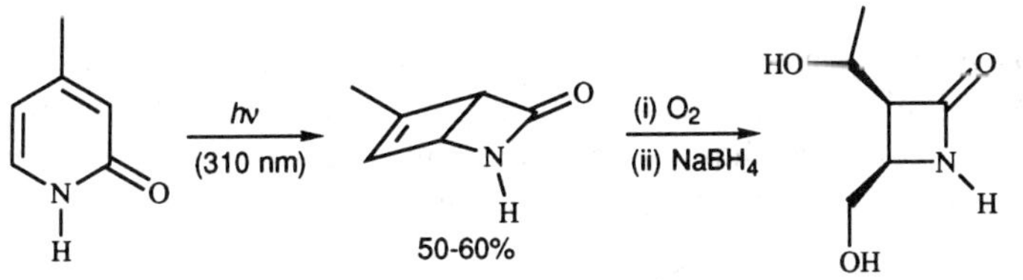

(149)

the olivanic acid antibiotics. By contrast, 3-oxidopyridinium systems give rise to dimers and a 6-azabicyclo[3,1,0]hexenone (**150**)[217]. A rather

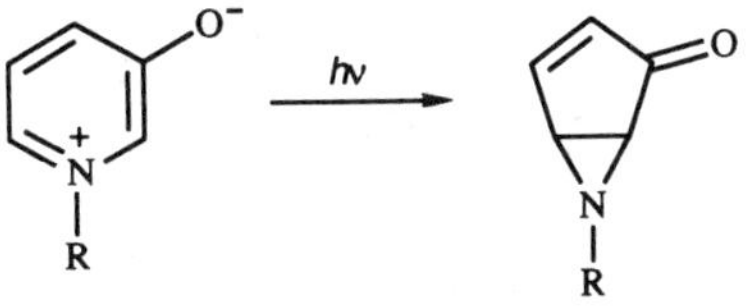

(150)

different modification of the basic isomerization reaction is that in which the side-chain-substituted 2-methylpyridine forms an *ortho*-substituted aniline

[218] by a pathway that involves a 2-methylene-2-azabicyclohexene (**151**).

(X = CN, COOR, Ph)

44% (X = CN)
$\phi = 0.068$

(151)

The bicyclic intermediate can be isolated if the irradiation is carried out in aqueous alkali, and subsequent irradiation with a high pressure mercury arc converts this intermediate to the aniline in high yield (greater than 90%).

Irradiation of pyridine gives 1-azabicyclohexadiene (**152**) [219], and

$h\nu$ (254 nm)

(152)

perfluoroalkylpyridines behave in a similar way [220], giving azabicyclohexadienes or azaprismanes depending on the wavelength (**153**). A study

$(C_2F_5)_5$ — $h\nu$ (Pyrex) → 95%; $h\nu$ (quartz) → 91%

(153)

of a pyridine with different substituents shows [221] that the 2-azabicyclohexadienes are not isolated because they are photochemically labile and undergo a 1,3-shift to an isomeric 1-aza-compound before cyclizing to an azaprismane (**154**). In the nitrogen heterocyclic series there is a strong

$(R^1 = CF_3, R^2 = C_2F_5, R^3 = C_3F_8)$

(154)

tendency to form *Dewar* isomers rather than azabenzvalenes.

## REFERENCES


[1] A. Padwa, *Chem. Rev.*, 1977, **77**, 37.

[2] A. C. Pratt, *Chem. Soc. Rev.*, 1977, **6**, 63.

[3] P. S. Mariano, *Tetrahedron*, 1983, **6**, 3845.

[4] G. Wettermark, *The Chemistry of the Carbon–Nitrogen Double Bond*, ed. S. Patai, Wiley-Interscience, 1969, 580.

[5] E. J. Poziomek, *J. Pharm. Sci.*, 1965, **54**, 333.

[6] A. C. Pratt and Q. Abdul-Majid, *J. Chem. Soc., Perkin Trans. 1*, 1986, 1691.

[7] G. Wettermark, *The Chemistry of the Carbon–Nitrogen Double Bond*, ed. S. Patai, Wiley-Interscience, 1969, 574.

[8] M. A. Quilliam, B. E. McCarry, K. H. Hoo, D. R. McCalla and S. Vaitekunas, *Can. J. Chem.*, 1987, **65**, 1128.

[9] D. Schulte-Frohlinde, *Justus Liebigs Ann. Chem.*, 1959, **622**, 47.

[10] J. Dale and L. Zechmeister, *J. Am. Chem. Soc.*, 1953, **75**, 2379.

[11] See *Acc. Chem. Res.*, 1975, **8**, 81–112.

[12] H. Rau and E. Lüddecke, *J. Am. Chem. Soc.*, 1982, **104**, 1616.

[13] (a) G. M. Badger, C. P. Joshua and G. E. Lewis, *Tetrahedron Lett.*, 1964, 3711.
(b) M. P. Cava and R. H. Schlessinger, *Tetrahedron Lett.*, 1964, 2109.

[14] T. Onaka, Y. Kanda and M. Natsume, *Tetrahedron Lett.*, 1974, 1179.

[15] F. B. Mallory and C. S. Wood, *Tetrahedron Lett.*, 1965, 2643.
[16] (a) D. Armesto, M. G. Gallego and W. M. Horspool, *J. Chem. Soc., Perkin Trans. 1*, 1989, 1623.
(b) J. Glinka, *Pol. J. Chem.*, 1979, **53**, 2143.
[17] V. H. M. Elferink and H. J. T. Bos, *J. Chem. Soc., Chem. Commun.*, 1985, 882.
[18] D. Armesto, M. G. Gallego, M. J. Ortiz, S. Romano and W. M. Horspool, *J. Chem. Soc., Perkin Trans. 1*, 1989, 1343.
[19] S. Arai, T. Takeuchi, M. Ishikawa, T. Takeuchi, M. Yamazari and M. Hida, *J. Chem. Soc., Perkin Trans. 1*, 1987, 481.
[20] D. Armesto, W. M. Horspool, M. Apoita, M. G. Gallego and A. Ramos, *J. Chem. Soc., Perkin Trans. 1*, 1989, 2035.
[21] T. Tschamber, H. Fritz and J. Streith, *Helv. Chim. Acta*, 1985, **68**, 1359.
[22] S. Hirokami, T. Takahashi, K. Kurosawa and M. Nagata, *J. Org. Chem.*, 1985, **50**, 166.
[23] T. Nishio, S. Kameyama and Y. Omote, *J. Chem. Soc., Perkin Trans. 1*, 1986, 1147.
[24] E. Sato, Y. Ikeda, Y. Kanaoka and H. Okajima, *Heterocycles*, 1987, **26**, 1611.
[25] J-C. Cuevas, J. de Mendoza and P. Prados, *Tetrahedron Lett.*, 1988, **29**, 4315.
[26] J. Kurita, K. Iwata and T. Tsuchiya, *J. Chem. Soc., Chem. Commun.*, 1986, 1188.
[27] H. Dürr, *Angew. Chem. Int. Ed. Engl.*, 1989, **28**, 413.
[28] A. Padwa, W. Bergmark, D. Pashayan, *J. Am. Chem. Soc.*, 1968, **90**, 4458.
[29] (a) O. Tsuge, T. Tashiro and K. Oe, *Tetrahedron Lett.*, 1968, 3971.
(b) O. Tsuge, K. Oe and T. Tashiro, *Tetrahedron*, 1973, **29**, 41.
[30] T. Nishio and Y. Omote, *J. Chem. Soc., Perkin Trans. 1*, 1987, 2611.
[31] S. Futamura, H. Ohta and Y. Kamiya, *Bull. Chem. Soc. Jpn*, 1982, **55**, 2190.
[32] N. Katagiri, H. Watanabe and C. Kaneko, *Chem. Pharm. Bull.*, 1988, **36**, 3354.
[33] T. H. Koch and K. H. Howard, *Tetrahedron Lett.*, 1972, 4035.
[34] (a) T. H. Koch and R. M. Rodehorst, *Tetrahedron Lett.*, 1972, 4039.
(b) R. M. Rodehorst and T. H. Koch, *J. Am. Chem. Soc.*, 1975, **97**, 7298.
[35] J. S. Swenton and J. A. Hyatt, *J. Am. Chem. Soc.*, 1974, **96**, 4879.
[36] T. Kumagai, Y. Kawamura and T. Mukai, *Chem. Lett.*, 1983, 1357.
[37] T. Nishio, T. Nishiyama and Y. Omote, *Liebigs Ann. Chem.*, 1988, 441.

[38] E. Malamidou-Xenikaki and D. N. Nicolaides, *Tetrahedron*, 1986, **42**, 5081.
[39] T. Nishio and Y. Omote, *J. Org. Chem.*, 1986, **50**, 1370.
[40] S. Fery-Forges and N. Paillous, *J. Org. Chem.*, 1986, **51**, 672.
[41] D. Lawrenz, S. Mohr and B. Wendländer, *J. Chem. Soc., Chem. Commun.*, 1984, 863.
[42] P. Margaretha, *Helv. Chim. Acta*, 1982, **65**, 290.
[43] P. S. Mariano, *Org. Photochem.*, 1987, **9**, 1.
[44] (a) S. S. Hixson, P. S. Mariano and H. E. Zimmerman, *Chem. Rev.*, 1973, **73**, 531.
(b) H. E. Zimmerman, *Rearrangements in Ground and Excited States*, ed. P. De Mayo, Academic Press, 1980, **3**, 131.
[45] D. Armesto, J. A. F. Martin, R. Perez-Ossorio and W. M. Horspool, *Tetrahedron Lett.*, 1982, **23**, 2149.
[46] M. Nitta, I. Kashara and T. K. Kobayashi, *Bull. Chem. Soc. Jpn*, 1981, **54**, 1275.
[47] D. Armesto, F. Langa, J. A. F. Martin, R. Perez-Ossorio and W. M. Horspool, *J. Chem. Soc., Perkin Trans. 1*, 1987, 743.
[48] D. Armesto, W. M. Horspool, J. A. F. Martin and R. Perez-Ossorio, *J. Chem. Research (S)*, 1986, 46.
[49] (a) D. Armesto, W. M. Horspool and F. Langa, *J. Chem. Soc., Perkin Trans. 2*, 1987, 1039.
(b) D. Armesto, W. M. Horspool and F. Langa, *J. Chem. Soc., Perkin Trans. 2*, 1989, 903.
[50] K. N. Houk, *Chem. Rev.*, 1976, **76**, 1.
[51] A. C. Pratt and Q. Abdul-Majid, *J. Chem. Soc., Perkin Trans. 1*, 1987, 359; 1986, 1691.
[52] D. Armesto, W. M. Horspool and F. Langa, *J. Chem. Soc., Chem. Commun.*, 1987, 1874.
[53] D. Armesto, W. M. Horspool, F. Langa and A. Ramos, *J. Chem. Soc., Perkin Trans. 1*, 1991, 223.
[54] (a) D. Armesto, M. G. Gallego and W. M. Horspool, *Tetrahedron Lett.*, 1990, **31**, 2475.
(b) D. Armesto, M. G. Gallego and W. M. Horspool, *Tetrahedron*, 1990, **46**, 6185.
[55] D. Armesto, A. R. Agarrabeitia, W. M. Horspool and M. G. Gallego, *J. Chem. Soc., Chem. Commun.*, 1990, 934.
[56] D. Armesto, W. M. Horspool, M. J. Mancheño and M. J. Ortiz, *J. Chem. Soc., Perkin Trans. 1*, 1990, 2348.
[57] H. Ohta and K. Tokumaru, *Chem. Lett.*, 1974, 1403.

[58] H. Ohta and K. Tokumaru, *Chem. Ind. (London)*, 1974, 1403.
[59] S. Prathapan, S. Loft and W. C. Agosta, *Tetrahedron Lett.*, 1988, **29**, 6853.
[60] Y. Hirai, H. Egawa, Y. Wakui and T. Yamazaki, *Heterocycles*, 1987, **25**, 201.
[61] A. Padwa, W. Bergmark and D. Pashayan, *J. Am. Chem. Soc.*, 1969, **91**, 2653.
[62] M. Fischer, *Tetrahedron Lett.*, 1966, 5273.
[63] N. Nata and M. Hokawa, *Chem. Lett.*, 1981, 507.
[64] I. Ono and N. Hata, *Bull. Chem. Soc. Jpn*, 1987, **60**, 2891.
[65] T. Nishio and Y. Omote, *J. Chem. Soc., Perkin Trans. 1*, 1988, 957.
[66] D. K. Buffel, C. McGuian and M. J. Robins, *J. Org. Chem.*, 1985, **50**, 2664.
[67] A. Erndt, A. Kostuch, A. Para and M. Fiedorowicz, *Liebigs Ann. Chem.*, 1985, 937.
[68] F. R. Stermitz, R. P. Seiber and D. E. Nicodem, *J. Org. Chem.*, 1968, **33**, 1136.
[69] M. Ochiai and K. Morita, *Tetrahedron Lett.*, 1967, 2349.
[70] S. Wake, Y. Takayama, Y. Otsuji and E. Imote, *Bull. Chem. Soc. Jpn*, 1974, **47**, 1257.
[71] P. S. Mariano, J. Stavinoha and E. Bay, *Tetrahedron*, 1981, **37**, 3385.
[72] Y. Kurauchi, H. Nobuhara and K. Ohga, *Bull. Chem. Soc. Jpn*, 1986, **59**, 897.
[73] H. Nozaki, M. Kato, R. Noyori and M. Kawanisi, *Tetrahedron Lett.*, 1967, 4259.
[74] H. Noyori, M. Kato, M. Kawanisi and H. Nozaki, *Tetrahedron*, 1969, **25**, 1125.
[75] Y. Karauchi, K. Ohga, H. Nobuhara and S. Morita, *Bull. Chem. Soc. Jpn*, 1985, **58**, 2711.
[76] A. Padwa, *Chem. Rev*, 1977, **77**, 37.
[77] A. Hassner and B. Fischer, *Tetrahedron*, 1989, **45**, 3535.
[78] K-H. Pfoertner, K. Bernauer, F. Kaufmann and E. Lorch, *Helv. Chim. Acta*, 1985, **68**, 584.
[79] T. Büchel, R. Prewo, J. H. Bieri and H. Heimgartner, *Helv. Chim. Acta*, 1984, **67**, 534.
[80] W. Stegmann, P. Uebelhart and H. Heimgartner, *Helv. Chim. Acta*, 1983, **66**, 2252.
[81] A. Padwa and J. Smolanoff, *Tetrahedron Lett.*, 1974, 33.
[82] A. Padwa, J. Smolanoff and A. Tremper, *Tetrahedron Lett.*, 1974, 29.
[83] (a) P. S. Mariano, *Org. Photochem.*, 1987, **9**, 1.

(b) P. S. Mariano, *Synthetic Organic Photochemistry*, ed. W. M. Horspool, Plenum Press, 1984, 145.

[84] J. L. Stavinoha and P. S. Mariano, *J. Am. Chem. Soc.*, 1981, **103**, 3136.

[85] J. L. Stavinoha, P. S. Mariano, A. Leone-Bay, R. Swanson and C. Bracken, *J. Am. Chem. Soc.*, 1981, **103**, 3148.

[86] T. Tiner-Harding, J. W. Ullrich, F. T. Chiu, S. F. Chen and P. S. Mariano, *J. Org. Chem.*, 1982, **47**, 3360.

[87] R. Borg, R. O. Heuckeroth, A. J. Y. Lan, S. L. Quillen and P. S. Mariano, *J. Am. Chem. Soc.*, 1987, **109**, 2728.

[88] G. Dai-Ho and P. S. Mariano, *J. Org. Chem.*, 1987, **52**, 704.

[89] R. Ahmed-Schofield and P. S. Mariano, *J. Org. Chem.*, 1987, **52**, 1478.

[90] J. W. Ullrich, F. T. Chiu, T. Tiner-Harding and P. S. Mariano, *J. Org. Chem.*, 1984, **49**, 220.

[91] K. Ohga, U. C. Yoon and P. S. Mariano, *J. Org. Chem.*, 1984, **49**, 213.

[92] A. L. Campbell and G. R. Lenz, *Synthesis*, 1987, 421.

[93] C. Bochu, A. Couture, P. Grandclaudon and A. Lablache-Combier, *J. Chem. Soc., Chem. Commun.*, 1986, 839.

[94] C. Bochu, A. Couture, P. Grandclaudon and A. Lablache-Combier, *Tetrahedron*, 1988, **44**, 1959.

[95] A. Couture and P. Grandclaudon, *Synthesis*, 1986, 576.

[96] A. Bhattacharjya, R. Mukhopadhyay, R. R. Sinha, E. Ali and S. C. Pakrashi, *Tetrahedron*, 1988, **44**, 3477.

[97] T. Naito, E. Doi, O. Miyata and I. Ninomiya, *Heterocycles*, 1986, **24**, 903.

[98] T. Naito, N. Kojima, O. Miyata and I. Ninomiya, *J. Chem. Soc., Chem. Commun.*, 1985, 1611.

[99] A. I. Scott, *Ultraviolet Spectra and Natural Products*, Pergamon Press, 1964, 39.

[100] K. Tsujimoto, T. Fujimori and M. Ohashi, *J. Chem. Soc., Chem. Commun.*, 1986, 304.

[101] J. Mattay, J. Rusink, R. Heckendron and T. Winkler, *Tetrahedron*, 1987, **43**, 5781.

[102] N. C. Yang, B. Kim, W. Chiang and T. Hamada, *J. Chem. Soc., Chem. Commun.*, 1976, 729.

[103] T. S. Cantrell, *J. Org. Chem.*, 1977, **42**, 4238.

[104] N. Al-Jalal, *J. Chem. Research*, 189, 110.

[105] M. Tada, H. Hamazaki and H. Hirano, *Bull. Chem. Soc. Jpn*, 1982, **55**, 3865.

[106] M. B. Robin, *The Chemistry of the Hydrazo, Azo and Azoxy Groups*, ed. S. Patai, Wiley-Interscience, 1975, 1.

[107] P. S. Engel and D. B. Gerth, *J. Am. Chem. Soc.*, 1981, **103**, 7689.
[108] J. G. Schantl and P. Margaretha, *Helv. Chim. Acta*, 1981, **64**, 2492.
[109] H. Rau and Y. Shen. *J. Photochem. Photobiol. A*, 1988, **42**, 321.
[110] S. Shinkai, M. Miyazaki and O. Manabe, *J. Chem. Soc., Perkin Trans. 1*, 1987, 449.
[111] F. Fariña, M. V. Martin, M. C. Paredes and A. Tito, *Heterocycles*, 1988, **27**, 365.
[112] H. Quast and G. Meichsner, *Chem. Ber.*, 1987, **120**, 1049.
[113] B. Carboni, F. Tonnard and R. Carrie, *Bull. Soc. Chim. Fr.*, 1987, 525.
[114] W. Adam and M. A. Miranda, *J. Org. Chem.*, 1987, **52**, 5498.
[115] W. Adam, M. Dörr, J. Kron and R. J. Rosenthal, *J. Am. Chem. Soc.*, 1987, **109**, 7074.
[116] Z. Majerski, Z. Hamersak and R. Sarac-Arneri, *J. Org. Chem.*, 1988, **53**, 5053.
[117] J. E. Jackson, N. Soundrarajan, M. S. Platz and M. T. H. Liu, *J. Am. Chem. Soc.*, 1988, **110**, 5597.
[118] R. A. Moss, G. Kmiecik-Lawryniwicz and D. P. Cox, *Synth. Commun.*, 1984, **14**, 21.
[119] R. A. Moss, W. Guo, D. Z. Denney, K. N. Houk and N. G. Rondan, *J. Am. Chem. Soc.*, 1981, **103**, 6164.
[120] B. Albert, N. Berning, C. Burschka, S. Huening and F. Prokschy, *Chem. Ber.*, 1984, **117**, 1465.
[121] K. Beck and S. Hünig, *Angew. Chem. Int. Ed. Engl.*, 1986, **25**, 187.
[122] C. H. Kuo and N. L. Wender, *Tetrahedron Lett.*, 1984, **25**, 2291.
[123] U. Burger, Y. Mentha and P. J. Thorel, *Helv. Chim. Acta*, 1986, **69**, 670.
[124] W. Ando, *The Chemistry of Diazonium and Diazo Groups, Part 1*, ed. S. Patai, Wiley-Interscience, 1978, 350.
[125] O. S. Mohamed, H. Dürr, M. T. Ismail and A. A. Abdel-Wahab, *Tetrahedron Lett.*, 1989, **30**, 1935.
[126] X. Creary and M. E. Mehrsheikh-Mohammadi, *Tetrahedron Lett.*, 1988, **29**, 749.
[127] A. Padwa, P. S. Carter, N. Nimmesgern and P. D. Stull, *J. Am. Chem. Soc.*, 1988, **110**, 2894.
[128] S. Ghost, I. Datta, R. Chakrabaory, T. K. Das, J. Sengupta and D. C. Sarkar, *Tetrahedron*, 1989, **45**, 1441.
[129] G. Homberger, A. E. Dorigo, W. Kirmse and K. N. Houk, *J. Am. Chem. Soc.*, 1989, **111**, 475.
[130] W. E. Lee, J. G. Calvert and E. W. Malmberg, *J. Am. Chem. Soc.*, 1961, **83**, 1928.

[131] K. Takahashi, K. L. Kirk and L. A. Cohen, *J. Org. Chem.*, 1984, **49**, 1951.
[132] T. Kamatani, T. Sugahara and K. Fukumoto, *Tetrahedron*, 1971, **27**, 5374.
[133] E. P. Kyba, S.-T. Liu, K. Chocklingam and B. R. Reddy, *J. Org. Chem.*, 1988, **53**, 3513.
[134] E. Leyva, M. S. Platz, G. Persy and J. Witz, *J. Am. Chem. Soc.*, 1986, **108**, 3783.
[135] H. Sawanishi, A. Fujii and T. Tsuchiya, *Chem. Pharm. Bull.*, 1987, **35**, 4101.
[136] H. Sawanishi, A. Fujii and T. Tsuchiya, *Chem. Pharm. Bull.*, 1987, **35**, 4110.
[137] H. Suschitzky, W. Kramer, R. Neidlein and H. Uhl, *J. Chem. Soc., Perkin Trans. 1*, 1988, 983.
[138] J. Frank, G. Stoll and H. Musso, *Liebigs Ann. Chem.*, 1986, 1990.
[139] D. M. B. Hickey, C. J. Moody and C. W. Rees, *J. Chem. Soc., Perkin Trans. 1*, 1986, 1119.
[140] Y. Naruta, T. Yokota, N. Nagai and K. Maruya, *J. Chem. Soc., Chem. Commun.*, 1986, 972.
[141] M. Mitani, O. Tachizawa, H. Takeuchi and K. Koyama, *Chem. Lett.*, 1987, 1029.
[142] O. Klingler and H. Prinzbach, *Angew. Chem. Int. Ed. Engl.*, 1987, **26**, 566.
[143] K. Banert, *Chem. Ber.*, 1987, **120**, 1891.
[144] (a) M. Akhtar, *Adv. Photochem.*, 1964, **2**, 263.
(b) C. N. R. Rao, *The Chemistry of Nitro and Nitroso Groups, Part 1*, ed. H. Feuer, Wiley-Interscience, 1969, 147.
[145] Z. Cekovic and D. Ilijev, *Tetrahedron Lett.*, 1988, **29**, 1414.
[146] (a) D. H. R. Barton, J. M. Beaton, L. E. Geller and M. M. Pechet, *J. Am. Chem. Soc.*, 1960, **82**, 2640.
(b) D. H. R. Barton, J. M. Beaton, L. E. Geller and M. M. Pechet, *J. Am. Chem. Soc.*, 1961, **83**, 4076.
[147] H. Suginome, N. Maeda and M. Kaji, *J. Chem. Soc., Perkin Trans. 1*, 1982, 111.
[148] S. W. Baldwin and H. R. Blomquist, *J. Am. Chem. Soc.*, 1982, **104**, 4990.
[149] C. N. R. Rao, *The Chemistry of Nitro and Nitroso Groups, Part 1*, ed. H. Feuer, Wiley-Interscience, 1969, 91.
[150] H. Suginome, K. Takakuwa and K. Orito, *Chem. Lett.*, 1982, 1357.
[151] (a) H. Suginome and Y. Kurokawa, *Bull. Chem. Soc. Jpn*, 1989, **62**, 1107.

(b) H. Suginome and Y. Kurokawa, *J. Org. Chem.*, 1989, **54**, 5945.
[152] O. L. Chapman, P. G. Cleveland and E. D. Hoganson, *J. Chem. Soc., Chem. Commun.*, 1966, 101.
[153] B. J. Wilkins, G. J. Gainsford and D. E. Moore, *J. Chem. Soc., Perkin Trans, 1*, 1987, 1817.
[154] R. D. Grant and J. T. Pinhey, *Aust. J. Chem.*, 1989, **37**, 1231.
[155] S. T. Reid, J. K. Thompson and C. F. Mushambi, *Tetrahedron Lett.*, 1983, **24**, 2209.
[156] R. Dessauer and J. P. Paris, *Adv. Photochem.*, 1963, **1**, 275 and references therein.
[157] J. R. Peyser and T. W. Flechtner, *J. Org. Chem.*, 1987, **52**, 4645.
[158] R. W. Binkley and T. W. Flechtner, *Synthetic Organic Photochemistry*, ed. W. M. Horspool, Plenum Press, 1984, 375.
[159] J. Hebert and D. Gravel, *Can. J. Chem.*, 1971, **52**, 187.
[160] H. Takeuchi and M. Machida, *Synthesis*, 1989, 206.
[161] K. Yamada, S. Tanaka, K. Naruchi and M. Yamamoto, *J. Org. Chem.*, 1982, **47**, 5283.
[162] K. Yamada, K. Kishikawa and M. Yamamoto, *J. Org. Chem.*, 1987, **52**, 2327.
[163] G. J. Edge, S. H. Iman and B. A. Marples, *J. Chem. Soc., Perkin Trans. 1*, 1984, 2319.
[164] Y. Ogata, K. Takgi and M. Mizuno, *J. Org. Chem.*, 1982, **47**, 3684.
[165] H. Suginome, K. Furukawa and K. Orito, *J. Chem. Soc., Perkin Trans. 1*, 1987, 1004.
[166] Q. Chen and X. Wang, *Xuaxue Xuebao*, 1987, **45**, 340 (*Chem. Abstr.*, 1987, **107**, 236396).
[167] M. L. M. Pennings, G. Okay, D. N. Reinhoudt, S. Harkema and G. J. van Hummel, *J. Org. Chem.*, 1982, **47**, 4413.
[168] D. S. C. Black and L. M. Johnstone, *Aust. J. Chem.*, 1984, **37**, 577.
[169] F. Bellamy and J. Streith, *Heterocycles*, 1976, **4**, 1391.
[170] A. Albini, E. Fasani and O. Buchardt, *Tetrahedron Lett.*, 1982, **23**, 4849.
[171] S. Yamada and C. Kaneko, *Tetrahedron*, 1979, **35**, 1273.
[172] F. Bellamy and J. Streith, *J. Chem. Research*, 1979, 18.
[173] O. Buchardt, P. L. Kumler and C. Lohse, *Acta Chem. Scand.*, 1969, **37**, 159.
[174] O. Buchardt, C. L. Pederson and N. Harrit, *J. Org. Chem.*, 1972, **37**, 3592.
[175] A. Albini, G. F. Bettinetti and G. Minoli, *Org. Synth.*, 1983, **61**, 98.
[176] R. Oberti, A. Albini and E. Fasani, *J. Heterocyclic Chem.*, 1983, **20**, 1007.

[177] H. Kawata, S. Nizuma and H. Kokobun, *J. Photochem.*, 1978, **9**, 463.
[178] O. Buchardt, J. J. Christensen, P. E. Nielson, R. R. Koganty, L. Finsen, C. Lohse and J. Becher, *Acta Chem. Scand.*, 1980, **B34**, 31.
[179] J. Becher, L. Finsen, I. Winckelmann, R. R. Koganty and O. Buchardt, *Tetrahedron*, 1981, **37**, 789.
[180] G. Serra-Errante and P. G. Sammes, *J. Chem. Soc., Chem. Commun.*, 1975, 579; M. N. Akhtar, D. R. Boyd, J. T. Neill and D. M. Jerina, *J. Chem. Soc., Perkin Trans. 1*, 1980, 1693; Y. Ogawa, S. Iwasaki and S. Okuda, *Tetrahedron Lett.*, 1981, **22**, 2277.
[181] J. Streith, J. P. Luttringer and M. Nastasi, *J. Org. Chem.*, 1971, **36**, 2962; J. Streith, *Pure Appl. Chem.*, 1977, **49**, 305; T. Kiguchi, J.-L. Schuppiser, J.-C. Schweller and J. Streith, *J. Org. Chem.*, 1980, **45**, 5095.
[182] T. Tsuchiya, S. Okajima, M. Enkaku and J. Kurita, *Chem. Pharm. Bull.*, 1980, **30**, 3757.
[183] T. Tsuchiya, M. Enkaku, J. Kurita and H. Sawanishi, *J. Chem. Soc., Chem. Commun.*, 1979, 534.
[184] T. J. Barton and G. P. Hussmann, *J. Org. Chem.*, 1985, **50**, 5881.
[185] H. Hiraoka, *Tetrahedron*, 1973, **29**, 2955.
[186] J. A. Barltrop, A. C. Day, P. D. Moxon and R. W. Ward, *J. Chem. Soc., Chem. Commun.*, 1975, 786.
[187] J. A. Barlrop, A. C. Day and R. W. Ward, *J. Chem. Soc., Chem. Commun.*, 1978, 131.
[188] J. A. Barltrop, A. C. Day, A. G. Mack, A. Shahrisa and S. Wakamatsu, *J. Chem. Soc., Chem. Commun.*, 1981, 604.
[189] H. Tiefenthaler, W. Dorscheln, H. Goth and H. Schmid, *Helv. Chim. Acta*, 1967, **50**, 2244.
[190] P. Beak and W. Messer, *Tetrahedron*, 1969, **25**, 3287.
[191] R. H. Good and G. Jones, *J. Chem. Soc. (C)*, 1971, 1196.
[192] A. Padwa, E. Chen and A. Ku, *J. Am. Chem. Soc.*, 1975, **97**, 6484.
[193] D. A. Murature and M. M. de Boertorello, *An. Asoc. Quim. Argent.*, 1981, **69**, 177 (*Chem. Abstr.*, 1981, **95**, 149 527).
[194] J. P. Ferris and F. R. Antonucci, *J. Am. Chem. Soc.*, 1974, **96**, 2014.
[195] M. Maeda and M. Kojima, *J. Chem. Soc., Perkin Trans. 1*, 1977, 239.
[196] C. Riou, J. C. Poite, G. Vernin and J. Metzger, *Tetrahedron*, 1974, **30**, 879.
[197] M. Maeda and M. Kojima, *J. Chem. Soc., Perkin Trans. 1*, 1978, 685.
[198] M. Maeda and M. Kojima, *J. Chem. Soc., Chem. Commun.*, 1973, 539.
[199] B. Singh and E. F. Ullman, *J. Am. Chem. Soc.*, 1967, **89**, 6911; E. F. Ullman, *Acc. Chem. Res.*, 1968, **1**, 353.

[200] J. P. Ferris and R. W. Trimmer, *J. Org. Chem.*, 1976, **41**, 13; K. H. Grellmann and E. Trauer, *J. Photochem.*, 1977, **6**, 365.
[201] S. Wakamatsu, J. A. Barltrop and A. C. Day, *Chem. Lett.*, 1982, 667.
[202] C. Parkanyi, *Bull. Soc. Chim. Belg.*, 1981, **90**, 599.
[203] K. Yoshida, *J. Chem. Soc., Chem. Commun.*, 1978, 1108.
[204] A. Gilbert and S. Krestonosich, *J. Chem. Soc., Perkin Trans. 1*, 1980, 2531.
[205] M. Terashima, K. Seki, C. Yoshida and Y. Kanaoka, *Heterocycles*, 1981, **15**, 1075.
[206] (a) A. P. Komin and J. F. Wolfe, *J. Org. Chem.*, 1977, **42**, 2481.
(b) M. P. Moon, A. P. Komin, J. W. Wolfe and G. F. Morris, *J. Org. Chem.*, 1983, **48**, 2392.
[207] R. Schutz and V. L. Ivanov, *J. Prakt. Chem.*, 1978, **320**, 667.
[208] T. Sugiyama, K. Yagi, Y. Ito and A. Sugimori, *Chem. Lett.*, 1982, 917.
[209] S. Caplain and A. Lablache-Combier, *J. Chem. Soc., Chem. Commun.*, 1970, 1247.
[210] R. D. Chambers, J. M. H. McBride, J. R. Maslakiewicz and K. C. Srivastava, *J. Chem. Soc., Perkin Trans. 1*, 1975, 396; M. A. Fox, D. M. Lemal, D. W. Johnson and J. R. Holman, *J. Org. Chem.*, 1982, **47**, 398.
[211] R. D. Chambers, J. R. Maslakiewicz and K. C. Srivastava, *J. Chem. Soc., Perkin Trans. 1*, 1975, 1130.
[212] N. Ivanoff, F. Lahmani, M. Magat and M. P. Pileni, (*Chem. Abstr.*, 1973, **79**, 131 331).
[213] J. A. Barltrop and A. C. Day, *J. Chem. Soc., Chem. Commun.*, 1975, 177.
[214] Y. Kobayashi, I. Kumadaki, A. Ohsawa and Y. Sekine, *Heterocycles*, 1977, **6**, 1587; Y. Kobayashi, K. Kawada, A. Ando and I. Kumadaki, *Heterocycles*, 1983, **12**, 174.
[215] T. Takagi and Y. Ogata, *J. Chem. Soc., Perkin Trans. 2*, 1979, 402.
[216] J. Brennan, *J. Chem. Soc., Chem. Commun.*, 1981, 880.
[217] A. R. Katritzky and H. Wilde, *J. Chem. Soc., Chem. Commun.*, 1975, 770.
[218] K. Takagi and Y. Ogata, *J. Chem. Soc., Perkin Trans. 2*, 1977, 1148; Y. Ogata and K. Takagi, *J. Org. Chem.*, 1978, **43**, 944.
[219] K. E. Wilzbach and D. J. Rausch, *J. Am. Chem. Soc.*, 1970, **92**, 2178.
[220] M. G. Barlow, R. N. Haszeldine and J. G. Dingwall, *J. Chem. Soc., Perkin Trans. 1*, 1973, 1542.
[221] R. D. Chambers and R. Middleton, *J. Chem. Soc., Perkin Trans. 1*, 1977, 1500.

# 6

# Halogen-containing compounds

This chapter deals with the photochemical reactions of organohalogen compounds and the photoreactivity of the compounds in this chapter is therefore dominated by the fission of a C—halogen bond. The halogen compounds discussed have the halo group in a variety of environments such as attached to a saturated group as in the alkyl halides, to an alkene as in vinyl halides or to an aryl group as with the aryl halides.

## 6.1 ALKYL HALIDES

### 6.1.1 Spectroscopic data

The longest-wavelength band in the ultraviolet spectra of alkyl halides is a continuum and corresponds to an $n\sigma^*$ transition from a non-bonding *p*-orbital on the halogen atom to an antibonding $\sigma^*$ orbital of the C—halogen bond. The position of the absorption is dependent on the electronegativity of the halogen atom changing from fluorine to iodine. Typical examples are $CH_3Cl$, $\lambda_{max}$ 173 nm $\varepsilon = 200$ dm$^3$ mol$^{-1}$ cm$^{-1}$, $CH_3Br$, $\lambda_{max}$ 202 nm $\varepsilon = 264$ dm$^3$ mol$^{-1}$ cm$^{-1}$, and $CH_3I$, $\lambda_{max}$ 258 nm $\varepsilon = 378$ dm$^3$ mol$^{-1}$ cm$^{-1}$. As can be seen, the extinction coefficients are usually low (*ca.* 300) but increase with increasing halogen content, e.g. $CH_2Br_2$ $\lambda_{max}$ 220 nm $\varepsilon = 1100$ dm$^3$ mol$^{-1}$ cm$^{-1}$, $CHBr_3$, $\lambda_{max}$ 224 nm $\varepsilon = 2130$ dm$^3$ mol$^{-1}$ cm$^{-1}$; $CH_2I_2$, $\lambda_{max}$ 290 nm $\varepsilon = 1320$ dm$^3$ mol$^{-1}$ cm$^{-1}$; $CHI_3$, $\lambda_{max}$ 394 nm $\varepsilon = 2170$ dm$^3$ mol$^{-1}$ cm$^{-1}$ [1]. Each of the iodoalkanes has a second strong absorption in the vacuum ultraviolet around 194 nm. This is localized on the halogen atom and has been assigned to a $p \rightarrow s$ transition where a non-bonding electron on iodine is promoted to the next higher orbital. Such transitions have been identified as of Rydberg type [2]. Higher-energy bands in the spectrum of chloromethane have also been assigned as Rydberg transitions [3].

### 6.1.2 Photochemical reactions of alkyl halides

The photochemistry of the simple haloalkanes is a simple efficient dissociative process when irradiation of the compound is carried out into the first absorption band. In flash photolysis experiments the free halogen has been detected indirectly by use of scavengers [4]. The free methyl radical has also been observed following the flash photolysis of bromo- and iodomethane [5]. In all the cases where quantitative measurements have been carried out, the C—halogen fission process has had a quantum yield of unity. The alkyl radical formed by this process has a large share of the available energy. This makes the radical 'hot' and it is rich in electronic, vibrational and translational energy [6]. This reactive radical becomes involved in secondary processes such as abstraction of an atom from another molecule or from solvent [7], or alkene formation [8]. As with the simple haloalkanes where photolysis leads to the fission of the weakest bond, the same rule can be applied to polyhaloalkanes, with irradiation into the first absorption band leading to the production of free radicals. Thus trifluoroiodomethane on irradiation at $\lambda > 400$ nm affords iodine atoms and the trifluoromethyl radical [9]. Similar behaviour is observed for other polyhalogenated compounds, and when irradiations are carried out in the presence of alkenes, the result is addition of the polyhaloalkane to the alkene [10]. Such reactions are well-documented [11] and some examples of synthetic value are shown in **1** [12–14]. Interestingly the irradiation of bromoform yields the $CHBr_2$

$$CH_3CH{=}CH_2 \xrightarrow[CCl_4]{h\nu} CH_3CHClCH_2CCl_3 \quad 55\% \qquad \text{Ref [12a]}$$

$$CH_3CH_2OCH{=}CH_2 \xrightarrow[CCl_4]{h\nu} CCl_3CCH_2CHClOCH_2CH_3 \quad 92\% \qquad \text{Ref [12b]}$$

$$CH_3(CH_2)_5CH{=}CH_2 \xrightarrow[CBr_4]{h\nu} CH_3(CH_2)_5CHBrCH_2CBr_3 \quad 88\% \qquad \text{Ref [13]}$$

$$CH_3CH{=}CH_2 \xrightarrow[Br_3CCl]{h\nu} CH_3CHBrCH_2CCl_3 \quad 62\% \qquad \text{Ref [14a]}$$

$$CH_2{=}CHCH_2Cl \xrightarrow[Br_3CCl]{h\nu} CBr_3CH_2CHClCH_2Cl \quad 82\% \qquad \text{Ref [14b, c]}$$

(1)

$$\text{cyclopentene} \xrightarrow[Br_3CCl]{h\nu} \text{2-chloro-1-(tribromomethyl)cyclopentane (62\%)}$$

Ref [14b, d]

$$CF_3CF{=}CF_2 \xrightarrow[ICF_3]{h\nu} CF_3CFICF_2CF_3$$

$$CF_3CH{=}CF_2 \xrightarrow[ICF_3]{h\nu} (CF_3)_2CHCF_2I$$

Ref [14e]

(1 continued)

radical by fission of a C—Br bond (**2**), while the irradiation of chloroform

$$CHBr_3 \xrightarrow{h\nu} \dot{C}HBr_2 + Br\bullet$$

$$\dot{C}HBr_2 \xrightarrow{RCH=CH_2} RCHBrCH_2CHBr_2$$

(2)

brings about fission of the C—H bond affording the trichloromethyl radical (**3**) [15]. Addition to alkynes is also of synthetic value, and some examples

$$HCCl_3 \xrightarrow{h\nu} \dot{C}Cl_3 + H\bullet$$

$$\dot{C}Cl_3 \xrightarrow{RCH=CH_2} RCH_2CH_2CCl_3$$

(3)

of this are shown in **4** [16–19].

$$CH_3(CH_2)_4C{\equiv}CH \xrightarrow[BrCCl_3]{h\nu} CH_3(CH_2)_4CBr{=}CHCCl_3$$

$$CF_3C{\equiv}CH \xrightarrow[ICCl_3]{h\nu} CF_3CI{=}CHCCl_3$$

(4)

Convincing evidence has been obtained in some cases that irradiation of the halo compound does not always yield free radicals and in these cases the carbon—halogen bond undergoes heterolysis. This is the case in the irradiation of 2-iodonorbornane which affords the two products shown (**5a**) [20a]. The formation of the rearrangement product is evidence for the participation of a carbocation intermediate. The carbocation could be the result of electron transfer within the initially formed radical pair following homolysis of the C—I bond. When the reaction of 4-iodonorbornane (**5b**) is carried out in methanol the methyl ether is formed, again confirming the intervention of a carbocation. This product is accompanied by norbornane formed by hydrogen abstraction by the norbornyl radical [20b]. The bromo derivative (**5c**) is also reactive, but the carbocation path only contributes in

hν, 254 nm, Ether: 81% + 19% [5a]

hν, MeOH, 254 nm: MeO product n = 1, 76%; n = 2, 90% + n = 1, 12%; n = 2, 2% [5b]

hν, MeOH: MeO product 30% + 55% [5c]

(5)

a small way to the reaction and the principal product from the process is norbornane formed by a radical path [20b]. Similar observations have been made with the diiodoadamantane (**6**) where irradiation in the presence of

hν, MeOH: 5% + 95%

(6)

methanol affords the dimethoxyadamantane derivative as the principal product [21]. Monoiodoadamantane (**7a**) behaves in a similar manner via a

I hν MeOH/NEt$_3$ OMe + + +

38% 30% 7% 15%

(7a)

cation. Rearrangement of this occurs when the irradiation is carried out in the presence of triethylamine when, in addition to the methoxy ether, and the reduction product, adamantane, the rearrangement products shown in **7a** are formed [22]. Other examples (**7b**) of the photoreactivity of haloalkanes

I I hν MeOH I H + I + MeO OMe (via cation)

3% 8% 45% Ref [23b]

$CH_2I$ hν via cation Ref [23c]

hν THF/Et$_3$N + +

I 56% 21% 20% via cation Ref [24b]

hν n-pentane I 70% + + +

4% 46% 45% 2% via cation Ref [24a]

(7b)

$CH_2I_2$ + $R^1R^3C{=}CR^2R^4$ $\xrightarrow[CH_2Cl_2]{h\nu}$ cyclopropane ($R^1$, $R^2$, $R^3$, $R^4$)

$R^1 = R^2 = H, R^3 = R^4 = (CH_2)_4$ 83% Ref [23b]
$R^1 = R^2 = Et, R^3 = R^4 = H$ 83%
$R^1 = R^4 = Et, R^2 = R^3 = H$ 84%
$R^1 = R^2 = R^3 = R^4 = Me$ 83%
$R^1 = Bu^t, R^2 = R^3 = R^4 = H$ 80%

(8)

have been reported which yield products by the intervention of carbocations [23, 24]. Within this group of reactions it was originally reported [23a] that irradiation of 1-iodo-4-phenylbutane failed to undergo formation of a cation but this was subsequently shown to be erroneous [24b]. Another point of interest is the ring expansion reaction involving the norbornane derivative. Processes such as this and the cyclization reactions are clear evidence for the intervention of cations. Additionally the irradiation of *gem*-diiodides (**8**) which cannot undergo rearrangement is worthy of mention. Here, the irradiation of diiodomethane in the presence of alkenes affords an efficient and stereospecific route to cyclopropanes. The reaction is not affected by steric factors. It is also of interest that the irradiation of the diodocubane (**9**)

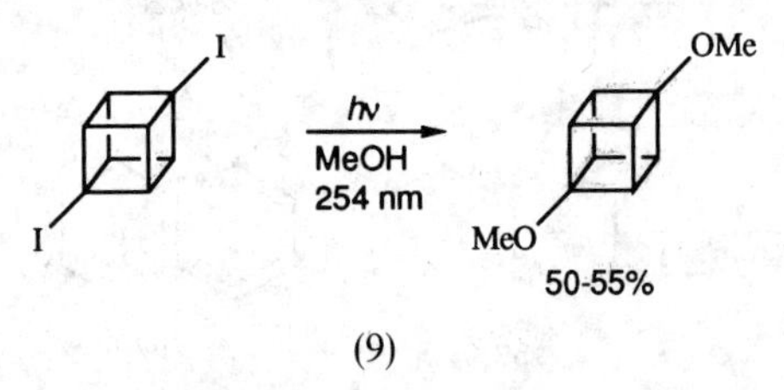


(9)

in methanol affords the corresponding dimethoxycubane, presumably via a carbocationic path [25]. Aromatic cyclohexylation occurs on irradiation of cyclohexyl iodide in benzene again, providing confirmation of the involvement of a carbocation [26].

## 6.2 VINYL HALIDES

With the introduction of a halogen at a vinyl site in an alkene, the $\pi\pi^*$ absorption band of the alkene moves to longer wavelengths [27]. However, in the simpler compounds the lowest transition is still of the $\pi\pi^*$ type, with an excited state of intermediate character where both the C—halogen and the C—C bonds have antibonding character [1].

### 6.2.1 Photochemical reactions of vinyl halides

Two reaction paths have been identified in the gas phase irradiation of a simple vinyl halide such as 1,2-dichloroethene (**10**). These processes are (a)

$$ClCH{=}CHCl \xrightarrow{h\nu} \dot{C}HCHCl + Cl\bullet \longrightarrow CH{\equiv}CH + Cl\bullet$$

$$ClCH{=}CHCl \xrightarrow{\lambda > 220\ nm} CH{\equiv}CCl + HCl$$

(10)

C—Cl fission to yield a chlorine atom and a chlorovinyl radical and (b) molecular elimination of HCl to yield chloroethyne. Thus with broad-spectrum light these two processes occur in a ratio of 9:1 while with $\lambda > 200$ nm the ratio is 3.3:1 [28, 29]. In the pure liquid this alkene undergoes *cis–trans*-isomerization when irradiated into the band at 313–366 nm [30]. (see Chapter 2 for further examples of this isomerization process). A laser flash study of the photochemical behaviour of the halovinyl compounds (**11**)

MeO, OMe, OMe, Br $\xrightarrow[347.1\ nm]{h\nu}$ MeO—C$^{+}$=, OMe, OMe

(11)

has shown that irradiation at 347.1 nm affords transients which have been identified as the carbocationic species [31].

This heterolytic process of vinyl halides has been exploited as a method for the generation of vinyl cations and a typical example of this is found with the iodoalkenes (**12**). Irradiation at 254 nm of these alkenes in methanol

hν
254 nm, MeOH
27%
OMe
65%
hν
254 nm
$HOCH_2CH_2OH$
50%
O O
30%

(12)

affords the products shown [32]. The route to these products involves the homolytic fission of the C—I bond followed by electron transfer to afford a vinyl cation which is subsequently trapped to afford the products depicted. However, radical paths can also be operative and result in the formation of reduction products. Thus the compounds studied in **12** give products from both the radical and the ionic paths. In general, for the examples where an ion is generated within a ring system, there is evidence for dependence on ring size, and nucleophilic trapping occurs in competition with deprotonation.

Vinyl cations are also important intermediates in the photolysis of 1,1-diphenyl-2-haloethenes **13** [33], where the dependence on the halogen

Ph, Ph, H, F $\xrightarrow{h\nu}$ Ph, Ph, H, F + Ph, F, H, Ph

$Et_2O$, 13% 9%
$C_6H_{12}$, 28% 23%

Ph, Ph, H, Cl $\xrightarrow{h\nu}$

Ph, Ph, H (•) ⟶ Ph, Ph + Ph, Ph, Ph, Ph

$Et_2O$, 28% 10%
$C_6H_{12}$, - -

Ph, Ph, H (+) ⟶ Ph—C≡C—Ph

$Et_2O$, 25%
$C_6H_{12}$, 16%

Ph, Ph, H, Br $\xrightarrow{h\nu}$

Ph, Ph, H (•) ⟶ Ph, Ph + Ph, Ph, Ph, Ph

$Et_2O$, 34% 10%
$C_6H_{12}$, - -

Ph, Ph, H (+) ⟶ Ph—C≡C—Ph + Ph, Ph

$Et_2O$, 23% -
$C_6H_{12}$, 3% 12%

\+ [phenanthrene]

\-
5%

Ph, Ph, H, I $\xrightarrow{h\nu}$

Ph, Ph, H (•) ⟶ Ph, Ph + Ph, Ph, Ph, Ph

$Et_2O$, 51% 6%
$C_6H_{12}$, 51% 6%

Ph, Ph, H (+) ⟶ Ph—C≡C—Ph

$Et_2O$, 19%
$C_6H_{12}$, 9%

(13)

involved was studied, and 1,1-diaryl-2-halo-propenes (**14**) [34]. The 2-

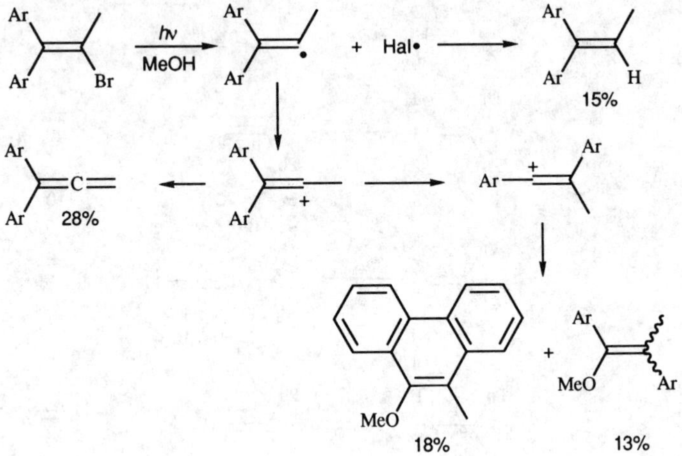

(14)

halogeno-1,1-diphenylethenes (**13**) afford a variety of products dependent upon the halogen and the solvent in which the reactions are carried out. This work has highlighted the fact that the C—F bond does not undergo fission and only rearrangement occurs. The absence of C—F bond fission is also demonstrated in example **15** where irradiation yields products of

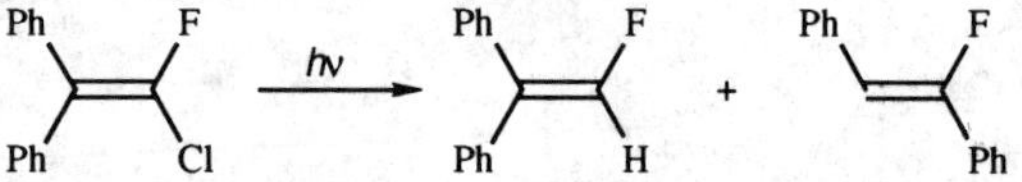

(15)

dechlorination and rearrangement. Both radical and cationic intermediates are involved in the formation of products [35]. The chloro-, bromo-, and iodo-derivatives in (**13**) all give substantial amounts of products derived from a cationic intermediate implicating an electron transfer after homolysis of the C—halogen bond. The electron transfer occurs while the radical pair are in close proximity. Electron transfer then yields the cation and the corresponding halide. The preference for which path is followed is dependent upon a variety of factors such as the halide and the nature of the substitution in the aryl groups. The reduction product 1,1-diphenylethylene arises from the radical path, whereas the formation of the allenes, stilbenes and phenanthrenes comes from the intermediacy of a vinyl cation, deprotonation affording the allenes, 1,2-phenyl migration leading to the stilbenes and electrophilic substitution affording the phenanthrene products. Similar reactions are observed in **14**. The influence of ring size on the mode of fission of vinyl bromides (**16**) has also been evaluated and in the seven-ring system,

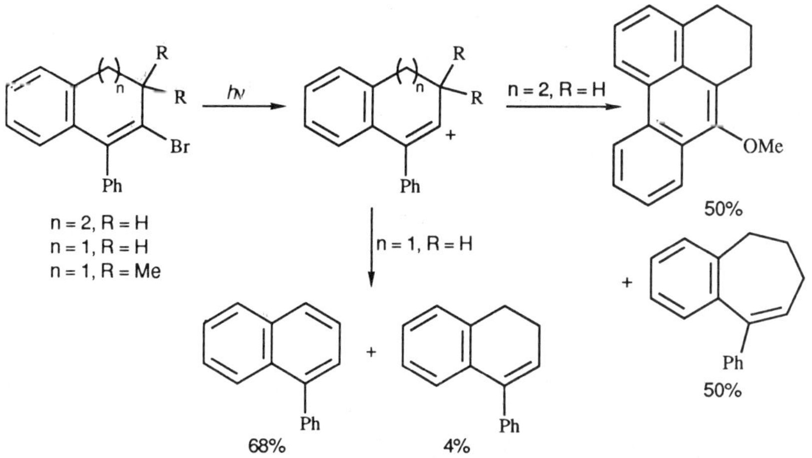

(16)

where there is more flexibility, both radical and cationic paths are operative to an equal extent, but in the more rigid six-membered ring examples, the ionic path predominates [36]. Other studies have shown that the percentage of products derived from a cationic intermediate can be enhanced greatly if the irradiation is carried out in the presence of copper(II) acetate (**17**) where

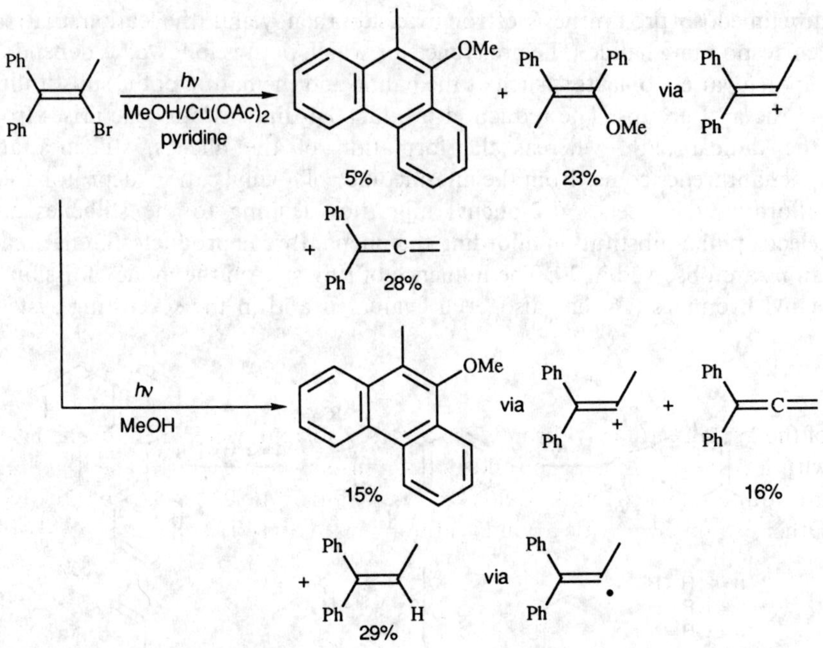

(17)

vinylcopper intermediates are proposed [37]. In the presence of copper(II) acetate no reduction products are obtained.

Examples of some synthetic value for this reaction type involving the production of cations by irradiation of vinyl halides have been reported. Thus the irradiation of **18** in the presence of azide ion affords the corre-

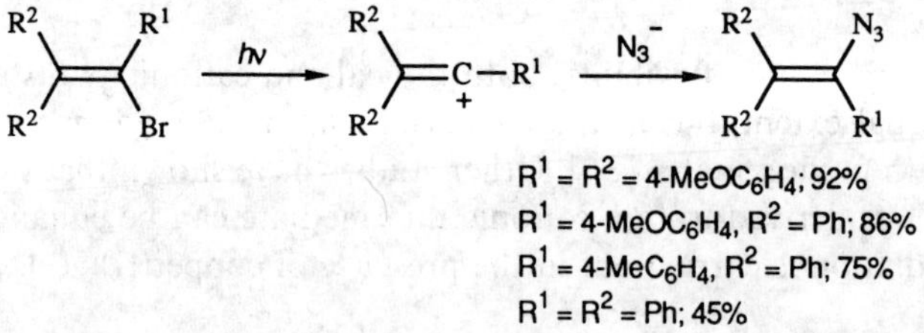

$R^1 = R^2 = 4\text{-}MeOC_6H_4$; 92%
$R^1 = 4\text{-}MeOC_6H_4$, $R^2 = Ph$; 86%
$R^1 = 4\text{-}MeC_6H_4$, $R^2 = Ph$; 75%
$R^1 = R^2 = Ph$; 45%

(18)

sponding azides [38]. The formation of products by the ionic path is favoured by the presence of electron-donating groups [35]. Another example has been developed for the synthesis of isoquinolinones (**19**) and involves irradiation

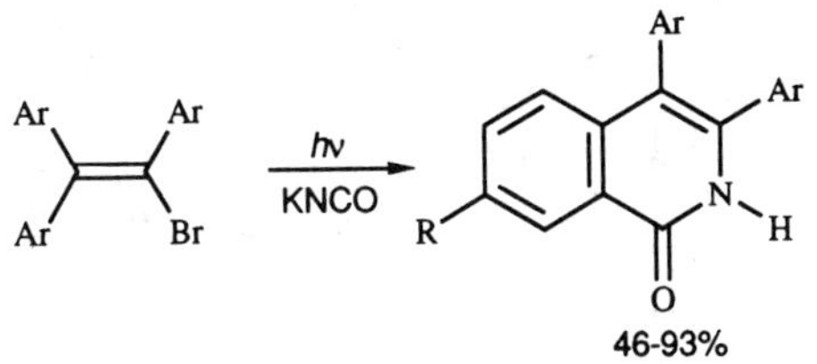

(19)

of the appropriate vinyl bromide in a mixed solvent, water/dichloromethane with a phase-transfer agent and potassium isocyanate [39]. The reactions are quite efficient and yields in the range 46–93% are obtainable. Other examples provide a route to thiophenes (**20**) [40] and furans (**21**) [41].

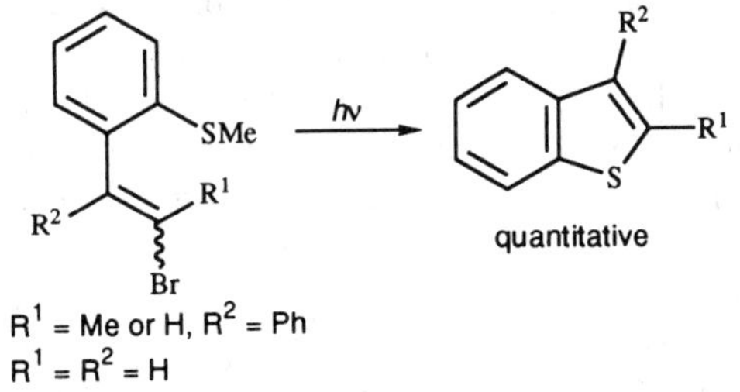

$R^1$ = Me or H, $R^2$ = Ph
$R^1$ = $R^2$ = H

quantitative

(20)

> 90%

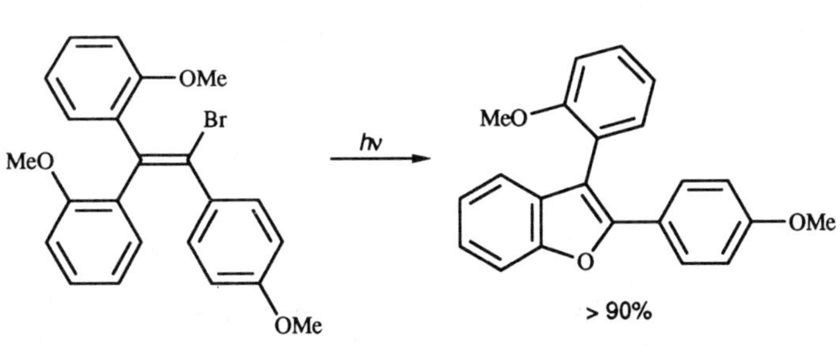

(21)

The vinyl bromide (**22**) is photochemically reactive (Pyrex filter) in a

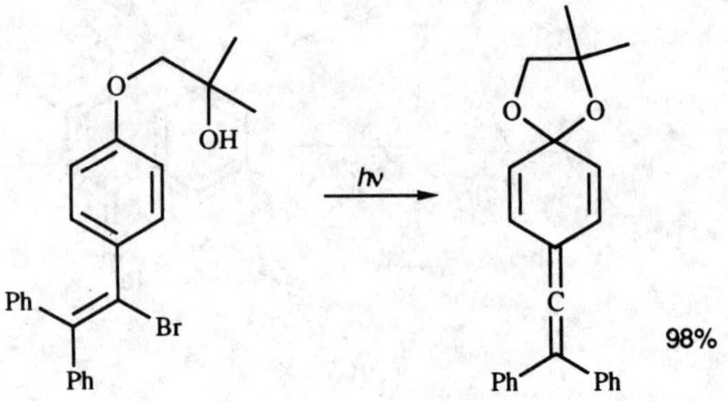

(22)

propan-2-ol/dichloromethane/sodium isopropoxide mixture and yields spiroallenes [42].

Some interest has also been shown in the photochemical debromination of 1,2-dibromo-1,2-diphenylethane and related compounds as a route to alkenes [43] and phase-transfer agents have proved useful to afford high yields. 1,2-Dichloroalkenes have been treated similarly [44]. Other work has shown that dihalides (**23**) are also reactive in methanol, resulting in reduction

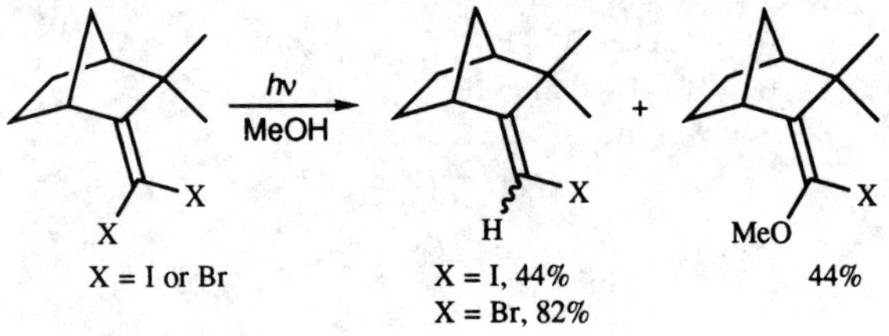

(23)

(a radical path) and ether formation (a cationic path). Lowering the temperature of the reaction enhances the yield of the ether product from the irradiation of the diiodo derivative but had the reverse effect with the dibromo compound. No ionic products were obtained from the irradiation of the monohalo compounds formed in these reactions [45].

## 6.3 ARYL HALIDES

### 6.3.1 Spectroscopic properties

Like the majority of benzenoid compounds (see Chapter 2) the halo-substituted compounds have two principal absorption bands in the ultraviolet region of the spectrum with the primary band at around 203 nm (iodobenzene 207 nm, $\varepsilon = 7000\ dm^3\ mol^{-1}\ cm^{-1}$; chlorobenzene 209.5 nm, $\varepsilon = 7400\ dm^3\ mol^{-1}\ cm^{-1}$; and bromobenzene 210 nm, $\varepsilon = 7900\ dm^3\ mol^{-1}\ cm^{-1}$) and the secondary band ($^1L_b$) around 254 nm (iodobenzene 257 nm, $\varepsilon = 700\ dm^3\ mol^{-1}\ cm^{-1}$, chlorobenzene 263.5 nm, $\varepsilon = 190\ dm^3\ mol^{-1}\ cm^{-1}$, and bromobenzene 261 nm, $\varepsilon = 192\ dm^3\ mol^{-1}\ cm^{-1}$) [46, 47]. The lower-energy band in the spectra arises from a $\pi\pi^*$ transition superimposed on a $\sigma\sigma^*$ transition associated with the halogen which has a tail, in the bromo and iodo cases, extending beyond 254 nm. Thus direct irradiation at 254 nm leads to dissociation of the C—halogen bond and the formation of aryl radicals and halo atoms. There is some doubt, however, as to which of the transitions, $\pi\pi^*$ or $\sigma\sigma^*$, is lower in energy, although evidence has been collected which implicates the $\sigma^*$ antibonding orbital as being lower [48].

### 6.3.2 Photochemistry of aryl halides

Irradiation of halo-aromatics at around 254 nm results usually in homolysis of the C—halogen bond to afford aryl radicals and halo atoms. Among the earliest examples of this behaviour is the photochemically induced fission of iodobenzene (**24**), which has been known for about 30 years [49]. Bromoben-

R—C₆H₄—I $\xrightarrow[254\ nm]{h\nu}$ R—C₆H₄• + I•

(24)

zene [50] and chlorobenzene [51] are also photoreactive in the same manner and the photochemistry of these compounds has been reviewed extensively [52].

Chlorobenzene is, however, somewhat different from the bromo and iodo derivatives in that it does not have a low-lying $\sigma^*$ orbital. However, irradiation of chlorobenzene does bring about fission of the C—halogen bond via a triplet $\pi\pi^*$ state [53]. Fluoro-substituted arenes are also photoreactive, but usually the C—F bond remains intact and other reaction paths are operative [54]. Substantiation of the proposal that the triplet state was responsible for

C—chlorine fission was obtained from various studies using sensitizers such as benzene [53] and alkyl ketones [55]. However, there is some doubt as to the nature of the energy-transfer process, and alternative mechanisms may be operative. One such alternative is the involvement of an exciplex. This was considered and discarded when it was demonstrated that 1-chlorophenyl-3-phenylpropane (**25**) did show exciplex emission but the dechlorination

$$\mathrm{Ph{-}(CH_2)_3{-}C_6H_4{-}Cl} \xrightarrow{h\nu} [\text{exciplex}]$$

(25)

reaction was inefficient and that the exciplex formation was an energy-wasting path [56]. Regardless, a variety of chlorobenzenes (**26**) have been shown to

$$\mathrm{R{-}C_6H_4{-}Cl} \xrightarrow[C_6H_6]{h\nu} \mathrm{R{-}C_6H_4^{\bullet}} + \mathrm{Cl^{\bullet}} \longrightarrow \mathrm{R{-}C_6H_4{-}C_6H_5}$$

(26)

undergo C—halogen fission in benzene to provide a route to biphenyls [57].

Dechlorination is an important path for many benzene derivatives. Thus dichloro-(**27**) [58, 59] and trichloro-(**28**) [60, 61] benzene compounds undergo

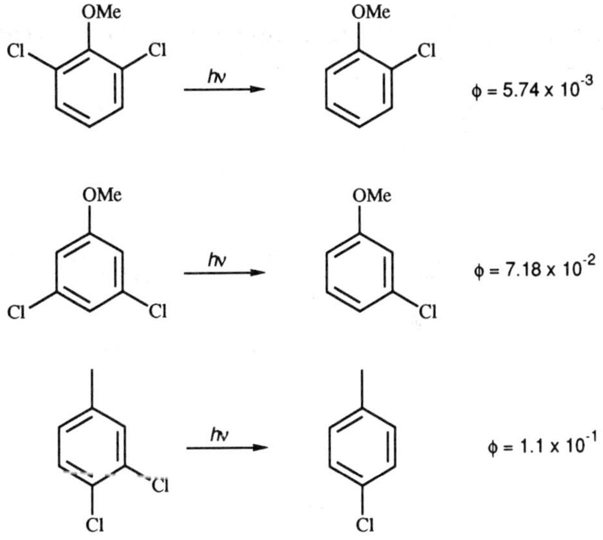

OMe Cl Cl $\xrightarrow{h\nu}$ OMe Cl $\phi = 5.74 \times 10^{-3}$

OMe Cl Cl $\xrightarrow{h\nu}$ OMe Cl $\phi = 7.18 \times 10^{-2}$

Cl Cl $\xrightarrow{h\nu}$ Cl $\phi = 1.1 \times 10^{-1}$

(27)

Cl Cl Cl $\xrightarrow{h\nu}$ Cl Cl + Cl Cl

Cl Cl Cl $\xrightarrow[MeOH]{h\nu}$ Cl Cl + Cl Cl + Cl

72.3% 13.3% 8%

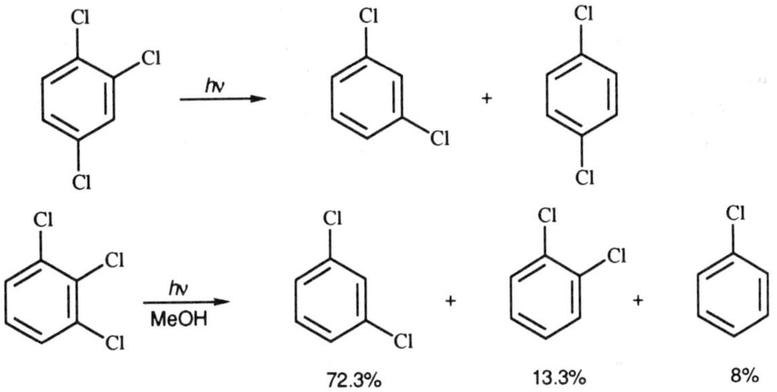

(28)

monodechlorination in a variety of solvents to yield the corresponding mono- and di-chloro derivatives as illustrated. The photochemical behaviour of the chlorinated biphenyls are also of interest and likewise undergo dechlorination from the triplet state [62, 63]. The reactions shown in **29** are typical of the

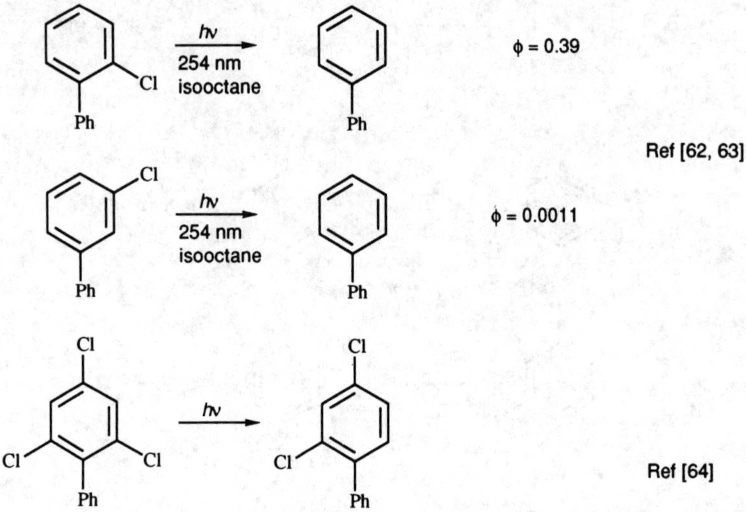

$\phi = 0.39$

$h\nu$, 254 nm, isooctane

Ref [62, 63]

$\phi = 0.0011$

$h\nu$

Ref [64]

(29)

reactions encountered where there is preferential loss of the *ortho*-chlorine atom, a feature also exhibited by 1,2,3-trichlorobenzene [64].

Some chlorobenzene derivatives, e.g. 4-chlorobenzonitrile, do not undergo dechlorination on direct irradiation in methanol. However, it is possible to bring about the reaction on irradiation in the presence of triethylamine. The reaction arises as shown in **30** and involves an electron-transfer reaction

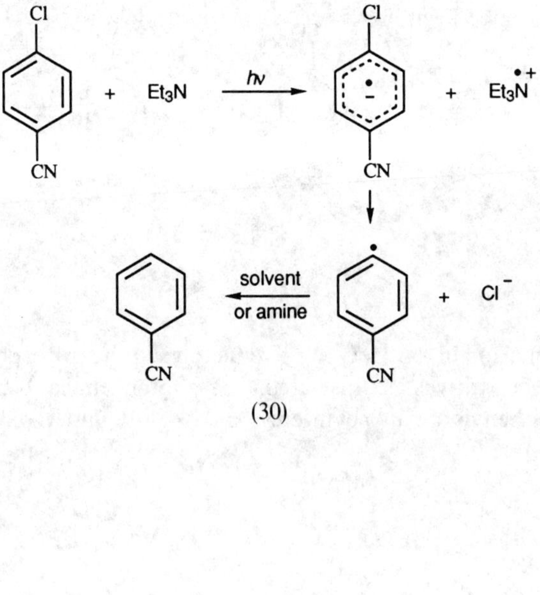

$+ Et_3N \xrightarrow{h\nu} + Et_3N^{\bullet +}$

solvent or amine

$+ Cl^-$

(30)

followed by expulsion of chloride [65a]. There is good evidence that electron-transfer mechanisms are involved in aryl halide photochemistry when irradiation is carried out in the presence of electron donors such as tertiary amines. Other studies have demonstrated that 1-chloronaphthalene undergoes photoreduction in the presence of donors such as triethylamine with a seven-fold enhancement of the quantum efficiency [65b]. Similar behaviour is observed for 1-chlorobiphenyl [66]. Evidence also indicates that the reduction of 1-chloronaphthalene to naphthalene is concentration-dependent and that the efficiency of the reduction increases with increasing concentration of reactant. The use of ammonia or amines as the electron-donating reagent has been made in a variety of systems. Thus triethylamine induces the photodehalogenation of chlorinated biphenyls [62], chlorinated *p*-terphenyls [67], and bromobiphenyls [66].

Other reactions which involve the formation of the radical anion of the aromatic compound have been reported [68] Typical of this is the reaction of iodobenzene with thiophenolate in liquid ammonia which gives diphenyl sulphide (**31**) [69]. When alkyl thiolates are used, there is a competing reaction

S⁻ + I → hν → S 94%

(31)

in which the alkyl-sulphur bond is cleaved, and one product from the reaction is the arylthiolate. A related reaction with diethyl phosphite has been shown to give arylphosphonates in good yield (**32**), and the quantum

I, Cl + $(EtO)_2PO^-$ → hν, 350 nm → $PO(OEt)_2$, Cl 91% ($\phi$>> 1)

(32)

yield for their formation is much greater than unity, indicating that a radical chain process is operative [70].

One of the most useful reactions in this category is that in which the nucleophile is an enolate anion. The basic reaction mechanism is illustrated in **33** [71]. Other examples of the same general process are shown in **34** [72].

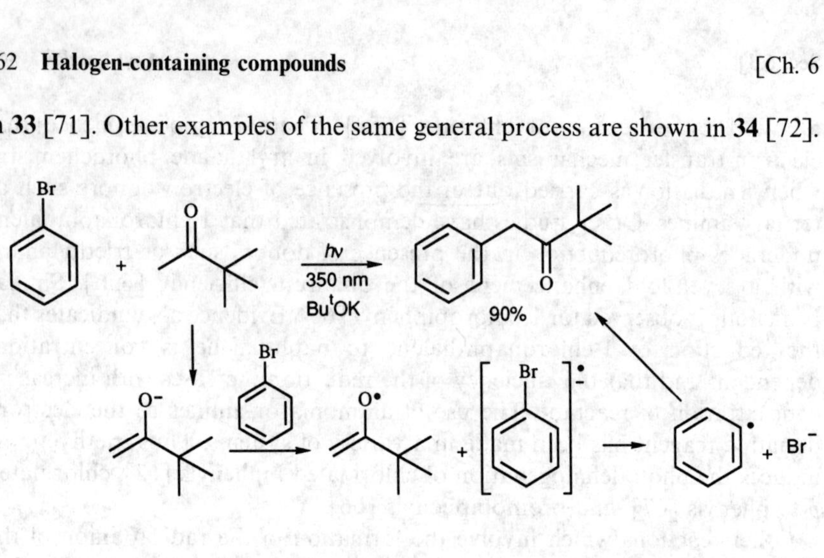

(33)

| | | | | |
|---|---|---|---|---|
| PhBr | + | $MeCH_2COCH_2Me$ | ⟶ | $PhCH(Me)COCH_2Me$ 80% |
| PhI | + | $Me_2CHCOCHMe_2$ | ⟶ | $PhC(Me)_2COCHMe_2$ 72% |

(34)

There have been a number of applications of this *modus operandi* leading to intramolecular cyclization and the synthesis of heterocyclic compounds (**35**)

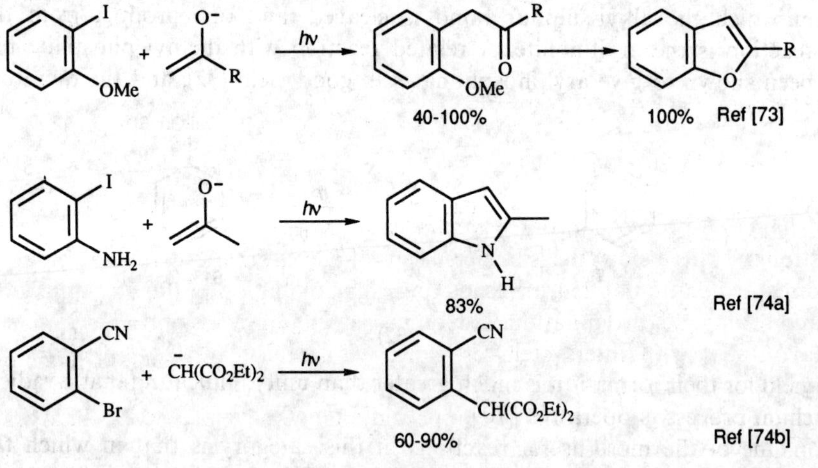

40-100% 100% Ref [73]

83% Ref [74a]

60-90% Ref [74b]

(35)

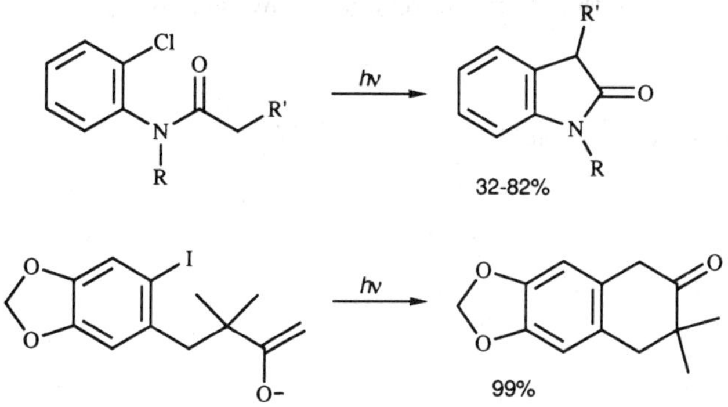

32-82% Ref [75]

99% Ref [76]

(35 continued)

[73–76]. An extension of the intramolecular version of this reaction has been put to good use in the synthesis of cephalotoxin (**36**) [77].

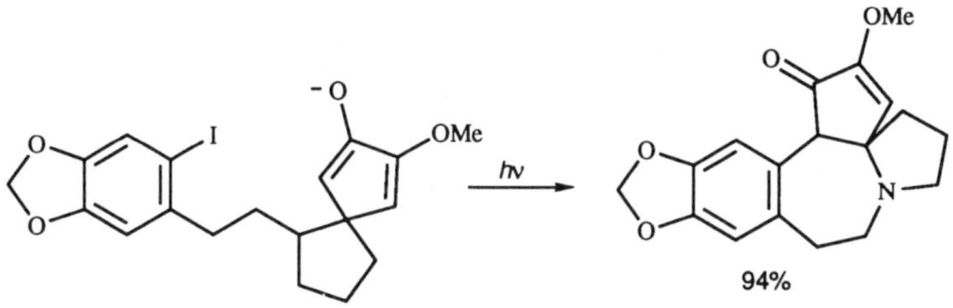

94%

(36)

The aryl-bromo bonds are weaker than aryl-chloro bonds and the bond strength falls below the singlet energy of the aromatic compounds. Thus direct irradiation at 254 nm leads to direct population of the $\sigma\sigma^*$ transition and resultant debromination. The singlet-state bond-fission process competes favourably with intersystem crossing. While this can also apply to the aryl-iodo compounds, the energy of the C-iodine bond is lower than the triplet energy of the aromatic compound and so the triplet state will also be operative. The earliest studies in this showed clearly that light-induced homolysis of halo-arene bonds had synthetic potential [49]. A typical example of the value is the synthesis of *p*-terphenyl (**37**) by irradiation of

Ph—C$_6$H$_4$—I $\xrightarrow[\text{254 nm, } C_6H_6]{h\nu}$ Ph—C$_6$H$_4$—Ph 90%

(37)

4-iodobiphenyl in benzene [78]. The reaction lends itself to a variety of applications, such as the intramolecular cyclization to yield phenanthrene (**38**) [79] or fluoranthrene (**39**) [80]. Considerable use has been made of the

$\xrightarrow{h\nu}$ 90%

(38)

$\xrightarrow{h\nu}$ 72%

(39)

intramolecular process as a route to compounds of synthetic value as shown in **40** [81].

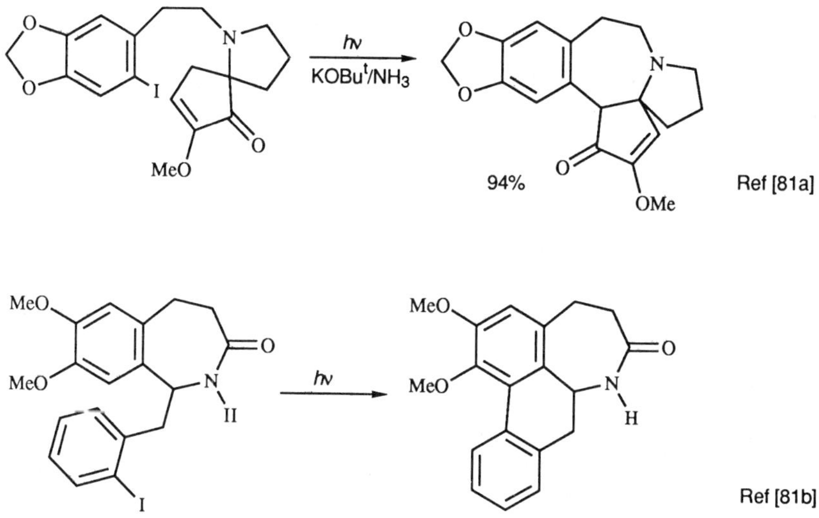

94% Ref [81a]

Ref [81b]

(40)

Reference has already been made in Chapter 2 to stilbene photocyclization that leads to a phenanthrene without the need for oxidation, owing to ready elimination of HX. In principle, reactions of *o*-iodoaromatic compounds such as that of **41**, leading to the methyl ester of the anti-tumour agent *aristolochic*

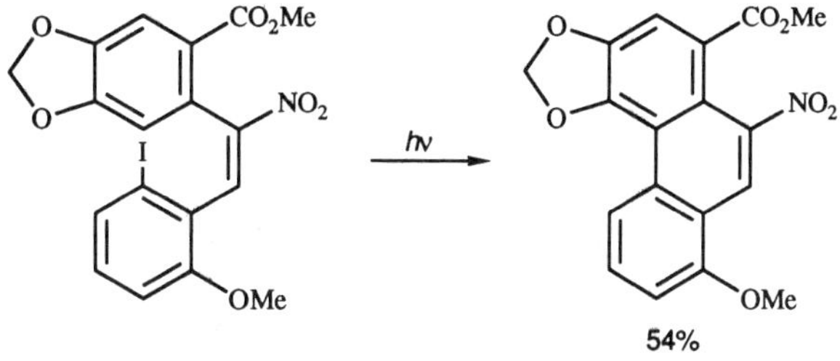

54%

(41)

acid [82] might fit into this category. However, it is more likely that the primary photochemical process is the cleavage of the carbon—halogen bond. Related reactions starting with 1-aroylisoquinolines have been used in the synthesis of aporphine alkaloids (e.g. **42**) [83], and *o*-halobenzoylenamides

MeO MeO I MeO OMe N O $h\nu$ MeO MeO MeO OMe N O 38%

(42)

undergo similar reactions [84]. *o*-Iodobenzanilides usually undergo photo-cyclization only inefficiently but a study [85] of *o*-chlorobenzanilides (**43**)

O N H Cl R $h\nu$ O N H R 67-74%

(43)

suggests that bond homolysis is assisted by the second aryl group and the chloro-compounds are more useful substrates than the bromo- or iodo-analogues in this system.

A variety of reactions have been reported that involve cyclization to an aromatic ring from a position (often in another aromatic ring) bearing a halogen substituent. These reactions cannot occur by a six-electron path because there are saturated carbon atoms in the bridging unit. Many of these have been used in alkaloid synthesis, and a typical example is shown in reaction **44** leading to the aporphine alkaloid *oliveroline* from an

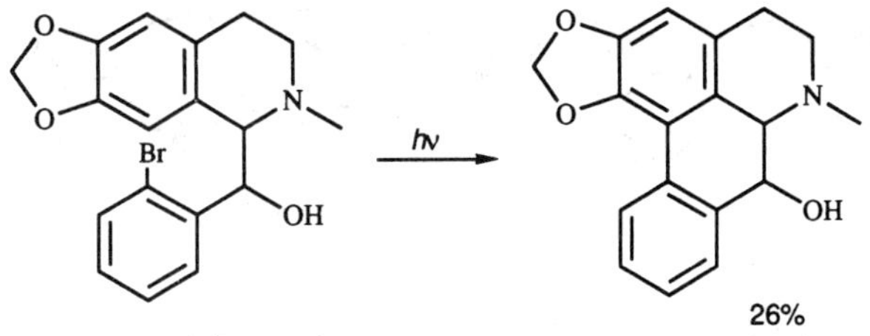

hν

26%

(44)

(*o*-bromobenzyl) tetrahydroisoquinoline [86]. Illustration **45** [87–91] contains a selection of processes of this type. It is of interest that in systems

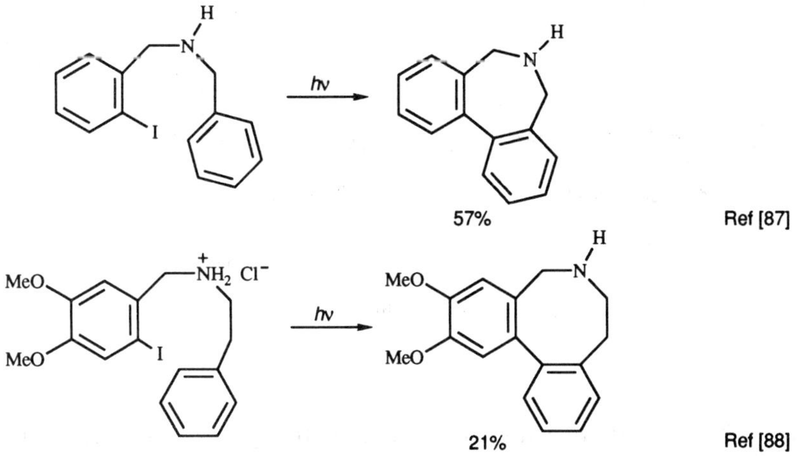

hν

57% Ref [87]

hν

21% Ref [88]

(45)

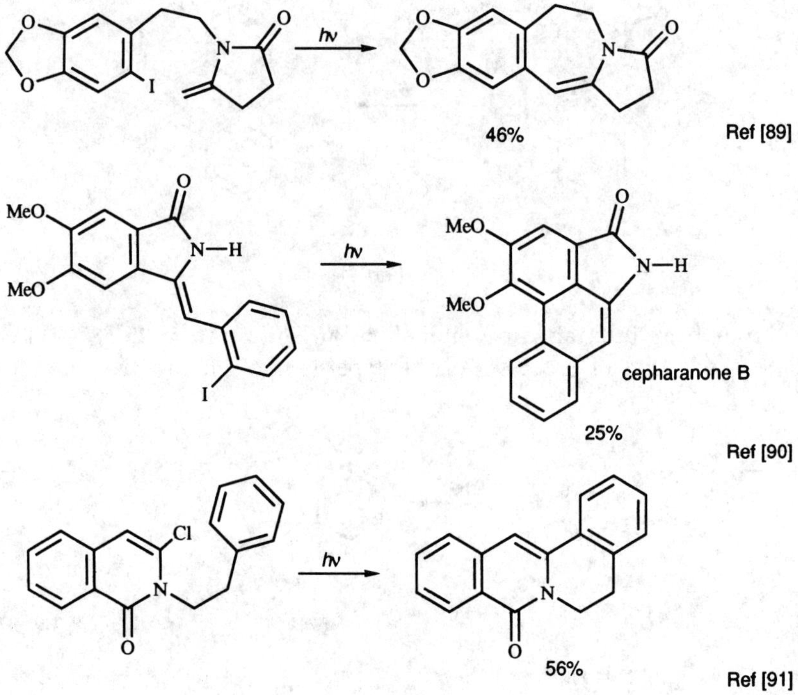

hν

46% Ref [89]

MeO MeO N—H hν MeO MeO N—H cepharanone B

25% Ref [90]

Cl N hν N

56% Ref [91]

(45 continued)

where the aromatic ring that is attacked by the (presumed) radical carries a hydroxy group, a major product can arise (e.g. **46**) by way of attack at the ring position *para* to this group [92]. The primary product, a

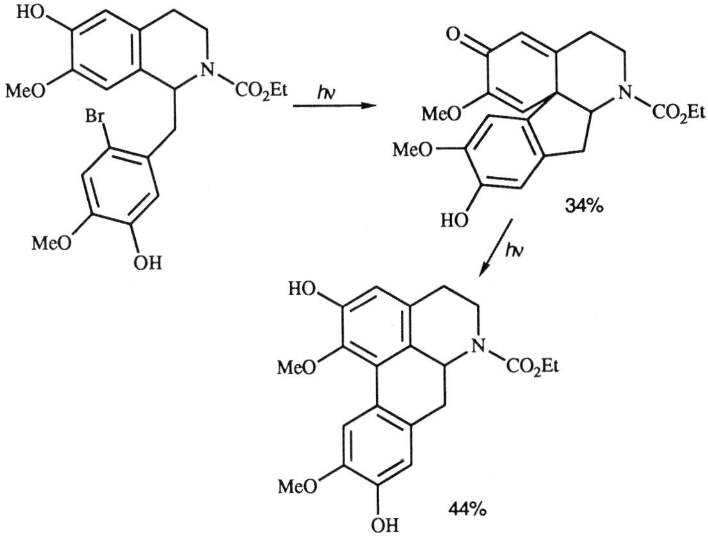

(46)

cyclohexadienone, is photochemically labile as shown (**46**). The yield of the latter compound is enhanced if the irradiation is carried out under alkaline conditions.

The use of *N*-chloroacetyl amines as substrates for photocyclization provides methods for making a variety of nitrogen heterocyclic systems [93]. The route is especially useful for making medium-ring lactams (**47**) [94],

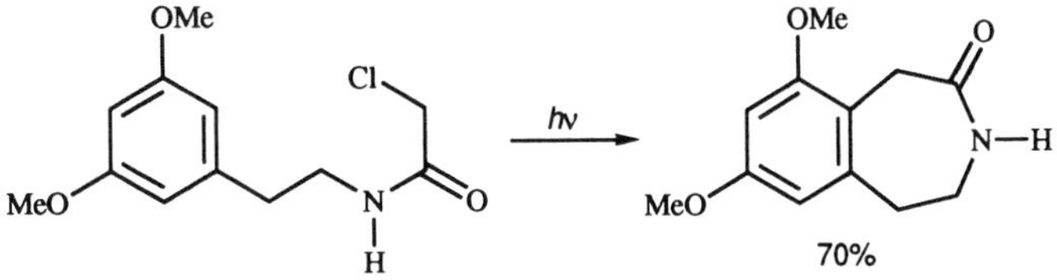

(47)

although it also works for smaller or larger rings. With certain patterns of substitution, different types of product, such as the cyclohexadienones, can be formed, and the mechanism is thought to involve initial electron transfer to give radical ions. Quite complex polycyclic structures can be constructed with moderate efficiency. One example is that of a catharanthine analogue (**48**) [95].

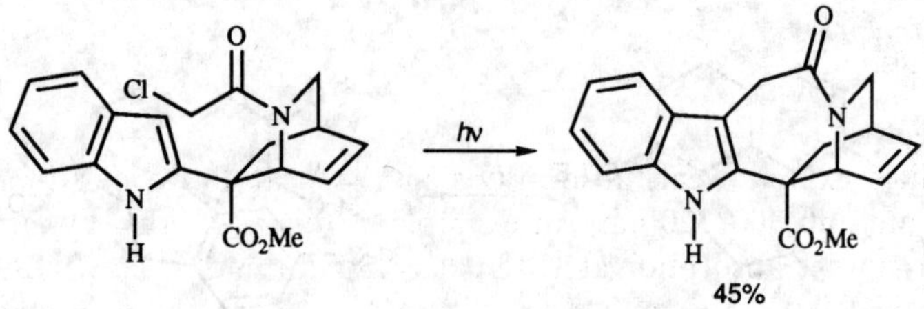

(48)

Electron transfer probably plays a role in the formation of a [3,1]metacyclophane by photocyclization of a benzyl alcohol linked to a dimethylaniline (**49**) [96]. Two examples of the formation of new carbon-heteroatom

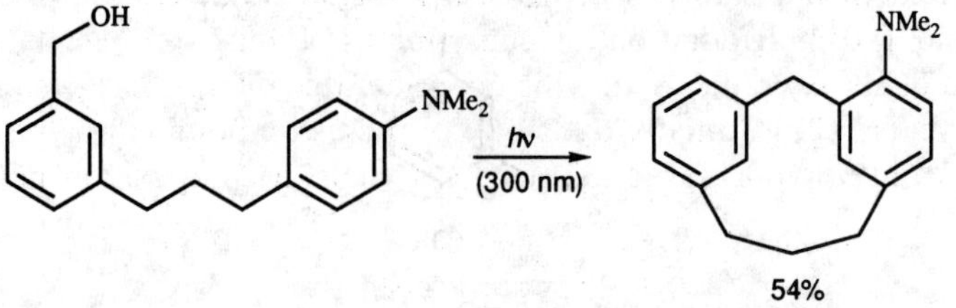

(49)

bonds are illustrated in the photocyclization of *N*-thioacetyl-*o*-haloanilines (**50**) [97], and the unusual reaction (**51**) in which 2,6-dichlorocinnamate esters

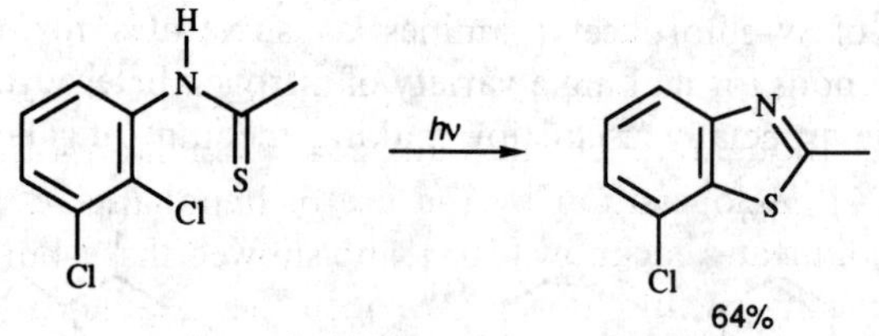

(50)

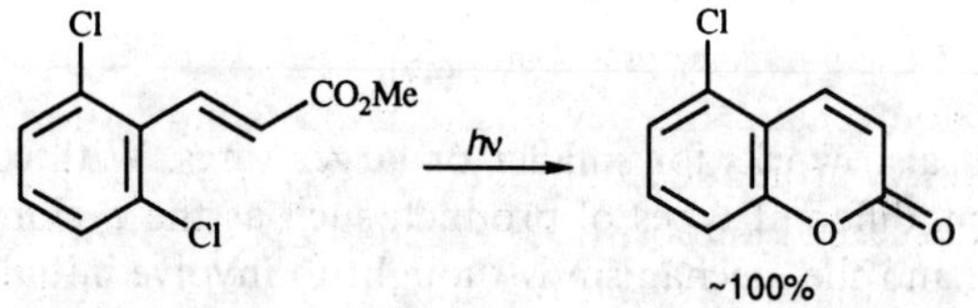

(51)

or amides give a coumarin [98].

## 6.4 HYPOHALITES

### 6.4.1 Spectra

Hypohalites exhibit weak absorptions in the near ultraviolet in regions 250–260 nm and 300–320 nm. An example of this is of ethyl hypochlorite, which has a weak absorption at 310 nm ($\varepsilon = 30\ dm^3\ mol^{-1}\ cm^{-1}$) [99, 100].

### 6.4.2 Photochemistry

In Chapter 5 the photochemistry of nitrites was described. This reaction type involves the irradiation of a compound with a weak O—X bond which in the nitrite example was an O—NO bond. Hypohalites fall into this reaction category and their photolysis involves the fission of an O—Cl, O—Br or O—I bond [101]. Irradiation of compounds of this sort results in the formation of alkoxy radicals, in which considerable interest has been shown over the years [102]. Initially the study of these compounds was restricted solely to *t*-butyl hypochlorite (**52**) where a chain decomposition path was

$$Bu^tOCl \xrightarrow{h\nu} Bu^tO\bullet + Cl\bullet$$

$$Bu^tO\bullet \longrightarrow CH_3COCH_3 + CH_3\bullet$$

$$CH_3\bullet + Bu^tOCl \longrightarrow CH_3Cl + Bu^tO\bullet$$

(52)

suggested [103], but *t*-butyl hypobromite has received much less attention, although it is thought to be mechanistically similar to the hypochlorite reactions [104]. Halogenation by the use of hypohalites was first demonstrated by Walling and Jacknow [105] who showed that *t*-butyl hypochlorite was a synthetically useful chlorinating agent for alkanes and alkylbenzene derivatives in the liquid phase. Usually, the conditions employed for chlorination with *t*-butyl hypochlorite involve high concentrations of substrate and moderate temperatures; under these conditions the reaction is dominated by the *t*-butoxy chain steps (**52**) and there is little incursion from a chlorine atom chain. Studies of the photo-chlorination of alkanes and alkyl-substituted aromatic compounds have shown that, although some substitution occurs at all positions, selectivity is shown with a preference for the formation of the more stable radical, i.e. primary $<$ secondary $<$ tertiary [105].

As far as organic photochemistry is concerned the intramolecular reactivity exhibited by hypochlorites is of importance. Several examples of tertiary hypochlorites with long alkyl chains undergo photochemical conversion into chloroalcohols, as illustrated in **53** [106, 107]. As with *t*-butyl hypochlorite,

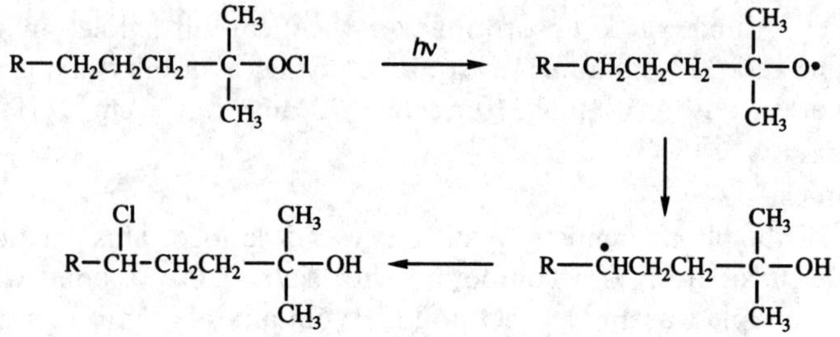
R—CH₂CH₂CH₂—C(CH₃)₂—OCl
hν
R—CH₂CH₂CH₂—C(CH₃)₂—O•
R—ĊHCH₂CH₂—C(CH₃)₂—OH
R—CH(Cl)—CH₂CH₂—C(CH₃)₂—OH

(53)

fission of the alkoxy radical to yield acetone can be an important competing reaction in the long-chain hypochlorites. However, this is dependent upon the nature of the compound and abstraction of the $\gamma$-hydrogen, the key reaction step in the formation of the chloroalcohol, is more efficient when the hydrogen is tertiary [106, 107]. When the site of attack is less heavily substituted, bond fission is a competing reaction. Thus the hypochlorite photolysis route provides a method of remote functionalization of a normally unreactive site, as in the conversion of the steroidal hypochlorite (**54**) into the ether [108, 109].

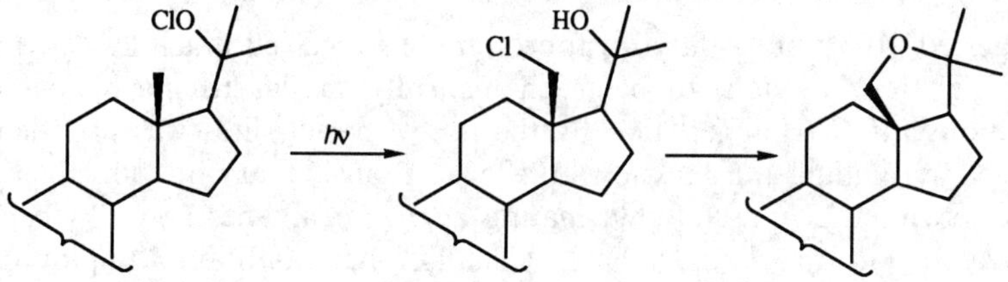
ClO
hν
HO
Cl
O

(54)

Hypoiodites, which are thermally unstable, have also been reported as useful intermediates which undergo irradiative conversion. The synthesis of the hypoiodites is carried out *in situ* with mercuric oxide and iodine or lead tetraacetate and iodine [110]. Irradiation of the resultant hypoiodite e.g. (**55**)

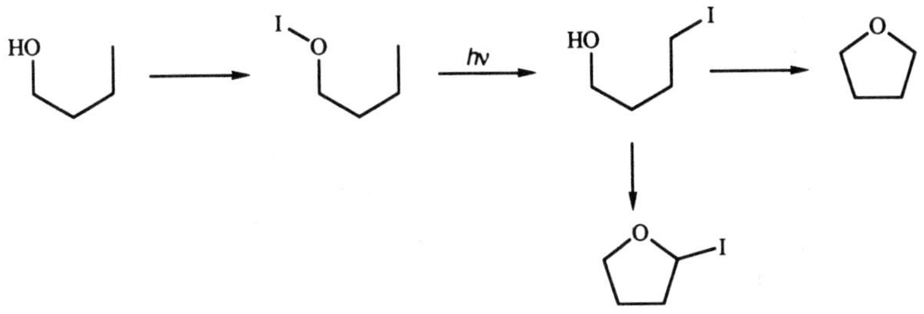

(55)

brings about the conversion shown [110]. Again this reaction path is analogous to the reactivity of the chlorites and bromites and involves fission of the O—I bond followed by reaction of the resultant alkoxy radical either by trapping by iodine atoms to afford the final product or undergoing fission of an adjacent C—C bond. The reaction found favour in remote functionalization procedures applied to steroidal molecules, as in **56** [110]. Many

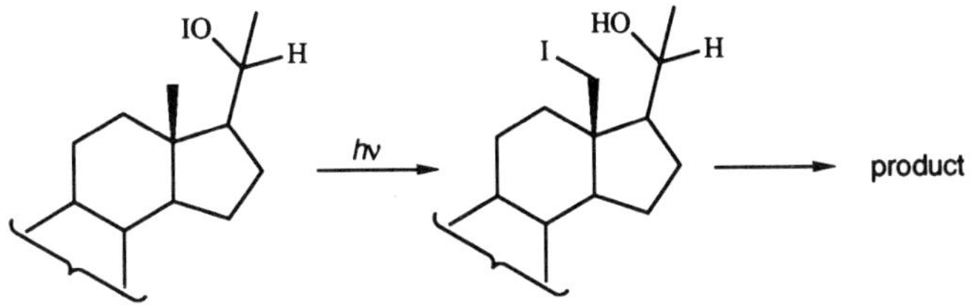

(56)

uses have been made since the initial discovery of the process such as the facile generation of an alkoxy radical which can undergo ring cleavage reactions. Some examples are shown in **57** [111].

$R^1 = H, R^2 = CO_2Me$
$R^1 = H, R^2 = CH_2OAc$
$R^1 = CH_2OAc, R^2 = H$

hν

c. 50%

Ref [111a]

35%

Ref [111b]

hν

43%

Ref [111c]

59%

(57)

Ref [111d]

## 6.5 PHOTOREACTIONS OF HALOGENS AND HYDROGEN HALIDES

### 6.5.1 Halogens

The photochemical reactions of halogens with alkanes and alkenes are not true examples of organic photochemistry since the process involves the excitation of the halogen molecule. Nevertheless these reactions feature prominently in industrial applications and in many laboratory procedures. The reactions involved are free radical chain processes involving initiation, propagation and termination steps and are typified by that shown in **58**;

$$Cl_2 \xrightarrow{h\nu} 2Cl\bullet \quad \text{initiation}$$

$$\left.\begin{array}{l} Cl\bullet + CH_2{=}CH_2 \longrightarrow ClCH_2\dot{C}H_2 \\ ClCH_2\dot{C}H_2 + Cl_2 \longrightarrow ClCH_2CH_2Cl + Cl\bullet \end{array}\right\} \text{propagation}$$

$$\left.\begin{array}{l} ClCH_2\dot{C}H_2 + ClCH_2\dot{C}H_2 \longrightarrow ClCH_2CH_2CH_2CH_2Cl \\ ClCH_2\dot{C}H_2 + Cl\bullet \longrightarrow ClCH_2CH_2Cl \end{array}\right\} \text{termination}$$

(58)

many examples have been recorded in the literature [11]. Usually the addition reactions involve chlorine and bromine. Iodine is seldom used in such processes and reactions involving iodination are prone to destructive decomposition. Some typical examples of the synthetic utility of the process with chlorine and bromine for some simple molecules are shown in **59** [112].

$$CH_2{=}CF_2 + Cl_2 \xrightarrow{h\nu} CH_2ClCF_2Cl \;\; 98\% \quad \text{Ref [112a]}$$

$$trans,\ ClCH{=}CHCl + Cl_2 \xrightarrow{h\nu} Cl_2CHCHCl_2 \;\; 80\% \quad \text{Ref [112b]}$$

$$CF_3CH{=}CH_2 + Cl_2 \xrightarrow{h\nu} CF_3CHClCH_2Cl \;\; 80\% \quad \text{Ref [112c]}$$

$$CH_2{=}CHCl + Br_2 \xrightarrow{h\nu} CH_2BrCHBrCl \quad \text{Ref [112d]}$$

$$CF_2{=}CHF + Br_2 \xrightarrow{h\nu} CF_2BrCHBrF \;\; 82\% \quad \text{Ref [112e]}$$

(59)

### 6.5.2 Hydrogen halides

As far as the synthetic aspects of this reaction are concerned the process is restricted to the addition of hydrogen bromide. Again the reaction is of the free radical chain type with a bromine atom as the chain carrier, as shown in **60** and

$$HBr \xrightarrow{h\nu} H\bullet + Br\bullet$$

$$Br\bullet + CH_2{=}CH_2 \longrightarrow BrCH_2\dot{C}H_2$$

$$BrCH_2\dot{C}H_2 + HBr \longrightarrow BrCH_2CH_3 + Br\bullet$$

(60)

is initiated by light of $\lambda < 290$ nm which is absorbed by hydrogen bromide. In some cases the hydrogen-abstracting properties of excited acetone, using 330 nm light, can be used as a photoinitiator. Again examples of synthetic value are illustrated in **61** [112, 113]. The addition of hydrogen bromide to

$$CH_2{=}CCl_2 + HBr \xrightarrow{h\nu} BrCH_2CHCl_2 \quad 62\% \qquad \text{Ref [112a]}$$

$$CF_3CH{=}CH_2 + HBr \xrightarrow{h\nu} CF_3CH_2CH_2Br \quad 90\% \qquad \text{Ref [112a]}$$

$$CH_3CH{=}CH_2 + HBr \xrightarrow{h\nu} CH_3CH_2CH_2Br \quad 87\% \qquad \text{Ref [113]}$$

$$CH_3CH_2CH{=}CH_2 + HBr \xrightarrow{h\nu} CH_3CH_2CH_2CH_2Br \quad 92\% \qquad \text{Ref [11]}$$

$$CH_3C{\equiv}CH + HBr \xrightarrow{h\nu} \textit{cis}\ CH_3CH{=}CHBr \quad 88\% \qquad \text{Ref [11]}$$

(61)

cyclic alkenes also shows synthetic utility (**62**) [114].

Br Br + HBr $h\nu$ Br Br Ref [114a]

Cl + HBr $h\nu$ Cl Br + Cl Br Bu$^t$ Bu$^t$ 5% Bu$^t$ 95% Ref [114b]

+ HBr $h\nu$ Br Br + Br Br Br Ref [114c]

total yield 70-97%

(62)

## REFERENCES

[1] J. G. Calvert and J. N. Pitts, jun., *Photochemistry*, Wiley, New York, 1966, p. 522.

[2] K. Kimura and S. Nagakura, *Spectrochim. Acta*, 1961, **17**, 166

[3] B. R. Russell, L. O. Edwards and J. W. Raymonda, *J. Am. Chem. Soc.*, 1973, **95**, 2129.

[4] J. R. Majer and J. P. Simons, *Adv. Photochem.*, 1964, **2**, 137; A. P. Zeelenberg, *Nature*, 1958, **181**, 42; J. F. McKeller and R. G. W. Norrish, *Proc. Roy. Soc. (London)*, 1961, **A263**, 51.

[5] G. Herzberg, *Proc. Chem. Soc. (London)*, 1959, 116.

[6] D. L. Bunberry, R. R. Williams, jun. and W. H. Hamill, *J. Am. Chem. Soc.*, 1956, **78**, 6228.

[7] J. J. Dannenberg and K. Dill, *Tetrahedron Lett.*, 1972 1571.

[8] D. K. Bakale and H. A. Gillis, *J. Phys. Chem.*, 1970, **74**, 2074.

[9] J. Banus, H. J. Emelius and R. N. Haszeldine, *J. Chem. Soc.*, 1950. 3041.

[10] R. N. Haszeldine and B. R. Steele, *J. Chem. Soc.*, 1955, 3005; R. N. Haszeldine and K. Leedham, *J. Chem. Soc.*, 1952, 3483.

[11] G. Sosnovsky, *Free Radical Reactions in Preparative Organic Chem-*

*istry*, Macmillan, New York, 1964; C. Walling and E. S. Huyser, *Org. Reactions*, 1963, **13**, 91; D. Elad, *Org. Photochem.*, 1969, **2**, 168.

[12] (a) E. C. Kooyman, *Rec. Trav. Chim. Pays-Bas*, 1951, **70**, 684, 867.
(b) M. F. Shostakovskii, A. V. Bogdanova, M. M. Zverov and G. I. Plotnikova, *Izv. Akad. Nauk SSSR Ser. Khim.*, 1956, 1236 (*Chem. Abstr.*, 1957,**51**, 5730).

[13] (a) M. S. Kharasch, E. V. Jensen and W. H. Urrey, *J. Am. Chem. Soc.*, 1946, **68**, 154.
(b) M. S. Kharasch, E. V. Jensen and W. H. Urrey, *J. Am. Chem. Soc.*, 1947, **69**, 1100.

[14] (a) M. S. Kharasch, O. Reinmuth and W. H. Urrey, *J. Am. Chem. Soc.*, 1947, **69**, 1105.
(b) E. A. I. Heiba and L. C. Anderson, *J. Am. Chem. Soc.*, 1957, **79**, 4940.
(c) M. S. Kharasch and M. Sage, *J. Org. Chem.*, 1949, **14**, 537.
(d) M. S. Kharasch and H. N. Friedlander, *J. Org. Chem.*, 1949, **14**, 239.
(e) R. N. Haszeldine and B. R. Steele, *J. Chem. Soc.*, 1955, 3005.

[15] C. Walling, *Free Radicals in Solution*, Wiley, New York, 1957, p. 256.

[16] E. I. Heiba and R. M. Dessau, *J. Am. Chem. Soc.*, 1967, **89**, 3772; R. N. Haszeldine and K. Leedham, *J. Chem. Soc.*, 1952, 3483.

[17] R. N. Haszeldine, *J. Chem. Soc.*, 1953, 922.

[18] R. N. Haszeldine and B. R. Steele, *J. Chem. Soc.*, 1955, 1199.

[19] R. N. Haszeldine and B. R. Steele, *J. Chem. Soc.*, 1955, 1592.

[20] (a) P. J. Kropp, T. H. Jones and G. S. Poindexter, *J. Am. Chem. Soc.*, 1973, **95**, 5420.
(b) G. S. Poindexter and P. J. Kropp, *J. Am. Chem. Soc.*, 1974, **96**, 7142; P. J. Kropp, P. R. Worsham, R. I. Davidson and T. H. Jones, *J. Am. Chem. Soc.*, 1982, **104**, 3972.
(c) P. J. Kropp, G. S. Poindexter, N. J. Pienta and D. C. Hamilton, *J. Am. Chem. Soc.*, 1976, **98**, 8135.

[21] R. R. Perkins and R. E. Pincock, *Tetrahedron Lett.*, 1975, 943.

[22] P. J. Kropp, J. R. Gibson, J. J. Snyder and G. S. Poindexter, *Tetrahedron Lett.*, 1978, 207.

[23] (a) J. L. Charlton, G. J. Williams and G. N. Lypka, *Can, J. Chem.*, 1980, **58**, 1271; J. L. Charlton and G. J. Williams, *Tetrahedron Lett.*, 1977, 1473.
(b) N. J. Pienta and P. J. Kropp, *J. Am. Chem. Soc.*, 1978, **100**, 655.
(c) N. Takaishi, N. Miyamoto and Y. Inamoto, *Chem. Lett.*, 1978, 1251.

[24] (a) K. M. Saplay, R. Sahni, N. P. Damodaran and S. Dev, *Tetrahedron*, 1980, **36**, 1455.
(b) K. V. Subbarao, N. P. Damodaran and S. Dev, *Tetrahedron*, 1987, **43**, 2543.
[25] D. S. Reddy, G. P. Sollott and P. E. Eaton, *J. Org. Chem.*, 1989, **55**, 722.
[26] M. Kurz and M. Rodgers, *J. Chem. Soc., Chem. Commun.*, 1985, 1227.
[28] H. E. Mahncke and W. A. Noyes, jun., *J. Am. Chem. Soc.*, 1936, **58**, 932.
[29] M. H. J. Wijnen, *J. Am. Chem. Soc.*, 1961, **83**, 4109.
[30] Z. R. Grabowski and A. Bylina, *Trans. Farad. Soc.*, 1964, **60**, 1131.
[31] W. Schnabel, I. Naito, T. Kitamura, S. Kobayashi and H. Taniguchi, *Tetrahedron* 1980, **36**, 3229.
[32] P. J. Kropp, S. A. McNeely and R. D. Davis, *J. Am. Chem. Soc.*, 1983, **105**, 6907; S. A. McNeely and P. J. Kropp, *J. Am. Chem. Soc.*, 1976, **98**, 4319.
[33] B. Sket and M. Zupan, *J. Chem. Soc., Perkin Trans. 1*, 1979, 752.
[34] T. Kitamura, S. Kobayashi and H. Taniguchi, *J. Org. Chem.*, 1982, **47**, 2323.
[35] A. Gregorcic and M. Zupan, *J. Fluorine Chem.*, 1987, **34**, 313 (*Chem. Abstr.*, 1987, **107**, 133961).
[36] T. Kitamura, T. Muta, T. Tahara, S. Kobayashi and H. Taniguchi, *Chem Lett.*, 1986, 759.
[37] T. Kitamura, S. Kobayashi and H. Taniguchi, *Chem. Lett.*, 1978, 1223 (*Chem. Abstr.*, 1979, **90**, 195); T. Kitamura, S. Kobayashi and H. Taniguchi, *J. Am. Chem. Soc.*, 1986, **108**, 2641.
[38] T. Kitamura, S. Kobayashi and H. Taniguchi, *Tetrahedron Lett.*, 1979, 1619; T. Kitamura, S. Kobayashi and H. Taniguchi, *Chem. Lett.*, 1984, 1523; T. Kitamura, S. Kobayashi and H. Taniguchi, *J. Org. Chem.*, 1984, **49**, 4755.
[39] T. Kitamura, S. Kobayashi and H. Taniguchi, *J. Org. Chem.*, 1990, **55**, 1801.
[40] T. Kitamura, S. Kobayashi and H. Taniguchi, *Chem. Lett.*, 1988, 1637.
[41] T. Suzuki, T. Kitamura, T. Sonada, S. Kobayashi and H. Taniguchi, *J. Org. Chem.*, 1981, **46**, 5324.
[42] T. Kitamura, I. Nakamura, S. Kobayashi and H. Taniguchi. *J. Chem. Soc., Chem. Commun.*, 1989, 1154.
[43] Z. Goran and I. Willner, *J. Am. Chem. Soc.*, 1983, **105**, 7764; R. Maidan and I. Willner, *J. Am. Chem. Soc.*, 1986, **108**, 1080.
[44] Y. Izawa, M. Takeuchi and H. Tomioka, *Chem. Lett.*, 1983, 1297.
[45] H. R. Sonawane, B. S. Nanjundiah and M. D. Panse, *Tetrahedron*

*Lett.*, 1985, **26**, 3507.

[46] H. H. Jaffe and M. Orchin, *Theory and Application of Ultraviolet Spectroscopy*, Wiley, New York, 1962, p. 257.

[47] J. G. Calvert and J. N. Pitts. jun., *Photochemistry*, Wiley, New York, 1966 p. 264.

[48] C. P. Andrieux, C. Blocman, J. M. Dumas-Bouchiat and J. M. Saveant, *J. Am. Chem. Soc.*, 1979, **101**, 3431.

[49] J. McD. Blair, D. Bryce-Smith and B. W. Pengilly, *J. Chem, Soc.*, 1959, 3174; J. McD. Blair and D. Bryce-Smith *J. Chem. Soc.*, 1965, 1788; W. Wolff and N. Kharasch, *J. Org. Chem.*, 1961, **26**, 283; (d) *ibid.*, 1960, **30**, 2493.

[50] T. Matsuura and K. Omura, *Bull. Chem. Soc. Jpn*, 1966, **39**, 944; G. R. Lappin and J. S. Zanucci, *Tetrahedron Lett.*, 1969, 5085.

[51] N. J. Bunce, Y. Kumar and B. G. Brownlee, *Chemosphere*, 1978, **7**, 155.

[52] J. Grimshaw and A. P. de Silva, *Chem. Soc. Reviews*, 1981, **10**, 181; R. S. Davidson, J. W. Goodwin and G. Kemp, *Adv. Phys. Org. Chem.*, 1984, **20**, 191.

[53] N. J. Bunce, J. P. Bergsma, M. D. Bergsma, W. De Graaf, Y. Kumar and L. Ravanal, *J. Org. Chem.*, 1980, **45**, 3708.

[54] See Chapter 2

[55] W. Augustyniak, *Ser. Chem.*, 1980, **39**, 72.

[56] N. J. Bunce and L. Ravanal, *J. Am. Chem. Soc.*, 1977, **99**, 4150.

[57] G. E. Robinson and J. M. Vernon, *J. Chem. Soc., Chem. Commun.*, 1969, 977; *J. Chem. Soc. (C)*, 3363.

[58] M. Mansour, S. Wawrik, H. Parlar and F. Korte, *Chem.-Ztg.*, 1980, **104**, 339.

[59] K. Nojima and S. Kanno, *Chemosphere*, 1980, **9**, 437.

[60] B. Akermark, P. Baeckstrom, U. E. Westlin, R. Gothe and C. Wachtmesiter, *Acta Chem. Scand.*, 1976, **B30**, 49.

[61] G. G. Choudhry, A. A. M. Roof and O. Hutzinger, *Tetrahedron Lett.*, 1979, 2059.

[62] N. J. Bunce, Y. Kumar, L. Ravanal and S. Safe, *J. Chem. Soc., Perkin Trans. 2*, 1978, 880.

[63] L. O. Ruzo, M. J. Zabik and R. D. Schuetz, *J. Am. Chem. Soc.*, 1974, **96**, 3809.

[64] T. Nishiwaki, *Yuki Gosei Kagaku Kyokai Shi*, 1981, **39**, 226; T. Nishiwaki, M. Usui and K. Anda, *Tokyo-Toritsu Kogyo Gijutsu Senta Kenkyu Hokoku*, 1980, 133 (Chem. Abstr., 1980, **93**, 113 514).

[65] (a) R. S. Davidson and J. W. Goodin, *Tetrahedron Lett.*, 1981, **22**, 163.
(b) L. O. Ruzo. N. J. Bunce and S. Safe, *Can. J. Chem.*, 1975, **53**, 688.

[66] N. J. Bunce, S. Safe and L. O. Ruzo, *J. Chem. Soc., Perkin Trans. 1*, 1975, 1607.
[67] B. Chittim, S. Safe, N. J. Bunce, L. O. Ruzo, K. Olie and O. Hutzinger, *Can. J. Chem.*, 1978, **56**, 1253.
[68] R. A. Rossi, *Acc. Chem. Res.*, 1982, **15**, 164.
[69] R. A. Rossi and S. M. Palacios, *J. Org. Chem.*, 1981, **46**, 5300.
[70] J. E. Bunnett and S. J. Shafer, *J. Org. Chem.*, 1978, **43**, 1873, 1877.
[71] J. F. Bunnett, *Acc. Chem. Res.*, 1978, **11**, 413.
[72] R. A. Rossi and J. F. Bunnett, *J. Org. Chem.*, 1973, **38**, 1407.
[73] R. Beugelmans and H. Ginsburg, *J. Chem. Soc., Chem. Commun.*, 1980, 508.
[74] (a) R. Beugelmans and G. Rossi, *Tetrahedron*, 1981, **37**, (supplement 9), 393.
(b) R. Beugelmans, M. Bois-Choussy, and B. Boudet, *Tetrahedron*, 1982, **38**, 3479.
[75] J. F. Wolfe, M. C. Sleevi and R. R. Goehring, *J. Am. Chem. Soc.*, 1980, **102**, 3646.
[76] M. F. Semmelhack and T. M. Barger, *J. Org. Chem.*, 1977, **42**, 1481.
[77] M. F. Semmelhack, B. P. Chong, R. D. Stauffer, T. D. Rogerson, A. Chong and L. D. Jones, *J. Am. Chem. Soc.*, 1975, **97**, 2507.
[78] W. Wolff and N. Kharasch, *J. Org. Chem.*, 1965, **30**, 2493; M. Sainsbury, *Tetrahedron*, 1980, **36**, 3327.
[79] S. M. Kupchan and H. C. Wormser, *Tetrahedron Lett.*, 1965, 359.
[80] W. A. Henderson, R. Lopresti and A. Zweig, *J. Am. Chem. Soc.*, 1969, **91**, 6049.
[81] (a) M. F. Semmelhack and T. M. Barger, *J. Org. Chem.*, 1977, **42**, 1481.
(b) B. L. Jensen and K. Chockalingham, *J. Heterocyclic Chem.*, 1986, **23**, 343.
[82] S. M. Kupchan and H. C. Wormser, *Tetrahedron Lett.*, 1965, 359.
[83] L. Trifonov and A. Orakhovats, *Izv. Khim.*, 1978, **11**, 297 (Chem. Abstr., 1980, **92**, 164129); T. Kametani, R. Nitadori, H. Terasawa, K. Takahashi, M. Ihara and K. Fukumoto, *Tetrahedron*, 1977, **33**, 1069.
[84] G. R. Lenz, *Tetrahedron Lett.*, 1973, 1963.
[85] J. Grimshaw and A. P. de Silva, *J. Chem. Soc., Perkin Trans. 2*, 1982, 857.
[86] S. V. Kessar, Y. P. Gupta, V. S. Yadav, M. Narula and T. Mohammad, *Tetrahedron Lett.*, 1980, 3307.
[87] P. W. Jeffs, J. F. Hanse and G. A. Brine, *J. Org. Chem.*, 1975, **40**, 2283.

[88] M. Kihara and S. Kobayashi, *Chem. Pharm, Bull.*, 1978, **26**, 155.

[89] I. Tse and V. Snieckus, *J. Chem. Soc., Chem. Commun.*, 1976, 505.

[90] L. Castedo, E. Guitian, J. M. Saa and R. Suau, *Heterocycles*, 1982, **19**, 279.

[91] C. Kaneko, T. Naito and C. Miwa, *Chem. Pharm. Bull.*, 1982, **30**, 752.

[92] S. M. Kupchan, C.-K. Kim and K. Miyano, *J. Chem. Soc., Chem. Commun.*, 1976, 91.

[93] R. J. Sundberg, *Org. Photochem.*, 1983, **6**, p. 121.

[94] Y. Ikuno, K. Hemmi and O. Yonemitsu, *Chem Pharm. Bull.*, 1972, **20**, 1164.

[95] R. J. Sundberg and J. D. Bloom, *J. Org. Chem.*, 1980, **45**, 3382.

[96] C.-I. Len, P. Singh, M. Maddox and E. F. Ullman, *J. Am. Chem. Soc.*, 1980, **102**, 3261.

[97] R. Paramasivan, R. Palaniappan and V. T. Ramakrishnan, *J. Chem. Soc., Chem. Commun.*, 1979, 260.

[98] R. Arad-Yellin, B. S. Green and K. A. Muszkat, *J. Org. Chem.*, 1983, **43**, 2578.

[99] J. G. Calvert and J. N. Pitts, jun., *Photochemistry*, Wiley, New York, 1966, p. 526.

[100] M. Anbar and I. Dostrovsky, *J. Chem. Soc.*, 1954, 1105.

[101] F. D. Chattaway and O. G. Backeberg, *J. Chem. Soc.*, 1923, 2999.

[102] M. Akhtar, *Adv. Photochem.*, 1964, **2**, 263.

[103] C. Walling, *Free Radicals in Solution*, Wiley, New York, 1957, p. 388.

[104] C. Walling and A. Padwa, *J. Org. Chem.*, 1962, **27**, 2976.

[105] C. Walling and B. B. Jacknow, *J. Am. Chem. Soc.*, 1960, **82**, 6108; C. Walling and B. B. Jacknow *J. Am. Chem. Soc.*, 1960, **82**, 6113.

[106] F. D. Greene, M. L. Savitz, H. H. Lau, D. Osterholtz and W. N. Smith, *J. Am. Chem. Soc.*, 1961, **83**, 2196; F. D. Greene, M. L. Savitz, H. H. Lau, D. Osterholtz, W. N. Smith and P. M. Zanet, *J. Org Chem.*, 1963, **28**, 55.

[107] C. Walling and A. Padwa, *J. Am. Chem. Soc.*, 1961, **83**, 2207.

[108] M. Akhtar and D. H. R. Barton, *J. Am. Chem. Soc.*, 1961, **83**, 2213.

[109] E. L. Jenner, *J. Org. Chem.*, 1962, **27**, 1013.

[110] Ch. Meystre, K. Heusler, J. Kalvoda, P. Wieland, G. Anner and A. Wettstein, *Experientia*, 1961, **17**, 475; Ch. Meystre, K. Heusler, J. Kalvoda, P. Wieland, G. Anner and A. Wettstein, *Helv. Chim. Acta*, 1962, **45**, 1317.

[111] (a) H. Suginome, M. Itoh, and K. Kobayashi, *J. Chem. Soc., Perkin Trans. 1*, 1988, 491
(b) H. Suginome, C. F. Liu, S. Seko and K. Kobayashi, *J. Org. Chem.*,

1988, **53**, 5952.

(c) H. Suginome, M. Itoh and K. Kobayashi, *Chem. Lett.*, 1987, 1527.

(d) H. Suginome and S. Yamada *Tetrahedron Lett.*, 1987, **28**, 3963.

[112] (a) R. N. Haszeldine and B. R. Steele, *J. Chem. Soc.*, 1957, 2800.

(b) K. L. Mueller and H. J. Schumacher, *Z. Physik. Chem. (Leipzig)*, 1937, **B35**, 285.

(c) R. N. Haszeldine, *J. Chem. Soc.*, 1951, 2495

(d) R. Schmitz and H. J. Schumacher, *Z. Physik. Chem. (Leipzig)*, 1942, **B52**, 72.

(e) J. D. Park, W. R. Lycan and J. R. Lacher, *J. Am. Chem. Soc.*, 1951, **73**, 711.

[113] A. L. Henne and M. Nager, *J. Am. Chem. Soc.*, 1951, **73**, 5527.

[114] (a) H. L. Goering, P. I. Abell and B. F. Aycock, *J. Am. Chem. Soc.*, 1952, **74**, 3588.

(b) P. D. Readio and P. S. Skell, *J. Org. Chem.*, 1966, **31**, 753.

(c) N. A. Lebel, *J. Am. Chem. Soc.*, 1960, **82**, 623.

# 7

# Experimental techniques

The previous chapters have outlined the many different reactions which can be brought about by the irradiation of organic compounds. Most of the work described has involved the solution phase, and this chapter is devoted to the more common experimental techniques which can be used to irradiate solutions of organic compounds.

In general, organic photochemical research has followed two main lines of study, where the principal aims are either the elucidation of the mechanistic steps of the reaction in question or the use of the reaction for synthetic purposes. Within this latter area, as has been well exemplified in the previous chapters, the photochemical approach to synthesis provides a profusion of new routes to a variety of molecules. Clearly the demands of the two research areas can involve different types of equipment. Thus the mechanistic photochemist often requires elaborate equipment involving the use of a monochromatic source of light, a filter train, a beam collimator, and means of measuring both the incident light and the light absorbed by the reactants. On the other hand the synthetic organic photochemist is primarily interested in high-intensity light sources abundant in the wavelengths absorbed by the compound under study. In this latter situation the apparatus should be easy to use and should have a variety of flask sizes so that preparative runs do not need redesign of the apparatus.

It is extremely difficult to give comprehensive coverage since the literature abounds with descriptions of apparatus which has been designed for specific tasks and often these pieces of apparatus have no applicability beyond that. Thus the apparatus described in this chapter is commercially available, in the main, although some reference will be made to specifically designed systems which can be readily constructed and which have some general applicability.

## 7.1 HAZARDS

Prior to a description of the apparatus the operator should be aware of the potential hazards involved and take the necessary safety precautions [1]. Thus, proper standards of electrical safety are required since high voltages are used to operate the discharge lamps. The lamps, particularly the high-pressure lamps, operate at high internal pressures and precautions have to be taken to guard against breakage. The medium-pressure and high-pressure lamps operate at high temperature. Care has to be taken to avoid fire if the apparatus is left unattended and also that the lamps have cooled sufficiently before handling. Short-wavelength ultraviolet light interacts with atmospheric oxygen producing ozone (TLV of 0.1 ppm). Thus adequate ventilation is required.

Further to the above it should be noted that the eyes and the skin can be damaged by exposure to u.v. irradiation[1–3]. Since individuals vary in their reaction, no general rule can be given apart from the need to wear safety goggles or spectacles as routine to protect the eyes. Facial skin should be protected by the use of a face mask and hands by wearing suitable gloves. In general the apparatus should be in a screened area with all of the necessary hazard warning signs and warning lights.

## 7.2 MERCURY VAPOUR LAMPS [1, 4–6]

The light source to be used for a given process is determined by the absorption spectrum of the compound under study; details of this have been included in the earlier chapters. The most widely used light source for preparative photochemistry are those based on mercury. These lamps are extremely convenient to use, are dependable and are capable of supplying steady illumination of both u.v. and visible light covering the range from 200 nm (599 kJ $mol^{-1}$) to 750 nm (159 kJ $mol^{-1}$) and above. The main categories are low-pressure discharge tubes with or without a phosphor coating, medium-pressure discharge tubes with or without a metal halide dopant, and high-pressure mercury lamps.

### 7.2.1 Low-pressure or resonance lamps

These lamps have a mercury pressure of 0.005–0.1 torr and operate at or slightly above room temperature with emission mainly (about 90%) at

**Table 7.1.** Output from a low pressure Hg lamp

| Wavelength/nm | Relative output |
|---|---|
| 248.2 | 0.01 |
| 253.7 | 100.00 |
| 265.2–265.5 | 0.05 |
| 275.3 | 0.03 |
| 280.4 | 0.02 |
| 289.4 | 0.04 |
| 296.7 | 0.20 |
| 302.2–302.8 | 0.06 |
| 312.6–313.2 | 0.60 |
| 334.1 | 0.03 |
| 365.0–366.3 | 0.54 |
| 404.5–407.8 | 0.39 |
| 435.8 | 1.00 |
| 546.1 | 0.88 |
| 577.0–579.0 | 10.14 |

253.7 and 184.9 nm. The output from these lamps is quite low and typically an arc of 40 cm in length has a power rating of 16 W (0.4 W $cm^{-1}$)[1]. The higher energy emission at 184.9 nm is responsible for ozone production by dissociation of atmospheric oxygen. However, the 184 nm line is not transmitted unless ultra-pure quartz is used for the lamp envelope and for the reaction vessel. The fractions of light emitted by such a source are tabulated (Table 7.1). These figures are given only for guidance, since the actual amounts are dependent upon the manufacturer of the lamp. The output from such lamps can be altered by coating the internal surface with a phosphor. By this technique the 254 nm line can be converted to longer wavelength radiation, as in fluorescent lighting tubes, and will have the spectral characteristics of the particular phosphor. This is illustrated in Fig. 7.1.

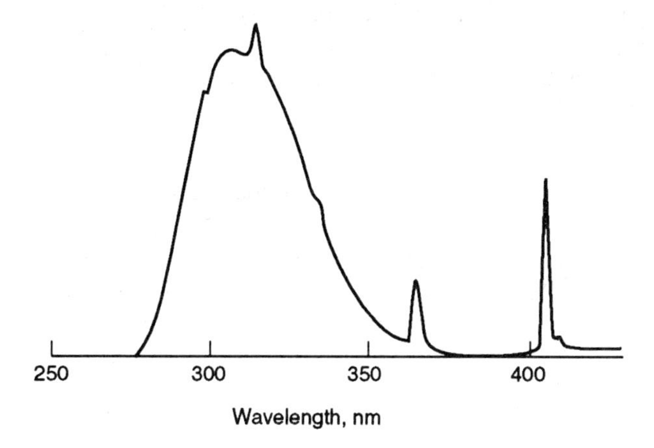


Fig. 7.1. Spectral output of a low-pressure phosphor-coated lamp. Bandwidth 15 mm (290–335 mm). Reproduced from *Photochemistry in Organic Synthesis*, ed. J. D. Coyle, The Royal Society of Chemistry with permission of the copyright holder.

### 7.2.2 Medium-pressure lamps

These lamps operate in the 1–10 atm ($10^5$–$10^6$ Pa) range and run at relatively high temperatures (600–800 °C). Consequently the lamp requires a few minutes to warm up to operating temperature, at which point the output is reasonably stable. The emission from such lamps is usually a weak continuum with superimposed spectral lines (Fig. 7.2). The higher operating pressures and

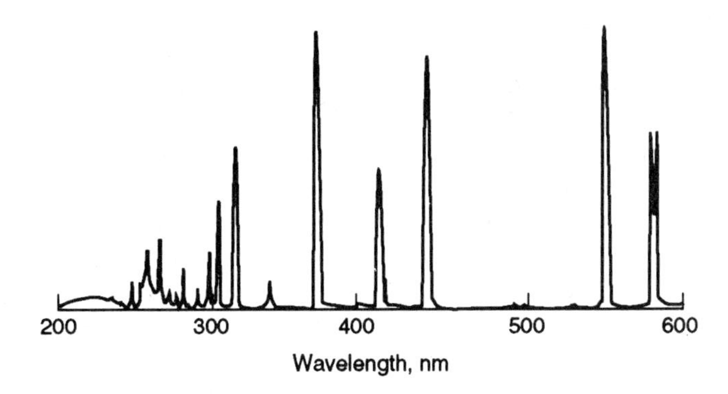


Fig. 7.2. Spectral output of undoped medium-pressure mercury lamp. Reproduced from *Photochemistry in Organic Synthesis*, ed. J. D. Coyle, The Royal Society of Chemistry with permission of the copyright holder.

temperatures result in diminished intensity of the 253.7 and 184.9 nm lines; at the higher mercury pressures what is known as reversed radiation occurs at 253.7 nm, with this line missing owing to self-absorption. The spectral lines at 265.4, 310, and 365 nm (Table 7.2) have reasonably high intensity

**Table 7.2.** Output from a medium-pressure Hg lamp.

| Wavelength/nm | Relative output |
|---|---|
| 222.4 | 14.0 |
| 232.0 | 8.0 |
| 236.0 | 6.0 |
| 238.0 | 8.6 |
| 240.0 | 7.3 |
| 248.2 | 8.6 |
| 253.7 | 16.6 |
| 257.1 | 6.0 |
| 265.2–265.5 | 15.3 |
| 270.0 | 4.0 |
| 275.3 | 2.7 |
| 284.0 | 9.3 |
| 289.4 | 6.0 |
| 296.7 | 16.6 |
| 302.2–302.8 | 23.9 |
| 312.6–313.2 | 49.9 |
| 334.1 | 9.3 |
| 365.0–366.3 | 100.0 |
| 404.5–407.8 | 42.2 |
| 435.8 | 77.5 |
| 546.1 | 93.0 |
| 577.0–579.0 | 76.5 |
| 1014.0 | 40.6 |
| 1128.7 | 12.6 |
| 1367.3 | 15.3 |

and are convenient for direct irradiation since they are in a useful part of the spectrum for the excitation of organic molecules.

Metal halide-doped lamps are also useful and produce additional lines characteristic of the metal halide (e.g. gallium, iron, magnesium or thallium) employed. A typical example of this is the use of thallium iodide, which produces enhanced emission at 535 nm.

### 7.2.3 High-pressure lamps

These lamps operate in the range 10–200 atm ($10^6$–$2 \times 10^7$ Pa). The lamps can be filled either with mercury alone or with a mercury/noble gas (such as xenon) mixture. The large increase in pressure and working temperature introduces many more lines and so a stronger continuum is obtained. The increase in pressure requires the lamps to be in a strong housing to protect the operator from explosion and from the intense irradiation in the 200–600 nm region. The output below 280 nm is very weak. The arc length in such lamps is only a few millimetres, but the output, which is virtually a point source, can range from tens of watts to several kilowatts. The high output from such lamps makes them especially useful for quantitative photochemical work.

## 7.3 LAMPS IN CONJUNCTION WITH FILTERS [1, 4–6]

### 7.3.1 Glass filters

The choice of lamp required for a reaction is dependent upon two main considerations. The one of major importance is the need for spectral overlap between the output of the lamp and the compound to be irradiated. Ideally the aim should be for the maximum overlap between the output of the lamp and the compound. This condition can be achieved in most cases by the use of a medium-pressure lamp which, as was seen in the last section, has a reasonable continuum of energy covering most of the wavelengths between 250 and 500 nm. A greater degree of selectivity may be required if irradiation into one of the absorption bands of the molecule is desired, or if the product from the irradiation is sensitive to a wavelength different from the one used to excite the starting material. This can be achieved by either a monochromatic source (a mercury vapour lamp and a diffraction grating) or less expensively by a system of filters. The simplest filter which can be used is a glass sleeve (quartz, Vycor, Corex or Pyrex) around the light source or a vessel constructed of such glasses. Approximate values for the wavelength of transmission of these are tabled (Table 7.3).

**Table 7.3.** Light transmission for several glasses

| | Percentage transmission for a thickness of 1 mm | | |
|---|---|---|---|
| Glass | 20% | 50% | 90% |
| Quartz | <200 nm | <200 nm | 240 nm |
| Vycor | 200 nm | 220 nm | 280 nm |
| Corex | 270 nm | 290 nm | 360 nm |
| Pyrex | 290 nm | 300 nm | 360 nm |

Other narrow band-pass glass filters are also available from a variety of suppliers.

### 7.3.2 Solution filters

The alternative to glass filters, which can only provide a minimum selectivity, is to use solution filters, many of which can be prepared from readily available chemicals. These solution filters fall into main classes: those which slightly improve on glass filters and those which provide narrow band-pass irradiation suitable for quantum yield measurements. The first type are often classed as short and long cut-off filters which serve the same purpose as glasses. A wide range of filter solutions are described in the literature and some are listed in Table 7.4.

Some of these filter systems have been developed for specific apparatus, such as the unit devised by Zimmerman and his coworkers known as the 'Black Box', which can be used for preparative-run solution-phase photochemistry.

**Table 7.4.** Short and long cut-off filters [7]

| Wavelength (nm) of cut-off | Chemical composition |
|---|---|
| Below 250 | $Na_2WO_4$ |
| Below 305 | $SnCl_2$ in HCl (0.1 M in 2:3 $HCl/H_2O$) |
| Below 330 | 2 M $Na_3VO_4$ |
| Below 355 | $BiCl_3$ in HCl |
| Below 400 | KH phthalate + $KNO_2$ in glycol at pH 11 |
| Below 460 | 0.1 M $K_2CrO_4$ in $NH_4OH/NH_4Cl$ at pH 10 |
| Above 360 | 1 M $NiSO_4$ + 1 M $CuSO_4$ in 5% $H_2SO_4$ |
| Above 450 | $CoSO_4$ + $CuSO_4$ |

With this system three component cells are used where each cell has a 2.4 cm pathlength and the whole gives bandwidths of about 50 nm. Details of the filter solutions used are available in the literature [7].

## 7.4 PREPARATIVE PHOTOCHEMICAL REACTORS

Preparative organic photochemistry can be carried out using two main types of apparatus: either with a solution surrounding the lamps or with a battery of lamps surrounding a solution of the reactant.

### 7.4.1 Immersion-well apparatus

The former method, the solution surrounding the lamp, allows for high light capture and is therefore more economical in terms of energy input. A typical example of this is shown in Fig. 7.3. These reactors are highly efficient, since the lamp is surrounded by the solution of the reacting substances, assuring maximum light capture. The lamps are contained in a double-walled vessel known as an immersion well, which is constructed of either quartz or borosilicate glass. The well is fitted with a standard taper joint allowing it to be used with a variety of outer vessels. Pyrex is less expensive than quartz but, as pointed out earlier, it can only be used with light of $\lambda > 280$ nm. Quartz on the other hand is transparent to 200 nm and is consequently more versatile over a greater range of wavelengths. The use of a quartz immersion well can be made in conjunction with filter sleeves which can be fitted around the lamp, thus permitting the exclusion of unwanted wavelengths. The double surface of the immersion well permits the circulation of cooling water to insulate the reaction medium from the heat generated by the medium-pressure arc lamp. The wavelength employed for the irradiation can also be varied by using a cooled filter solution circulated through the jacket of the immersion well. Again, this allows for selection of the wavelength in use for a given set of conditions. The construction of the immersion well permits the use of either a medium- or a low-pressure Hg lamp. It can be an advantage to pass nitrogen through the central portion of the immersion well since this eliminates the formation of ozone and also protects the lamp from corrosion of the electrode seals.

The outer vessel of such an apparatus is usually made of Pyrex and can be varied in size and in the number of outlets and inlets attached to it. The various reaction flasks which can be constructed enable irradiation to be carried out under aerobic and anaerobic conditions or a constant low or high temperature. It is usually essential, especially with large reactors, to stir the reaction mixture, since the light will not penetrate through the entire

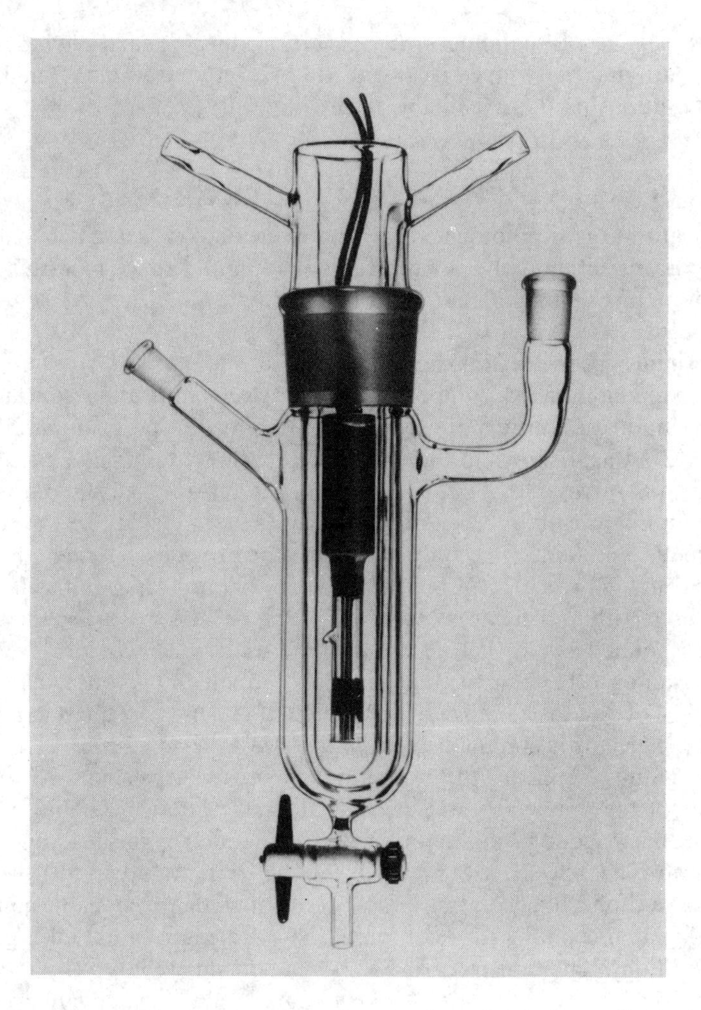

Fig. 7.3. Immersion well for solution-phase photochemistry. Originally supplied by Applied Photophysics Ltd but now manufactured by PHOTOCHEMICAL REACTORS LTD, Haleacre Works, Watchet Lane, Little Kingshill, Buckinghamshire, HP16 0DR, U.K.

solution. Either a mechanical stirrer or a nitrogen gas stream is sufficient to effect adequate mixing, with the added advantage that the use of nitrogen also excludes oxygen from the reaction. This is particularly important if the process is oxygen-sensitive. One drawback of such a method of stirring is that there can be a loss of volatile products and reactants by vaporization on the gas stream. However, stirring, by whatever means chosen, has the effect of constantly replacing the layer of solution closest to the lamp, and this can eliminate the build-up of wall deposits.

The design of the reaction vessel for an immersion well apparatus is obviously open to the ingenuity of the researcher and also to the needs of a given experiment. Indeed there are many reactors described in the literature which have been constructed for a special purpose. One of these that deserves special mention is designed for low-temperature reactions [8]. At low temperatures medium-pressure Hg arc lamps fail to strike and consequently, if the temperature for the reaction is much lower than that of cold water, difficulties will occur. These can be surmounted by using an unmirrored Dewar as the immersion well (Fig. 7.4). This effectively insulates the lamp from the cold reaction mixture and the coolant, and so reactions can be carried out successfully.

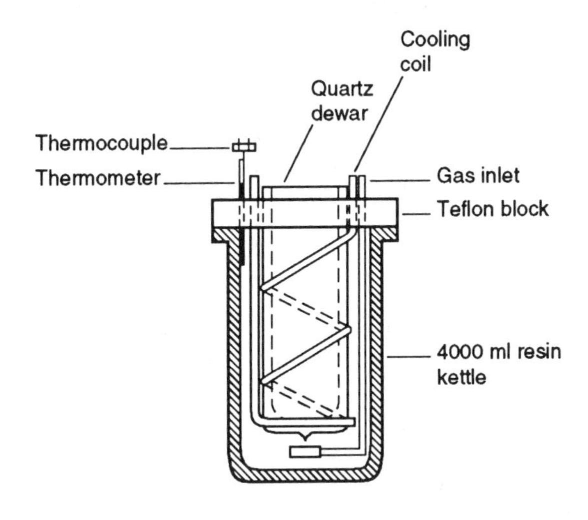


Fig. 7.4.

### 7.4.2 External irradiation

The alternative mode for irradiating solutions is with an external lamp. In this type of arrangement the solution to be irradiated is surrounded by a battery of lamps within a chamber. Obviously it is possible to construct these reactors oneself, but commercial models are available such as the one shown

Fig. 7.5. External irradiation. Originally supplied by Applied Photophysics Ltd but now manufactured by PHOTOCHEMICAL REACTORS LTD, Haleacre Works, Watchet Lane, Little Kingshill, Buckinghamshire, HP16 0DR, U.K.

in Fig. 7.5. This example of a multilamp reactor can be used with low-pressure, medium-pressure or phosphor-coated lamps. The space within the housing is cooled by a circulating fan. However, if lower temperatures are required, a Dewar (quartz or borosilicate glass) can be used.

### 7.4.3 Reactors for quantitative work

Quantitative studies can be carried out using apparatus composed of an immersion-well apparatus surrounded by a merry-go-round or carousel assembly which can take a number of sample tubes (Fig. 7.6). Such an arrangement is of considerable use either for carrying out preliminary studies on a variety of compounds or for determination of the quantum yield of a given reaction. The one essential feature for the use of such a merry-go-round

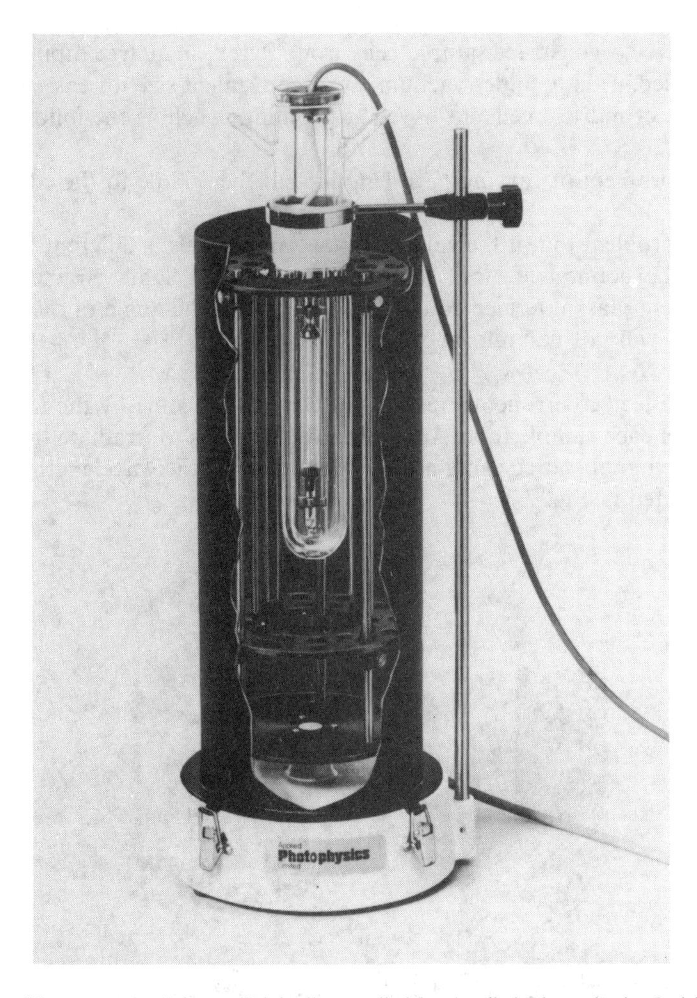


Fig. 7.6. Merry-go-round, large. Originally supplied by Applied Photophysics Ltd but now manufactured by PHOTOCHEMICAL REACTORS LTD, Haleacre Works, Watchet Lane, Little Kingshill, Buckinghamshire, HP16 0DR, U.K.

assembly is to ensure that the light incident upon the sample or the actinometer is completely absorbed. This usually means that solutions have to have an optical density greater than 5. The need to ensure total light capture arises from the fact that, if light capture is not complete, then a method for measuring the fraction of light passing through the sample has to be devised. One further experimental drawback is the use of cylindrical sample cells. Such cells are often used for economy, since it is relatively inexpensive to construct sample cells from Pyrex or quartz tubing. These, when sealed, usually under vacuum, are a convenient size for easy handling. However, cylindrical cells do have shortcomings, such as the following:

(1) the problem of varying path length from the middle to the edge of the tube;
(2) the problem that if the solvent in the sample cells is different from that in the actinometer cell, then the amount of light refracted at the solvent-glass interface could be different and the angle of incidence of light will not be uniform.

These can lead to erroneous results but can be overcome by the use of slits in front of each sample tube. An alternative to this is to irradiate the sample with an external source using a filter train to select the wavelength required as illustrated in Fig. 7.7.

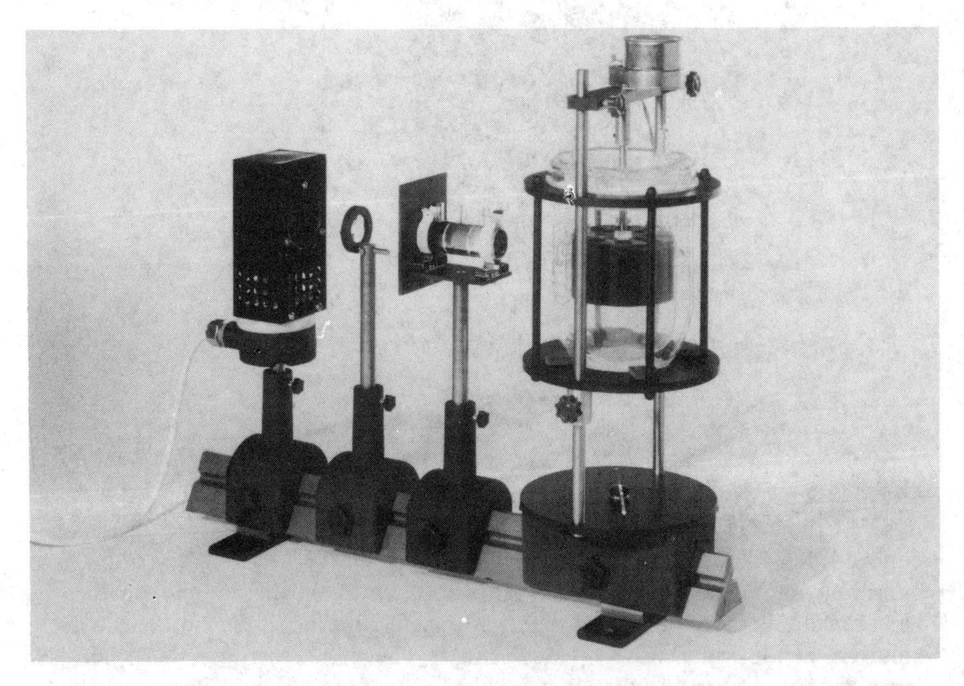

Fig. 7.7. Merry-go-round, small. Originally supplied by Applied Photophysics Ltd but now manufactured by PHOTOCHEMICAL REACTORS LTD, Haleacre Works, Watchet Lane, Little Kingshill, Buckinghamshire, HP16 0DR, U.K.

The light capture by the methods described above is poorer than for the previous methods, owing to the amount of light which escapes and is not absorbed by the samples. A more efficient utilization of the available light energy can be achieved by focusing the light from the lamp by means of a suitable mirror. Such an apparatus has been described in the literature for the irradiation of samples in the 5–50 g range. This uses a 1 kW arc lamp with the light focused by an aluminium parabolic mirror which provides a collimated beam of light 14 cm in diameter. A narrow band-pass can be achieved using combinations of filter solutions in a three-stage filter cell (Fig. 7.8). For quantum yield determinations, the light incident on the face of the sample cell can be determined by the use of a beam splitter to deflect about 10% of the light to an actinometer cell or photomultiplier. While it is possible to make such apparatus oneself, commercially available lamps and mirror assemblies are available from a variety of suppliers.

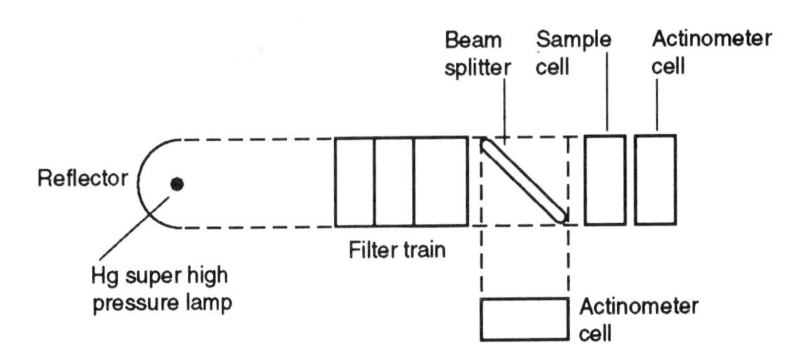


Fig. 7.8. Apparatus using focused light beam and filter train.

### 7.4.4 Thin-film reactors

Yet another mode of irradiation, that of a thin film, has found application in the irradiation of small volumes of reactant or for the irradiation of concentrated solutions where the incident light penetrates only a fraction of a millimetre. This system, shown in Fig. 7.9, pumps the solution to be irradiated from a reservoir through a jet which allows a thin film of the solution to fall under gravity over a quartz or Pyrex tube containing the lamp. Re-collection and re-circulation allows for eventual high conversions.

Details of other methods for the irradiation of compounds on a large scale can be found in a recent monograph [9].

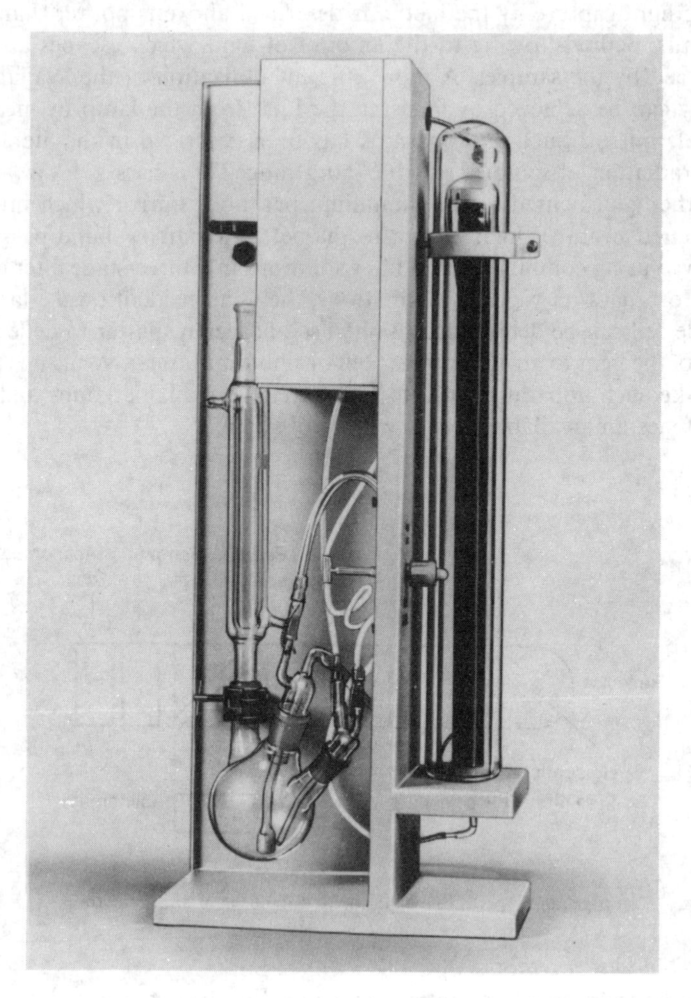

Fig. 7.9. Falling-film irradiation. Originally supplied by Applied Photophysics Ltd but now manufactured by PHOTOCHEMICAL REACTORS LTD, Haleacre Works, Watchet Lane, Little Kingshill, Buckinghamshire, HP16 0DR, U.K.

## 7.5 ACTINOMETRY

The accurate measurement of the efficiency of a photochemical reaction, i.e. its quantum yield, depends mainly on the accuracy of the actinometer cmployed, although it is also important to be able to measure accurately the product or the reactant consumed during irradiation.

Ideally a chemical actinometer should have the following features: thermal stability, availability, reproducibility, uniform response over a large wavelength range, and ease of analysis of chemical change. In the 250–450 nm range, the normal range for solution-phase photochemistry, the ferrioxalate (potassium trioxalatoferrate(III)) actinometer has reigned supreme [10, 11]. This actinometer, used in aqueous solution as $6 \times 10^{-3}$ M solutions without the need for degassing, reacts with light to produce $Fe^{2+}$ (as iron(II) oxalate

**Table 7.5.** Quantum yield of $Fe^{2+}$ ion production for 0.006 M Solution of potassium ferrioxalate

| $\lambda$ (nm) | $\Phi$ |
|---|---|
| 254 | 1.25 |
| 267 | 1.24 |
| 312 | 1.24 |
| 334 | 1.24 |
| 361 | 1.21 |
| 365 | 1.21 |

which does not absorb light) and $CO_2$. This reaction involves only one photon but the oxalyl radical anion produces, by a dark reaction, another molecule of trisoxalatoferrate(II) ion. Thus the observed quantum yield is twice that for the primary photoprocess. The actinometer is monitored by measuring the amount of $Fe^{2+}$ formed by complexation with 1,10-phenanthroline, which produces a red complex analysed spectrophotometrically at 510 nm. The formation of $Fe^{2+}$ is almost constant at short wavelengths (Table 7.5) and full details of the system can be found elsewhere [10]. More recent studies on this actinometer system indicate that an improvement in performance can be obtained using deoxygenated solutions (0.15 M). Other studies [12] have shown that care has to be exercised in the use of the 1,10-phenanthroline since these solutions can suffer photo-oxidation under fluorescent lighting.

Many other systems can be used as actinometers, and often it is convenient to use a reaction whose quantum yield has been determined with reference to ferrioxalate. The Norrish Type II elimination of valerophenone is one such [13]. This reaction produces acetophenone with a quantum yield of 0.33 when valerophenone (**1**) is irradiated at 313 nm in vacuum-degassed benzene. The hexan-2-one system (**2**) is of the same type [14], undergoing

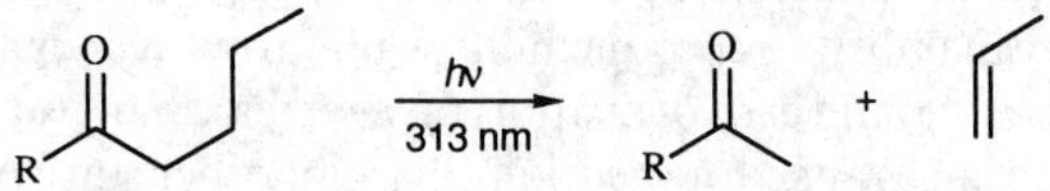

(1) R = Ph
(2) R = Me

fragmentation at 313 nm to afford acetone and propene. Both these products are volatile and can be measured quantitatively by g.l.c, although it is more usual to quantify the amount of acetone which is produced with a quantum yield of 0.22. Such systems have the advantage that they require long irradiation times and so can be used throughout such an irradiative conversion, thus accommodating any variation in the light intensity.

A new attractive system has recently appeared on the market and is operative in the 310–370 nm and 436–545 nm ranges [15]. The system is based on the photochromic compound known as Aberchrome 540 (**3**). This

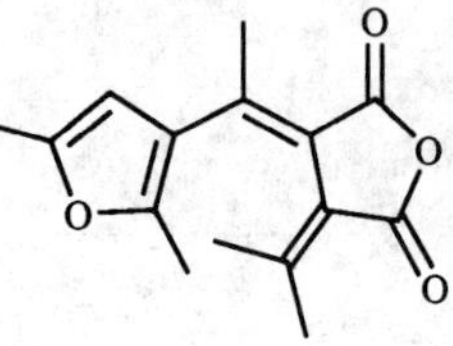

(3)

compound, on irradiation at 310–370 nm, is converted into an isomer (**4**).

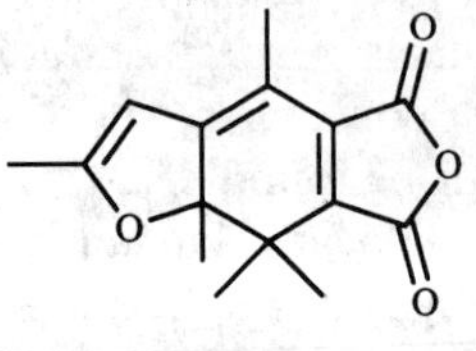

(4)

The use of a magnetically stirred solution of Aberchrome 540 in toluene in a cuvette allows the compound to be used as an actinometer with analysis in the visible at 494 nm for the isomer. The solution can then be irradiated with white light to reverse the reaction and it is then ready for re-usc. Provided the photostability of such systems is beyond doubt, reagents such as Aberchrome 540 are obviously of great utility.

In recent years, electronic means of measuring incident and transmitted light have become popular. These systems are particularly useful in a focused-beam apparatus, where one has the problem of determining the splitting ratio each time the apparatus is used. Automatic or semi-automatic electronic systems have obvious advantages in simplicity of use.

## 7.6 PURITY OF GASES AND SOLVENTS

Solvents may play an important part in the photoreaction of organic molecules. Primarily, dilution of a pure compound prevents the accumulation of high concentrations of excited species which, if produced near the walls of the containing vessel, can lead to side reactions such as polymerization. Dilution also allows for reasonable penetration of the irradiating light into the solution, and so reasonable reaction rates can be achieved. The optimum concentration for a reactant can only be found by trial and error and is a balance between a solution which is too dilute, where most of the incident light passes through unabsorbed, and too concentrated a solution, where side reactions such as dimerization might become important. Many solvents are used in irradiations and the spectral transmissions of some of them are shown in Table 7.6. From these data and more detailed information available from other sources, it is possible to select the best solvent with appropriate optical transparency at the wavelength to be used for the reaction.

**Table 7.6.** Transmission characteristics of various solvents

| Solvent | 10% transmission | 100% transmission |
|---|---|---|
| Acetone | 329 | 366 |
| Acetonitrile | 190 | 313 |
| Benzene | 280 | 366 |
| Carbon tetrachloride | 265 | 313 |
| Cyclohexane | 205 | 254 |
| Diethyl ether | 215 | 313 |
| Dimethyl sulphoxide | 262 | 366 |
| Ethanol | 205 | 313 |
| Hexane | 195 | 254 |
| Propan-2-ol | 205 | 313 |
| Tetrahydrofuran | 233 | 366 |

Hydrocarbons such as cyclohexane are relatively trouble-free since they do not have low-lying excited states which can interfere with the reaction of the solute and they are essentially chemically unreactive (apart from hydrogen abstraction). They are also free from impurities, which can often be a problem when ethers and alcohols are used. Care has to be taken to ensure that ethers are free of peroxides and that alcohols are acid-free, since these impurities could lead to undesirable reactions. The exclusion of air ($O_2$) from the photoreaction is essential, particularly when a triplet-state reaction is suspected. Table 7.7 shows the concentration of $O_2$ in non-degassed solvents. Although these amounts are small, there is often sufficient dissolved oxygen to quench a triplet state. Thus it is essential to outgas the solvent prior to irradiation.

**Table 7.7.** Concentration of oxygen in non-degassed solvents

| Solvent | Oxygen concentration (M) |
|---|---|
| Acetone | 0.0024 |
| Benzene | 0.0019 |
| Carbon tetrachloride | 0.0026 |
| Cyclohexane | 0.0023 |
| Diethyl ether | 0.0040 |
| Ethanol | 0.0021 |
| Hexane | 0.0031 |
| Propan-2-ol | 0.0021 |

Degassing by vacuum techniques can be effective for small samples, but for qualitative and preparative work such degassing is impractical. The degassing technique most commonly used in such situations is the use of an inert gas such as helium, argon, or nitrogen. For most purposes nitrogen will suffice, but nitrogen, even of the highest purity supplied in large cylinders, contains appreciable quantities of $O_2$ and water. Thus it is often necessary to remove these impurities by passing the gas through a scrubbing chain. A few methods are available for the removal of oxygen from nitrogen, the simplest of which makes use of vanadium(II) sulphate. Degassing is carried out by passing nitrogen through the solution of reactant for 0.5–1 h, which is usually sufficient time to reduce the $O_2$ level to below $10^{-4}$ M. The apparatus is then sealed or else kept under a positive pressure of nitrogen throughout the irradiation.

## REFERENCES

[1] J. Hutchison, in *Photochemistry in Organic Synthesis*, ed. J. D. Coyle, Royal Society of Chemistry, Specialist Publication 57, pp 19–38.
[2] D. Hughes, in *Hazards of Occupational Exposure of Ultraviolet Irradiation*, University of Leeds Industrial Service, 1978.
[3] G. Beddard, in *Light Chemical Change and Life: A Source Book in Photochemistry*, eds J. D. Coyle, R. R. Hill and D. Roberts, the Open University Press, 1982, pp 178–194.
[4] S. L. Murov, *Handbook of Photochemistry*, Marcel Dekker, New York, 1973.
[5] W. M. Horspool, *Aspects of Organic Photochemistry*, Academic Press, 1976.
[6] W. M. Horspool, in *Synthetic Organic Photochemistry*, ed. W. M. Horspool, Plenum Press, New York and London, 1984 pp 489–509.
[7] H. E. Zimmerman, *Mol. Photochem.*, 1971, **3**, 281.
[8] D. C. Owsley and J. J. Bloomfield, *Org. Prep. Proc. Internat.*, 1971, **3**, 61.
[9] A. Braun, M.-T. Maurette and E. Oliveros, *Technologie Photochimique*, Presses Polytechniques Romandes, Lausanne, 1986, pp. 540.
[10] J. G. Calvert and J. N. Pitts jun., *Photochemistry*, Wiley, New York, 1966.
[11] C. G. Hatchard and C. A. Parker, *Proc. Roy. Soc. London*, 1965, **A235**, 518.
[12] D. E. Nicodem, M. L. P. F. Cabral, and J. C. N. Ferreira, *Mol. Photochem.*, 1977, **8**, 213.
[13] P. J. Wagner and A. E. Kemppainer, *J. Am. Chem. Soc.*, 1968, **90**, 5896.
[14] P. J. Wagner, *Tetrahedron Lett.*, 1968, 5385.
[15] H. G. Heller and J. R. Langan, *E. P. A. Newsletter*, 1981, 71.

# Index

OXFORD